T0350074

MODERN ANALYSIS OF AUTOMORPHIC FORMS BY EXAMPLE, VOLUME 2

This is Volume 2 of a two-volume book that provides a self-contained introduction to the theory and application of automorphic forms, using examples to illustrate several critical analytical concepts surrounding and supporting the theory of automorphic forms. The two-volume book treats three instances, starting with some small unimodular examples, followed by adelic GL2, and finally GLn. Volume 2 features critical results, which are proven carefully and in detail, including automorphic Green's functions, metrics and topologies on natural function spaces, unbounded operators, vector-valued integrals, vector-valued holomorphic functions, and asymptotics. Volume 1 features discrete decomposition of cuspforms, meromorphic continuation of Eisenstein series, spectral decomposition of pseudo-Eisenstein series, and automorphic Plancherel theorem. With numerous proofs and extensive examples, this classroom-tested introductory text is meant for a second-year or advanced graduate course in automorphic forms, and also as a resource for researchers working in automorphic forms, analytic number theory, and related fields.

Paul Garrett is Professor of Mathematics at the University of Minnesota–Minneapolis. His research focuses on analytical issues in the theory of automorphic forms. He has published numerous journal articles as well as five books.

CAMBRIDGE STUDIES IN ADVANCED MATHEMATICS

Editorial Board:
B. Bollobás, W. Fulton, F. Kirwan, P. Sarnak, B. Simon, B. Totaro

All the titles listed below can be obtained from good booksellers or from Cambridge University Press.
For a complete series listing visit: www.cambridge.org/mathematics.

Modern Analysis of Automorphic Forms by Example, Volume 2

PAUL GARRETT

University of Minnesota–Minneapolis

CAMBRIDGE
UNIVERSITY PRESS

CAMBRIDGE
UNIVERSITY PRESS

University Printing House, Cambridge CB2 8BS, United Kingdom

One Liberty Plaza, 20th Floor, New York, NY 10006, USA

477 Williamstown Road, Port Melbourne, VIC 3207, Australia

314-321, 3rd Floor, Plot 3, Splendor Forum, Jasola District Centre, New Delhi - 110025, India

79 Anson Road, #06-04/06, Singapore 079906

Cambridge University Press is part of the University of Cambridge.

It furthers the University's mission by disseminating knowledge in the pursuit of
education, learning and research at the highest international levels of excellence.

www.cambridge.org
Information on this title: www.cambridge.org/9781108473842
DOI: 10.1017/9781108571814

© Paul Garrett 2018

First published 2018

A catalogue record for this publication is available from the British Library

ISBN 978-1-108-47384-2 Hardback

Contents

Introduction and Historical Notes

The aim of this book is to offer persuasive proof of several important analytical results about automorphic forms, among them spectral decompositions of spaces of automorphic forms, discrete decompositions of spaces of cuspforms, meromorphic continuation of Eisenstein series, spectral synthesis of automorphic forms, a Plancherel theorem, and various notions of convergence of spectral expansions. Rather than assuming prior knowledge of the necessary analysis or giving extensive external references, this text provides customized discussions of that background, especially of ideas from 20th-century analysis that are often neglected in the contemporary standard curriculum. Similarly, I avoid assumptions of background that would certainly be useful in studying automorphic forms but that beginners cannot be expected to have. Therefore, I have kept external references to a minimum, treating the modern analysis and other background as a significant part of the discussion.

Not only for reasons of space, the treatment of automorphic forms is deliberately neither systematic nor complete, but instead provides three families of examples, in all cases aiming to illustrate aspects beyond the introductory case of $SL_2(\mathbb{Z})$ and its congruence subgroups.

The first three chapters set up the three families of examples, proving essential preparatory results and many of the basic facts about automorphic forms, while merely stating results whose proofs are more sophisticated or difficult. The proofs of the more difficult results occupy the remainder of the book, as in many cases the arguments require various ideas not visible in the statements.

The first family of examples is introduced in Chapter 1, consisting of *waveforms* on quotients having dimensions $2, 3, 4, 5$ with a single *cusp*, which is just a *point*. In the two-dimensional case, the space on which the functions live is the usual quotient $SL_2(\mathbb{Z})\backslash\mathfrak{H}$ of the complex upper half-plane $\mathfrak{H}$. The three-dimensional case is related to $SL_2(\mathbb{Z}[i])$, and the four-dimensional and five-dimensional cases are similarly explicitly described. Basic discussion of the physical spaces themselves involves explication of the groups acting on them, and decompositions of these groups in terms of subgroups, as well as the expression of the physical spaces as G/K for K a maximal compact subgroup of G. There are natural invariant measures and integrals on G/K and on $\Gamma\backslash G/K$ whose salient properties can be described quickly, with proofs deferred to a later point. Similarly, a natural Laplace-Beltrami operator Δ on G/K and $\Gamma\backslash G/K$ can be described easily, but with proofs deferred. The first serious result specific to automorphic forms is about *reduction theory*, that is, determination of a nice set in G/K that *surjects* to the quotient $\Gamma\backslash G/K$, for specific *discrete* subgroups Γ of G. The four examples in this simplest scenario all admit very simple sets of representatives, called *Siegel sets*, in every case a product of a ray and a box, with Fourier expansions possible along the box-coordinate, consonant with a decomposition of part of the group G (Iwasawa decomposition). This greatly simplifies both statements and proofs of fundamental theorems.

In the simplest family of examples, the space of *cuspforms* consists of those functions on the quotient $\Gamma\backslash G/K$ with 0th Fourier coefficient identically 0. The basic theorem, quite nontrivial to prove, is that the space of cuspforms in $L^2(\Gamma\backslash G/K)$ has a basis consisting of eigenfunctions for the invariant Laplacian Δ. This result is one form of the *discrete decomposition of cuspforms*. We delay its proof, which uses many ideas not apparent in the statement of the theorem. The orthogonal complement to cuspforms in $L^2(\Gamma\backslash G/K)$ is readily characterized as the space of *pseudo-Eisenstein series*, parametrized here by test functions on $(0, +\infty)$. However, these simple, explicit automorphic forms are never eigenfunctions for Δ. Rather, via Euclidean Fourier-Mellin inversion, they are expressible as *integrals* of (genuine) Eisenstein series, the latter eigenfunctions for Δ, but unfortunately not in $L^2(\Gamma\backslash G/K)$. Further, it turns out that the best expression of pseudo-Eisenstein series in terms of genuine Eisenstein series E_s involves the latter with complex parameter outside the region of convergence of the defining series. Thus arises the need to *meromorphically continue* the Eisenstein series in that complex parameter. Genuine proof of meromorphic continuation, with control over the behavior of the meromorphically continued function, is another basic but nontrivial result, whose proof is delayed. Granting those postponed proofs, a *Plancherel theorem* for the space of pseudo-Eisenstein series follows from their expansion in terms of genuine Eisenstein series, together with attention to integrals as *vector-valued* (rather than merely numerical), with the important corollary that such integrals commute with continuous operators on the vector space. This and other aspects of vector-valued integrals are treated at length in an appendix. Then we obtain the Plancherel theorem for the whole space of L^2 waveforms. Even for the simplest examples, these few issues illustrate the goals of this book: discrete decomposition of spaces of cuspforms, meromorphic continuation of Eisenstein series, and a Plancherel theorem.

In Chapter 2 is the second family of examples, *adele groups* GL_2 over number fields. These examples subsume classical examples of quotient $\Gamma_0(N)\backslash\mathfrak{H}$ with several cusps, reconstituting things so that operationally there is a *single* cusp. Also, examples of Hilbert modular groups and Hilbert modular forms are subsumed by rewriting things so that the vagaries of class numbers and unit groups become irrelevant. Assuming some basic algebraic number theory, we prove p-adic analogues of the group decomposition results proven earlier in Chapter 1 for the purely archimedean examples. Integral operators made from C_c^o functions on the p-adic factor groups, known as *Hecke operators*, are reasonable p-adic analogues of the archimedean factors' Δ, although the same integral operators do make the same sense on archimedean factors. Again, the first serious result for these examples is that of *reduction theory*, namely, that there is a single nice set, an adelic form of a *Siegel set*, again nearly the product of a ray and a box, that surjects to the quotient $Z^+GL_2(k)\backslash GL_2(\mathbb{A})$, where Z^+ is itself a ray in the center of the group. The first serious analytical result is again about *discrete decomposition of spaces of cuspforms*, where now relevant operators are both the invariant Laplacians and the Hecke operators. Again, the deferred proof is much more substantial than the statement and needs ideas not visible in the assertion itself. The orthogonal complement to cuspforms is again describable as the L^2 span of *pseudo-Eisenstein series*, now with a discrete parameter, a Hecke character (grossencharacter) of the ground field, in addition to the test function on $(0, +\infty)$. The pseudo-Eisenstein series are never eigenfunctions for invariant Laplacians or for Hecke operators. Within each family, indexed by Hecke characters, every pseudo-Eisenstein series again decomposes via Euclidean Fourier-Mellin inversion as an integral of (genuine) Eisenstein series with the same discrete parameter. The genuine Eisenstein series are eigenfunctions for invariant Laplacians and are eigenfunctions for Hecke operators at almost all finite places, but are not square-integrable. Again, the best assertion of spectral decomposition requires a meromorphic continuation of the genuine Eisenstein series in the continuous parameter. Then a Plancherel theorem for pseudo-Eisenstein series for each discrete parameter value follows from the integral representation in terms of genuine Eisenstein series and general properties of vector-valued integrals. These are assembled into a Plancherel theorem for all L^2 automorphic forms. An appendix computes *periods* of Eisenstein series along copies of $GL_1(\tilde{k})$ of quadratic field extensions $\tilde{k}$ of the ground field.

Chapter 3 treats the most complicated of the three families of examples, including automorphic forms for $SL_n(\mathbb{Z})$, both purely archimedean and adelic.

Again, some relatively elementary set-up regarding group decompositions is necessary and carried out immediately. Identification of invariant differential operators and Hecke operators at finite places is generally similar to that for the previous example GL_2. A significant change is the proliferation of types of *parabolic subgroups* (essentially, subgroups conjugate to subgroups containing upper-triangular matrices). This somewhat complicates the notion of *cuspform*, although the general idea, that zeroth Fourier coefficients vanish, is still correct, if suitably interpreted. Again, the space of square-integrable cuspforms decomposes discretely, although the complexity of the proof for these examples increases significantly and is again delayed. The increased complication of parabolic subgroups also complicates the description of the orthogonal complement to cuspforms in terms of pseudo-Eisenstein series. For purposes of spectral decomposition, the discrete parameters now become more complicated than the GL_2 situation: *cuspforms* on the Levi components (diagonal blocks) in the parabolics generalize the role of Hecke characters. Further, the continuous complex parametrizations need to be over larger-dimensional Euclidean spaces. Thus, I restrict attention to the two extreme cases: minimal parabolics (also called *Borel subgroups*), consisting exactly of upper-triangular matrices, and *maximal proper* parabolics, which have exactly two diagonal blocks. The minimal parabolics use no cuspidal data but for $SL_n(\mathbb{Z})$ have an $(n-1)$-dimensional complex parameter. The maximal proper parabolics have just a one-dimensional complex parameter but typically need two cuspforms on smaller groups, one on each of the two diagonal blocks. The general qualitative result that the L^2 orthogonal complement to cuspforms is spanned by pseudo-Eisenstein series of various types does still hold, and the various types of pseudo-Eisenstein series are integrals of genuine Eisenstein series with the same discrete parameters. Again, the best description of these integrals requires the meromorphic continuation of the Eisenstein series. For nonmaximal parabolics, Bochner's lemma (recalled and proven in an appendix) reduces the problem of meromorphic continuation to the maximal proper parabolic case, with cuspidal data on the Levi components. Elementary devices such as Poisson summation, which suffice for meromorphic continuation for GL_2, as seen in the appendix to Chapter 2, are inadequate to prove meromorphic continuation involving the nonelementary cuspidal data. I defer the proof. Plancherel theorems for the spectral fragments follow from the integral representations in terms of genuine Eisenstein series, together with properties of vector-valued integrals.

The rest of the book gives proofs of those foundational analytical results, discreteness of cuspforms and meromorphic continuation of Eisenstein series, at various levels of complication and by various devices. Perhaps surprisingly, the required analytical underpinnings are considerably more substantial than an unsuspecting or innocent bystander might imagine. Further, not everyone interested in the *truth* of foundational analytical facts about automorphic forms will necessarily care about their *proofs*, especially upon discovery that that burden is greater than anticipated. These obvious points reasonably explain the compromises made in many sources. Nevertheless, rather than either gloss over the analytical issues, refer to encyclopedic treatments of modern analysis on a scope quite unnecessary for our immediate interests, or give suggestive but misleading neoclassical heuristics masquerading as adequate arguments for what is truly needed, the remaining bulk of the book aims to discuss analytical issues at a technical level truly sufficient to convert appealing heuristics to persuasive, genuine proofs. For that matter, one's own lack of interest in the proofs might provide all the more interest in knowing that things widely believed are in fact provable by standard methods.

Chapter 4 explains enough Lie theory to understand the invariant differential operators on the ambient archimedean groups G, both in the simplest small examples and, more generally, determining the invariant Laplace-Beltrami operators explicitly in coordinates on the four simplest examples.

Chapter 5 explains how to integrate on quotients, without concern for explicit sets of representatives. Although in very simple situations, such as quotients $\mathbb{R}/\mathbb{Z}$ (the circle), it is easy to manipulate sets of representatives (the interval $[0, 1]$ for the circle), this eventually becomes infeasible, despite the traditional example of the explicit fundamental domain for $SL_2(\mathbb{Z})$ acting on the upper half-plane $\mathfrak{H}$. That is, much of the picturesque detail is actually inessential, which is fortunate because that level of detail is also unsustainable in all but the simplest examples.

Chapter 6 introduces natural actions of groups on spaces of functions on *physical* spaces on which the groups act. In some contexts, one might make a more elaborate *representation theory* formalism here, but it is possible to reap many of the benefits of the ideas of representation theory without the usual superstructure. That is, the *idea* of a linear action of a topological group on a topological vector space of functions on a physical space is the beneficial notion, with or without *classification*. It is true that at certain technical moments, classification results are crucial, so although we do not prove either the Borel-Casselman-Matsumoto classification in the p-adic case [Borel 1976], [Matsumoto 1977], [Casselman 1980], nor the subrepresentation theorem [Casselman 1978/1980], [Casselman-Miličić 1982] in the archimedean case, ideally the roles of these results are made clear. Classification results per se, although difficult and interesting problems, do not necessarily affect the foundational analytic aspects of automorphic forms.

Chapter 7 proves the discreteness of spaces of cuspforms, in various senses, in examples of varying complexity. Here, it becomes apparent that genuine proofs, as opposed to heuristics, require some sophistication concerning topologies on natural function spaces, beyond the typical Hilbert, Banach, and Fréchet spaces. Here again, there is a forward reference to the extended appendix on function spaces and classes of topological vector spaces necessary for practical analysis. Further, even less immediately apparent, but in fact already needed in the discussion of decomposition of pseudo-Eisenstein series in terms of genuine Eisenstein series, we need a coherent and effective theory of *vector-valued integrals*, a complete, succinct form given in the corresponding appendix, following Gelfand and Pettis, making explicit the most important corollaries on uniqueness of invariant functions, differentiation under integral signs with respect to parameters, and related.

Chapter 8 fills an unobvious need, proving that automorphic forms that are of *moderate growth* and are eigenfunctions, for Laplacians have asymptotics given by their *constant terms*. In the smaller examples, it is easy to make this precise. For SL_n with $n \geq 3$, some effort is required for an accurate statement. As corollaries, L^2 cuspforms that are eigenfunctions are of rapid decay, and Eisenstein series have relatively simple asymptotics given by their constant terms. Thus, we discover again the need to prove that Eisenstein series have vector-valued meromorphic continuations, specifically as moderate-growth functions.

Chapter 9 carefully develops ideas concerning unbounded symmetric operators on Hilbert spaces, thinking especially of operators related to Laplacians Δ, and especially those such that $(\Delta - \lambda)^{-1}$ is a compact-operator-valued meromorphic function of $\lambda \in \mathbb{C}$. On one hand, even a naive conception of the general behavior of Laplacians is fairly accurate, but this is due to a subtle fact that needs proof, namely, the *essential self-adjointness* of Laplacians on natural spaces such as $\mathbb{R}^n$, multi-toruses $\mathbb{T}^n$, spaces G/K, and even spaces $\Gamma \backslash G/K$. This has a precise sense: the (invariant) Laplacian restricted to *test functions* has a *unique* self-adjoint extension, which then is necessarily its *graph-closure*. Thus, the naive presumption, implicit or explicit, that the graph closure is a (maximal) self-adjoint extension is *correct*. On the other hand, the proof of meromorphic continuation of Eisenstein series in [Colin de Verdière 1981, 1982/1983] makes essential use of some quite counterintuitive features of (Friedrichs's) self-adjoint extensions of *restrictions* of self-adjoint operators, which therefore merit careful attention. In this context, the basic examples are the usual Sobolev spaces on $\mathbb{T}$ or $\mathbb{R}$ and the quantum harmonic oscillator $-\Delta + x^2$ on $\mathbb{R}$. An appendix recalls the proof of the spectral theorem for compact, self-adjoint operators.

Chapter 10 extends the idea from [Lax-Phillips 1976] to prove that larger spaces than spaces of cuspforms decompose *discretely* under the action of self-adjoint extensions $\widetilde{\Delta}_a$ of suitable restrictions Δ_a of Laplacians. Namely, the space of *pseudo-cuspforms* L_a^2 at cutoff height a is specified, not by requiring constant terms to vanish *entirely*, but by requiring that all constant terms vanish above height a. The discrete decomposition is proven, as expected, by showing that the resolvent $(\widetilde{\Delta}_a - \lambda)^{-1}$ is a meromorphic compact-operator-valued function of λ, and invoking the spectral theorem for self-adjoint compact operators. The compactness of the resolvent is a Rellich-type compactness result, proven by observing that $(\widetilde{\Delta}_a - \lambda)^{-1}$ maps L_a^2 to a Sobolev-type space $\mathfrak{B}_a^1$ with a finer topology on $\mathfrak{B}_a^1$ than the subspace topology and that the inclusion $\mathfrak{B}_a^1 \to L_a^2$ is compact.

Chapter 11 uses the discretization results of Chapter 10 to prove meromorphic continuations and functional equations of a variety of Eisenstein series, following [Colin de Verdière 1981, 1982/1983]'s application of the discreteness result in [Lax-Phillips 1976]. This is carried out first for the four simple examples, then for maximal proper parabolic Eisenstein series for $SL_n(\mathbb{Z})$, with cuspidal data. In both the simplest cases and the higher-rank examples, we identify the *exotic eigenfunctions* as being certain truncated Eisenstein series.

Chapter 12 uses several of the analytical ideas and methods of the previous chapters to reconsider automorphic Green's functions, and solutions to other differential equations in automorphic forms, by spectral methods. We prove a *pretrace formula* in the simplest example, as an application of a comparably simple instance of a *subquotient theorem*, which follows from asymptotics of solutions of second-order ordinary differential equations, recalled in a later appendix. We recast the pretrace formula as a demonstration that an automorphic Dirac δ-function lies in the expected *global automorphic Sobolev space*. The same argument gives a corresponding result for any compact automorphic *period*. Subquotient/subrepresentation theorems for groups such as $G = SO(n, 1)$ (rank-one groups with abelian unipotent radicals) appeared in [Casselman-Osborne 1975], [Casselman-Osborne 1978]. For higher-rank groups $SL_n(\mathbb{Z})$, the corresponding subrepresentation theorem is [Casselman 1978/1980], [Casselman-Miličić 1982]. Granting that, we obtain a corresponding pretrace formula for a class of compactly supported automorphic distributions, showing that these distributions lie in the expected global automorphic Sobolev spaces.

Chapter 13 is an extensive appendix with many examples of natural spaces of functions and appropriate topologies on them. One point is that too-limited types of topological vector spaces are inadequate to discuss natural function spaces arising in practice. We include essential standard arguments characterizing locally convex topologies in terms of families of seminorms. We prove the *quasi-completeness* of all natural function spaces, weak duals, and spaces of maps between them. Notably, this includes spaces of distributions.

Chapter 14 proves existence of Gelfand-Pettis vector-valued integrals of compactly supported continuous functions taking values in locally convex, quasi-complete topological vector space. Conveniently, the previous chapter showed that all function spaces of practical interest meet these requirements. The fundamental property of Gelfand-Pettis integrals is that for V-valued f, $T : V \to W$ continuous linear,

$$T\left(\int f\right) = \int T \circ f$$

at least for f continuous, compactly supported, V-valued, where V is quasi-complete and locally convex. That is, continuous linear operators pass inside the integral. In suitably topologized natural function spaces, this situation includes differentiation with respect to a parameter. In this situation, as corollaries we can easily prove uniqueness of invariant distributions, density of smooth vectors, and similar.

Chapter 15 carefully discusses holomorphic V-valued functions, using the Gelfand-Pettis integrals as well as a variant of the Banach-Steinhaus theorem. That is, weak holomorphy implies (strong) holomorphy, and the expected Cauchy integral formulas and Cauchy-Goursat theory apply almost *verbatim* in the vector-valued situation. Similarly, we prove that for f a V-valued function on an interval $[a, b]$, $\lambda \circ f$ being C^k for all $\lambda \in V^*$ implies that f itself is C^{k-1} as a V-valued function.

Chapter 16 reviews basic results on asymptotic expansions of integrals and of solutions to second-order ordinary differential equations. The methods are deliberately general, rather than invoking specific features of special functions, to illustrate methods that are applicable more broadly. The simple subrepresentation theorem in Chapter 12 makes essential use of asymptotic expansions.

Our coverage of modern analysis does not aim to be either systematic or complete but well-grounded and adequate for the aforementioned issues concerning automorphic forms. In particular, several otherwise-apocryphal results are treated carefully. We want a sufficient viewpoint so that attractive heuristics, for example, from physics, can become succinct, genuine proofs. Similarly, we do *not* presume familiarity with Lie theory, nor algebraic groups, nor representation theory, nor algebraic geometry, and certainly not with classification of representations of Lie groups or p-adic groups. All these are indeed useful, in the long run, but it is unreasonable to demand mastery of these before thinking about analytical issues concerning automorphic forms. Thus, we directly develop some essential ideas in these supporting topics, sufficient for immediate purposes here. [Lang 1975] and [Iwaniec 2002] are examples of the self-supporting exposition intended here.

Naturally, any novelty here is mostly in the presentation, rather than in the facts themselves, most of which have been known for several decades. Sources and origins can be most clearly described in a historical context, as follows.

The reduction theory in [1.5] is merely an imitation of the very classical treatment for $SL_2(\mathbb{Z})$, including some modern ideas, as in [Borel 1997]. The subtler versions in [2.2] and [3.3] are expanded versions of the first part of [Godement 1962–1964], a more adele-oriented reduction theory than [Borel 1965/1966b], [Borel 1969], and [Borel-HarishChandra 1962]. Proofs [1.9.1], [2.8.6], [3.10.1-2], [3.11.1] of convergence of Eisenstein series are due to Godement use similar ideas, reproduced for real Lie groups in [Borel 1965/1966a]. Convergence arguments on larger groups go back at least to [Braun 1939]'s treatment of convergence of Siegel Eisenstein series. Holomorphic Hilbert-Blumenthal modular forms were studied by [Blumenthal 1903/1904]. What would now be called degenerate Eisenstein series for GL_n appeared in [Epstein 1903/1907]. [Picard 1882, 1883, 1884] was one of the earliest investigations beyond the elliptic modular case. Our notion of *truncation* is from [Arthur 1978] and [Arthur 1980].

Eigenfunction expansions and various notions of convergence are a pervasive theme here and have a long history. The idea that periodic functions should be expressible in terms of sines and cosines is at latest from [Fourier 1822], including what we now call the Dirichlet kernel, although [Dirichlet 1829] came later. Somewhat more generally, eigenfunction expansions for Sturm-Liouville problems appeared in [Sturm 1836] and [Sturm 1833a,b,1836a,b] but were not made rigorous until [Bôcher 1898/1899] and [Steklov 1898] (see [Lützen 1984]). Refinements of the spectral theory of ordinary differential equations continued in [Weyl 1910], [Kodaira 1949], and others, addressing issues of non-compactness and unboundedness echoing complications in the behavior of Fourier transform and Fourier inversion on the line [Bochner 1932], [Wiener 1933]. Spectral theory and eigenfunction expansions for integral equations, which we would now call compact operators [9.A], were recognized as more tractable than direct treatment of differential operators soon after 1900: [Schmidt 1907], [Myller-Lebedev 1907], [Riesz 1907], [Hilbert 1909], [Riesz 1910], [Hilbert 1912]. Expansions in spherical harmonics were used in the 18th century by S.P. Laplace and J.-L. Lagrange, and eventually subsumed in the representation theory of compact Lie groups [Weyl 1925/1926], and in eigenfunction expansions on Riemannian manifolds and Lie groups, as in [Minakshisundaram-Pleijel 1949], [Povzner 1953], [Avakumović 1956], [Berezin 1956], and many others.

Spectral decomposition and synthesis of various types of automorphic forms is more recent, beginning with [Maaß 1949], [Selberg 1956], and [Roelcke 1956a, 1956b]. The spectral decomposition for automorphic forms on general reductive groups is more complicated than might have been anticipated by the earliest pioneers. Subtleties are already manifest in [Gelfand-Fomin 1952], and then in [Gelfand-Graev 1959], [Harish-Chandra 1959], [Gelfand-PiatetskiShapiro 1963], [Godement 1966b], [Harish-Chandra 1968], [Langlands 1966], [Langlands 1967/1976], [Arthur 1978], [Arthur 1980], [Jacquet 1982/1983], [Moeglin-Waldspurger 1989], [Moeglin-Waldspurger 1995], [Casselman 2005], [Shahidi 2010]. Despite various formalizations, spectral synthesis of automorphic forms seems most clearly understood in fairly limited scenarios: [Godement 1966a], [Faddeev 1967], [Venkov 1971], [Faddeev-Pavlov 1972], [Arthur 1978], [Venkov 1979], [Arthur 1980], [Cogdell-PiatetskiShapiro 1990], largely due to issues of convergence, often leaving discussions in an ambiguous realm of (nevertheless interesting) heuristics.

Regarding meromorphic continuation of Eisenstein series: our proof [2.B] for the case [2.9] of GL_2 is an adaptation of the Poisson summation argument from [Godement 1966a]. The essential idea already occurred in [Rankin 1939] and [Selberg 1940]. [Elstrodt-Grunewald-Mennicke 1985] treated examples including our example $SL_2(\mathbb{Z}[i])$, and in that context [Elstrodt-Grunewald-Mennicke 1987] treats special cases of the *period* computation of [2.C]. For Eisenstein series in rank one groups, compare also [Cohen-Sarnak 1980], which treats a somewhat larger family including our simplest four examples, and then [Müller 1996]. The minimal-parabolic example in [3.12] using Bochner's lemma [3.A] essentially comes from an appendix in [Langlands 1967/1976]. The arguments for the broader class of examples in Chapter 11 are adaptations of [Colin de Verdière 1981, 1982/1983], using discretization effects of pseudo-Laplacians from Chapter 10, which adapts the idea of [Lax-Phillips 1976]. Certainly one should compare the arguments in [Harish-Chandra 1968], [Langlands 1967/1976], [Wong 1990], and [Moeglin-Waldspurger 1995], the last of which gives a version of Colin de Verdière's idea due to H. Jacquet.

The discussion of group actions on function spaces in Chapter 6 is mostly very standard. Apparently the first occurrence of the Gelfand-Kazhdan criterion idea is in [Gelfand 1950]. An extension of that idea appeared in [Gelfand-Kazhdan 1975].

The arguments for discrete decomposition of cuspforms in Chapter 11 are adaptations of [Godement 1966b]. The discrete decomposition examples for larger spaces of pseudo-cuspforms in Chapter 10 use the idea of [Lax-Phillips 1976]. The idea of this decomposition perhaps goes back to [Gelfand-Fomin 1952] and, as with many of these ideas, was elaborated on in the iconic sources [Gelfand-Graev 1959], [Harish-Chandra 1959], [Gelfand-PiatetskiShapiro 1963], [Godement 1966b], [Harish-Chandra 1968], [Langlands 1967/1976], and [Moeglin-Waldspurger 1989].

Difficulties with pointwise convergence of Fourier series of continuous functions, and problems in other otherwise-natural Banach spaces of functions, were well appreciated in the late 19th century. There was a precedent for constructs avoiding strictly pointwise conceptions of functions in the very early 20th century, when B. Levi, G. Fubini, and D. Hilbert used Hilbert space constructs to legitimize Dirichlet's minimization principle, in essence that a nonempty closed convex set should have a (unique) point nearest a given point not in that set. The too-general form of this principle is false, in that both existence and uniqueness easily fail in Banach spaces, in natural examples, but the principle is correct in Hilbert spaces. Thus, natural Banach spaces of pointwise-valued functions, such as continuous functions on a compact set with sup norm, do not support this minimization principle. Instead, Hilbert-space versions of continuity and differentiability are needed, as in [Levi 1906]. This idea was systematically developed by [Sobolev 1937, 1938, 1950]. We recall the L^2 Sobolev spaces for circles in [9.5] and for lines in [9.7] and develop various (global) automorphic versions of Sobolev spaces in Chapters 10, 11, and 12.

For applications to analytic number theory, automorphic forms are often constructed by *winding up* various simpler functions containing parameters, forming *Poincaré series* [Cogdell-PiatetskiShapiro 1990] and [Cogdell-PiatetskiShapiro-Sarnak 1991]. Spectral expansions are the standard device for demonstration of meromorphic continuation in the parameters, if it exists at all, which is a nontrivial issue [Estermann 1928], [Kurokawa 1985a,b]. For the example of automorphic Green's functions, namely, solutions to equations $(\Delta - s(s - 1))u = \delta_w^{\text{afc}}$ with invariant Laplacian Δ on $\mathfrak{H}$ and automorphic Dirac δ on the right, [Huber 1955] had considered such matters in the context of lattice-point problems in hyperbolic spaces, and, independently, [Selberg 1954] had addressed this issue in lectures in Göttingen. [Neunhöffer 1973] carefully considers the convergence and meromorphic continuation of a solution of that equation formed by winding up. See also [Elstrodt 1973]. The complications or failures of pointwise convergence of the spectral synthesis expressions can often be avoided entirely by considering convergence in suitable global automorphic Sobolev spaces described in Chapter 12. See [DeCelles 2012] and [DeCelles 2016] for developments in this spirit.

Because of the naturality of the issue and to exploit interesting idiosyncrasies, we pay considerable attention to invariant Laplace-Beltrami operators and their eigenfunctions. To have genuine proofs, rather than heuristics, Chapter 9 attends to rigorous notions of unbounded operators on Hilbert spaces [vonNeumann 1929], with motivation toward [vonNeumann 1931], [Stone 1929, 1932], [Friedrichs 1934/1935], [Krein 1945], [Krein 1947]. In fact, [Friedrichs 1934/1935]'s special construction [9.2] has several useful idiosyncrasies, exploited in Chapters 10 and 11. Incidentally, the apparent fact that the typically naive treatment of many natural Laplace-Beltrami operators without boundary conditions does not lead to serious mistakes is a corollary of their *essential self-adjointness* [9.9], [9.10]. That is, in many situations, the naive form of the operator admits a *unique* self-adjoint extension, and this extension is the *graph closure* of the original. Thus, in such situations, a naive treatment is provably reasonable. However, the Lax-Phillips discretization device, and Colin de Verdière's use of it to prove meromorphic continuation of Eisenstein series and also to convert certain inhomogeneous differential equations to homogeneous ones, illustrate the point that *restrictions* of essentially self-adjoint operators need not remain essentially self-adjoint. With hindsight, this possibility is already apparent in the context of Sturm-Liouville problems [9.3].

The *global automorphic Sobolev spaces* of Chapter 12 already enter in important auxiliary roles as the spaces $\mathfrak{B}^1$, $\mathfrak{B}^1_a$ in Chapter 10's proofs of discrete decomposition of spaces of pseudo-cuspforms, and $\mathfrak{C}^1$ and $\mathfrak{C}^1_a$ in [11.7-11.11] proving meromorphic continuation of Eisenstein series. The basic estimate called a *pretrace formula* occurred as a precursor to *trace formulas*, as in [Selberg 1954], [Selberg 1956], [Hejhal 1976/1983], and [Iwaniec 2002]. The notion of global automorphic Sobolev spaces provides a reasonable context for discussion of automorphic Green's functions, other automorphic distributions, and solutions of partial differential equations in automorphic forms. The heuristics for Green's functions [Green 1828], [Green 1837] had repeatedly shown their utility in the 19th century. Differential equations $(-\Delta - \lambda)u = \delta$ related to Green's functions had been used by physicists [Dirac 1928a/1928b, 1930], [Thomas 1935], [Bethe-Peierls 1935], with excellent corroboration by physical experiments and are nowadays known as *solvable models*. At the time, and currently, in physics contexts they are rewritten as $((-\Delta + \delta) - \lambda)u = 0$, viewing $-\Delta + \delta$ as a perturbation of $-\Delta$ by a *singular potential* δ, a mathematical idealization of a very short-range force. This was treated rigorously in [Berezin-Faddeev 1961]. The necessary systematic estimates on eigenvalues of integral operators use a *subquotient theorem*, which we prove for the four simple examples, as in that case the issue is about asymptotics of solutions of second-order differential equations, classically understood as recalled in an appendix (Chapter 16). The general result is the *subrepresentation theorem* from [Casselman 1978/1980], [Casselman-Miličić 1982], improving the *subquotient theorem* of [HarishChandra 1954]. In [Varadarajan 1989] there are related computations for $SL_2(\mathbb{R})$.

In the discussion of natural function spaces in Chapter 13, in preparation for the vector-valued integrals of the following chapter, the notion of *quasi-completeness* proves to be the correct general version of completeness. The *incompleteness* of weak duals has been known at least since [Grothendieck 1950], which gives a systematic analysis of completeness of various types of duals. This larger issue is systematically discussed in [Schaefer-Wolff 1966/1999], pp. 147–148 and following. The significance of the compactness of the closure of the convex hull of a compact set appears, for example, in the discussion of vector-valued integrals in [Rudin 1991], although the latter does not make clear that this condition is fulfilled in more than Fréchet spaces and does not mention quasi-completeness. To apply these ideas to distributions, one might cast about for means to prove the compactness condition, eventually hitting on the hypothesis of quasi-completeness in conjunction with ideas from the proof of the Banach-Alaoglu theorem. Indeed, in [Bourbaki 1987] it is shown (by apparently different methods) that quasi-completeness implies this compactness condition. The fact that a bounded subset of a countable strict inductive limit of closed subspaces must actually be a bounded subset of one of the subspaces, easy to prove once conceived, is attributed to Dieudonne and Schwartz in [Horvath 1966]. See also [Bourbaki 1987], III.5 for this result. Pathological behavior of uncountable colimits was evidently first exposed in [Douady 1963].

In Chapter 14, rather than *constructing* vector-valued integrals as limits following [Bochner 1935], [Birkhoff 1935], et alia, we use the [Gelfand 1936]-[Pettis 1938] *characterization* of integrals, which has good functorial properties and gives a forceful reason for *uniqueness*. The issue is *existence*. Density of smooth vectors follows [Gårding 1947]. Another of application of holomorphic and meromorphic vector-valued functions is to *generalized functions*, as in [Gelfand-Shilov 1964], studying *holomorphically parametrized families* of distributions. A hint appears in the discussion of holomorphic vector-valued functions in [Rudin 1991]. A variety of developmental episodes and results in the Banach-space-valued case is surveyed in [Hildebrandt 1953]. Proofs and application of many of these results are given in [Hille-Phillips 1957]. (The first edition, authored by Hille alone, is sparser in this regard.) See also [Brooks 1969] to understand the viewpoint of those times.

Ideas about vector-valued holomorphic and differentiable functions, in Chapter 15, appeared in [Schwartz 1950/1951], [Schwartz 1952], [Schwartz 1953/1954], and in [Grothendieck 1953a,1953b].

The asymptotic expansion results of Chapter 16 are standard. [Blaustein-Handelsman 1975] is a standard source for asymptotics of integrals. *Watson's lemma* and *Laplace's method* for integrals have been used and rediscovered repeatedly. Watson's lemma dates from at latest [Watson 1918], and Laplace's method at latest from [Laplace 1774]. [Olver 1954] notes that Carlini [Green 1837] and [Liouville 1837] investigated relatively simple cases of asymptotics at irregular singular points of ordinary differential equations, without complete rigor. According to [Erdélyi 1956] p. 64, there are roughly two proofs that the standard argument produces genuine asymptotic expansions for solutions of the differential equation. Poincaré's approach, elaborated by J. Horn, expresses solutions as Laplace transforms and invokes Watson's lemma to obtain asymptotics. G.D. Birkhoff and his students constructed auxiliary differential equations from partial sums of the asymptotic expansion, and compared these auxiliary equations to the original [Birkhoff 1908], [Birkhoff 1909], [Birkhoff 1913]. Volterra integral operators are important in both approaches, insofar as asymptotic expansions behave better under *integration* than under *differentiation*. Our version of the Birkhoff argument is largely adapted from [Erdélyi 1956].

Many parts of this exposition are adapted and expanded from [Garrett vignettes], [Garrett mfms-notes], [Garrett fun-notes], and [Garrett alg-noth-notes]. As is surely usual in book writing, many of the issues here had plagued me for decades.

9

Unbounded Operators on Hilbert Spaces

This is preparation for eigenfunction decompositions of Hilbert spaces by operators closely related to *invariant Laplacians*.

Amazingly, *resolvents* $R_\lambda = (T - \lambda)^{-1}$ can exist, as *everywhere-defined*, continuous linear maps on a Hilbert space, even for T *unbounded* and only *densely defined*. Further hypotheses on T are needed, but these hypotheses are met in useful situations occurring in practice. In particular, we need that T is *symmetric*, in the sense that $\langle Tv, w \rangle = \langle v, Tw \rangle$ for v, w in the domain D_T of T, and *semibounded* in the sense that there is a constant C such that either $\langle Tv, v \rangle \geq C \cdot \langle v, v \rangle$ for all v in D_T or $\langle Tv, v \rangle \leq C \langle v, v \rangle$ for all v in D_T. In that circumstance, T has a self-adjoint *Friedrichs extension*, with several good features, described explicitly in what follows.

In practice, anticipating that a given unbounded operator is self-adjoint *when extended suitably*, a simple version of the operator is defined on an easily described, small, dense domain, specifying a *symmetric* operator. Then a self-adjoint *extension* is shown to exist, as shown subsequently in Friedrichs's theorem.

For example, [9.5] gives a simple application, recovering the standard fact that the Hilbert space $L^2(\mathbb{T})$ on the circle $\mathbb{T} = \mathbb{R}/2\pi\mathbb{Z}$ has an *orthogonal Hilbert-space basis* of exponentials e^{inx} with $n \in \mathbb{Z}$, using ideas still applicable to situations *lacking* analogues of Dirichlet or Fejér kernels. These exponentials are *eigenfunctions* for the Laplacian $\Delta = d^2/dx^2$, so it would suffice to show that $L^2(\mathbb{T})$ has an orthogonal basis of eigenfunctions for Δ. Two technical issues must be overcome: the most awkward is that Δ does not map $L^2(\mathbb{T})$ to itself. Second, there is no guarantee that infinite-dimensional Hilbert spaces have Hilbert-space bases of eigenfunctions for a given linear operator. Indeed, reasonable operators on infinite-dimensional spaces may fail to have *any* eigenvectors. For example, on $L^2[a, b]$, the multiplication operator $Tf(x) = x \cdot f(x)$ is continuous, possesses the symmetry property $\langle Tf, g \rangle = \langle f, Tg \rangle$, but has no eigenvectors. That is, the *spectrum* of operators on infinite-dimensional Hilbert spaces typically includes more than *eigenvalues*.

Natural operators like d^2/dx^2 on $L^2[a, b]$ are not *bounded*, that is, not *continuous* operators. Not-necessarily-bounded operators are called *unbounded*, despite the inconsistency of language.

Self-adjoint operators on Hilbert spaces generally do *not* give orthogonal Hilbert-space bases of eigenvectors, but an important special class *does* have a spectral theory imitating finite-dimensional spectral theory: the *compact self-adjoint* operators, which *always* give an orthogonal Hilbert-space basis of eigenvectors, as in [9.A].

Genuinely unbounded operators such as Δ are never *continuous*, much less *compact*, but in happy circumstances they may have *compact resolvent* $(1 - \Delta)^{-1}$. Often, this compactness can be proven by a Sobolev imbedding lemma, a Rellich compactness lemma, and Friedrichs's construction, recovering a good spectral theory, as in examples below.

9.1 Unbounded Symmetric Operators on Hilbert Spaces

The natural differential operator $\Delta = \frac{d^2}{dx^2}$ on $\mathbb{R}$ has no sensible definition as mapping all of the Hilbert space $L^2(\mathbb{R})$ to itself, whatever else can be said. The possibility of thinking of Δ as differentiating L^2 functions *distributionally* is useful but sacrifices information if it abandons the L^2 behavior too completely. There is substantial motivation to accommodate discontinuous (*unbounded*) linear maps on Hilbert spaces, under some reasonable technical hypotheses, as apply to operators like Δ. At the most cautious, Δ certainly maps $C_c^\infty(\mathbb{R})$ to itself, and, by integration by parts,

$$\langle \Delta f, g \rangle = \langle f, \Delta g \rangle \qquad \text{(for both } f, g \in C_c^\infty(\mathbb{R}))$$

This is *symmetry* of Δ.

A not-necessarily continuous, that is, not-necessarily *bounded*, linear operator T, defined on a *dense* subspace D_T of a Hilbert space V, is called an *unbounded operator on V*, even though it is likely not defined or definable on all of V. We consider mostly *symmetric* unbounded operators T, meaning that $\langle Tv, w \rangle = \langle v, Tw \rangle$ for v, w in the domain D_T of T. For example, the Laplacian is symmetric on $C_c^\infty(\mathbb{R})$, by integration by parts.

For unbounded operators on V, description of the *domain* is critical: an unbounded operator T on V is a subspace D of V and a linear map $T : D \longrightarrow V$. Nevertheless, explicit naming of the domain of an unbounded operator is often suppressed, instead writing $T_1 \subset T_2$ when T_2 is an *extension* of T_1, in the sense that the domain of T_2 contains that of T_1, and the restriction of T_2 to the domain of T_1 agrees with T_1. Unlike self-adjoint operators on finite-dimensional spaces, and unlike self-adjoint *bounded* operators on Hilbert spaces, symmetric *unbounded* operators, even when densely defined, usually need to be *extended* to behave more like self-adjoint operators in finite-dimensional and bounded-operator situations.

An operator T', D' is a *subadjoint* to a symmetric operator T, D when

$$\langle Tv, w \rangle = \langle v, T'w \rangle \qquad \text{(for } v \in D, w \in D')$$

For *dense* domain D, for given D' there is *at most* one T' meeting the subadjointness condition.

In various useful circumstances there is a *unique maximal* element, in terms of domain, among all subadjoints to T, the *adjoint* T^* of T. *Uniqueness* of a maximal subadjoint is proven subsequently for T symmetric. Perhaps surprisingly, we will see that the adjoint T^* of a symmetric operator T is *not* symmetric *unless* already T is self-adjoint, that is, unless $T = T^*$. In particular, existence of adjoints for symmetric, densely defined operators T does not immediately imply existence of $(T^*)^*$. Paraphrasing the notion of *symmetry*: a densely defined operator T is *symmetric* when $T \subset T^*$, and *self-adjoint* when $T = T^*$. These comparisons refer to the *domains* of these not-everywhere-defined operators. In the following claim and its proof, the domain of a map S on V is incorporated in a reference to its *graph*

$$\text{graph } S = \{v \oplus Sv : v \in \text{domain } S\} \subset V \oplus V$$

The direct sum $V \oplus V$ is a Hilbert space with natural inner product $\langle v \oplus v', w \oplus w' \rangle = \langle v, v' \rangle + \langle w, w' \rangle$. Define an isometry $U : V \oplus V \to V \oplus V$ by $v \oplus w \to -w \oplus v$.

[9.1.1] Claim: Given T with *dense* domain D, there is a unique *maximal* T^*, D^* among all subadjoints to T, D. The adjoint T^* is *closed*, in the sense that its *graph* is closed in $V \oplus V$. In fact, the adjoint is *characterized* by its graph, the orthogonal complement in $V \oplus V$ to the image of the graph of T under U, namely,

$$\text{graph } T^* = \text{orthogonal complement of } U(\text{graph } T)$$

Proof: The adjointness condition $\langle Tv, w \rangle = \langle v, T^*w \rangle$ for given $w \in V$ is an orthogonality condition

$$\langle w \oplus T^*w, \ U(v \oplus Tv) \rangle = 0 \qquad \text{(for all } v \text{ in the domain of } T\text{)}$$

The graph of *any* sub-adjoint is a subset of $X = U(\text{graph } T)^\perp$. Since T is densely defined, for given $w \in V$ there is *at most* one possible value w' such that $w \oplus w' \in X$, so this orthogonality condition determines a well-defined function T^* on a subset of V, by $T^*w = w'$ if there exists $w' \in V$ such that $w \oplus w' \in X$. Linearity of T^* is immediate. It is maximal among subadjoints to T because the graph of any subadjoint is a subset of the graph of T^*. Orthogonal complements are closed, so T^* has a closed graph.   ///

[9.1.2] Corollary: For $T_1 \subset T_2$ with dense domains, $T_2^* \subset T_1^*$.   ///

[9.1.3] Corollary: $T \subset T^{**}$ for densely defined, *symmetric* T.

Proof: Since T is symmetric, and by uniqueness of the adjoint, $T^* \supset T$. In particular, T^* is densely defined. Hence, from earlier, T^* has an adjoint T^{**}. The description of the adjoint in terms of orthogonality in $V \oplus V$ shows that $T^{**} \supset T$.   ///

[9.1.4] Corollary: A densely defined self-adjoint operator has a closed graph.

Proof: Self-adjointness of densely defined T includes equality of domains $T = T^*$. Again, since the graph of T^* is an orthogonal complement, it is closed. ///

Closed-ness of the graph of a self-adjoint operator is essential in proving existence of *resolvents*.[1]

---

[1] In general, the graph-closure of an operator need not be the graph of an operator! An operator whose graph-closure *is* the graph of an operator is *closeable*. A broader discussion of unbounded operators would consider such issues at greater length, but such a discussion is not necessary here.

[9.1.5] Corollary: The adjoint T^* of a symmetric densely defined operator T is *also* symmetric if and only if $T = T^*$. ///

The use of the term *symmetric* in this context is potentially misleading. The notation $T = T^*$ allows an inattentive reader to forget nontrivial assumptions on the *domains* of the operators. The equality of domains of T and T^* is critical for legitimate computations.

[9.1.6] Proposition: Eigenvalues for symmetric operators T, D are *real*.

Proof: Suppose $0 \neq v \in D$ and $Tv = \lambda v$. Then

$$\lambda \langle v, v \rangle = \langle \lambda v, v \rangle = \langle Tv, v \rangle = \langle v, T^*v \rangle$$

because $v \in D \subset D^*$. Because T^* agrees with T on D, $\langle v, T^*v \rangle = \langle v, \lambda v \rangle = \overline{\lambda} \langle v, v \rangle$. Thus, $\lambda \langle v, v \rangle = \overline{\lambda} \langle v, v \rangle$. ///

The *resolvent* of T is $R_\lambda = (T - \lambda)^{-1}$ for $\lambda \in \mathbb{C}$, when this inverse exists as a *continuous, everywhere-defined* linear operator on V.

[9.1.7] Theorem: Let T be self-adjoint with dense domain D. For $\lambda \in \mathbb{C}$, $\lambda \notin \mathbb{R}$, the image $(T - \lambda)D$ is the whole Hilbert space V. The resolvent R_λ exists. For T *positive*, for $\lambda \notin [0, +\infty)$, the image $(T - \lambda)D$ is the whole space V, and R_λ exists.

Proof: For $\lambda = x + iy$ off the real line and v in the domain of T,

$$|(T - \lambda)v|^2 = |(T - x)v|^2 + \langle (T - x)v, iyv \rangle + \langle iyv, (T - x)v \rangle + y^2 |v|^2$$

$$= |(T - x)v|^2 - iy\langle (T - x)v, v \rangle + iy\langle v, (T - x)v \rangle + y^2 |v|^2$$

By the symmetry of T, and the fact that the domain of T^* contains that of T, $\langle v, Tv \rangle = \langle T^*v, v \rangle = \langle Tv, v \rangle$. Thus,

$$|(T - \lambda)v|^2 = |(T - x)v|^2 + y^2 |v|^2 \geq y^2 |v|^2$$

For $y \neq 0$ and $v \neq 0$, $(T - \lambda)v \neq 0$, so $T - \lambda$ is *injective*. We must show that $(T - \lambda)D$ is the whole Hilbert space V. If

$$0 = \langle (T - \lambda)v, w \rangle \qquad \text{(for all } v \in D)$$

then the adjoint of $T - \lambda$ can be defined on w simply as $(T - \lambda)^*w = 0$, since

$$\langle (T - \lambda)v, w \rangle = 0 = \langle v, 0 \rangle \qquad \text{(for all } v \in D)$$

Thus, $T^* = T$ is defined on w, and $Tw = \overline{\lambda}w$. For λ not real, this implies $w = 0$. Thus, $(T - \lambda)D$ is *dense* in V.

Since T is self-adjoint, it is *closed*, so $T - \lambda$ is closed. The equality

$$|(T - \lambda)v|^2 = |(T - x)v|^2 + y^2 |v|^2$$

gives

$$|(T - \lambda)v|^2 \gg_y |v|^2$$

Thus, for fixed $y \neq 0$, the map

$$F : v \oplus (T - \lambda)v \longrightarrow (T - \lambda)v$$

respects the metrics, in the sense that

$$|(T - \lambda)v|^2 \leq |(T - \lambda)v|^2 + |v|^2 \ll_y |(T - \lambda)v|^2$$

for fixed $y \neq 0$. The graph of $T - \lambda$ is *closed*, so is a *complete* metric subspace of $V \oplus V$. Since F respects the metrics, it preserves completeness. Thus, the metric space $(T - \lambda)D$ is *complete*, so is a closed subspace of V. Since the closed subspace $(T - \lambda)D$ is dense, it is V. Thus, for $\lambda \notin \mathbb{R}$, R_λ is everywhere-defined. Its norm is bounded by $1/|\mathrm{Im}\,\lambda|$, so it is a continuous linear operator on V.

Similarly, for T *positive*, for $\mathrm{Re}\,(\lambda) < 0$,

$$|(T - \lambda)v|^2 = |Tv|^2 - \lambda\langle Tv, v\rangle - \overline{\lambda}\langle v, Tv\rangle + |\lambda|^2 \cdot |v|^2$$

$$= |Tv|^2 + 2|\mathrm{Re}\,\lambda|\langle Tv, v\rangle + |\lambda|^2 \cdot |v|^2 \geq |\lambda|^2 \cdot |v|^2$$

Then the same argument proves the existence of an everywhere-defined inverse $R_\lambda = (T - \lambda)^{-1}$, with $\|R_\lambda\| < 1/|\lambda|$ for $\mathrm{Re}\,\lambda < 0$.   ///

[9.1.8] Theorem: (Hilbert) For T self-adjoint, for points λ, μ off the real line, or, for T *positive* self-adjoint and λ, μ off $[0, +\infty)$,

$$R_\lambda - R_\mu = (\lambda - \mu)R_\lambda R_\mu$$

For the operator-norm topology, $\lambda \to R_\lambda$ is *holomorphic* at such points.

Proof: From the previous theorem, for such T, λ, the image $(T - \lambda)D$ is the whole Hilbert space V. Applying R_λ to

$$1_V - (T - \lambda)R_\mu = \big((T - \mu) - (T - \lambda)\big)R_\mu = (\lambda - \mu)R_\mu$$

gives

$$R_\lambda(1_V - (T - \lambda)R_\mu) = R_\lambda\big((T - \mu) - (T - \lambda)\big)R_\mu = R_\lambda(\lambda - \mu)R_\mu$$

Then

$$\frac{R_\lambda - R_\mu}{\lambda - \mu} = R_\lambda R_\mu$$

For holomorphy, with $\lambda \to \mu$,

$$\frac{R_\lambda - R_\mu}{\lambda - \mu} - R_\mu^2 = R_\lambda R_\mu - R_\mu^2 = (R_\lambda - R_\mu)R_\mu = (\lambda - \mu)R_\lambda R_\mu R_\mu$$

Taking operator norm, using $\|R_\lambda\| \le 1/|\mathrm{Im}\,\lambda|$ from the previous computation,

$$\left\| \frac{R_\lambda - R_\mu}{\lambda - \mu} - R_\mu^2 \right\| \le \frac{|\lambda - \mu|}{|\mathrm{Im}\,\lambda| \cdot |\mathrm{Im}\,\mu|^2}$$

Thus, for $\mu \notin \mathbb{R}$, as $\lambda \to \mu$, this operator norm goes to 0, demonstrating the holomorphy.

For *positive* T, the estimate $\|R_\lambda\| \le 1/|\lambda|$ for $\mathrm{Re}\,\lambda \le 0$ yields holomorphy on the negative real axis by the same argument.   ///

9.2 Friedrichs's Self-Adjoint Extensions of Semibounded Operators

Semibounded operators are more tractable than general unbounded symmetric operators. A densely defined symmetric operator T, D is *positive* (or *nonnegative*), denoted $T \ge 0$, when $\langle Tv, v \rangle \ge 0$ for all $v \in D$. All the eigenvalues of a positive operator are nonnegative real. Similarly, T is *negative* when $\langle Tv, v \rangle \le 0$ for all v in the (dense) domain of T. Generally, if there is a constant $c \in \mathbb{R}$ such that $\langle Tv, v \rangle \ge c \cdot \langle v, v \rangle$ (written $T \ge c$), or $\langle Tv, v \rangle \le c \cdot \langle v, v \rangle$ (written $T \le c$), say T is *semi-bounded*. The following argument for *positive* operators can easily be adapted to the general semibounded situation.

For *positive*, symmetric T on V with dense domain D, define a Hermitian form $\langle , \rangle_1$ and corresponding norm $| \cdot |_1$ by[2]

$$\langle v, w \rangle_1 = \langle v, w \rangle + \langle Tv, w \rangle = \langle v, (1+T)w \rangle = \langle (1+T)v, w \rangle$$

for $v, w \in D$. The symmetry and positivity of T make $\langle , \rangle_1$ positive-definite Hermitian on D, and $\langle v, w \rangle_1$ has sense whenever at least one of v, w is in D. Let V^1 be the Hilbert-space completion of D with respect to the metric d_1 induced by the norm $| \cdot |_1$ on D. The completion V^1 continuously injects to V: for v_i a $| \cdot |_1$-Cauchy sequence in D, v_i is Cauchy in V in the original topology, since $|v_i - v_j| \le |v_i - v_j|_1$. For two sequences v_i, w_j with the same $| \cdot |_1$-limit v, the d_1-limit of $v_i - w_i$ is 0, so $|v_i - w_i| \le |v_i - w_i|_1 \to 0$. We identify V^1 with its natural image inside V, noting that V^1 has a finer topology than would be induced from V.

[9.2.1] Theorem: (Friedrichs) A *positive*, densely defined, symmetric operator T with domain D dense in Hilbert space V has a positive *self-adjoint* extension

---

[2] This is a slightly abstracted version of a Sobolev norm, as in [9.5] and [9.7].

$\widetilde{T}$ with domain $\widetilde{D} \subset V^1$, characterized by

$$\langle (1 + T)v, (1 + \widetilde{T})^{-1}w \rangle = \langle v, w \rangle \qquad \text{(for } v \in D \text{ and } w \in V)$$

The bound $\langle \widetilde{T}v, v \rangle \geq 0$ for v in the domain $\widetilde{D}$ of $\widetilde{T}$ is preserved. The resolvent $(1 + \widetilde{T})^{-1} : V \to V^1$ is continuous with the finer topology on V^1.

Proof: For $h \in V$ and $v \in V^1$, the functional $\lambda_h : v \to \langle v, h \rangle$ has a bound

$$|\lambda_h v| \leq |v| \cdot |h| \leq |v|_1 \cdot |h|$$

so the norm of the functional λ_h on V^1 is at most $|h|$. By Riesz-Fréchet, there is unique Bh in the Hilbert space V^1 with $|Bh|_1 \leq |h|$, such that $\lambda_h(v) = \langle v, Bh \rangle_1$ for $v \in V^1$, and then $|Bh| \leq |Bh|_1 \leq |h|$. The map $B : V \to V^1$ is verifiably linear. There is an obvious *symmetry* of B:

$$\langle Bv, w \rangle = \lambda_w(Bv) = \langle Bv, Bw \rangle_1 = \overline{\langle Bw, Bv \rangle_1}$$

$$= \overline{\lambda_v(Bw)} = \overline{\langle Bw, v \rangle} = \langle v, Bw \rangle \qquad \text{(for } v, w \in V)$$

Positivity of B is similar:

$$\langle v, Bv \rangle = \lambda_v(Bv) = \langle Bv, Bv \rangle_1 \geq \langle Bv, Bv \rangle \geq 0$$

B is *injective*: for $Bw = 0$, for all $v \in V^1$

$$0 = \langle v, 0 \rangle_1 = \langle v, Bw \rangle_1 = \lambda_w(v) = \langle v, w \rangle$$

Since V^1 is dense in V, this gives $w = 0$. The image of B is *dense* in V^1: if $w \in V^1$ is such that $\langle Bv, w \rangle_1 = \lambda_v(w) = 0$ for all $v \in V$, taking $v = w$ gives

$$0 = \lambda_w(w) = \langle w, Bw \rangle_1 = \langle Bw, Bw \rangle$$

and by injectivity $w = 0$. Thus, $B : V \to V^1 \subset V$ is bounded, symmetric, positive, injective, with dense image. In particular, B is self-adjoint.

Thus, B has a possibly *unbounded* positive, symmetric inverse A. Since B injects V to a dense subset V^1, necessarily A *surjects* from its domain (inside V^1) to V. We claim that A is *self-adjoint*. Let $S : V \oplus V \to V \oplus V$ by $S(v \oplus w) = w \oplus v$. Then graph $A = S(\text{graph } B)$. In computing orthogonal complements $X^\perp$, clearly

$$(SX)^\perp = S(X^\perp)$$

From the obvious $U \circ S = -S \circ U$, compute

$$\text{graph } A^* = (U \text{ graph } A)^\perp = (U \circ S \text{ graph } B)^\perp$$

$$= (-S \circ U \text{ graph } B)^\perp = -S((U \text{ graph } B)^\perp)$$

$$= -\text{ graph } A = \text{ graph } A$$

since the domain of B^* is the domain of B. Thus, A is self-adjoint.

We claim that for v in the domain of A, $\langle Av, v \rangle \geq \langle v, v \rangle$. Indeed, letting $v = Bw$,

$$\langle v, Av \rangle = \langle Bw, w \rangle = \lambda_w Bw = \langle Bw, Bw \rangle_1 \geq \langle Bw, Bw \rangle = \langle v, v \rangle$$

Similarly, with $v' = Bw'$, and $v \in V^1$,

$$\langle v, Av' \rangle = \langle v, w' \rangle = \lambda_{w'} v = \langle v, Bw' \rangle_1 = \langle v, v' \rangle_1$$

for $v \in V^1$, v' in the domain of A.

Last, prove that A is an extension of $S = 1 + T$. On one hand, as earlier,

$$\langle v, Sw \rangle = \lambda_{Sw} v = \langle v, BSw \rangle_1 \qquad \text{(for } v, w \in D)$$

On the other hand, by definition of $\langle , \rangle_1$,

$$\langle v, Sw \rangle = \langle v, w \rangle_1 \qquad \text{(for } v, w \in D)$$

Thus,

$$\langle v, w - BSw \rangle_1 = 0 \qquad \text{(for all } v, w \in D)$$

Since D is d_1-dense in V^1, $BSw = w$ for $w \in D$. Thus, $w \in D$ is in the range of B, so is in the domain of A, and

$$Aw = A(BSw) = Sw$$

Thus, the domain of A contains that of S and extends S, so the domain of A is dense in V^1 in the d_1-topology. In fact, $B = (1 + \widetilde{T})^{-1}$ maps $V \to V^1$ continuously even with the finer $\langle , \rangle_1$-topology on V^1: the relation $\langle v, Bw \rangle_1 = \langle v, w \rangle$ for $v \in V^1$ with $v = Bw$ gives

$$|Bw|_1^2 = \langle Bw, Bw \rangle_1 = \langle Bw, w \rangle \leq |Bw| \cdot |w| \leq |Bw|_1 \cdot |w|$$

The resulting $|Bw|_1 \leq |w|$ is continuity in the finer topology.   ///

Continuity of $(1 + \widetilde{T})^{-1} : V \to V^1$ with the finer topology on V^1 is a useful property of Friedrichs's self-adjoint extensions not shared by the other self-adjoint extensions of a given symmetric operator. It has an important corollary:

[9.2.2] Corollary: When the inclusion $V^1 \to V$ is *compact*, the resolvent $(1 + \widetilde{T})^{-1} : V \to V$ is compact.

Proof: In the notation of the proof of the theorem, $B : V \to V^1 \to V$ is the composition of this continuous map with the injection $V^1 \to V$ where V^1 has the finer topology. The composition of a continuous linear map with a compact operator is compact, so compactness of $V^1 \to V$ with the finer topology on V^1 suffices to prove compactness of the resolvent.   ///

9.3 Example: Incommensurable Self-adjoint Extensions

The differential operator $\frac{d^2}{dx^2}$ on $L^2[a, b]$ or $L^2(\mathbb{R})$ is a prototypical natural *unbounded operator*. It is undeniably *not continuous* in the L^2 topology: on $L^2[0, 1]$ the norm of $f(x) = x^n$ is $1/\sqrt{2n+1}$, and the second derivative of x^n is $n(n-1)x^{n-2}$, so the ratio of L^2 norms $|(x^n)''|/|x^n|$ goes to $+\infty$ as $n \to +\infty$. Since the operator is unbounded on polynomials, it certainly has no bounded *extension* to $L^2[0, 1]$.

Just below, we exhibit a *continuum* of mutually incomparable self-adjoint extensions of the restriction T of $-\frac{d^2}{dx^2}$ to smooth functions on $[a, b]$ vanishing at the endpoints. As this will illustrate, it is unreasonable to expect naturally occurring positive/negative, symmetric operators to have *unique* self-adjoint extensions. In brief, for unbounded operators arising from differential operators, varying *boundary conditions* gives mutually incomparable self-adjoint extensions. In that situation, the *graph-closure* $\overline{T} = T^{**}$ is *not* self-adjoint. Equivalently, T^* is *not* symmetric, proven as follows.

In general, the graph-closure of an unbounded operator need not be the graph of an operator, but this potential problem does not exist for a *semibounded* operator T, since the Friedrichs self-adjoint extension $\widetilde{T}$ exists, and the graph of $\widetilde{T}$ contains the graph-closure of T.

Suppose positive, symmetric, densely defined T has positive, symmetric extensions A, B *admitting no common symmetric extension*. Let $\overline{A} = A^{**}$, $\overline{B} = B^{**}$ be the graph-closures of A, B. Friedrichs's construction $T \to \widetilde{T}$ applies to T, A, B. The inclusion-reversing property of $S \to S^*$ gives a diagram of extensions, where ascending lines indicate extensions:

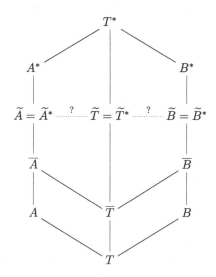

Since T^* *is* a common extension of A, B, but A, B have no common *symmetric* extension, T^* cannot be symmetric. Thus, any such situation gives an example of *nonsymmetric adjoints* of symmetric operators. Equivalently, $\overline{T}$ cannot be self-adjoint because its adjoint is T^*, which cannot be *symmetric*.

Further, although the graph closures $\overline{A}$ and $\overline{B}$ *are* (not necessarily *proper*) extensions of $\overline{T}$, neither of their Friedrichs extensions can be directly comparable to that of $\overline{T}$ without being *equal* to it, since *comparable* self-adjoint densely defined operators are necessarily equal: a densely defined self-adjoint operator cannot be a proper extension of another such: for $S \subset T$ with $S = S^*$ and $T = T^*$, the inclusion-reversing property gives $T = T^* \subset S^* = S$. By hypothesis, A, B have no common symmetric extension, so *both* equalities cannot hold.

Let $V = L^2[a, b]$, $T = -d^2/dx^2$, with domain

$$D_T = \{f \in C_c^\infty[a, b] : f \text{ vanishes at } a, b\}$$

The sign on the second derivative makes T *positive*: using the boundary conditions, integrating by parts,

$$\langle Tv, v \rangle = -\langle v'', v \rangle = -v'(b)\overline{v}(b) + v'(a)\overline{v}(a) + \langle v', v' \rangle = \langle v', v' \rangle \geq 0$$

for $v \in D_T$. Integration by parts *twice* proves *symmetry*:

$$\langle Tv, w \rangle = -\langle v'', w \rangle = -v'(b)\overline{w}(b) + v'(a)\overline{w}(a) + \langle v', w' \rangle = \langle v', w' \rangle$$

$$= v(b)\overline{w}'(b) - v(a)\overline{w}'(a) - \langle v, w'' \rangle = \langle v, Tw \rangle \qquad \text{(for } v, w \in D_T\text{)}$$

For each pair α, β of complex numbers, an extension $T_{\alpha,\beta} = -d^2/dx^2$ of T is defined by taking a larger domain, by relaxing the boundary conditions:

$$D_{\alpha,\beta} = \{f \in C_c^\infty[a, b] : f(a) = \alpha \cdot f(b), \ f'(a) = \beta \cdot f'(b)\}$$

Integrating by parts,

$$\langle T_{\alpha,\beta}v, w \rangle = v'(b)\overline{w}(b) \cdot (1 - \beta\overline{\alpha}) + v(b)\overline{w}'(b) \cdot (1 - \alpha\overline{\beta}) + \langle v, T_{\alpha,\beta}w \rangle$$

for v, $w \in D_{\alpha,\beta}$. The values $v'(b)$, $v(b)$, $w(b)$, and $w'(b)$ can be arbitrary, so the extension $T_{\alpha,\beta}$ is *symmetric* if and only if $\alpha\overline{\beta} = 1$, and in that case T is *positive*, since again

$$\langle T_{\alpha,\beta}v, v \rangle = -\langle v'', v \rangle = \langle v', v' \rangle \geq 0$$

for $\alpha\overline{\beta} = 1$ and $v \in D_{\alpha,\beta}$. For two values α, α', taking $\beta = 1/\overline{\alpha}$ and $\beta' = 1/\overline{\alpha}'$, for the symmetric extensions $T_{\alpha,\beta}$ and $T_{\alpha',\beta'}$ to have a *common symmetric extension* $\widetilde{T}$ requires that the domain of $\widetilde{T}$ include both $D_{\alpha,\beta} \cup D_{\alpha',\beta'}$. The integration by parts computation gives

$$\langle \widetilde{T}v, w \rangle = v'(b)\overline{w}(b) \cdot (1 - \beta\overline{\alpha}) + v(b)\overline{w}'(b) \cdot (1 - \alpha\overline{\beta}) + \langle v, T_{\alpha,\beta}w \rangle$$

$$= v'(b)\overline{w}(b)(1 - \beta\overline{\alpha}') + v(b)\overline{w}'(b) \cdot (1 - \alpha\overline{\beta}') + \langle v, \widetilde{T}w \rangle$$

for $v \in D_{\alpha,\beta}$, $w \in D_{\alpha',\beta'}$. Thus, the required symmetry $\langle \widetilde{T}v, w \rangle = \langle v, \widetilde{T}w \rangle$ holds only for $\alpha = \alpha'$ and $\beta = \beta'$. That is, the original operator T has a continuum of distinct symmetric extensions, no two of which admit a common symmetric extension. In particular, no two of these symmetric extensions have a common *self-adjoint* extension. Yet each does have at least the *Friedrichs* positive, self-adjoint extension. Thus, T has infinitely many distinct positive, self-adjoint extensions.

For example, the two similar boundary-value problems on $L^2[0, 2\pi]$

$$\begin{cases} u'' = \lambda \cdot u & \text{and} \quad u(0) = u(2\pi), \ u'(0) = u'(2\pi) \\ u'' = \lambda \cdot u & \text{and} \quad u(0) = 0 = u(2\pi) \end{cases}$$

(provably) have eigenfunctions and eigenvalues indexed by $n = 1, 2, 3, \ldots$:

$$\begin{cases} 1, \ \sin(nx), \ \cos(nx) & \text{eigenvalues} \quad 0, \ 1, \ 1, \ 4, \ 4, \ 9, \ 9, \ldots \\ \sin(\frac{nx}{2}) & \text{eigenvalues} \quad \frac{1}{4}, \ 1, \ \frac{9}{4}, \ 4, \ \frac{25}{4}, \ 9, \ \frac{49}{4}, \ldots \end{cases}$$

Half the eigenfunctions and eigenvalues are common, while the other half of eigenvalues of the first are shifted upward for the second. Both collections of eigenfunctions give orthogonal bases for $L^2[0, 2\pi]$. Expressions of the unshared eigenfunctions of one in terms of those of the other are not trivial, despite considerable mythology suggesting the contrary.

9.4 Unbounded Self-adjoint Operators with Compact Resolvents

The following unsurprising claim and its proof are standard:

[9.4.1] Claim: For a not-necessarily-bounded self-adjoint operator T, if T^{-1} exists and is *compact*, then $(T - \lambda)^{-1}$ exists and is a compact operator for λ off a *discrete* set in $\mathbb{C}$, and is *meromorphic* in λ. Further, the spectrum of T and nonzero spectrum of T^{-1} are in the bijection $\lambda \leftrightarrow \lambda^{-1}$.

Proof: The set of *eigenvalues* or *point spectrum* of a possibly unbounded operator T consists of $\lambda \in \mathbb{C}$ such that $T - \lambda$ fails to be *injective*. The *continuous* spectrum consists of λ with $T - \lambda$ *injective* and with *dense* image, but *not surjective*. Further, for possibly unbounded operators, we require a *bounded* (= continuous) inverse $(T - \lambda)^{-1}$ on $(T - \lambda)D_T$ for λ to be in the continuous spectrum. The *residual spectrum* consists of λ with $T - \lambda$ injective, but $(T - \lambda)D_T$ not dense.

The description of *continuous spectrum* simplifies for *closed* T, that is, for T with closed graph: we claim that for $(T - \lambda)^{-1}$ densely defined and continuous, $(T - \lambda)D_T$ is the whole space, so $(T - \lambda)^{-1}$ is *everywhere* defined, so λ cannot be in the residual spectrum. Indeed, the continuity gives a constant C such that $|x| \leq C \cdot |(T - \lambda)x|$ for all $x \in D_T$. Then $(T - \lambda)x_i$ Cauchy implies x_i Cauchy, and T closed implies $T(\lim x_i) = \lim Tx_i$. Thus, $(T - \lambda)D_T$ is *closed*. Then *density* of $(T - \lambda)D_T$ implies it is the whole space.

Now prove that for T^{-1} *compact*, the resolvent $(T - \lambda)^{-1}$ exists and is compact for λ off a discrete set, and is meromorphic in λ. The nonzero spectrum of the compact self-adjoint operator T^{-1} is *point spectrum*, from basic spectral theory for such operators, as in [9.A]. We claim that the spectrum of T and nonzero spectrum of T^{-1} are in the obvious bijection $\lambda \leftrightarrow \lambda^{-1}$. From the algebraic identities

$$T^{-1} - \lambda^{-1} = T^{-1}(\lambda - T)\lambda^{-1} \qquad T - \lambda = T(\lambda^{-1} - T^{-1})\lambda$$

failure of either $T - \lambda$ or $T^{-1} - \lambda^{-1}$ to be *injective* forces the failure of the other, so the point spectra are identical.

For (nonzero) λ^{-1} not an eigenvalue of *compact* T^{-1}, $T^{-1} - \lambda^{-1}$ is injective *and* has a continuous, everywhere-defined inverse. That $S - \lambda$ is *surjective* for compact self-adjoint S and $\lambda \neq 0$ not an eigenvalue follows from the spectral theorem for self-adjoint compact operators. For such λ, inverting the relation $T - \lambda = T(\lambda^{-1} - T^{-1})\lambda$ gives

$$(T - \lambda)^{-1} = \lambda^{-1}(\lambda^{-1} - T^{-1})^{-1}T^{-1}$$

from which $(T - \lambda)^{-1}$ is continuous and everywhere-defined. That is, λ is *not* in the spectrum of T. Finally, $\lambda = 0$ is not in the spectrum of T, because T^{-1} exists and is continuous. This establishes the bijection.

Thus, for T^{-1} compact self-adjoint, the spectrum of T is *countable*, with no accumulation point in $\mathbb{C}$. Letting $R_\lambda = (T - \lambda)^{-1}$, the resolvent relation

$$R_\lambda = (R_\lambda - R_0) + R_0 = (\lambda - 0)R_\lambda R_0 + R_0 = (\lambda R_\lambda + 1) \circ R_0$$

expresses R_λ as the composition of a continuous operator with a compact operator, proving its compactness. ////

As earlier, continuity is immediate from Hilbert's relation

$$(T - \lambda)^{-1}(\lambda - \mu)(T - \mu)^{-1} = (T - \lambda)^{-1}\Big((T - \mu) - (T - \lambda)\Big)(T - \mu)^{-1}$$

$$= (T - \lambda)^{-1} - (T - \mu)^{-1}$$

Then dividing through by $\lambda - \mu$ gives

$$\frac{(T-\lambda)^{-1} - (T-\mu)^{-1}}{\lambda - \mu} = (T-\lambda)^{-1}(T-\mu)^{-1}$$

proving differentiability.

9.5 Example: Δ on $L^2(\mathbb{T})$ and Sobolev Spaces

There are many ways to understand and prove that exponentials $x \to e^{i\xi x}$ give an orthogonal basis for L^2 of the circle. Here, we use Friedrichs extensions and compact resolvents.

On the circle $\mathbb{T} = \mathbb{R}/2\pi\mathbb{Z}$ or $\mathbb{R}/\mathbb{Z}$, there are no boundary terms in integration by parts, so Δ has the *symmetry*

$$\langle \Delta f, g \rangle = \langle f, \Delta g \rangle$$

with usual $\langle f, g \rangle = \int_{\mathbb{T}} f \cdot \bar{g}$, for $f, g \in C^\infty(\mathbb{T})$. Friedrichs's *self-adjoint extension* $\widetilde{\Delta}$ of Δ is essentially described by the relation

$$\langle (1-\widetilde{\Delta})^{-1}x, (1-\Delta)y \rangle = \langle x, y \rangle \quad (x \in L^2(\mathbb{T}), y \in C^\infty(\mathbb{T}))$$

The *compactness* of the *resolvent* $(\widetilde{\Delta} - z)^{-1}$, proven subsequently, and the *spectral theorem* for compact, self-adjoint operators, yield an orthogonal Hilbert-space basis for $L^2(\mathbb{T})$ consisting of $\widetilde{\Delta}$-eigenfunctions. Further, these eigenfunctions will be proven eigenfunctions for Δ itself.

The compactness of the resolvent will follow from Friedrichs' construction of $\widetilde{\Delta}$ via the continuous linear map $(1-\widetilde{\Delta})^{-1}$, itself a continuous linear map $L^2(\mathbb{T}) \longrightarrow H^1(\mathbb{T})$, where the Sobolev space $H^1(\mathbb{T})$ is the completion of $C^\infty(\mathbb{T})$ with respect to the Sobolev norm

$$|f|_{H^1(\mathbb{T})} = \left(|f|_{L^2(\mathbb{T})}^2 + |f'|_{L^2(\mathbb{T})}^2 \right)^{\frac{1}{2}} = \langle (1-\Delta)f, f \rangle^{\frac{1}{2}}$$

Compactness of the resolvent is *Rellich's lemma:* the inclusion $H^1(\mathbb{T}) \to L^2(\mathbb{T})$ is *compact*. The map $(1-\widetilde{\Delta})^{-1}$ of $L^2(\mathbb{T})$ to itself is compact because it is the composition of the continuous map $(1-\widetilde{\Delta})^{-1} : L^2(\mathbb{T}) \to H^1(\mathbb{T})$ and the compact inclusion $H^1(\mathbb{T}) \to L^2(\mathbb{T})$.

The eigenfunctions for the extension $\widetilde{\Delta}$ certainly *include* the Δ-eigenfunctions $\psi_n(x) = e^{inx}$ with $n \in \mathbb{Z}$, but the issue is exactly to show that there are no *other* eigenfunctions for $\widetilde{\Delta}$ than for Δ. The genuine possibility of *exotic* eigenfunctions is illustrated in the following section. That is, while we can solve the differential equation $\Delta u = \lambda \cdot u$ on $\mathbb{R}$ and identify λ having $2\pi\mathbb{Z}$-periodic solutions, more must be done to ensure that there are no *further*

$\widetilde{\Delta}$-eigenfunctions in the orthogonal basis promised by the spectral theorem. We *hope* that the natural heuristic, of straightforward solution of the differential equation $\Delta u = \lambda \cdot u$, gives the whole orthogonal basis, but this is exactly the issue.

The kth *Sobolev space* $H^k(\mathbb{T})$ is the Hilbert space completion of $C^\infty(\mathbb{T})$ with respect to k^{th} Sobolev norm given by

$$|f|^2_{H^k(\mathbb{T})} = \langle (1 - \Delta)^k f, f \rangle_{L^2} \qquad \text{(for } 0 \leq k \in \mathbb{Z}, \text{ for } f \in C^\infty(\mathbb{T}))$$

There are other useful, slightly different, expressions for a kth Sobolev norm, such as

$$\left(|f|^2_{L^2(\mathbb{T})} + |f'|^2_{L^2(\mathbb{T})} + \cdots + |f^{(k)}|^2 \right)^{\frac{1}{2}}$$

The two are *comparable*: with *uniform* implied constants,[3]

$$\left(|f|^2_{L^2(\mathbb{T})} + |f'|^2_{L^2(\mathbb{T})} + \cdots + |f^{(k)}|^2 \right)^{\frac{1}{2}} \asymp \langle (1 - \Delta)^k f, f \rangle^{\frac{1}{2}}$$

for $0 \leq k \in \mathbb{Z}$, but they are *not* constant multiples of each other. In fact, precise comparison of constants between the two versions of the Sobolev norms proves irrelevant, as we will see that the exponentials $\psi_n(x) = e^{inx}$ give an orthogonal basis for any/all versions of the Hilbert-space structure.

The *Sobolev imbedding theorem* [9.5.4] shows that $H^{k+1}(\mathbb{T})$ is inside the Banach space $C^k(\mathbb{T})$ with norm

$$|f|_{C^k(\mathbb{T})} = \sup_{0 \leq i \leq k} \sup_{x \in \mathbb{T}} |f^{(i)}(x)|$$

by showing the dominance relation

$$|f|_{H^k(\mathbb{T})} \ll |f|_{C^k(\mathbb{T})} \ll |f|_{H^{k+1}(\mathbb{T})}$$

giving

$$H^{k+1}(\mathbb{T}) \subset C^k(\mathbb{T}) \subset H^k(\mathbb{T})$$

The inclusions $C^k(\mathbb{T}) \subset H^k(\mathbb{T})$ follow from the density of $C^\infty(\mathbb{T})$ in every $C^k(\mathbb{T})$. Letting $H^\infty(\mathbb{T}) = \lim_k H^k(\mathbb{T})$, the intersection $C^\infty(T)$ of Banach spaces $C^k(\mathbb{T})$ is an intersection of Hilbert spaces

$$H^\infty(\mathbb{T}) = \bigcap_k H^k(\mathbb{T}) = \bigcap_k H^{k+1}(\mathbb{T})$$

$$\subset \bigcap_k C^k(\mathbb{T}) = C^\infty(\mathbb{T}) \subset H^\infty(\mathbb{T})$$

---

[3] Here $a \asymp b$ means that there are positive, finite constants c, c' such that $c \cdot a \leq b \leq c' \cdot b$.

For $f \in C^\infty(\mathbb{T})$, let $f \to \overline{f}$ be the value-wise conjugation. Extend this by continuity to $f \to \overline{f}$ on $L^2(\mathbb{T})$, and let $(\Lambda f)(g) = \langle g, \overline{f} \rangle_{L^2}$. From the Riesz-Fréchet theorem, Λ is an isomorphism of $L^2(\mathbb{T})$ to itself. For $1 \le k \in \mathbb{Z}$, let $H^{-k}(\mathbb{T})$ be the Hilbert-space dual of $H^k(\mathbb{T})$, *not* identified with $H^k(\mathbb{T})$ itself via Riesz-Fréchet. With $i : H^1(\mathbb{T}) \to L^2(\mathbb{T})$ the inclusion, let $i^* : L^2(\mathbb{T})^* \to H^{-1}(\mathbb{T})$ be the adjoint. Similarly, the adjoint of the inclusion $H^{k+1}(\mathbb{T}) \to H^k(\mathbb{T})$ is $H^{-k}(\mathbb{T}) \to H^{-(k+1)}(\mathbb{T})$. Thus, suppressing the reference to $\mathbb{T}$, we have

$$H^\infty \quad \cdots \longrightarrow H^1 \xrightarrow{\ i\ } L^2 \xrightarrow[\Lambda]{\ \approx\ } L^2 \xrightarrow{\ i^*\ } H^{-1} \longrightarrow \cdots$$

Let $H^{-\infty}(\mathbb{T}) = \bigcup_{k \ge 0} H^{-k}(\mathbb{T}) = \operatorname{colim}_k H^{-k}(\mathbb{T})$. Assuming Sobolev imbedding [9.5.4], we have

[9.5.1] Corollary: The dual space $C^\infty(\mathbb{T})^*$ to $C^\infty(\mathbb{T})$ of *distributions* on $\mathbb{T}$ is $H^{-\infty}(\mathbb{T})$.

Proof: $C^\infty(\mathbb{T}) = H^\infty(\mathbb{T})$ by Sobolev. In general, the dual of a limit is *not* the colimit of the duals of the limitands, but when the limitands are *normed* and the transition maps have dense images, as with $H^{k+1}(\mathbb{T}) \to H^k(\mathbb{T})$, [13.14.4] shows that $(\lim_k H^k)^* = \operatorname{colim}_k (H^k)^* = \operatorname{colim}_k H^{-k}$. ///

[9.5.2] Corollary: For f in the domain of $\widetilde{\Delta}$, the image $\widetilde{\Delta}f$ is the genuine distributional derivative Δf, and the domain of $\widetilde{\Delta}$ is $H^2(\mathbb{T})$.

Proof: For $g \in C^\infty(\mathbb{T})$ and f in the domain of $\widetilde{\Delta}$, by the characterization of the Friedrichs extension $f = (1 - \widetilde{\Delta})^{-1}F$ for some F in L^2. The characterization $\langle (1 - \widetilde{\Delta})^{-1}F, g \rangle_{H^1} = \langle F, g \rangle_{L^2}$ gives $\langle f, g \rangle_{H^1} = \langle (1 - \widetilde{\Delta})f, g \rangle_{L^2}$. Then

$$\langle (1 - \widetilde{\Delta})f, g \rangle_{L^2} = \langle f, g \rangle_{H^1} = \langle f, (1 - \Delta)g \rangle_{L^2}$$
$$= \langle f, (1 - \Delta)g \rangle_{H^{-\infty} \times H^\infty} = \langle (1 - \Delta)f, g \rangle_{H^{-\infty} \times H^\infty}$$

where we restrict the Hermitian pairing on $L^2 \times L^2$ to a Hermitian pairing $L^2 \times H^\infty$ and then extend it to $H^{-\infty} \times H^\infty$. This holds for all $g \in C^\infty(\mathbb{T})$, so $(1 - \widetilde{\Delta})f = (1 - \Delta)f$ as distributions. Thus, for $f \in H^1(\mathbb{T})$ in the domain of $\widetilde{\Delta}$, $\Delta f \in L^2(\mathbb{T})$. Thus, $\langle (1 - \Delta)^2 f, f \rangle_{L^2}$ is finite, so $f \in H^2(\mathbb{T})$. ///

[9.5.3] Corollary: $\widetilde{\Delta}$-eigenvectors are *smooth*, and are Δ-eigenvectors.

Proof: A λ-eigenfunction u for $\widetilde{\Delta}$ is in the domain of $\widetilde{\Delta}$, and by the previous $\widetilde{\Delta}u = \Delta u$, so $\Delta u = \lambda u$. Thus, $(1 - \Delta)u = (1 - \lambda)u$. By design, $(1 - \Delta)$ is a continuous map $H^{k+2} \to H^k$: for $f \in C^\infty(\mathbb{T})$,

$$|(1 - \Delta)f|^2_{H_k} = \langle (1 - \Delta)^k(1 - \Delta)f, \, (1 - \Delta)f \rangle_{L^2}$$

$$= \langle (1 - \Delta)^{k+2}f, \, f \rangle_{L^2} = |f|^2_{H^{k+2}}$$

Each H^k is the completion of $C^\infty(\mathbb{T})$, so $(1 - \Delta)$ is an isomorphism $H^{k+2} \to H^k$, and $(1 - \Delta)^{-1}$ is an isomorphism $H^k \to H^{k+2}$, so

$$u = (1 - \Delta)^{-1}(1 - \lambda)u \in (1 - \Delta)^{-1}H^2(\mathbb{T}) \subset H^4(\mathbb{T})$$

By induction, $u \in H^\infty(\mathbb{T}) = C^\infty(\mathbb{T})$.   ///

Of course, granting the foregoing, to find smooth Δ-eigenfunctions, solve the differential equation

$$\lambda \cdot u = \widetilde{\Delta}u = \Delta u = u''$$

on $\mathbb{R}$: for $\lambda \neq 0$, the solutions are linear combinations of $u(x) = e^{\pm\sqrt{\lambda} \cdot x}$; for $\lambda = 0$, the solutions are linear combinations of $u(x) = 1$ and $u(x) = x$. The $2\pi\mathbb{Z}$-periodicity is equivalent to $\lambda \in i\mathbb{Z}$ in the former case, and eliminates $u(x) = x$ in the latter. As usual, uniqueness is proven via the mean-value theorem. This would prove that the exponentials $\{e^{inx} : n \in \mathbb{Z}\}$ are a Hilbert-space basis for $L^2(\mathbb{T})$.

Now we prove Sobolev imbedding and Rellich compactness.

[9.5.4] Theorem: *(Sobolev imbedding)* $H^{k+1}(\mathbb{T}) \subset C^k(\mathbb{T})$.

Proof: This is the fundamental theorem of calculus and the Cauchy-Schwarz-Bunyakowsky inequality. The case $k = 0$ adequately illustrates the causality: prove that the H^1 norm dominates the C^o norm, namely, sup norm, on $C^\infty(\mathbb{T})$. Use coordinates from the real line, with $\mathbb{T} = \mathbb{R}/\mathbb{Z}$. For $0 \le x \le y \le 1$, the difference between maximum and minimum values of $f \in C^\infty[0, 1]$ is constrained:

$$|f(y) - f(x)| = \left| \int_x^y f'(t) \, dt \right| \le \int_0^1 |f'(t)| \, dt$$

$$\le \left(\int_0^1 |f'(t)|^2 \, dt \right)^{1/2} \cdot \left(\int_x^y 1 \, dt \right)^{1/2} = |f'|_{L^2} \cdot |x - y|^{\frac{1}{2}}$$

Let $y \in [0, 1]$ be such that $|f(y)| = \min_x |f(x)|$. Using the previous inequality,

$$|f(x)| \le |f(y)| + |f(x) - f(y)| \le \int_0^1 |f(t)| \, dt + |f(x) - f(y)|$$

$$\le \int_0^1 |f| \cdot 1 + |f'|_{L^2} \cdot 1 \le |f|_{L^2} + |f'|_{L^2} \ll 2(|f|^2 + |f'|^2)^{1/2} = 2|f|_{H^1}$$

Thus, on $C_c^\infty(\mathbb{T})$ the H^1 norm dominates the sup norm. Thus, this comparison holds on the H^1 completion $H^1(\mathbb{T})$, and $H^1(\mathbb{T}) \subset C^o(\mathbb{T})$. The same argument applies to the H^1 norm and C^o norm of successive derivatives.   ///

The space $C^\infty(\mathbb{T})$ of smooth functions on $\mathbb{T}$ is the nested intersection of the spaces $C^k(\mathbb{T})$, an instance of a *(projective) limit* of Banach spaces:

$$C^\infty(\mathbb{T}) = \bigcap_{k=0}^\infty C^k(\mathbb{T}) = \lim_k C^k(\mathbb{T})$$

so it has a uniquely determined *Fréchet* space topology, as in [13.5]. Similarly,

$$H^\infty(\mathbb{T}) = \bigcap_{k=0}^\infty H^k(\mathbb{T}) = \lim_k H^k(\mathbb{T})$$

[9.5.5] Corollary: $C^\infty(\mathbb{T}) = H^\infty(\mathbb{T})$.

Proof: The *interlacing* property $C^{k+1}(\mathbb{T}) \subset H^{k+1}(\mathbb{T}) \subset C^{k+1}(\mathbb{T})$ gives (dashed) compatible maps from $H^\infty(\mathbb{T})$ to the spaces $C^k(\mathbb{T})$ inducing a unique (dotted) map to the limit $C^\infty(\mathbb{T})$:

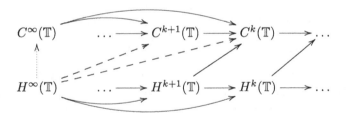

Oppositely, we obtain a unique (dotted) map of $C^\infty(\mathbb{T})$ to $\lim_k H^k(\mathbb{T})$:

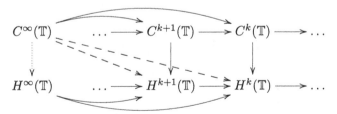

Thus, the two dotted maps must be mutual inverses.   ///

Rellich's lemma on $\mathbb{T}$ uses some finer details from the discussion just above, namely, the Lipschitz property $|f(x) - f(y)| \ll |x - y|^{\frac{1}{2}}$ for $|f|_{H^1} \le 1$, and the related fact that the map $H^1(T) \to C^o(\mathbb{T})$ has operator norm at most 2.

[9.5.6] Theorem: *(Rellich compactness)* The inclusion $H^{k+1}(\mathbb{T}) \to H^k(\mathbb{T})$ is *compact.*

Proof: The causality is adequately illustrated by the $k = 0$ case, showing that the unit ball in $H^1(\mathbb{T})$ is *totally bounded* in $L^2(\mathbb{T})$. Approximate $f \in H^1(\mathbb{T})$ in $L^2(\mathbb{T})$ by piecewise-constant functions

$$
F(x) = \begin{cases}
c_1 & \text{for } 0 \leq x < \frac{1}{n} \\[4pt]
c_2 & \text{for } \frac{1}{n} \leq x < \frac{2}{n} \\[4pt]
\cdots & \\[4pt]
c_n & \text{for } \frac{n-1}{n} \leq x \leq 1
\end{cases}
$$

The sup norm of $|f|_{H^1} \leq 1$ is bounded by 2, so we only need c_i in the range $|c_i| \leq 2$.

Given $\varepsilon > 0$, take N large enough such that the disk of radius 2 in $\mathbb{C}$ is covered by N disks of radius less than ε, with centers C. Given $f \in H^1(\mathbb{T})$ with $|f|_1 \leq 1$, choose constants $c_k \in C$ such that $|f(k/n) - c_k| < \varepsilon$. Then

$$
|f(x) - c_k| \leq |f\left(\frac{k}{n}\right) - c_k| + \left|f(x) - f\left(\frac{k}{n}\right)\right| < \varepsilon + \left|x - \frac{k}{n}\right|^{\frac{1}{2}} \leq \varepsilon + \frac{1}{\sqrt{n}}
$$

for $\frac{k}{n} \leq x \leq \frac{k+1}{n}$. Then

$$
\int_0^1 |f - F|^2 \leq \sum_{k=1}^n \int_{k/n}^{(k+1)/n} \left(\varepsilon + \frac{1}{\sqrt{n}}\right)^2
$$

$$
\leq n \cdot \frac{1}{n} \cdot \left(\varepsilon + \frac{1}{\sqrt{n}}\right)^2 = \left(\varepsilon + \frac{1}{\sqrt{n}}\right)^2
$$

For ε small and n large, this is small. Thus, the image in $L^2(\mathbb{T})$ of the unit ball in $H^1(\mathbb{T})$ is *totally bounded*, so has compact closure. This proves that the inclusion $H^1(\mathbb{T}) \subset L^2(\mathbb{T})$ is *compact*. ///

This completes the arguments for Sobolev imbedding and Rellich compactness, which are used to prove that a $\widetilde{\Delta}$-eigenfunction is actually smooth, so is in the natural domain $\mathbb{C}^\infty(\mathbb{T})$ of Δ, justifying determination of the orthogonal basis for $L^2(\mathbb{T})$ by solving the differential equations $u'' = \lambda \cdot u$ in classical (nondistributional terms).

Because the exponentials are *smooth* eigenfunctions for this self-adjoint extension of Δ,

[9.5.7] Corollary: The exponentials $\psi_n(x) = e^{inx}$ are an orthogonal basis for every Sobolev space $H^k(\mathbb{T})$.

Proof: The exponentials are smooth, so are inside every $H^k(\mathbb{T})$. The H^k norm of ψ_n is $(1 + n^2)^{k/2}$ times the L^2 norm:

$$|\psi_n|^2_{H^k} = \langle (1 - \Delta)^k \psi_n, \psi_n \rangle_{L^2} = \langle (1 + n^2)^k \psi_n, \psi_n \rangle_{L^2}$$
$$= (1 + n^2)^k \langle \psi_n, \psi_n \rangle_{L^2}$$

They are mutually orthogonal in every $H^k(\mathbb{T})$:

$$\langle \psi_m, \psi_n \rangle_{H^k} = \langle (1 - \Delta)^k \psi_m, \psi_n \rangle_{L^2} = \langle (1 + m^2)^k \psi_m, \psi_n \rangle_{L^2}$$
$$= (1 + m^2)^k \cdot \langle \psi_m, \psi_n \rangle_{L^2}$$

For $f \in H^k(\mathbb{T}) \subset L^2(\mathbb{T})$, orthogonality

$$0 = \langle \psi_n, f \rangle_{H^k} = \langle (1 - \Delta)^k \psi_n, f \rangle_{L^2} = \langle (1 + n^2))^k \psi_n, f \rangle_{L^2}$$
$$= (1 + n^2))^k \langle \psi_n, f \rangle_{L^2}$$

gives orthogonality in L^2, so completeness of the exponentials in L^2 gives completeness in H^k.   ///

Granting that the exponentials are an orthogonal basis for every $H^k(\mathbb{T})$, we have the *spectral* characterization of $H^k(\mathbb{T})$:

[9.5.8] Corollary: For $f \in C^\infty(\mathbb{T})$, for every $0 \le k \in \mathbb{Z}$,

$$|f|^2_{H^k} = \frac{1}{2\pi} \sum_{n \in \mathbb{Z}} |\langle f, \psi_n \rangle_{L^2}|^2 \cdot (1 + n^2)^k$$

The Fourier series $\frac{1}{2\pi} \sum_n \langle F, \psi_n \rangle_{H^k} \cdot \psi_n$ of a function F in $H^k(\mathbb{T})$ converges to F in the H^k topology, and for $u \in H^k(\mathbb{T})^* = H^{-k}(\mathbb{T})$,

$$u\Big(\frac{1}{2\pi} \sum_n \langle F, \psi_n \rangle_{H^k} \cdot \psi_n \Big) = \frac{1}{2\pi} \sum_n \langle F, \psi_n \rangle_{H^k} \cdot u(\psi_n)$$

Proof: Starting with Plancherel for Hilbert spaces,

$$|f|^2_{H^k} = \frac{1}{2\pi} \sum_n \frac{|\langle f, \psi_n \rangle_{H^k}|^2}{\langle \psi_n, \psi_n \rangle_{H^k}} = \frac{1}{2\pi} \sum_{n \in \mathbb{Z}} \frac{|\langle f, (1 + n^2)^k \psi_n \rangle_{L^2}|^2}{(1 + n^2)^k}$$
$$= \frac{1}{2\pi} \sum_{n \in \mathbb{Z}} |\langle f, \psi_n \rangle_{L^2}|^2 \cdot (1 + n^2)^k$$

Completing, the same holds for $F \in H^k$. In particular, the partial sums of the Fourier series of F converge to F in H^k. Thus, by the continuity of u in the dual,

$$u(F) = u\Big(\lim_{M,N} \frac{1}{2\pi} \sum_{-M \le n \le N} \langle F, \psi_n \rangle \cdot \psi_n \Big)$$
$$= \lim_{M,N} \frac{1}{2\pi} \sum_{-M \le n \le N} \langle F, \psi_n \rangle \cdot u(\psi_n) = \frac{1}{2\pi} \sum_n \langle F, \psi_n \rangle \cdot u(\psi_n)$$

a convergent series.   ///

The spectral characterization allows extension to a definition of the sth Sobolev norm for *real s*:

$$|f|^2_{H^s} = \frac{1}{2\pi} \sum_{n \in \mathbb{Z}} |\langle f, \psi_n \rangle_{L^2}|^2 \cdot (1 + n^2)^s \qquad (\text{for } f \in C^\infty(\mathbb{T}))$$

It would be natural to declare that $H^s(\mathbb{T})$ is the completion of $C^\infty(\mathbb{T})$ with respect to the sth norm, for all $s \in \mathbb{R}$, but one potential issue is that $H^{-k}(\mathbb{T})$ is already described as the dual of $H^k(\mathbb{T})$, so there is consistency to be checked. First, for duality, the L^2 Plancherel assertion

$$\langle f, g \rangle_{L^2} = \frac{1}{2\pi} \sum_n \langle f, \psi_n \rangle \cdot \overline{\langle g, \psi_n \rangle}$$

can be *desymmetrized*:

[9.5.9] Corollary: With Sobolev spaces described as completions of $C^\infty(\mathbb{T})$, for $f \in H^s(\mathbb{T})$ and $g \in H^{-s}(\mathbb{T})$, the pairing $\langle f, g \rangle_{H^s \times H^{-s}}$ defined by

$$\langle f, g \rangle_{H^s \times H^{-s}} = \frac{1}{2\pi} \sum_n \langle f, \psi_n \rangle_{H^s} \cdot \overline{\langle g, \psi_n \rangle_{H^{-s}}}$$

for $f, g \in C^\infty(\mathbb{T})$, gives a (conjugate-linear) isomorphism of H^{-s} to the Hilbert-space dual of H^s, for all real s.

Proof: It suffices to prove that this pairing matches the L^2 pairing for $f, g \in C^\infty(\mathbb{T})$, since the smooth functions are dense in every $H^s(\mathbb{T})$. Indeed, for $f, g \in C^\infty(\mathbb{T})$,

$$\langle f, g \rangle_{H^s \times H^{-s}} = \frac{1}{2\pi} \sum_n \langle f, \psi_n \rangle_{H^s} \cdot \overline{\langle g, \psi_n \rangle_{H^{-s}}}$$

$$= \frac{1}{2\pi} \sum_n (1 + n^2)^s \langle f, \psi_n \rangle_{L^2} \cdot \overline{(1 + n^2)^{-s} \langle g, \psi_n \rangle_{L^2}}$$

$$= \frac{1}{2\pi} \sum_n \langle f, \psi_n \rangle_{L^2} \cdot \overline{\langle g, \psi_n \rangle_{L^2}} = \langle f, g \rangle_{L^2}$$

by ordinary Plancherel for L^2.   ///

Granting that the Sobolev spaces defined as completions of $C^\infty(\mathbb{T})$ are in suitable duality,

[9.5.10] Claim: For a *distribution u*, if $\sum_n |u(\psi_n)|^2 \cdot (1 + n^2)^{-s} < \infty$ then $u \in H^{-s}(\mathbb{T})$, and we have the Fourier expansion in $H^s(\mathbb{T})$:

$$u = \frac{1}{2\pi} \sum_n u(\psi_n) \cdot \psi_n = \frac{1}{2\pi} \sum_n \langle u, \psi_n \rangle_{H^s \times H^{-s}} \cdot \psi_n$$

Proof: For $f \in H^s(\mathbb{T})$,

$$u(f) = \lim_{M,N} u\Big(\frac{1}{2\pi} \sum_{-M \le n \le N} \langle f, \psi_n \rangle_{H^s} \cdot \psi_n \Big)$$

$$= \lim_{M,N} \frac{1}{2\pi} \sum_{-M \le n \le N} \langle f, \psi_n \rangle_{H^s} \cdot u(\psi_n) = \frac{1}{2\pi} \sum_n \langle f, \psi_n \rangle_{H^s} \cdot u(\psi_n)$$

and by Cauchy-Schwarz-Bunyakowsky

$$\Big| \frac{1}{2\pi} \sum_n \langle f, \psi_n \rangle_{H^s} \cdot u(\psi_n) \Big|^2$$

$$= \Big| \frac{1}{2\pi} \sum_n \langle f, \psi_n \rangle_{L^2} \cdot (1+n^2)^{s/2} \cdot \frac{u(\psi_n)}{(1+n^2)^{s/2}} \Big|^2$$

$$\le \frac{1}{2\pi} \sum_n |\langle f, \psi_n \rangle|_{L^2}^2 \cdot (1+n^2)^s \cdot \frac{1}{2\pi} \sum_n \frac{|u(\psi_n)|^2}{(1+n^2)^s}$$

$$= |f|_{H^s}^2 \cdot \Big| \sum_n u(\psi_n) \cdot \psi_n \Big|_{H^{-s}}^2$$

Thus, u is a continuous linear functional on H^s, so is in the dual H^{-s}. Accommodation of complex conjugation by $u(\psi_n) = \langle u, \psi_{-n} \rangle_{H^{-s} \times H^s}$ and replacing n by $-n$ gives the second form of the Fourier expansion.   ///

Sobolev imbedding becomes simpler and sharper on the spectral side:

[9.5.11] Corollary: For $s > k + \frac{1}{2}$, $H^s(\mathbb{T}) \subset C^k(\mathbb{T})$.

Proof: As earlier, it suffices to treat $k = 0$, and prove that the C^k-norm is dominated by the H^s-norm, on C^∞-functions f. Indeed,

$$|f|_{C^o} = \sup_x |f(x)| = \sup_x \Big| \frac{1}{2\pi} \sum_n \langle f, \psi_n \rangle_{L^2} \cdot e^{inx} \Big|$$

$$\le \frac{1}{2\pi} \sum_n |\langle f, \psi_n \rangle_{L^2}|$$

$$= \frac{1}{2\pi} \sum_n |\langle f, \psi_n \rangle_{L^2}|(1+n^2)^{s/2} \cdot (1+n^2)^{-s/2}$$

$$\le \frac{1}{2\pi} \Big(\sum_n |\langle f, \psi_n \rangle_{L^2}|^2 (1+n^2)^s \Big)^{\frac{1}{2}} \cdot \Big(\sum_n (1+n^2)^{-s} \Big)^{\frac{1}{2}}$$

by Cauchy-Schwarz-Bunyakowsky. For $s > \frac{1}{2}$, the latter sum is finite. Similarly, for general k,

$$|f|_{C_k} \ll_s |f|_{H^s} \qquad \text{(for any } s > \frac{1}{2} + k, \text{ for } f \in C^\infty(\mathbb{T}))$$

Thus, the completion $H^s(\mathbb{T})$ has a canonical continuous inclusion to $C^k(\mathbb{T})$. ///

Similarly, *granting* that the exponentials form an orthonormal basis for every H^s, proof of an extended form of Rellich's lemma becomes easier:

[9.5.12] Corollary: For real $t > s$, the inclusion $H^t(\mathbb{T}) \subset H^s(\mathbb{T})$ is *compact*. For $t > s + \frac{1}{2}$, this inclusion is *Hilbert-Schmidt*, and for $t > s + 1$ it is *trace-class*.

Proof: The inclusion maps one orthogonal basis to another, but the lengths change. That is, ignoring the constant 2π, $|\psi_n|_{H^s} = (1 + n^2)^{s/2}$. Thus, letting $e_n^s = \psi_n/(1 + n^2)^{s/2}/\sqrt{2\pi}$ be an orthonormal basis for H^s, the inclusion map maps $e_n^t \to e_n^s \cdot (1 + n^2)^{s-t}$.

Generally, a map $T : V \to W$ of Hilbert spaces with orthonormal bases $\{e_n\}, \{f_n\}$, of the form $Te_n = \lambda_n f_n$, is *compact* when $\lambda_n \to 0$, is Hilbert-Schmidt when $\sum_n |\lambda_n|^2 < \infty$, and is trace-class when $\sum_n |\lambda_n| < \infty$. ///

[9.5.13] Corollary: $H^\infty(\mathbb{T}) = C^\infty(\mathbb{T})$ is *nuclear Fréchet*, in the sense that it is a (projective) limit of Hilbert spaces V_n, with transition maps $V_n \to V_{n-1}$ Hilbert-Schmidt. ///

Sobolev imbedding and Rellich compactness on $\mathbb{T}^n$ can be proven either by reducing to the case of a single circle $\mathbb{T}$, or by repeating analogous arguments directly on $\mathbb{T}^n$. Sobolev norms and spaces $H^s(\mathbb{T}^n)$ can be defined for real s. The index-shift in the Sobolev imbedding is easy to understand in the spectral form:

[9.5.14] Theorem: *(Sobolev imbedding)* $H^s(\mathbb{T}^n) \subset C^k(\mathbb{T}^n)$ for $s > k + \frac{n}{2}$.

Proof: Index the characters on $\mathbb{T}^n$ by $\psi_\xi(x) = e^{i\xi \cdot x}$, where $\xi \cdot x$ is the usual pairing on $\mathbb{R}^n \times \mathbb{R}^n$. For $k = 0$,

$$|f|_{C^0} = \sup_x |f(x)| = \sup_x \left| \frac{1}{(2\pi)^n} \sum_{\xi \in \mathbb{Z}^n} \langle f, \psi_\xi \rangle_{L^2} \cdot \psi_\xi(x) \right|$$

$$\leq \frac{1}{(2\pi)^n} \sum_\xi |\langle f, \psi_\xi \rangle_{L^2}|$$

$$= \frac{1}{(2\pi)^n} \sum_\xi |\langle f, \psi_\xi \rangle_{L^2}|(1 + |\xi|^2)^{s/2} \cdot (1 + |\xi|^2)^{-s/2}$$

$$\leq \frac{1}{(2\pi)^n} \left(\sum_\xi |\langle f, \psi_\xi \rangle_{L^2}|^2 (1 + |\xi|^2)^s \right)^{\frac{1}{2}} \cdot \left(\sum_\xi (1 + |\xi|^2)^{-s} \right)^{\frac{1}{2}}$$

by Cauchy-Schwarz-Bunyakowsky, where $|\xi| = (\xi \cdot \xi)^{\frac{1}{2}}$. For $s > \frac{n}{2}$, the latter sum is finite.   ///

[9.5.15] Theorem: *(Rellich compactness)* $H^t(\mathbb{T}^n) \subset H^s(\mathbb{T}^n)$ is *compact* for $t > s$, *Hilbert-Schmidt* for $t > s + \frac{n}{2}$, and *trace-class* for $t > s + n$.   ///

[9.5.16] Corollary: $H^\infty(\mathbb{T}^n) = C^\infty(\mathbb{T}^n)$ is *nuclear Fréchet*.   ///

The relevance of the *nuclearity* property is the *Schwartz kernel theorem* [Schwartz 1950] for this situation: every continuous linear map $T : C^\infty(\mathbb{T}^m) \to C^\infty(\mathbb{T}^n)^*$ from smooth functions to *distributions* has a Schwartz kernel $K(x, y) \in C^\infty(\mathbb{T}^{m+n})^*$, meaning that

$$(T\varphi)(\psi) = K(\varphi \otimes \psi) \qquad (\text{for } \varphi \in C^\infty(\mathbb{T}^m) \text{ and } \psi \in C^\infty(\mathbb{T}^n))$$

9.6 Example: Exotic Eigenfunctions on $\mathbb{T}$

In this example, the *exotic eigenfunctions* are not truly exotic, but do illustrate the possibility that eigenfunctions for a Friedrichs self-adjoint extension of a slight *restriction* of an otherwise natural differential operator like Δ may fail to be *smooth*.

Let δ be the Dirac delta distribution at 0 in $\mathbb{T} \approx \mathbb{R}/2\pi\mathbb{Z}$. The Sobolev imbedding $H^1(\mathbb{T}) \subset C^o(\mathbb{T})$ essentially shows that $\delta \in H^{-s}$ for every $s > \frac{1}{2}$:

$$\left| \sum_n \delta \psi_n \cdot \psi_n \right|^2_{H^{-s}} = \sum_n \frac{|\delta \psi_n|^2}{(1 + n^2)^s} = \sum_n \frac{1}{(1 + n^2)^s} < +\infty$$

Thus, $\ker \delta$ is a closed subspace of $H^1(\mathbb{T})$. Let S be the restriction of Δ to the space D_S of smooth functions f such that $\delta f = 0$. A functional on $L^2(\mathbb{T})$ is continuous if and only if its kernel is closed, so D_S is *dense* in $L^2(\mathbb{T})$. Let $\widetilde{S}$ be its Friedrichs extension.

[9.6.1] Claim: The domain of $\widetilde{S}$ is

$$\{f \in H^1(\mathbb{T}) : \Delta f \in L^2(\mathbb{T}) + \mathbb{C} \cdot \delta, \text{ and } \delta f = 0\}$$

and $\widetilde{S}f = u$ if and only if $\Delta f = u + c \cdot \delta$ for some $c \in \mathbb{C}$, and $\delta f = 0$. In particular, $f \in H^1(\mathbb{T})$ such that $(\Delta - \lambda)f = \delta$ and $\delta f = 0$ is an eigenfunction for $\widetilde{S}$.

Proof: Since S is nonpositive, by the Friedrichs characterization, a function in the domain of $\widetilde{S}$ is expressible as $f = (1 - \widetilde{S})^{-1}F$ for some $F \in L^2(\mathbb{T})$. The H^1 completion of the domain D_S is the H^1-closed subspace $\ker \delta|_{H^1}$. The characterization $\langle (1 - \widetilde{S})^{-1}F, g \rangle_{\ker \delta|_{H^1}} = \langle F, g \rangle_{L^2}$ gives $\langle f, g \rangle_{H^1} = \langle (1 - \widetilde{S})f, g \rangle_{L^2}$.

Then

$$\langle (1 - \widetilde{S})f, g \rangle_{L^2} = \langle f, g \rangle_{H^1} = \langle f, (1 - \Delta)g \rangle_{L^2}$$
$$= \langle f, (1 - \Delta)g \rangle_{H^{-\infty} \times H^{\infty}} = \langle (1 - \Delta)f, g \rangle_{H^{-\infty} \times H^{\infty}}$$

where we restrict the Hermitian pairing on $L^2 \times L^2$ to a Hermitian pairing $L^2 \times H^{\infty}$, and then extend it to $H^{-\infty} \times H^{\infty}$. Thus, $\widetilde{S} - \Delta = 0$ on D_S. This does not quite imply that $\widetilde{S} - \Delta = 0$ as a *distribution*, since D_S is the kernel of δ on $C^{\infty}(\mathbb{T})$. Recall [13.14.5] that a continuous linear functional vanishing on the kernel of a (continuous) linear function is a scalar multiple of it. Thus, for f in the domain of $\widetilde{S}$, there is a constant c_f such that

$$\langle (\widetilde{S} - \Delta)f, g \rangle_{H^{-\infty} \times H^{\infty}} = c_f \cdot \delta g \qquad \text{(for all } g \in C^{\infty}(\mathbb{T}))$$

which gives $(\widetilde{S} - \Delta)f = c_f \cdot \delta$. Thus, $\widetilde{S}f = u$ if and only if $\Delta f + c_f \delta = u$, which is $\Delta f = u - c_f \delta$.

On the other hand, if $\Delta f = u + c \cdot \delta$ and $\delta f = 0$, then $(1 - \Delta)f = f - u - c\delta$, and the Friedrichs characterization of $f = (1 - \widetilde{S})^{-1}(f - u)$ is satisfied: essentially running the earlier computation in reverse, for $g \in D_S$,

$$\langle f, g \rangle_{H^1} = \langle f, (1 - \Delta)g \rangle_{L^2} = \langle f, (1 - \Delta)g \rangle_{H^{-\infty} \times H^{\infty}}$$
$$= \langle (1 - \Delta)f, g \rangle_{H^{-\infty} \times H^{\infty}} = \langle f - u - c\delta, g \rangle_{H^{-\infty} \times H^{\infty}}$$
$$= \langle f - u, g \rangle_{H^{-\infty} \times H^{\infty}} - c\langle \delta, g \rangle_{H^{-\infty} \times H^{\infty}} = \langle f - u, g \rangle_{H^{-\infty} \times H^{\infty}} + 0$$
$$= \langle f - u, g \rangle_{L^2}$$

The condition $\delta f = 0$ is necessary for f to be in the H^1-closure of the original domain D_S. If $\delta f \neq 0$, then f cannot be $(1 - \widetilde{S})^{-1}F$ for any $F \in L^2(\mathbb{T})$.   ///

Explicitly, while the Δ-eigenfunctions $u_n(x) = \sin nx$ for $n = 0, 1, 2, \ldots$ satisfy $\delta u_n = 0$, the Δ-eigenfunctions $\cos nx$ for $n > 0$ do not, so cannot be $\widetilde{S}$-eigenfunctions. In effect, they are replaced by *exotic* $\widetilde{S}$-eigenfunctions, essentially $\sin(nx/2)$ with $n = 1, 3, 5, 7, \ldots$ made 2π-periodic by force. That is, $v_n(x) = \sin(nx/2)$ for $x \in [0, 2\pi]$, and made 2π-periodic. That is, given $x \in \mathbb{R}$, for $\ell \in \mathbb{Z}$ such that $x - \ell \cdot 2\pi \subset [0, 2\pi)$, put $[x] = x - \ell \cdot 2\pi$, and $v_n(x) = \sin(n[x]/2)$. These are mildly exotic since the forced 2π-periodicity gives their graphs *corners* at $2\pi\mathbb{Z}$ on $\mathbb{R}/2\pi\mathbb{Z}$, and

$$\Delta v_n = -\left(\frac{n}{2}\right)^2 \cdot v_n + n \cdot \delta$$

The differential equation $(\Delta - \lambda)u = \delta$ can be solved *by division*, using Fourier expansions of distributions on $\mathbb{T}$: let $u = \frac{1}{2\pi} \sum_n c_n \psi_n$,

$$\frac{1}{2\pi} \sum_n (-n^2 - \lambda)\psi_n = (\Delta - \lambda)u = \delta = \frac{1}{2\pi} \sum_n \psi_n$$

convergent in $H^{-1}(\mathbb{T})$. Thus,

$$u = \sum_n \frac{\psi_n}{-n^2 - \lambda} \qquad \text{(convergent in } H^1(\mathbb{T}))$$

For $\lambda \neq -n^2$ this has a unique solution $u \in H^1$. The condition $\delta u = 0$ is

$$0 = \delta u = \frac{1}{2\pi} \sum_n \frac{\delta \psi_n}{-n^2 - \lambda} = \frac{1}{2\pi} \sum_n \frac{1}{-n^2 - \lambda}$$

It is perhaps not obvious that this has solutions exactly for $\lambda = -(n/2)^2$ for $n = 1, 3, 5, \ldots$.

9.7 Example: Usual Sobolev Spaces on $\mathbb{R}$

In contrast to Δ on $\mathbb{T}$, there are no square-integrable Δ-eigenfunctions on $\mathbb{R}$: the eigenfunction condition $u'' = \lambda \cdot u$ is an explicitly solvable constant-coefficient differential equation, whose solutions are all linear combinations of $e^{\pm\sqrt{\lambda} \cdot x}$, and none of these is square-integrable on $\mathbb{R}$. There cannot be an *orthogonal basis* for $L^2(\mathbb{R})$ consisting of Δ-eigenfunctions, although Fourier transform and inversion

$$\widehat{f}(\xi) = \int_{\mathbb{R}} e^{-i\xi x} f(x)\, dx \qquad f(x) = \frac{1}{2\pi} \int_{\mathbb{R}} e^{i\xi x}\, \widehat{f}(\xi)\, d\xi$$

for f in the Schwartz space $\mathscr{S}(\mathbb{R})$ ([13.7], [13.15]) do express functions as *superpositions* of Δ-eigenfunctions. For $0 \leq k \in \mathbb{Z}$, the usual Sobolev spaces are completions of $C_c^\infty(\mathbb{R})$ with respect to the the kth Sobolev norm given by

$$|f|_{H^k}^2 = \int_{\mathbb{R}} (1 - \Delta)^k f(x) \cdot \overline{f}(x)\, dx \qquad \text{(for } f \in C_c^\infty(\mathbb{R}))$$

By the Plancherel theorem for the Fourier transform, this is

$$|f|_{H^k}^2 = \int_{\mathbb{R}} (1 - \Delta)^k f(x) \cdot \overline{f}(x)\, dx = \frac{1}{2\pi} \int_{\mathbb{R}} (1 + \xi^2)^k\, |\widehat{f}(\xi)|^2\, d\xi$$

for $f \in C_c^\infty(\mathbb{R})$. This also gives an sth Sobolev norm for all real s:

$$|f|_{H^s}^2 = \frac{1}{2\pi} \int_{\mathbb{R}} (1 + \xi^2)^s\, |\widehat{f}(\xi)|^2\, d\xi \qquad \text{(for } s \in \mathbb{R}, \text{ for } f \in C_c^\infty(\mathbb{R}))$$

The same sort of arguments as for $\mathbb{T}$ prove Sobolev imbedding here:

[9.7.1] **Claim:** $H^s(\mathbb{R}) \subset C_c^k(\mathbb{R})$ for any $s > \frac{1}{2} + k$, because the semi-norm $\nu_k(f) = \sup_{0 \leq i \leq k} \sup_{x \in \mathbb{R}} |f^{(i)}(x)|$ is dominated by the H^s-norm for $s > k + \frac{1}{2}$.

Proof: The seminorm comparison implies that the H^s-completion of $C_c^\infty(\mathbb{R})$ is contained in the ν_k-completion, which consists of C^k-functions whose k derivatives all vanish at ∞. First, the case $k = 0$ illustrates the key idea. By Cauchy-Schwarz-Bunyakowsky, for $f \in C_c^\infty(\mathbb{R})$, by Fourier inversion,

$$
\begin{aligned}
\sup_{x \in \mathbb{R}} |f(x)| &= \sup_{x \in \mathbb{R}} \left| \frac{1}{2\pi} \int_{\mathbb{R}} \widehat{f}(\xi) \cdot e^{i\xi x} \, d\xi \right| \leq \frac{1}{2\pi} \int_{\mathbb{R}} |\widehat{f}(\xi)| \, d\xi \\
&= \frac{1}{2\pi} \int_{\mathbb{R}} |\widehat{f}(\xi)| \cdot (1 + \xi^2)^{s/2} \cdot \frac{1}{(1 + \xi^2)^{s/2}} \, d\xi \\
&\leq \left(\frac{1}{2\pi} \int_{\mathbb{R}} |\widehat{f}(\xi)|^2 \cdot (1 + \xi^2)^s \, d\xi \right)^{\frac{1}{2}} \cdot \left(\frac{1}{2\pi} \int_{\mathbb{R}} \frac{1}{(1 + \xi^2)^s} \, d\xi \right)^{\frac{1}{2}} \\
&\ll_s |f|_{H^s}
\end{aligned}
$$

since for any $s > \frac{1}{2}$ the last integral is finite. For $k \geq 0$, use Gelfand-Pettis corollaries [14.3] to justify moving the differentiation inside the integral:

$$
\begin{aligned}
\sup_{x \in \mathbb{R}} |f^{(k)}(x)| &= \sup_{x \in \mathbb{R}} \left| \frac{\partial^k}{\partial x^k} \frac{1}{2\pi} \int_{\mathbb{R}} \widehat{f}(\xi) \cdot e^{i\xi x} \, d\xi \right| \\
&= \sup_{x \in \mathbb{R}} \left| \frac{1}{2\pi} \int_{\mathbb{R}} \frac{\partial^k}{\partial x^k} \widehat{f}(\xi) \cdot e^{i\xi x} \, d\xi \right| \\
&= \sup_{x \in \mathbb{R}} \left| \frac{1}{2\pi} \int_{\mathbb{R}} (i\xi)^k \widehat{f}(\xi) \cdot e^{i\xi x} \, d\xi \right| \\
&\leq \frac{1}{2\pi} \int_{\mathbb{R}} (1 + \xi^2)^{k/2} \cdot |\widehat{f}(\xi)| \, d\xi \\
&= \frac{1}{2\pi} \int_{\mathbb{R}} |\widehat{f}(\xi)| \cdot (1 + \xi^2)^{s/2} \cdot \frac{1}{(1 + \xi^2)^{(s-k)/2}} \, d\xi \\
&\leq \left(\frac{1}{2\pi} \int_{\mathbb{R}} |\widehat{f}(\xi)|^2 \cdot (1 + \xi^2)^s \, d\xi \right)^{\frac{1}{2}} \cdot \left(\frac{1}{2\pi} \int_{\mathbb{R}} \frac{1}{(1 + \xi^2)^{s-k}} \, d\xi \right)^{\frac{1}{2}} \\
&\ll_s |f|_{H^s}
\end{aligned}
$$

for $s - k > \frac{1}{2}$, again because the latter integral is convergent for $s - k > \frac{1}{2}$. ///

However, there is no *Rellich compactness* here: the inclusion $H^1(\mathbb{R}) \to L^2(\mathbb{R})$ is *not* compact, and the Friedrichs extension $\widetilde{\Delta}$ of the restriction of Δ to $\mathscr{S}(\mathbb{R})$ does not have compact resolvent. The noncompactness of $H^1(\mathbb{R}) \subset L^2(\mathbb{R})$ follows easily from the spectral characterization, together with Plancherel for Fourier transform on $L^2(\mathbb{R})$: letting V^1 be the image of $H^1(\mathbb{R})$ under Fourier transform, that is, the Hilbert space of f on $\mathbb{R}$ such that

$\int_{\mathbb{R}} |f(x)|^2 \cdot (1 + x^2)\, dx < \infty$, we have a commutative diagram, where vertical maps are isometries given by Fourier transform or inversion,

$$
\begin{array}{ccccc}
L^2(\mathbb{R}) & \xrightarrow{\;(1-\tilde{\Delta})^{-1}\;} & H^1(\mathbb{R}) & \xrightarrow{\;\text{inc}\;} & L^2(\mathbb{R}) \\
\approx \downarrow & & \uparrow \approx & & \downarrow \approx \\
L^2(\mathbb{R}) & \xrightarrow[1/(1+x^2)]{} & V^1 & \xrightarrow[\text{inc}]{} & L^2(\mathbb{R})
\end{array}
$$

where the map on the bottom left is multiplication by $1/(1 + x^2)$. If the inclusion $H^1(\mathbb{R}) \to L^2(\mathbb{R})$ were compact, then the composition

$$
\begin{array}{ccc}
H^1(\mathbb{R}) & \xrightarrow{\;\text{inc}\;} & L^2(\mathbb{R}) \\
\uparrow \approx & & \downarrow \approx \\
L^2(\mathbb{R}) \xrightarrow[1/(1+x^2)]{} V^1 & & L^2(\mathbb{R})
\end{array}
$$

would be compact, so the multiplication operator $L^2(\mathbb{R}) \to L^2(\mathbb{R})$ by multiplication by $1/(1 + x^2)$ would be compact. This operator is continuous and self-adjoint, but has no eigenvectors and has continuous spectrum $[0, 1]$, so cannot be compact.

9.8 Example: Discrete Spectrum of $-\Delta + x^2$ on $L^2(\mathbb{R})$

To obtain an operator on $\mathbb{R}$ related to Δ but with compact resolvent, add the *confining potential* x^2, construed as a *multiplication operator*, obtaining a *Schrödinger operator*

$$
S = -\Delta + x^2 = -\frac{d^2}{dx^2} + x^2 \qquad Sf(x) = -f''(x) + x^2 \cdot f(x)
$$

This operator is also called the *quantum harmonic oscillator*. We will see that the resolvent $\tilde{S}^{-1}$ of the Friedrichs extension $\tilde{S}$ of S *is* compact, so $\tilde{S}$ has entirely discrete spectrum.

The eigenfunctions for S are somewhat less well known than those for Δ, the latter easy to obtain by solving the constant-coefficient equation $u'' = \lambda u$. The standard device to obtain eigenfunctions is as follows. The relevant *Dirac operator* here[4] is

$$\mathbb{D} = i\frac{\partial}{\partial x} \qquad \text{so that} \qquad \mathbb{D}^2 = -\Delta$$

the factorization

$$-\Delta + x^2 = (\mathbb{D} - ix)(\mathbb{D} + ix) + [ix, \mathbb{D}] = (\mathbb{D} - ix)(\mathbb{D} + ix) + 1$$

with $[ix, \mathbb{D}] = ix \circ \mathbb{D} - \mathbb{D} \circ ix$, allows determination of many S-eigenfunctions, although proof that *all* are produced requires some effort. The eigenfunctions will be smooth but *not* compactly supported, so it is not optimal to declare the natural domain of the operator to be $C_c^\infty(\mathbb{R})$. Instead, we take the Schwartz functions $\mathscr{S}(\mathbb{R})$, as in [13.7], to be the domain.

The *raising* and *lowering* operators are

$$R = \text{raising} = \mathbb{D} - ix \qquad L = \text{lowering} = \mathbb{D} + ix$$

[9.8.1] Claim: The operator $S = -\Delta + x^2$ satisfies $S \geq 1$, in the sense that $\langle Sf, f \rangle \geq \langle f, f \rangle$ for $f \in \mathscr{S}(\mathbb{R})$.

Proof: This follows from the Dirac factorization:

$$\langle Sf, f \rangle = \langle ((\mathbb{D} - ix)(\mathbb{D} + ix) + 1)f, f \rangle = \langle (\mathbb{D} - ix)(\mathbb{D} + ix)f, f \rangle + \langle f, f \rangle$$

$$= \langle (\mathbb{D} + ix)f, (\mathbb{D} + ix)f \rangle + \langle f, f \rangle \geq \langle f, f \rangle$$

from the integration-by-parts fact $(\mathbb{D} - ix)^* = \mathbb{D} + ix$ on Schwartz functions. ///

Rather than attempting a direct solution of the differential equation $Su = \lambda u$, special features are exploited. First, a smooth function u annihilated by $\mathbb{D} + ix$ will be an eigenfunction for S with eigenvalue 1:

$$Su = \big((\mathbb{D} - ix)(\mathbb{D} + ix) + 1\big)u = (\mathbb{D} - ix)0 + u = 1 \cdot u$$

Dividing through by i, the equation $(\mathbb{D} + ix)u = 0$ is

$$\left(\frac{\partial}{\partial x} + x\right)u = 0$$

That is, $u' = -xu$ or $u'/u = -x$, so $\log u = -x^2/2 + C$ for arbitrary constant C. With $C = 1$

$$u(x) = e^{-x^2/2}$$

---

[4] Conveniently, the Dirac operator in this situation has *complex* coefficients. In two dimensions, Dirac operators have *Hamiltonian quaternion* coefficients, a special case of the general situation, that Dirac operators have coefficients in *Clifford algebras*.

Conveniently, this is in $L^2(\mathbb{R})$ and in fact is in the Schwartz space [13.7] $\mathscr{S}(\mathbb{R})$ on $\mathbb{R}$. The alternative factorization

$$S = -\Delta + x^2 = (\mathbb{D} + ix)(\mathbb{D} - ix) - [ix, \mathbb{D}] = (\mathbb{D} + ix)(\mathbb{D} - ix) - 1$$

does also lead to an eigenfunction $u(x) = e^{x^2/2}$, but this grows too fast. It is unreasonable to expect such luck in general, but here *the raising and lower operators map S-eigenfunctions to other eigenfunctions*: for $Su = \lambda u$, noting that $S = RL + 1 = LR - 1$,

$$S(Ru) = (RL + 1)(Ru) = RLRu + Ru$$

$$= R(LR)u + Ru = R(LR - 1)u + 2Ru = RSu + 2Ru$$

$$= R\lambda u + 2Ru = (\lambda + 2) \cdot Ru \qquad \text{(for } Su = \lambda u\text{)}$$

Similarly, $S(Lu) = (\lambda - 2) \cdot Lu$. Many eigenfunctions are produced by application of R^n to $u_1(x) = e^{-x^2/2}$:

$$R^n e^{-x^2/2} = (2n + 1) - \text{eigenfunction for } - \Delta + x^2$$

Repeated application of R to $e^{-x^2/2}$ produces polynomial[5] multiples of $e^{-x^2/2}$

$$R^n e^{-x^2/2} = H_n(x) \cdot e^{-x^2/2}$$

with polynomial $H_n(x)$ of degree n. The commutation relation shows that application of $LR^n u$ is just a multiple of $R^{n-1} u$, so application of L to the eigenfunctions $R^n u$ produces nothing new.

We can *almost* prove that the functions $R^n u$ are *all* the square-integrable eigenfunctions. We have seen that $\langle (-\Delta + x^2)f, f \rangle \geq |f|_{L^2}^2$, so an L^2 eigenfunction has *real* eigenvalue $\lambda \geq 1$. *Granting* that repeated application of L to a λ-eigenfunction u stays in $L^2(\mathbb{R})$, the function $L^n u$ has eigenvalue $\lambda - 2n$, and the requirement $\lambda - 2n \geq 1$ on $L^2(\mathbb{R})$ implies that $L^n u = 0$ for some n. Then $L(L^{n-1}u) = 0$, but we already have shown that the only $L^2(\mathbb{R})$-function in the kernel of L is $u_1(x) = e^{-x^2/2}$.

To make this discussion a proof requires some preparation, since in general a Friedrichs extension can have eigenvectors outside the original domain, as in [9.6].

[9.8.2] Sobolev Norms Associated with the Schrödinger Operator: A Friedrichs extension $\widetilde{S}$ requires specification of a *domain* for S. The space $C_c^\infty(\mathbb{R})$ of test functions is universally reasonable, but we have already seen the

---

[5] The polynomials H_n are the *Hermite polynomials*, but everything needed about them can be proven from this spectral viewpoint.

not-compactly-supported eigenfunctions for the differential operator S. Happily, those eigenfunctions are in the Schwartz space $\mathscr{S}(\mathbb{R})$, confirming specification of $\mathscr{S}(\mathbb{R})$ as the domain of S.

There is a hierarchy of Sobolev-like norms

$$|f|_{\mathfrak{B}^\ell} = \left\langle (-\Delta + x^2)^\ell f, f \right\rangle^{\frac{1}{2}}_{L^2(\mathbb{R})} \qquad \text{(for } f \in \mathscr{S}(\mathbb{R}))$$

with corresponding Hilbert-space completions

$$\mathfrak{B}^\ell = \text{completion of } \mathscr{S}(\mathbb{R}) \text{ with respect to } |f|_{\mathfrak{B}^\ell}$$

and $\mathfrak{B}^0 = L^2(\mathbb{R})$. The Friedrichs extension $\widetilde{S}$ is characterized via its *resolvent* $\widetilde{S}^{-1}$, the resolvent characterized by

$$\langle \widetilde{S}^{-1} f, Sg \rangle = \langle f, g \rangle \qquad \text{(for } f \in L^2(\mathbb{R}) \text{ and } g \in \mathscr{S}(\mathbb{R}))$$

and $\widetilde{S}^{-1}$ maps $L^2(\mathbb{R})$ continuously to $\mathfrak{B}^1$. Thus, an eigenfunction u for $\widetilde{S}$ is in $\mathfrak{B}^\infty = \bigcap_\ell \mathfrak{B}^\ell = \lim_\ell \mathfrak{B}^\ell$. We will see that

$$\mathfrak{B}^\infty = \mathscr{S}(\mathbb{R})$$

In particular, $\widetilde{S}$-eigenfunctions are in the presumed-natural domain $\mathscr{S}(\mathbb{R})$ of S, so evaluation of $\widetilde{S}$ on them is evaluation of S. Thus, $\widetilde{S}$-eigenfunctions are S-eigenfunctions. Further, repeated application of the lowering operator stabilizes $\mathscr{S}(\mathbb{R})$, so the foregoing *near-proof* becomes a *proof* that all eigenfunctions in $L^2(\mathbb{R})$ are of the form $R^n e^{-x^2/2}$.

To prove that these eigenfunctions are a Hilbert space basis for $L^2(\mathbb{R})$, we will prove that the resolvent is compact, so the eigenfunctions for the resolvent form an orthogonal Hilbert-space basis, and these are eigenfunctions for $\widetilde{S}$ itself, and then for S. That is, *there is an orthogonal basis for $L^2(\mathbb{R})$ consisting of S-eigenfunctions, all obtained as*

$$(2n + 1) - \text{eigenfunction} = R^n e^{-x^2/2} = \left(i\frac{\partial}{\partial x} - ix \right)^n e^{-x^2/2}$$

On $\mathbb{R}$, the compactness result depends on *both* smoothness and decay properties of the functions, in contrast to $\mathbb{T}$, where smoothness was the only issue.

[9.8.3] Theorem: *(Rellich compactness)* The injection $\mathfrak{B}^{\ell+1} \to \mathfrak{B}^\ell$ is compact.

Proof: The mechanism is well illustrated by the $\ell = 0$ case. We show compactness of $\mathfrak{B}^1 \to L^2(\mathbb{R})$ by showing *total boundedness* [14.7.1] of the image of the unit ball. Let φ be a smooth cutoff function, with

$$
\varphi_N(x) = \begin{cases} 1 & \text{(for } |x| \le N\text{)} \\ \text{smooth, between 0 and 1} & \text{(for } N \le |x| \le N+1\text{)} \\ 0 & \text{(for } |x| \ge N+1\text{)} \end{cases}
$$

The derivatives of φ_N in $N \le |x| \le N+1$ can easily be arranged to be independent of N. For $|f|_1 \le 1$, write $f = f_1 + f_2$ with

$$
f_1 = \varphi_N \cdot f \qquad f_2 = (1 - \varphi_N) \cdot f
$$

The function f_1 on $[-N-1, N+1]$ can be considered a function on a circle $\mathbb{T}$, by sticking $\pm(N+1)$ together. Then the Rellich compactness lemma on $\mathbb{T}$ [9.5.6] shows that the image of the unit ball from $\mathcal{B}^1$ is totally bounded in $L^2(\mathbb{T})$, which we can identify with $L^2[-N-1, N+1]$. The L^2 norm of the function f_2 is directly estimated

$$
|f_2|^2_{L^2(\mathbb{R})} = \int_{|x| \ge N} \varphi_N^2(x) \cdot |f_2(x)|^2 \, dx \le \frac{1}{N^2} \int_{|x| \ge N} |f_2(x) \cdot x|^2 \, dx
$$

$$
\le \frac{1}{N^2} \int_{\mathbb{R}} x^2 f(x) \cdot \overline{f}(x) \, dx \le \frac{1}{N^2} \int_{\mathbb{R}} (-\frac{d^2}{dx^2} + x^2) f(x) \cdot \overline{f}(x) \, dx
$$

$$
= \frac{1}{N^2} |f|_1^2 \le \frac{1}{N^2}
$$

Thus, given $\varepsilon > 0$, for N large the *tail* f_2 lies within a single ε-ball in $L^2(\mathbb{R})$. This proves total boundedness of the image of the unit ball, and compactness.   ///

[9.8.4] Corollary: The Friedrichs extension $\widetilde{S}$ of $S = -\frac{d^2}{dx^2} + x^2$ has compact resolvent.

Proof: The map $\widetilde{S}^{-1}$ of $L^2(\mathbb{R})$ to itself is compact because it is the composition of the continuous map $\widetilde{S}^{-1} : L^2(\mathbb{R}) \to \mathcal{B}^1$ and the compact inclusion $\mathcal{B}^1 \to \mathcal{B}^0 = L^2(\mathbb{R})$.   ///

[9.8.5] Corollary: The spectrum of $S = -\frac{d^2}{dx^2} + x^2$ is *discrete*. There is an orthonormal basis of $L^2(\mathbb{R})$ consisting of eigenfunctions for $\widetilde{S}$.

Proof: Self-adjoint compact operators have discrete spectrum with finite multiplicities for nonzero eigenvalues. From the preceding, these eigenfunctions are exactly the eigenfunctions for the Friedrichs extension $\widetilde{S}$. Since these eigenfunctions give an orthogonal Hilbert-space basis, $\widetilde{S}$ has no further spectrum.   ///

It remains to show that the eigenfunctions are in $\mathscr{S}(\mathbb{R})$, to know that they are eigenfunctions of S itself, rather than only of the extension $\widetilde{S}$.

[9.8.6] Theorem: $\mathfrak{B}^\infty = \mathscr{S}(\mathbb{R})$

Proof: It is clear that $\mathscr{S}(\mathbb{R}) \subset \mathfrak{B}^\infty$. The issue is the other containment. The *Weyl algebra* $A = A_1$ of operators, generated over $\mathbb{C}$ by the multiplication x and derivative $\partial = d/dx$, is also generated by $R = i\partial - ix$ and $L = i\partial + ix$. The Weyl algebra is *filtered* by degree in R and L: let $A^{\leq n}$ be the C-subspace of A spanned by all noncommuting monomials in R, L of total degree at most n, with $A^{\leq 0} = \mathbb{C}$. Note that R and L *commute* modulo $A^{\leq 0}$: as operators, $\partial \circ x = 1 + x \circ \partial$, and the commutation relation is obtained again, by

$$[R, L] = RL - LR = (i\partial - ix)(i\partial + ix) - (i\partial + ix)(i\partial - ix)$$

$$= -(\partial - x)(\partial + x) + (\partial + x)(\partial - x)$$

$$= -(\partial^2 - x\partial + \partial x - x^2) + (\partial^2 + x\partial - \partial x - x^2)$$

$$= 2(x\partial - \partial x) = -2$$

[9.8.7] Claim: For a monomial w_{2n} in R and L of degree $2n$,

$$|\langle w_{2n} \cdot f, f \rangle_{L^2(\mathbb{R})}| \ll_n |f|^2_{\mathfrak{B}^n} \qquad \text{(for } f \in C_c^\infty(\mathbb{R}))$$

Proof: Induction. First,

$$\langle RLf, f \rangle = \langle (RL + 1)f, f \rangle - \langle f, f \rangle \leq \langle (RL + 1)f, f \rangle$$

$$= \langle Sf, f \rangle = |f|^2_{\mathfrak{B}^1}$$

A similar argument applies to LR. For the length-two word L^2,

$$|\langle L^2 f, f \rangle| = |\langle Lf, Rf \rangle| \leq |Lf| \cdot |Rf| = \langle Lf, Lf \rangle^{\frac{1}{2}} \cdot \langle Rf, Rf \rangle^{\frac{1}{2}}$$

$$= \langle RLf, f \rangle^{\frac{1}{2}} \cdot \langle LRf, f \rangle^{\frac{1}{2}} \leq \langle Sf, f \rangle = |f|^2_{\mathfrak{B}^1}$$

A similar argument applies to R^2, completing the argument for $n = 1$. For the induction step, any word w_{2n} of length $2n$ is equal to $R^a L^b \bmod A^{\leq 2n-2}$ for some $a + b = 2n$, so, by induction,

$$|\langle w_{2n} f, f \rangle| = |\langle R^a L^b f, f \rangle| + |f|^2_{\mathfrak{B}^{n-1}}$$

In the case that $a \geq 1$ and $b \geq 1$, by induction

$$|\langle R^a L^b f, f \rangle| = |\langle R^{a-1} L^{b-1}(Lf), Lf \rangle| \ll_n$$

$$= |Lf|^2_{\mathfrak{B}^{n-1}} = \langle S^{n-1} Lf, Lf \rangle = \langle RS^{n-1} Lf, f \rangle$$

Since $RS^{n-1}L$ is $S^n \bmod A^{\leq 2n-2}$, by induction

$$\langle RS^{n-1} Lf, f \rangle \ll_n \langle S^n f, f \rangle + |f|^2_{\mathfrak{B}^{n-1}} = |f|^2_{\mathfrak{B}^n} + |f|^2_{\mathfrak{B}^{n-1}} \ll |f|^2_{\mathfrak{B}^n}$$

In the extreme case $a = 0$,

$$\langle L^{2n} f, f \rangle = \langle L^n f, R^n f \rangle \leq |L^n f| \cdot |R^n f|$$
$$= \langle L^n f, L^n f \rangle^{\frac{1}{2}} \cdot \langle R^n f, R^n f \rangle^{\frac{1}{2}} = \langle R^n L^n f, f \rangle^{\frac{1}{2}} \cdot \langle L^n R^n f, f \rangle^{\frac{1}{2}}$$

which brings us back to the previous case. The extreme case $b = 0$ is similar. ///

[9.8.8] Corollary: For a monomial w_n in R and L of degree n,

$$|\langle w_n \cdot f, f \rangle_{L^2(\mathbb{R})}| \ll_n |f|_{\mathfrak{B}^n} \cdot |f|_{L^2} \qquad (\text{for } f \in C_c^\infty(\mathbb{R}))$$

Proof: By Cauchy-Schwarz-Bunyakowsky and the claim,

$$|\langle w_n \cdot f, f \rangle_{L^2}| \leq |w_n f|_{L^2} \cdot |f|_{L^2} = \langle w_n^* w_n f, f \rangle^{\frac{1}{2}} \cdot |f|_{L^2} \leq |f|_{\mathfrak{B}^n} \cdot |f|_{L^2}$$

as claimed. ///

[9.8.9] Corollary: The limit $\mathfrak{B}^\infty = \lim_k \mathfrak{B}^k$ is contained in $\mathscr{S}(\mathbb{R})$.

Proof: This is the same idea as in [9.7.1] and its proof. Use the density of test functions in $\mathscr{S}(\mathbb{R})$, whose completion in the $\mathfrak{B}^k$ norm gives $\mathfrak{B}^k$, for every k. Thus, test functions are dense in $\mathfrak{B}^\infty$. For $f \in C_c^\infty(\mathbb{R})$, by Fourier inversion,

$$\sup_{x \in \mathbb{R}} |(1 + x^2)^m f^{(n)}(x)|$$

$$= \sup_{x \in \mathbb{R}} \left| \frac{1}{2\pi} \int_{\mathbb{R}} \left((1 + x^2)^m f^{(n)} \right)^\widehat{}(\xi) \cdot e^{i\xi x} \, d\xi \right|$$

$$= \sup_{x \in \mathbb{R}} \left| \frac{1}{2\pi} \int_{\mathbb{R}} \left((1 - \Delta)^m \cdot (-i\xi)^n \cdot \widehat{f}(\xi) \right) \cdot e^{i\xi x} \, d\xi \right|$$

$$\leq \frac{1}{2\pi} \int_{\mathbb{R}} \left| (1 - \Delta)^m \cdot \xi^n \cdot \widehat{f}(\xi) \right| \, d\xi$$

$$= \frac{1}{2\pi} \int_{\mathbb{R}} \left| (1 - \Delta)^m \cdot \xi^n \cdot \widehat{f}(\xi) \right| (1 + \xi^2)^{s/2} \cdot \frac{1}{(1 + \xi^2)^{s/2}} \, d\xi$$

$$\leq \frac{1}{2\pi} \left| (1 + \xi^2)^{s/2} \cdot (1 - \Delta)^m \cdot \xi^n \cdot \widehat{f}(\xi) \right|_{L^2} \cdot \left| \frac{1}{(1 + \xi^2)^{s/2}} \right|_{L^2}$$

$$= \frac{1}{2\pi} \left| (1 - \Delta)^{s/2} \cdot (1 + x^2)^m \cdot f^{(n)} \right|_{L^2} \cdot \left| \frac{1}{(1 + \xi^2)^{s/2}} \right|_{L^2}$$

taking $s \in 2\mathbb{Z}$, by Plancherel and Cauchy-Schwarz-Bunyakowsky. The L^2-norm of $1/(1 + \xi^2)^{s/2}$ is finite for large-enough s, and the L^2-norm of $(1 - \Delta)^{s/2}(1 + x^2)^m f^{(n)}(x)$ is dominated by a finite linear combination of the seminorms $\mu_w(f) = |\langle wf, f \rangle|^{\frac{1}{2}}$. ///

We return to the proof that all eigenfunctions of $\widetilde{S}$ are in $\mathscr{S}(\mathbb{R})$ and, therefore, in the domain of the original operator S. An eigenfunction u for $\widetilde{S}$ lies in $\mathfrak{B}^1$, by Friedrichs's construction. Friedrichs extensions preserve semiboundedness, so $\widetilde{S} \geq 1$, and the inverse $\widetilde{S}^{-1}$ exists as a bounded operator, and is self-adjoint. An eigenvector relation $\widetilde{S}u = \lambda \cdot u$ entails $\lambda \neq 0$, and gives $u = \lambda^{-1}\widetilde{S}^{-1}u$. Any $\widetilde{S}$-eigenfunction u is in the domain of $\widetilde{S}$, inside $\mathfrak{B}^1$, by construction. By induction, $u \in \mathfrak{B}^\infty = \mathscr{S}(\mathbb{R})$. ///

That is, again, the heuristic above that appears to determine an orthogonal basis of eigenfunctions for S does succeed: there is an orthogonal basis of eigenfunctions for $\widetilde{S}$, and we have shown that all eigenfunctions for $\widetilde{S}$ are actually in the domain of S.

Up to a constant, the nth *Hermite polynomial* $H_n(x)$ is characterized by

$$H_n(x) \cdot e^{-x^2/2} = R^n \, e^{-x^2/2} = (i\frac{\partial}{\partial x} - ix)^n \, e^{-x^2/2}$$

The preceding discussion shows that $H_0, H_1, H_2, \ldots$ are orthogonal on $\mathbb{R}$ with respect to the weight e^{-x^2}, and give an orthogonal basis for the weighted L^2-space

$$\{f : \int_{\mathbb{R}} |f(x)|^2 \cdot e^{-x^2} \, dx < \infty\}$$

9.9 Essential Self-adjointness

The simple examples that exhibit nonsymmetric adjoints to symmetric operators show that there can be *many* self-adjoint extensions, *incomparable* in the partial ordering on operators. While Friedrichs extensions are *canonical* positive self-adjoint extension $T_{\text{Fr}} \supset T$ of a *positive*, symmetric, densely defined operator T, we are interested in clarifying the conditions under which symmetric, densely defined T has a *unique* self-adjoint extension.

A symmetric, densely defined operator T with a *unique* self-adjoint extension is *essentially self-adjoint*. Although this use of *essential* is approximately compatible with the colloquial sense of the word, unfortunately there is some risk that its use in this context be mistaken for the more ambiguous colloquial sense.

In brief, for unbounded operators arising from differential operators, imposition of various *boundary conditions* often gives rise to mutually incomparable self-adjoint extensions. Thus, *free-space* situations, lacking boundary conditions, are the best candidates for essential self-adjointness.

Since a self-adjoint operator is (graph-) *closed*, any self-adjoint extension of symmetric T must extend the (graph-) closure $\overline{T}$.

As earlier, symmetric T has no nonreal complex eigenvalues λ, that is, that $T - \lambda$ is *injective* on D_T. This allows definition of an operator U on the image $(T - \lambda)D_T$ by

$$U = (T - \bar{\lambda})(T - \lambda)^{-1} \qquad \text{(for } \lambda \notin \mathbb{R}, \text{ on the image } (T - \lambda)D_T)$$

[9.9.1] Claim: The operator U defined on the image $(T - \lambda)D_T$ is *unitary*, in the sense that $\langle Uv, Uw \rangle = \langle v, w \rangle$ for v, w in the domain of U.

Proof: For v, w in the image $(T - \lambda)D_T$, let $v' = (T - \lambda)^{-1}v$ and $w' = (T - \lambda)^{-1}w$. Then

$$\langle Uv, Uw \rangle = \langle (T - \bar{\lambda})v', (T - \bar{\lambda})w' \rangle$$

while

$$\langle v, w \rangle = \langle (T - \lambda)v', (T - \lambda)w' \rangle$$

Thus, we want to show that

$$\langle (T - \bar{\lambda})v', (T - \bar{\lambda})w' \rangle = \langle (T - \lambda)v', (T - \lambda)w' \rangle$$

This follows from the symmetry of T.   ///

[9.9.2] Theorem: For (graph-) *closed*, symmetric, densely defined T, if for some nonreal λ both $(T - \lambda)D_T$ and $(T - \bar{\lambda})D_T$ are *dense*, then T is *self-adjoint*.

Proof: First, claim that for *(graph-) closed* and symmetric T, for nonreal λ the image $(T - \lambda)D_T$ is *closed*. To see this, let $(T - \lambda)v_i$ be Cauchy, with v_i in the domain of T. By the unitariness of U, the sequence

$$U((T - \lambda)v_i) = (T - \bar{\lambda})v_i$$

is also Cauchy. Subtracting one sequence from the other, $(\lambda - \bar{\lambda})v_i$ is Cauchy. Since $\lambda \notin \mathbb{R}$, v_i is Cauchy. Similarly, adding the two sequences, $(2T + \lambda + \bar{\lambda})v_i$ is Cauchy. Because v_i is Cauchy, $(\lambda + \bar{\lambda})v_i$ is Cauchy, so $2Tv_i$ and Tv_i are Cauchy. Since the graph of T is closed, the sequence $v_i \oplus Tv_i$ converges to some $v \oplus Tv$ *in the graph of* T. Thus, $(T - \lambda)v_i$ certainly converges to $(T - \lambda)v$, and verifies the claim that $(T - \lambda)D_T$ is *closed*. By hypothesis, the closed subspaces $(T - \lambda)D_T$ and $(T - \lambda)D_T$ are also *dense*, so each is the whole space V.

Given v in the domain D_{T^*} of the adjoint T^*, we show that $v \in D_T$. Since $(T - \lambda)D_T = V$, there is $v' \in D_T$ such that

$$(T - \lambda)v' = (T^* - \lambda)v$$

Thus,

$$\langle v', (T^* - \overline{\lambda})w \rangle = \langle (T - \lambda)v', w \rangle = \langle (T^* - \lambda)v, w \rangle = \langle v, (T - \overline{\lambda})w \rangle$$

for all $w \in D_T$. Since $(T - \overline{\lambda})D_T$ is dense, $v' = v$. That is, $v \in D_T$.   ///

[9.9.3] Corollary: For symmetric, densely defined T, suppose that for some nonreal λ both $(T - \lambda)D_T$ and $(T - \overline{\lambda})D_T$ are *dense*. Then the *closure* $\overline{T}$ of T is *self-adjoint* and is the *unique* self-adjoint extension of T. In particular, T is *essentially self-adjoint*.

Proof: The closure $\overline{T}$ extends T and is *symmetric* for symmetric T. Certainly $(\overline{T} - \lambda)D_{\overline{T}}$ contains $(T - \lambda)D$, so when the latter is dense the former is dense. Thus, $\overline{T}$ meets the hypothesis of the theorem and is self-adjoint.

Any self-adjoint extension $S = S^*$ of T is *closed*, since adjoints are closed. Thus, any self-adjoint extension S of T contains the closure $\overline{T} = T^{**}$, for topological reasons. Taking adjoints is inclusion-reversing, so, $S = S^* \subset (T^{**})^* = T^*$ from characterization of adjoints in terms of graphs as shown earlier. Therefore, $S = T^{**} = \overline{T}$.   ///

[9.9.4] Claim: For a symmetric, densely defined operator T, *density* of $(T - \lambda)D_T$ is equivalent to the assertion that T^* does *not* have eigenvalue $\overline{\lambda}$.

Proof: From earlier examples, T^* need not be *symmetric*, so the natural argument that eigenvalues of symmetric operators must be *real* does not apply to T^*. Apart from that, the argument is the natural one. The density of $(T - \lambda)D_T$ implies that $\langle (T - \lambda)v, w \rangle = 0$ for all $v \in D_T$ if and only if $w = 0$. If $(T^* - \lambda)v = 0$, then

$$0 = \langle (T^* - \overline{\lambda})v, w \rangle = \langle v, (T - \lambda)w \rangle \qquad \text{(for all } w \in D_T)$$

Since $(T - \lambda)D_T$ is *dense*, this implies $w = 0$. Conversely, if $(T - \lambda)D_T$ were *not* dense, then its closure would not be the whole space, so would be orthogonal to some $v \neq 0$. Then

$$0 = \langle v, (T - \lambda)w \rangle = \langle (T^* - \overline{\lambda})v, w \rangle$$

for every $w \in D_T$, for $v \in D_{T^*}$. Thus, since D_T is dense, we imagine that it would be consistent to *define* $T^* v = \overline{\lambda}v$. In fact, by [9.1.1], the graph of T^* is the orthogonal complement in $V \oplus V$ of the image of the graph of S under the isometry J, so $0 = \langle (T - \overline{\lambda})D_T, v \rangle$ implies that the pair $(v, \lambda v)$ is in the graph of the adjoint. Thus, if $(T - \lambda)D_T$ were not dense, then T^* would have eigenvalue $\overline{\lambda}$.   ///

Thus, we have a variant form of the criterion for the closure of T being self-adjoint:

[9.9.5] Corollary: For symmetric, densely defined T, if for some nonreal λ, neither λ nor $\bar{\lambda}$ is an *eigenvalue* for the adjoint T^*, then the *closure* $\bar{T}$ of T is the *unique* self-adjoint extension of T. In particular, T is *essentially self-adjoint.* ///

In the situation of the corollary, since $\bar{T} = T^{**}$, and $T^{***} = T^*$, in fact $\bar{T}^* = T^*$.

9.10 Example: Essentially Self-adjoint Operator

In fact, [9.5.2] proved that the domain of the Friedrichs extension $\tilde{S}$ of the restriction S of Δ to $C^\infty(\mathbb{T})$ is $H^2(\mathbb{T})$, which says that the graph-closure of S is its Friedrichs extension. Since every self-adjoint extension contains the graph-closure and distinct self-adjoint extensions are not comparable, the graph-closure must be the *only* self-adjoint extension. That is, the restriction of Δ to $C^\infty(\mathbb{T})$ is essentially self-adjoint. Nevertheless, we want to practice application of the criterion above.

Let $S = \frac{d^2}{dx^2}$ on $\mathbb{T} = \mathbb{R}/2\pi\mathbb{Z}$ with domain $D_S = C^\infty(\mathbb{T})$. Since S is nonpositive, there is *at least* one *meaningful* self-adjoint extension, the Friedrichs extension. However, we want the (graph-) *closure* $\bar{S}$ of S to be that self-adjoint extension, giving *uniqueness* in a strong, unambiguous fashion.

We do not directly characterize the domain D_{S^*} of S^*, apart from the fact that it contains the domain of S. It is convenient that S *stabilizes* D_S. The *translation action* of $\mathbb{T}$ on functions on $\mathbb{T}$ is $(R_x f)(y) = f(y + x)$. This action is *unitary*, and gives a (jointly) continuous map $\mathbb{T} \times L^2(\mathbb{T}) \longrightarrow L^2(\mathbb{T})$. A *constant-coefficient* differential operator such as S commutes with the translation action, at least on D_S: in symbols, $R_t \circ T = T \circ R_t$ for all $t \in \mathbb{T}$. Indeed, this invariance allows such operators to *descend* from $\mathbb{R}$ to the quotient $\mathbb{T}$. Certainly D_S is *stable* under translation.

[9.10.1] Claim: The domain D_{S^*} of S^* is *stable* under translation.

Proof: Let $J(x \oplus y) = -y \oplus x$ be the usual map on $L^2(\mathbb{T}) \oplus L^2(\mathbb{T})$. The map J is an isometry with respect to the usual inner product $\langle x + x', y + y' \rangle = \langle x, y \rangle + \langle x', y' \rangle$ on $L^2(\mathbb{T}) \oplus L^2(\mathbb{T})$. The graph of the adjoint is characterized as the orthogonal complement of the image by J of the graph of S. Thus, for $y \oplus S^*y$ in the graph of S^*, for all $x \oplus Sx$ in the graph of S, because S commutes with R_t on D_S,

$$\langle R_t y \oplus R_t S^* y, J(x \oplus Sx) \rangle = \langle R_t y \oplus R_t S^* y, -Sx \oplus x \rangle$$
$$= \langle y \oplus S^* y, -R_t^{-1} Sx \oplus R_t^{-1} x \rangle$$
$$= \langle y \oplus S^* y, -SR_t^{-1} x \oplus R_t^{-1} x \rangle = 0$$

because $R_t^{-1} x \in D_S$. Thus, $R_t y \in D_{S^*}$, as claimed. ////

The action of $\varphi \in C^\infty(\mathbb{T})$ on $L^2(\mathbb{T})$ is by the *integral operator*

$$R_\varphi v = \int_{\mathbb{T}} \varphi(t) \cdot R_t\, v\, dt$$

Since $t \to \varphi(t) \cdot R_t v$ is a compactly supported, continuous, $L^2(\mathbb{T})$-valued function on $\mathbb{T}$, it has a *Gelfand-Pettis* integral. Further, D_S is stable under this action, and the translation action

$$\mathbb{T} \times C^\infty(\mathbb{T}) \longrightarrow C^\infty(\mathbb{T})$$

is continuous with respect to the Fréchet-space topology. For $\varphi \in C^\infty(\mathbb{T})$, the corresponding integral operator R_φ maps $L^2(\mathbb{T})$ to $C^\infty(\mathbb{R})$, by smoothing of distributions [14.5].

[9.10.2] Claim: The operators R_φ for $\varphi \in C^\infty(\mathbb{T})$ commute with S^*.

Proof: Since the operators S on D_S and S^* on D_{S^*} are *not continuous* on $L^2(\mathbb{T})$, the properties of Gelfand-Pettis integrals must be used scrupulously. For $\varphi \in C^\infty(\mathbb{T})$, $v \in D_{S^*}$, and $w \in D_S$, using the commutativity of Gelfand-Pettis integrals with *continuous* maps, a sensible computation succeeds:

$$\langle R_\varphi S^* v, w \rangle = \langle \int_{\mathbb{T}} \varphi(t) R_t S^* v\, dt, w \rangle = \int_{\mathbb{T}} \langle \varphi(t) R_t S^* v, w \rangle\, dt$$

$$= \int_{\mathbb{T}} \varphi(t) \langle R_t S^* v, w \rangle\, dt = \int_{\mathbb{T}} \varphi(t) \langle S^* v, R_t^{-1} w \rangle\, dt$$

$$= \int_{\mathbb{T}} \varphi(t) \langle v, S R_t^{-1} w \rangle\, dt = \int_{\mathbb{T}} \varphi(t) \langle v, R_t^{-1} S w \rangle\, dt$$

$$= \int_{\mathbb{T}} \varphi(t) \langle R_t v, S w \rangle\, dt = \langle \int_{\mathbb{T}} \varphi(t) R_t v\, dt, S w \rangle$$

$$= \langle R_\varphi v, S w \rangle = \langle S^* R_\varphi v, w \rangle$$

This is the desired commutativity. ///

Now we can prove that S^* has *no* nonreal eigenvalues, so S meets the hypotheses of the theorem of the previous section, and its closure $\overline{S}$ is the unique self-adjoint extension of S. Suppose $v \in D_{S^*}$ and $(S^* - \lambda)v = 0$. Then, for any $\varphi \in C^\infty(\mathbb{T})$,

$$0 = R_\varphi \cdot 0 = R_\varphi(S^* - \lambda)v = (S^* - \lambda)R_\varphi v = (S - \lambda)R_\varphi v$$

the last equality because R_φ maps everything to $C^\infty(\mathbb{T})$, on which S^* acts by S. Although S^* is not guaranteed to be symmetric (unless $S^* = \overline{S} = S^{**}$, which is the sought-after essential self-adjointness of S itself!), the operator S *is* symmetric, so has no nonreal eigenvalues, giving $R_\varphi v = 0$. For given φ, taking φ

sufficiently far along in an *approximate identity* gives $R_\varphi v \neq 0$ for $v \neq 0$. Thus, we conclude that $v = 0$, and S^* has no nonreal eigenvalues.   ///

Thus, the closure $\overline{S}$ of S is the unique self-adjoint extension of S. Restricted to the graph of S, the metric on $L^2(\mathbb{T}) \oplus L^2(\mathbb{T})$ gives norm-squared

$$|x \oplus Sx|^2 = |x|^2 + |Sx|^2 \geq |x|^2 + |\langle Sx, x\rangle| + |Sx|^2 - \tfrac{1}{2} \cdot (|x| + |Sx|)^2$$

$$\geq \tfrac{1}{2} \cdot \left(|x|^2 + \langle -Sx, x\rangle + |Sx|^2\right)$$

since Cauchy-Schwarz-Bunyakowsky and $2ab \leq (a+b)^2$ give

$$|\langle Sx, x\rangle| \leq |Sx| \cdot |x| \leq \tfrac{1}{2} \cdot (|x| + |Sx|)^2$$

Thus, the completion $H^2(\mathbb{T})$ of D_S with respect to the norm attached to the Hermitian inner product

$$\langle x, y\rangle + \langle -Sx, y\rangle + \langle S^2 x, y\rangle \asymp \langle x, y\rangle_{H^2}   \text{(for } x, y \in D_S)$$

is exactly the domain of $\overline{S}$. In the $\langle, \rangle_{H^2}$ topology, S is *continuous* on D_S, and $\overline{S}$ is the *extension by continuity* to $H^2(\mathbb{T})$.

9.A Appendix: Compact Operators

The *spectrum* $\sigma(T)$ of a continuous linear operator $T : X \to X$ on a Hilbert space X is the collection of complex numbers λ such that $T - \lambda$ *does not have a continuous linear inverse*. The *discrete spectrum* $\sigma_{\text{disc}}(T)$ is the collection of complex numbers λ such that $T - \lambda$ fails to be *injective*. In other words, the discrete spectrum is the collection of *eigenvalues*. The *continuous spectrum* $\sigma_{\text{cont}}(T)$ is the collection of complex numbers λ such that $T - \lambda \cdot 1_X$ *is* injective, *does* have dense image, but fails to be *surjective*. The *residual spectrum* $\sigma_{\text{res}}(T)$ is everything else: neither discrete nor continuous spectrum. That is, the residual spectrum of T is the collection of complex numbers λ such that $T - \lambda \cdot 1_X$ *is* injective and *fails* to have dense image (so is certainly not surjective).

To see that there are no *other* possibilities for failure of existence of an inverse, note that the *closed graph theorem* [9.B.3] implies that a bijective, continuous, linear map $T : X \to Y$ of Banach spaces has continuous inverse. Indeed, granting that the inverse exists as a linear map, its graph is

$$\text{graph of } T^{-1} = \{(y, x) \in Y \times X : (x, y) \text{ in the graph of } T \subset X \times Y\}$$

Since the graph of T is closed, the graph of T^{-1} is closed, and by the closed graph theorem T^{-1} is continuous.

As usual, the *adjoint* T^* of a continuous linear map $T : X \to Y$ from one Hilbert space is defined by

$$\langle Tx, y \rangle_Y = \langle x, T^* y \rangle_Y$$

[9.A.1] Claim: An (bounded) *normal* operator $T : X \to X$, that is, with $TT^* = T^*T$, has empty residual spectrum. That is, for λ not an eigenvalue, $T - \lambda$ has *dense image*.

Proof: The adjoint of $T - \lambda$ is $T^* - \overline{\lambda}$, so consider $\lambda = 0$ to lighten the notation. Suppose that T does *not* have dense image. Then there is nonzero z such that

$$0 = \langle z, Tx \rangle = \langle T^*z, x \rangle \qquad \text{(for every } x \in X)$$

Therefore $T^*z = 0$, and the 0-eigenspace Z of T^* is nonzero. Since $T^*(Tz) = T(T^*z) = T(0) = 0$ for $z \in Z$, T^* stabilizes Z. That is, Z is both T and T^*-stable. Therefore, $T = (T^*)^*$ acts on Z by (the complex conjugate of) 0, and T has nontrivial 0-eigenvectors, contradiction. ///

A set in a topological space is *precompact* when its closure is compact. A linear operator $T : X \to Y$ on Hilbert spaces is *compact* when it maps the unit ball in X to a *precompact* set in Y. Equivalently, T is compact if and only if it maps *bounded* sequences in X to sequences in Y with *convergent subsequences*.

[9.A.2] Claim: An operator-norm limit of compact operators is compact. A compact operator $T : X \to Y$ with Y a Hilbert space is an operator norm limit of *finite rank* operators.

Proof: Let $T_n \to T$ in uniform operator norm, with compact T_n. Given $\varepsilon > 0$, let n be sufficiently large such that $|T_n - T| < \varepsilon/2$. Since $T_n(B)$ is precompact, there are finitely many $y_1, \ldots, y_t$ such that for any $x \in B$ there is i such that $|T_n x - y_i| < \varepsilon/2$. By the triangle inequality

$$|Tx - y_i| \leq |Tx - T_n x| + |T_n x - y_i| < \varepsilon$$

Thus, $T(B)$ is covered by finitely many balls of radius ε. ////

A continuous linear operator is of *finite rank* when its image is finite-dimensional. A finite-rank operator is *compact*, since all balls are precompact in a finite-dimensional Hilbert space.

[9.A.3] Theorem: A compact operator $T : X \to Y$ with Y a Hilbert space is an operator norm limit of *finite rank* operators.

Proof: Let B be the closed unit ball in X. Since $T(B)$ is precompact it is totally bounded, so for given $\varepsilon > 0$ cover $T(B)$ by open balls of radius ε centered at points $y_1, \ldots, y_n$. Let p be the orthogonal projection to the finite-dimensional subspace F spanned by the y_i and define $T_\varepsilon = p \circ T$. Note that for any $y \in Y$ and for any y_i

$$|p(y) - y_i| \leq |y - y_i|$$

since $y = p(y) + y'$ with y' orthogonal to all y_i. For x in X with $|x| \leq 1$, by construction there is y_i such that $|Tx - y_i| < \varepsilon$. Then

$$|Tx - T_\varepsilon x| \leq |Tx - y_i| + |T_\varepsilon x - y_i| < \varepsilon + \varepsilon$$

Thus, $T_\varepsilon \to T$ in operator norm as $\varepsilon \to 0$.   ///

Hilbert-Schmidt operators are an important concrete class of compact operators, as is verified in the claim that follows. Originally Hilbert-Schmidt operators on function spaces $L^2(X)$ arose as operators given by *integral kernels*: for X and Y σ-finite measure spaces, and for integral kernel $K \in L^2(X \times Y)$, the associated *Hilbert-Schmidt* operator $T : L^2(X) \longrightarrow L^2(Y)$ is

$$Tf(y) = \int_X K(x, y) f(x) \, dx$$

By Fubini's theorem and the σ-finiteness, for orthonormal bases φ_α for $L^2(X)$ and ψ_β for $L^2(Y)$, the collection of functions $\varphi_\alpha(x)\psi_\beta(y)$ is an orthonormal basis for $L^2(X \times Y)$. Thus, for some scalars c_{ij},

$$K(x, y) = \sum_{ij} c_{ij} \, \overline{\varphi_i}(x) \, \psi_j(y)$$

Square-integrability is

$$\sum_{ij} |c_{ij}|^2 = |K|^2_{L^2(X \times Y)} < \infty$$

The indexing sets may as well be countable, since an uncountable sum of positive reals cannot converge. Given $f \in L^2(X)$, the image Tf is in $L^2(Y)$, since

$$Tf(y) = \sum_{ij} c_{ij} \langle f, \varphi_i \rangle \, \psi_j(y)$$

has $L^2(Y)$ norm easily estimated by

$$|Tf|^2_{L^2(Y)} \leq \sum_{ij} |c_{ij}|^2 |\langle f, \varphi_i \rangle|^2 \, |\psi_j|^2_{L^2(Y)}$$

$$\leq |f|^2_{L^2(X)} \sum_{ij} |c_{ij}|^2 \, |\varphi_i|^2_{L^2(X)} \, |\psi_j|^2_{L^2(Y)}$$

$$= |f|^2_{L^2(X)} \sum_{ij} |c_{ij}|^2 = |f|^2_{L^2(X)} \cdot |K|^2_{L^2(X \times Y)}$$

The adjoint $T^* : L^2(Y) \to L^2(X)$ has kernel

$$K^*(y, x) = \overline{K(x, y)}$$

by computing

$$\langle Tf, g \rangle_{L^2(Y)} = \int_Y \left(\int_X K(x, y) f(x) \, dx \right) \overline{g(y)} \, dy$$

$$= \int_X f(x) \overline{\left(\int_Y \overline{K(x, y)} g(y) \, dy \right)} \, dx$$

The *intrinsic* characterization of Hilbert-Schmidt operators $V \to W$ on Hilbert spaces V, W is as the *completion* of the space of *finite-rank* operators $V \to W$ with respect to the *Hilbert-Schmidt norm*, whose square is

$$|T|^2_{\mathrm{HS}} = \mathrm{tr}(T^*T) \qquad \text{(for } T : V \to W \text{ and } T^* : W^* \to V^*\text{)}$$

The *trace* of a finite-rank operator from a Hilbert space to itself can be described in coordinates and then proven independent of the choice of coordinates, or trace can be described *intrinsically*, obviating need for proof of coordinate-independence. First, in coordinates, for an orthonormal basis e_i of V, and finite-rank $T : V \to V$, define

$$\mathrm{tr}(T) = \sum_i \langle Te_i, e_i \rangle$$

with reference to orthonormal basis $\{e_i\}$. With this description, one would need to show independence of the orthonormal basis. For the intrinsic description, consider the map from $V \otimes V^*$ to finite-rank operators on V induced from the bilinear map

$$v \times \lambda \longrightarrow \left(w \to \lambda(w) \cdot v \right) \qquad \text{(for } v \in V \text{ and } \lambda \in V^*\text{)}$$

Trace is easy to define in these terms $\mathrm{tr}(v \otimes \lambda) = \lambda(v)$, and

$$\mathrm{tr}\left(\sum_{v, \lambda} v \otimes \lambda \right) = \sum_{v, \lambda} \lambda(v) \qquad \text{(finite sums)}$$

Expression of *trace* in terms of an orthonormal basis $\{e_j\}$ is easily obtained from the intrinsic form: given a finite-rank operator T and an orthonormal basis $\{e_i\}$, let $\lambda_i(v) = \langle v, e_i \rangle$. We claim that $T = \sum_i Te_i \otimes \lambda_i$. Indeed,

$$\left(\sum_i Te_i \otimes \lambda_i \right)(v) = \sum_i Te_i \cdot \lambda_i(v) = \sum_i Te_i \cdot \langle v, e_i \rangle$$

$$= T\left(\sum_i e_i \cdot \langle v, e_i \rangle \right) = Tv$$

Then the trace is

$$\operatorname{tr} T = \operatorname{tr}\left(\sum_i T e_i \otimes \lambda_i \right) = \sum_i \operatorname{tr}(T e_i \otimes \lambda_i)$$

$$= \sum_i \lambda_i(T e_i) = \sum_i \langle T e_i, e_i \rangle$$

Similarly, *adjoints* $T^* : W \to V$ of maps $T : V \to W$ are expressible in these terms: for $v \in V$, let $\lambda_v \in V^*$ be $\lambda_v(v') = \langle v', v \rangle$, and for $w \in W$ let $\mu_w \in W^*$ be $\mu_w(w') = \langle w', w \rangle$. Then

$$(w \otimes \lambda_v)^* = v \otimes \mu_w \qquad (\text{for } w \in W \text{ and } v \in V)$$

since

$$\langle (w \otimes \lambda_v) v', w' \rangle = \langle \lambda_v(v')w, w' \rangle = \langle v', v \rangle \langle w, w' \rangle$$

$$= \langle v', \langle w', w \rangle \cdot v \rangle = \langle v', (v \otimes \mu_w) w' \rangle$$

Since it is defined as a completion, the collection of all Hilbert-Schmidt operators $T : V \to W$ is a Hilbert space, with the Hermitian inner product $\langle S, T \rangle = \operatorname{tr}(T^* S)$.

[9.A.4] Claim: The Hilbert-Schmidt norm $|\ |_{\mathrm{HS}}$ dominates the uniform operator norm $|\ |_{\mathrm{op}}$, so Hilbert-Schmidt operators are *compact*.

Proof: Given $\varepsilon > 0$, let e_1 be a vector with $|e_1| \leq 1$ such that $|T v_1| \geq |T|_{\mathrm{op}} - \varepsilon$. Extend $\{e_1\}$ to an orthonormal basis $\{e_i\}$. Then

$$|T|_{\mathrm{op}}^2 = \sup_{|v| \leq 1} |T v|^2 \leq |T v_1|^2 + \varepsilon \leq \varepsilon + \sum_j |T v_j|^2 = |T|_{\mathrm{HS}}^2$$

Thus, Hilbert-Schmidt-norm limits of finite-rank operators are operator-norm limits of finite-rank operators, so are compact. ///

It is already nearly visible that the $L^2(X \times Y)$ norm on kernels $K(x, y)$ is the same as the Hilbert-Schmidt norm on corresponding operators $T : V \to W$, yielding

[9.A.5] Claim: Operators $T : L^2(X) \to L^2(Y)$ given by integral kernels $K \in L^2(X \times Y)$ are Hilbert-Schmidt, that is, are Hilbert-Schmidt norm limits of finite-rank operators.

Proof: To prove properly that the $L^2(X \times Y)$ norm on kernels $K(x, y)$ is the same as the Hilbert-Schmidt norm on corresponding operators $T : V \to W$, T

should be expressed as a limit of finite-rank operators T_n in terms of kernels $K_n(x, y)$ which are finite sums of products $\varphi(x) \otimes \psi(y)$. Thus, first claim that

$$K(x, y) = \sum_i \overline{\varphi_i}(x) \, T\varphi_i(y) \qquad (\text{in } L^2(X \times Y))$$

Indeed, the inner product in $L^2(X \times Y)$ of the right-hand side against any $\varphi_i(x)\psi_j(y)$ agrees with the inner product of the latter against $K(x, y)$, and we have assumed $K \in L^2(X \times Y)$. With $K = \sum_{ij} c_{ij}\overline{\varphi_i} \otimes \psi_j$,

$$T\varphi_i = \sum_j c_{ij}\, \psi_j$$

Since $\sum_{ij} |c_{ij}|^2$ converges,

$$\lim_i |T\varphi_i|^2 = \lim_i \sum_j |c_{ij}|^2 = 0$$

and

$$\lim_n \sum_{i>n} |T\varphi_i|^2 = \lim_n \sum_{i>n} |c_{ij}|^2 = 0$$

so the infinite sum $\sum_i \overline{\varphi_i} \otimes T\varphi_i$ converges to K in $L^2(X \times Y)$. In particular, the truncations

$$K_n(x, y) = \sum_{1 \le i \le n} \overline{\varphi_i}(x) \, T\varphi_i(y)$$

converge to $K(x, y)$ in $L^2(X \times Y)$, and give finite-rank operators

$$T_n f(y) = \int_X K_n(x, y) \, f(x) \, dx$$

We claim that $T_n \to T$ in Hilbert-Schmidt norm. It is convenient to note that by a similar argument $\overline{K(x, y)} = \sum_i T^*\psi_i(x)\overline{\psi_i}(y)$. Then

$$|T - T_n|_{\text{HS}}^2 = \text{tr}\Big((T - T_n)^* \circ (T - T_n)\Big)$$

$$= \sum_{i,j>n} \text{tr}\Big(\big(T^*\psi_i \otimes \overline{\psi_i}\big) \circ \big(\overline{\varphi_j} \otimes T\varphi_j\big)\Big)$$

$$= \sum_{i,j>n} \langle T^*\psi_i, \varphi_j \rangle_{L^2(X)} \cdot \langle T\varphi_j, \psi_i \rangle_{L^2(Y)} = \sum_{i,j>n} |c_{ij}|^2 \longrightarrow 0$$

as $n \to \infty$, since $\sum_{ij} |c_{ij}|^2$ converges. Thus, $T_n \to T$ in Hilbert-Schmidt norm. ////

Now we come to the spectral theorem for self-adjoint compact operators. Again, the λ-eigenspace V_λ of a self-adjoint compact operator T on a Hilbert space T is

$$V_\lambda = \{v \in V : Tv = \lambda \cdot v\}$$

We have already shown that eigenvalues, if any, of self-adjoint T are *real*.

[9.A.6] Theorem: Let T be a self-adjoint compact operator on a nonzero Hilbert space V.

- The completion of $\oplus V_\lambda$ is all of V. In particular, there is an orthonormal basis of *eigenvectors*.
- For infinite-dimensional V, 0 is the only accumulation point of the set of eigenvalues.
- Every eigenspaces X_λ for $\lambda \neq 0$ is *finite-dimensional*. The 0-eigenspace may be {0}, finite-dimensional, or infinite-dimensional.
- *(Rayleigh-Ritz)* One or the other of $\pm |T|_{\mathrm{op}}$ is an eigenvalue of T, with operator norm $| \cdot |_{\mathrm{op}}$.

Proof: An alternative expression for the operator norm is needed:

[9.A.7] Lemma: $|T|_{\mathrm{op}} = \sup_{|x| \leq 1} |\langle Tx, x \rangle|$ for T a *self-adjoint* continuous linear operator on a Hilbert space.

Proof: Let s be that supremum. By Cauchy-Schwarz-Bunyakowsky, $s \leq |T|_{\mathrm{op}}$. For any x, y, by polarization

$$
\begin{aligned}
2|\langle Tx, y \rangle + \langle Ty, x \rangle| &= |\langle T(x+y), x+y \rangle - \langle T(x-y), x-y \rangle| \\
&\leq |\langle T(x+y), x+y \rangle| + |\langle T(x-y), x-y \rangle| \\
&\leq s|x+y|^2 + s|x-y|^2 \\
&= 2s(|x|^2 + |y|^2)
\end{aligned}
$$

With $y = t \cdot Tx$ with $t > 0$, because $T = T^*$,

$$\langle Tx, y \rangle = \langle Tx, t \cdot Tx \rangle = t \cdot |Tx|^2 \geq 0$$

for $y = t \cdot Tx$ with $t > 0$, and

$$\langle Ty, x \rangle = \langle t \cdot T^2 x, t \cdot x \rangle = t \cdot \langle Tx, Tx \rangle = t \cdot |Tx|^2 \geq 0$$

for $y = t \cdot Tx$ with $t > 0$. Thus,

$$|\langle Tx, y \rangle| + |\langle Ty, x \rangle| = \langle Tx, y \rangle + \langle Ty, x \rangle = |\langle Tx, y \rangle + \langle Ty, x \rangle|$$

for $y = t \cdot Tx$ with $t > 0$. From this, and from the polarization identity divided by 2,

$$|\langle Tx, y \rangle| + |\langle Ty, x \rangle| = |\langle Tx, y \rangle + \langle Ty, x \rangle| \leq s(|x|^2 + |y|^2)$$

with $y = t \cdot Tx$. Divide through by t to obtain

$$|\langle Tx, Tx \rangle| + |\langle T^2 x, x \rangle| \leq \frac{s}{t} \cdot (|x|^2 + |Tx|^2)$$

Minimize the right-hand side by taking $t = |x|/|Tx|$ and note that $\langle T^2 x, x \rangle = \langle Tx, Tx \rangle$, giving

$$2|\langle Tx, Tx \rangle| \leq 2s \cdot |x| \cdot |Tx| \leq 2s \cdot |x|^2 \cdot |T|_{\text{op}}$$

Thus, $|T|_{\text{op}} \leq s$. ///

The last assertion of the theorem is the starting point of the proof and uses $|T| = \sup_{|x| \leq 1} |\langle Tx, x \rangle|$ and the fact that any value $\langle Tx, x \rangle$ is *real*, by self-adjointness. Choose a sequence $\{x_n\}$ so that $|x_n| \leq 1$ and $|\langle Tx, x \rangle| \to |T|$. Replacing it by a subsequence if necessary, the sequence $\langle Tx, x \rangle$ of real numbers has a limit $\lambda = \pm|T|$. Then

$$0 \leq |Tx_n - \lambda x_n|^2 = \langle Tx_n - \lambda x_n, Tx_n - \lambda x_n \rangle$$
$$= |Tx_n|^2 - 2\lambda \langle Tx_n, x_n \rangle + \lambda^2 |x_n|^2$$
$$\leq \lambda^2 - 2\lambda \langle Tx_n, x_n \rangle + \lambda^2$$

The right-hand side goes to 0. By *compactness* of T, replace x_n by a subsequence so that Tx_n converges to some vector y. The previous inequality shows $\lambda x_n \to y$. For $\lambda = 0$, we have $|T| = 0$, so $T = 0$. For $\lambda \neq 0$, $\lambda x_n \to y$ implies $x_n \longrightarrow \lambda^{-1} y$. For $x = \lambda^{-1} y$, we have $Tx - \lambda x$ and x is the desired eigenvector with eigenvalue $\pm|T|$.

Now use induction. The completion Y of the sum of nonzero eigenspaces is T-stable. We claim that the orthogonal complement $Z = Y^\perp$ is T-stable, and the restriction of T to is a compact operator. Indeed, for $z \in Z$ and $y \in Y$, $\langle Tz, y \rangle = \langle z, Ty \rangle = 0$, proving stability. The unit ball in Z is a subset of the unit ball B in X, so has precompact image $TB \cap Z$ in X. Since Z is *closed* in X, the intersection $TB \cap Z$ of Z with the precompact TB is precompact, proving T restricted to $Z = Y^\perp$ is still compact. Self-adjoint-ness is clear.

By construction, the restriction T_1 of T to Z has no eigenvalues on Z, since any such eigenvalue would also be an eigenvalue of T on Z. Unless $Z = \{0\}$ this would contradict the previous argument, which showed that $\pm|T_1|$ is an eigenvalue on a *nonzero* Hilbert space. Thus, it must be that the completion of the sum of the eigenspaces is all of X.

To prove that eigenspaces V_λ for $\lambda \neq 0$ are finite-dimensional, and that there are only finitely many eigenvalues λ with $|\lambda| > \varepsilon$ for given $\varepsilon > 0$, let B be the unit ball in $Y = \sum_{|\lambda| > \varepsilon} X_\lambda$. The image of B by T contains the ball of radius ε in Y. Since T is compact, this ball is *pre*compact, so Y is finite-dimensional. Since the dimensions of the X_λ are positive integers, there can be only finitely many of them with $|\lambda| > \varepsilon$, and each is finite-dimensional. It follows that the only possible accumulation point of the set of eigenvalues is 0, and, for X infinite-dimensional, 0 *must* be an accumulation point.   ///

[9.A.8] Corollary: For a self-adjoint compact operator $T : X \to X$ on a Hilbert space X, for $\lambda \neq 0$ not an eigenvalue, $(T - \lambda)X = X$.

Proof: By the spectral theorem, $(T - \lambda)^{-1}$ exists.   ///

9.B Appendix: Open Mapping and Closed Graph Theorems

[9.B.1] Theorem: *(Open Mapping Theorem)* For a continuous linear *surjection* $T : X \to Y$ of Banach spaces, there is $\delta > 0$ such that for all $y \in Y$ with $|y| < \delta$ there is $x \in X$ with $|x| \leq 1$ such that $Tx = y$. In particular, T is an *open map*.

[9.B.2] Corollary: A *bijective* continuous linear map of Banach spaces is an *isomorphism*.   ///

Proof: In the corollary the nontrivial point is that T is *open*, which is the point of the theorem. The linearity of the inverse is easy.

For every $y \in Y$, there is $x \in X$ so that $Tx = y$. For some integer n we have $n > |x|$, so Y is the union of the sets $TB(n)$, with usual open balls

$$B(n) = \{x \in X : |x| < n\}$$

By Baire category [15.A], the *closure* of some one of the sets $TB(n)$ contains a nonempty open ball

$$V = \{y \in Y : |y - y_o| < r\}$$

for some $r > 0$ and $y_o \in Y$. Since we are in a metric space, the conclusion is that every point of V occurs as the limit of a Cauchy sequence consisting of elements from $TB(n)$. Certainly

$$\{y \in Y : |y| < r\} \subset \{y_1 - y_2 : y_1, y_2 \in V\}$$

Thus, every point in the ball B'_r of radius r centered at 0 in Y is the sum of two limits of Cauchy sequences from $TB(n)$. Thus, surely every point in B'_r is the limit of a single Cauchy sequence from the image $TB(2n)$ of the open ball $B(2n)$ of twice the radius. That is, the *closure* of $TB(2n)$ contains the ball $B'(r)$.

Using the linearity of T, the *closure* of $TB(\rho)$ contains the ball $B'(r\rho/2n)$ in Y.

Given $|y| < 1$, choose $x_1 \in B(2n/r)$ so that $|y - Tx_1| < \varepsilon$. Choose $x_2 \in B(\varepsilon \cdot \frac{2n}{r})$ so that $|(y - Tx_1) - Tx_2| < \varepsilon/2$. Choose $x_3 \in B(\frac{\varepsilon}{2} \cdot \frac{2n}{r})$ so that

$$|(y - Tx_1 - Tx_2) - Tx_3| < \varepsilon/2^2$$

Choose $x_4 \in B(\frac{\varepsilon}{2^2} \cdot \frac{2n}{r})$ so that

$$|(y - Tx_1 - Tx_2 - Tx_3) - Tx_4| < \varepsilon/2^3$$

and so on. The sequence

$$x_1, \ x_1 + x_2, \ x_1 + x_2 + x_3, \ldots$$

is Cauchy in X. Since X is complete, the limit x of this sequence exists in X, and $Tx = y$. We find that

$$x \in B(\frac{2n}{r}) + B(\varepsilon \frac{2n}{r}) + B(\frac{\varepsilon}{2} \cdot \frac{2n}{r}) + B(\frac{\varepsilon}{2^2} \cdot \frac{2n}{r}) + \cdots \subset B((1 + 2\varepsilon)\frac{2n}{r})$$

Thus,

$$TB((1 + \varepsilon)\frac{2n}{r}) \supset \{y \in Y : |y| < 1\}$$

This proves openness at 0.   ///

It is straightforward to show[6] that a *continuous* map $f : X \to Y$ of *Hausdorff* topological spaces has *closed graph*

$$\Gamma_f = \{(x, y) : f(x) = y\} \subset X \times Y$$

Similarly, a topological space X is Hausdorff if and only if the diagonal $X^\Delta = \{(x, x) : x \in X\}$ is closed in $X \times X$.[7]

[9.B.3] Theorem: *(Closed Graph Theorem)* A linear map $T : V \to W$ of Banach spaces is *continuous* if it has closed graph $\Gamma = \{(v, w) : Tv = w\}$.

Proof: The direct sum $V \oplus W$ with norm $|v \oplus w| = |v| + |w|$ is a Banach space. Since Γ is a closed subspace of $V \oplus W$, it is a Banach space itself with the restriction of this norm. The projection $\pi_V : V \oplus W \to V$ is a continuous

---

[6] To show that a continuous map $f : X \to Y$ of topological spaces with Y Hausdorff has closed graph Γ_f, show the complement is open. Take $(x, y) \notin \Gamma_f$. Let V_1 be a neighborhood of $f(x)$ and V_2 a neighborhood of y such that $V_1 \cap V_2 = \phi$, using Hausdorff-ness. By continuity of f, for x' in a suitable neighborhood U of x, the image $f(x')$ is inside V_1. Thus, the neighborhood $U \times V_2$ of (x, y) does not meet Γ_f.

[7] To show that closedness of the diagonal X^Δ in $X \times X$ implies X is Hausdorff, let $x_1 \neq x_2$ be points in X. Then there is a neighborhood $U_1 \times U_2$ of (x_1, x_2), with U_i a neighborhood of x_i, not meeting the diagonal. That is, $(x, x') \in U_1 \times U_2$ implies $x \neq x'$. That is, $U_1 \cap U_2 = \phi$.

linear map. The restriction $\pi_V|_\Gamma$ of π_V to Γ is still continuous and still *surjective* because T is an everywhere-defined function on V. By the open mapping theorem, $\pi_V|_\Gamma$ is *open*. Thus, the bijection $\pi_V|_\Gamma$ is a *homeomorphism*. Letting $\pi_W : V \oplus W \to W$ be the projection to W,

$$T = \pi_W \circ \left(\pi_V|_\Gamma\right)^{-1} : V \longrightarrow W$$

expresses T as a composition of continuous functions.   ///

9.C Appendix: Irreducibles of Compact Groups

As usual, now specifically for *compact* topological groups K, a *representation* of K on a quasi-complete, locally convex topological vector space V is a continuous map $K \times V \to V$ making K act by continuous linear maps on V. Such a representation V is (topologically) *irreducible* when there are no K-stable (topologically) closed subspaces of V except $\{0\}$ and V itself. A K-homomorphism $\varphi : V \to W$ of K-representations is a continuous linear map that respects the action of K: $\varphi(k \cdot v) = k \cdot \varphi(v)$.

[9.C.1] Claim: Every representation of compact K on a Hilbert space V is isomorphic to a *unitary* representation of K on V. That is, there is another inner product $\langle, \rangle'$ on V, comparable to the original inner product $\langle, \rangle$ in the sense that there are finite constants $0 < c_1, c_2$ such that

$$c_1 \cdot \langle v, v \rangle \leq \langle v, v \rangle' \leq c_2 \cdot \langle v, v \rangle$$

for all $v \in V$, and such that $\langle k \cdot v, k \cdot v \rangle = \langle v, v \rangle'$ for all $v \in V$ and $k \in K$.

Proof: The natural idea to *average* the original inner product by the action of K succeeds, because K is compact: let

$$\langle v, v \rangle' = \int_K \langle k \cdot v, k \cdot v \rangle \, dk$$

Since K is compact, for each $v \in V$ the orbit $K \cdot v = \{k \cdot v : k \in K\}$ is compact, so bounded. By Banach-Steinhaus (uniform boundedness) [13.12.3], the action of elements $k \in K$ are *uniformly equicontinuous*: given $\varepsilon > 0$, there is $\delta > 0$ such that

$$|v| < \delta \implies |k \cdot v| < \varepsilon$$

With $\varepsilon = 1$, $|v| \leq 1$ implies $|k \cdot v| < \delta^{-1}$. That is, the operator norm of $v \to k \cdot v$ is at most δ^{-1} for all $k \in K$. That is, $|k \cdot v| \leq \delta^{-1} \cdot |v|$ for all k, v. Replacing k by k^{-1} and v by $k \cdot v$, we similarly have $|v| \leq \delta^{-1}|k \cdot v|$, which gives $\delta|v| \leq$

$|k \cdot v|$. Thus, integrating,

$$\delta^2 \cdot |v|^2 \cdot \text{meas}\,(K) \le \int_K |k \cdot v|^2 \, dk \le \delta^{-2} \cdot |v|^2 \cdot \text{meas}\,(K)$$

Not every norm arises from an inner product. To see that the new norm-squared $|v|^2_{\text{new}} = \int_K |k \cdot v|^2 \, dk$ does arise from an inner product, it suffices to prove the polarization identity

$$|v + w|^2_{\text{new}} - |v - w|^2_{\text{new}} = 2|v|^2_{\text{new}} + 2|w|^2_{\text{new}}$$

This follows by integrating the polarization identity for the original norm. ///

The action of K on $L^2(K)$ by right translation is *unitary* because the measure is invariant. The continuity of this action follows by a simpler form of the argument of [6.1] and [6.2]. As there, K acts continuously on $C^o_c(K)$ by right translation. The density of $C^o_c(K)$ in $L^2(K)$, and the domination of the L^2 norm by the sup norm, give the continuity.

For $\lambda \in V^*$ and $v \in V$, let $c_{v,\lambda}(k) = \lambda(k \cdot v)$ for $k \in K$. The function $c_{v,\lambda}$ on K is a *(matrix) coefficient function*.

[9.C.2] Claim: Every Hilbert space irreducible V of K has a K-homomorphism to $L^2(K)$, by the map

$$v \longrightarrow c_{v,\lambda} \qquad \text{(for fixed } 0 \ne \lambda \in V^*)$$

Proof: Without loss of generality, we can assume V is unitary, by the previous. The function $k \times v \to k \cdot v$ is a (jointly) continuous function $K \times V \to V$, by assumption. Composing with λ gives a continuous function $K \times V \to \mathbb{C}$. We claim that $v \to (k \to c_{v,\lambda}(k))$ is a continuous $C^o(K)$-valued function on V: for $|v - v'| < \delta$,

$$|c_{v,\lambda}(k) - c_{v',\lambda}(k')| = |\lambda(k \cdot v - k' \cdot v')|$$

$$= |\lambda(k \cdot v - k' \cdot v) + \lambda(k' \cdot v - k' \cdot v')|$$

$$\le |\lambda|_{V^*} \cdot ((k \cdot v - k' \cdot v|_V + (k' \cdot v - k' \cdot v'|_V)$$

By unitariness, $|k' \cdot (v - v')| = |v - v'|$. By the continuity of the action of K on V, $|k \cdot v - k' \cdot v| < \varepsilon$ for given v for k' sufficiently close to k.

To see that $v \to c_{v,\lambda}$ is a K-homomorphism, for $x, y \in K$,

$$c_{x \cdot v, \lambda}(y) = \lambda(y \cdot (x \cdot v)) = \lambda((y \cdot x) \cdot v) = c_{v,\lambda}(yx)$$

This proves the claim. ///

As usual [14.1], $\varphi \in C_c^o(K)$ acts on a K-representation W by integral operators

$$\varphi \cdot w = \int_K \varphi(k) \, k \cdot w \, dk$$

Thus, such W becomes a $C_c^o(K)$-representation, as discussed in somewhat greater generality in [9.D]. Potential issues about *multiplicities* are clarified in [9.D.14].

[9.C.3] Claim: $L^2(K)$ is the completion of an orthogonal direct sum $\oplus_V m_V \cdot V$ of orthogonal sums $m_V \cdot V = \underbrace{V \oplus \cdots \oplus V}_{m_V}$ of $C_c^o(K)$-irreducibles V, each occurring with finite multiplicity m_V.

[9.C.4] Remark: This claim is an extreme case of [7.B]'s treatment of compact $\Gamma \backslash G$, where now $\Gamma = \{1\}$. The argument simplifies as well. Potential ambiguities about the notion of *multiplicity* are resolved in [9.D.14].

Proof: On $L^2(K)$ this is

$$(\varphi \cdot f)(x) = \int_K \varphi(y) f(xy) \, dy = \int_K \varphi(x^{-1}y) f(y) \, dy$$

The function $x \times y \to \varphi(x^{-1}y)$ is continuous on $K \times K$, so is in $L^2(K \times K)$ by the compactness of K. Thus, φ gives a Hilbert-Schmidt operator [9.A.5] on $L^2(K)$. The adjoint of the operator given by φ is easily determined and is again in $C_c^o(K)$. This action is *non-degenerate* in the sense that for given $f \in L^2(K)$, there is $\varphi \in C_c^o(K)$ such that $\varphi \cdot f \neq 0$, from [14.1.5]. That is, the ring of operators on L^2 is *adjoint-stable*, nondegenerate, and consists of compact operators, so [7.2.18] applies: $L^2(K)$ is the completion of a direct sum of irreducible $C_c^o(K)$-representations, each occurring with finite multiplicity.   ///

[9.C.5] Corollary: The $C_c^o(K)$-irreducible subrepresentations in $L^2(K)$ are exactly the K-irreducible subrepresentations. Thus, $L^2(K)$ is the completion of an orthogonal direct sum of K-irreducibles V, each occurring with finite multiplicity m_V.

Proof: This is a special case of [14.1.6] and [14.1.7]: irreducible $C_c^o(K)$-subrepresentations of a K-representation are irreducible K-subrepresentations.   ///

[9.C.6] Remark: With a little more effort, one can prove that $m_V = \dim_{\mathbb{C}} V$, and more (for example, *Schur inner-product relations*), but the assertion of the claim is all we need for our immediate purposes.

[9.C.7] Corollary: All Hilbert-space irreducibles of compact K are finite-dimensional.

Proof: A copy of every K-irreducible appears inside $L^2(K)$, where all irreducibles are finite-dimensional. ///

[9.C.8] Corollary: For two compact groups K_1 and K_2, the Hilbert-space irreducibles of $K_1 \times K_2$ are tensor products of Hilbert-space irreducibles of K_1 and of K_2.

Proof: Let V be an irreducible Hilbert-space representation of $K_1 \times K_2$. From the foregoing, without loss of generality, the representation is unitary. From the previous corollary, V is finite-dimensional. Forgetting the action of K_2, V is a finite-dimensional representation of K_1, so is a finite orthogonal direct sum of irreducibles.

For an irreducible W of K appearing in V, the *W-isotype* V^W of V is the (not necessarily direct) sum of all copies of W in V. By [9.D.14], this sum is expressible an orthogonal *direct* sum. We claim that K_2 stabilizes V^W. If not, the orthogonal projection from some image $k_2 \cdot V^W$ to some other isotype $V^{W'}$ would be nonzero. But the orthogonal projections to K_1-isotypes are K_1-homomorphisms, as are the orthogonal projections to copies of W inside V^W. The kernel and image of K_1-homomorphisms $W \to W'$ are subrepresentations, since in finite-dimensional spaces all subspaces are (topologically) closed. Thus, if the kernel is not all of W, the map is an injection, so has nonzero image, so is all of W', giving an *isomorphism* $W \to W'$, which is impossible for nonisomorphic irreducibles. Thus, K_2 stabilizes V^W. Thus, $K_1 \times K_2$ stabilizes V^W, so by irreducibility of V this (nonzero) isotype is all of V, that is, $V = V^W$.

In any case, $\mathrm{Hom}_{K_1}(W, V)$ has a K_2-representation structure given by postapplication of the action of k_2:

$$(k_2 \cdot \varphi)(w) = k_2 \cdot \varphi(w)$$

The map $W \otimes_{\mathbb{C}} \mathrm{Hom}_{K_1}(W, V) \longrightarrow V$ by $w \otimes \varphi \to \varphi(w)$ is a nonzero $K_1 \times K_2$-homomorphism to V, so must surject to V, by the irreducibility of V.

For the converse: let W_1, W_2 be unitary irreducibles of K_1, K_2, and claim that $V = W_1 \otimes W_2$ is an irreducible $K_1 \times K_2$-representation. For a $K_1 \times K_2$-subrepresentation $W \subset V$, the orthogonal projections to W and $W^\perp$ are $K_1 \times K_2$-homomorphisms. Thus, if V is reducible, then it has nonscalar $K_1 \times K_2$-endomorphisms. Proving that any endomorphism φ of V is scalar will prove that V is irreducible. For fixed $w_2 \in W_2$ and $\lambda_2 \in W_2^*$, we can map $W_1 \otimes W_2 \to W_1$ by $w_1 \otimes w_2 \longrightarrow \lambda_2(w_2) \cdot w_1$, and then consider

$$w_1 \longrightarrow w_1 \otimes w_2 \longrightarrow \varphi(w_1 \otimes w_2) \longrightarrow W_1$$

This is a K_1-homomorphism, so is a scalar c_{w_2,λ_2} by (the finite-dimensional version of) Schur's lemma. The map $W_2 \to W_2^{**} \approx W_2$ by $w_2 \to (\lambda_2 \to c_{w_2,\lambda_2})$ is a K_2-homomorphism, so by Schur's lemma there is a constant c such that $c_{w_2,\lambda_2} = c \cdot \lambda_2(w_2)$. Then, for all $\lambda_1 \in W_1^*$ and $\lambda_2 \in W_2^*$,

$$(\lambda_1 \otimes \lambda_2)(\varphi(w_1 \otimes w_2)) = \lambda_1(c_{w_2,\lambda_2} \cdot w_1) = \lambda_1(w_1) \cdot c \cdot \lambda_2(w_2)$$

$$= c \cdot (\lambda_1 \otimes \lambda_2)(w_1 \otimes w_2)$$

Thus, any $K_1 \times K_2$-endomorphism φ acts by a scalar, so V is irreducible.   ///

9.D Appendix: Spectral Theorem, Schur's Lemma, Multiplicities

A portion of a spectral theorem for bounded self-adjoint operators on Hilbert spaces is necessary to prove a form of *Schur's lemma* [9.D.12], itself used to remove ambiguities about *multiplicities* of irreducible representations [9.D.14].

The present discussion continues in the context of [9.A]. Let T be a continuous self-adjoint linear map $V \to V$ for a (separable) Hilbert space V, with *spectrum*

$$\sigma(T) = \{\lambda \in \mathbb{C} : (T - \lambda)^{-1} \text{ does not exist}\}$$

[9.D.1] Claim: For self-adjoint T, the spectrum $\sigma(T)$ is a nonempty compact subset of $\mathbb{R}$.

Proof: First, we show that $T - \lambda$ is invertible for $|\lambda| > |T|_{\text{op}}$. The natural heuristic expands a geometric series:

$$(T - \lambda)^{-1} = -\lambda^{-1} \cdot (1 - \frac{T}{\lambda})^{-1} = -\lambda^{-1} \cdot \left(1 + \frac{T}{\lambda} + (\frac{T}{\lambda})^2 + \cdots\right)$$

Since $|T/\lambda|_{\text{op}} < 1$, the latter infinite sum does converge in operator norm. Then, just as with geometric series of real or complex numbers, it is easy to check that this infinite sum converges to $(T - \lambda)^{-1}$.

To prove that $\sigma(T)$ is *closed*, show that $\mu \in \mathbb{C}$ sufficiently close to $\lambda \notin \sigma(T)$ is also not in $\sigma(T)$. Again, this uses geometric series expansions as a natural heuristic to obtain an expression for $(T - \mu)^{-1}$ as a convergent series:

$$(T - \mu)^{-1} = ((T - \lambda) - (\mu - \lambda))^{-1}$$

$$= \left(1 - (\mu - \lambda)(T - \lambda)^{-1}\right) \circ (T - \lambda)^{-1}$$

$$= \left(1 + (\mu - \lambda)(T - \lambda)^{-1} + ((\mu - \lambda)(T - \lambda)^{-1})^2 + \cdots\right)$$

$$\circ (T - \lambda)^{-1}$$

For $|\mu - \lambda|$ small enough that $|(\mu - \lambda) \cdot (T - \lambda)^{-1}| = |\mu - \lambda| \cdot |(T - \lambda)^{-1}|_{op} < 1$, the geometric series converges and is readily checked to give $(T - \mu)^{-1}$.

To show that $\sigma(T) \subset \mathbb{R}$, show that $T - \lambda$ is both injective and surjective for $\lambda \notin \mathbb{R}$. Then the open mapping theorem [9.B.1] shows that the inverse is continuous. For injectivity, note that $\langle Tv, v \rangle = \langle v, Tv \rangle = \overline{\langle Tv, v \rangle}$ implies that $\langle Tv, v \rangle$ is *real*. Then $(T - \lambda)v = 0$ with $v \neq 0$ implies $\langle (T - \lambda)v, v \rangle = 0$, from which $\lambda \in \mathbb{R}$. For surjectivity, suppose $\langle (T - \lambda)v, w \rangle = 0$ for some $w \neq 0$. In particular, $\langle (T - \lambda)w, w \rangle = 0$. Again using the fact that $0 \neq \langle Tw, w \rangle \in \mathbb{R}$, this would require that $\lambda \in \mathbb{R}$.

Liouville's theorem on bounded entire functions implies that the spectrum of a continuous linear operator on a Hilbert space is not empty, as follows. If a continuous $R_\lambda = (T - \lambda)^{-1}$ exists for every complex λ, then for $0 \neq v \in V$, $R_\lambda v \in V$ is never $0 \in V$. Take $w \in V$ such that $\langle R_{\lambda_o} v, w \rangle \neq 0$ for some $\lambda_o \in \mathbb{C}$. Then $f(\lambda) = \langle R_\lambda v, w \rangle$ is a not-identically 0 entire function. At the same time, for large $|\lambda|$, the operator norm of R_λ is small. Thus, $f(\lambda)$ is small for large $|\lambda|$, and must be identically 0, by Liouville, contradiction. ///

For a self-adjoint continuous operator S on V, write $S \geq 0$ when $\langle Sv, v \rangle \geq 0$ for all $v \in V$. For self-adjoint S, T, write $S \leq T$ when $T - S \geq 0$. At the outset, with $a \leq -|T|_{op}$ and $b \geq |T|_{op}$, we have, $\langle a \cdot v, v \rangle \leq \langle Tv, v \rangle \leq \langle b \cdot v, v \rangle$. That is, $a \leq T \leq b$, where the scalars refer to scalar operators on V. Here all functions are real-valued, and $C^o[a, b]$ refers to real-valued continuous functions on $[a, b]$.

[9.D.2] Theorem: The map $\mathbb{R}[x] \to \mathbb{R}[T]$ on polynomials given by $f \to f(T)$ is *continuous*, where $\mathbb{R}[x]$ has the sup norm on $[a, b]$ and $\mathbb{R}[T]$ has the uniform operator norm. Thus, by Weierstraß approximation, this map extends to a continuous map $C^o[a, b] \to \overline{\mathbb{R}[T]}$, the latter being the operator-norm completion of $\mathbb{R}[T]$. This map factors through $C^o(\sigma(T))$:

$$C^o[a, b] \longrightarrow C^o(\sigma(T)) \longrightarrow \overline{\mathbb{R}[T]}$$

and the map $C^o(\sigma(T)) \to \overline{\mathbb{R}[T]}$ is an *isometric isomorphism*, where $C^o(\sigma(T))$ has sup norm.

Proof: We claim that for $f \in \mathbb{R}[x]$ with $f(x) \geq 0$ on $[a, b]$, then $f(T) \geq 0$. From the following lemma on polynomials, f is expressible as a finite sum of the form

$$f = \sum_i P_i^2 + (x - a) \sum_j Q_j^2 + (b - x) \sum_k R_k^2$$

for polynomials P_i, Q_j, R_k in $\mathbb{R}[x]$. Incidentally,

[9.D.3] Lemma: For *commuting* self-adjoint S, T with $T \geq 0$, also $S^2 T \geq 0$.

Proof: $\langle S^2 T v, v \rangle = \langle T S v, S^* v \rangle = \langle T(Sv), (Sv) \rangle \geq 0$.   ///

Thus, since $a \leq T \leq b$, and all these operators commute (being polynomials in T), each $P_i^2(T) \geq 0$, each $(T - a)Q_j^2(T) \geq 0$, and $(b - T)R_k^2(T) \geq 0$. Thus, $f(T) \geq 0$, proving the claim.

Since $g(x) = \sup_{[a,b]} |f| \pm f(x) \geq 0$ on $[a, b]$, $\sup_{[a,b]} |f| \pm f(T) \geq 0$. That is, $-\sup_{[a,b]} |f| \leq f(T) \leq \sup_{[a,b]} |f|$, which gives

$$|f(T)|_{\text{op}} = \sup_{|v| \leq 1} |f(T)v| \leq \sup_{|v| \leq 1} |\sup_{[a,b]} |f| \cdot |v| = |\sup_{[a,b]} |f|$$

which is the desired inequality. Thus, we can extend by continuity to the sup-norm closure of $\mathbb{R}[x]$ in $C^o[a, b]$, which by Weierstraß is the whole $C^o[a, b]$, giving $C^o[a, b] \to \overline{\mathbb{R}[T]}$, the latter being the operator-norm closure of $\mathbb{R}[T]$, with $|f(T)|_{\text{op}} \leq |f|_{C^o[a,b]}$. Since $\mathbb{R}[x] \to \mathbb{R}[T]$ is a ring homomorphism, the extension by continuity is also a ring homomorphism.

[9.D.4] Corollary: (*Existence of square roots of positive operators*) For $T \geq 0$, there is $S \in \overline{\mathbb{R}[x]}$ such that $S \geq 0$ and $S^2 = T$.

Proof: Since $T \geq 0$, we can take $[a, b] = [0, b]$ in the previous discussion. The function $f(x) = \sqrt{x} \in C^o[0, b]$ is nonnegative on $[0, b]$, and $f(T)^2 = f^2(T) = T$. Take $S = f(T)$.   ///

[9.D.5] Corollary: (*Positivity of products of commuting positive operators*) For $S \geq 0$ and $T \geq 0$ with $ST = TS$, also $ST \geq 0$.

Proof: From the previous corollary, there is $R \in \overline{\mathbb{R}[S]}$ such that $R \geq 0$ and $R^2 = S$. Also, R commutes with T, by continuity. Thus,

$$\langle ST v, v \rangle = \langle R^2 T v, v \rangle = \langle RTR v, v \rangle = \langle TR v, R v \rangle \geq 0$$

because $T \geq 0$.   ///

The kernel I of $C^o[a, b] \to \overline{\mathbb{R}[T]}$ is an *ideal* in $C^o[a, b]$ and is (topologically) *closed* because $C^o[a, b] \to \overline{\mathbb{R}[T]}$ is continuous. Let $\tau(T) \subset [a, b]$ be the simultaneous zero-set of all the functions in I. Shortly, we will see that $\tau(T) = \sigma(T)$, but we cannot use this yet.

[9.D.6] Claim: The restriction map $C^o[a, b] \to C^o(\tau(T))$ has kernel I. That is, if $f|_{\tau(T)} = 0$, then $f(T) = 0$. More precisely, $f \geq 0$ on $\tau(T)$ if and only if $f(T) \geq 0$.

Proof: It suffices to show that $f(T) \geq 0$ implies $f \geq 0$ on $\tau(T)$. For f not nonnegative on $\tau(T)$, there is $x_o \in \tau(T)$ where $f(x_o) < 0$. Using the continuity of f, take a small neighborhood N of x_o in $[a, b]$ such that $f(x) < 0$ on N. Let $g \in C^o[a, b]$ be supported inside N, non-negative, and strictly positive at x_o. Then $fg \leq 0$, and $fg(x_o) < 0$, so $-fg(T) \geq 0$. But $f(T) \geq 0$ and $g(T) \geq 0$, so by the corollary on positivity of commuting positive operators, $fg(T) \geq 0$. Thus, $fg(T) = 0$, so $fg \in I$, and $fg|_{\tau(T)} = 0$, contradiction. Thus, $f \geq 0$ on $\tau(T)$. Thus, if $f = 0$ on $\tau(T)$, both $f \geq 0$ and $-f \geq 0$ on $\tau(T)$, so both $f(T) \geq 0$ and $-f(T) \geq 0$, so $f(T) = 0$, and $f \in I$.   ///

[9.D.7] Corollary: $C^o[a, b] \to \overline{\mathbb{R}[T]}$ factors through $C^o(\tau(T))$, giving a commutative diagram

$$C^o[a, b] \longrightarrow C^o(\tau(T)) \longrightarrow \overline{\mathbb{R}[T]}$$

The induced map $C^o(\tau(T)) \to \overline{\mathbb{R}[T]}$ is a *bijection*, and $|f(T)|_{\text{op}} \geq |f|_{C^o(\tau(T))}$.

Proof: By the Tietze-Urysohn-Brouwer extension theorem [9.E.1], every continuous function on $\tau(T)$ has an extension to a continuous function on $[a, b]$, with the same sup norm. This gives the surjectivity of $C^o[a, b] \to C^o(\tau(T))$. By the claim, $C^o(\tau(T)) \approx C^o[a, b]/I$, giving the injectivity to $\overline{\mathbb{R}[T]}$.

Given the positivity, since $|f(T)|_{\text{op}} \pm f(T) \geq 0$, from the previous claim $|f(T)|_{\text{op}} \pm f(x) \geq 0$ for $x \in \tau(T)$. Thus, $\sup_{x \in \tau(T)} |f(x)| \leq |f(T)|_{\text{op}}$.   ///

Now a refinement of the earlier argument gives the other inequality on norms:

[9.D.8] Corollary: The induced map $C^o(\tau(T)) \to \overline{\mathbb{R}[T]}$ is an *isometric isomorphism*. That is, the map is a bijection, and $|f(T)|_{\text{op}} = |f|_{C^o(\tau(T))}$.

Proof: For $f \geq 0$ on $\tau(T)$, again by Tietze-Urysohn-Brouwer, there is an extension $g \geq 0$ of f to $[a, b]$ with the same sup norm. The first claim of the proof showed that $|f(T)|_{\text{op}} \leq |g|_{C^o[a,b]}$, so

$$|f|_{C^o(\tau(T))} \leq |f(T)|_{\text{op}} \leq |g|_{C^o[a,b]} = |f|_{C^o(\tau(T))}$$

giving the isometry. In particular, for $f_n(T)$ a Cauchy sequence in the operator norm (for $f_n \in C^o(\tau(T))$), the sequence f_n is Cauchy in $C^o(\tau(T))$, so converges to some $f \in C^o(\tau(T))$. By the isometry, $f_n(T) \to f(T)$, giving the surjection to the closure.   ///

It remains to show $\tau(T) = \sigma(T)$.

First, we reprove the fact that $\sigma(T) \subset \mathbb{R}$. For $\lambda \in \mathbb{C}$ such that there is no $(T - \lambda)^{-1}$, the polynomial $f(x) = (x - \lambda)(x - \bar{\lambda})$ is nonzero on $\mathbb{R}$, so certainly on $\tau(T)$, so has an inverse $h(x) = 1/g(x) \in C^o(\tau(T))$. Then $h(T)(T - \bar{\lambda})$ would be an inverse for $T - \lambda$, contradiction. Thus, $\sigma(T) \subset \mathbb{R}$.

For λ real and not in $\tau(T)$, $x - \lambda$ is invertible on $\tau(T)$ with inverse $h \in C^o(\tau(T))$, so

$$h(T) \circ (T - \lambda) = (h \cdot (x - \lambda))(T) = 1(T) = 1$$

and similarly $(T - \lambda) \circ h(T) = 1$, so $T - \lambda$ is invertible. For $\lambda \in \tau(T)$, for $n > 0$, let $f_n(x) \in C^[a, b]$ be

$$f_n(x) = \begin{cases} N & (\text{for } |x - \lambda| \leq \frac{1}{N}) \\ \frac{1}{|x-\lambda|} & (\text{for } |x - \lambda| \geq \frac{1}{N}) \end{cases}$$

Thus, $|(x - \lambda) \cdot f_n|_{C^o(\tau(T))} \leq 1$, and $(T - \lambda)f_n(T)|_{op} \leq 1$. If $T - \lambda$ had an inverse S, then for all n

$$n \leq |f_n|_{C^o(\tau(T))} = |f_n(T)|_{op} = |1 \cdot f_n(T)|_{op} = |S \cdot (T - \lambda) \cdot f_n(T)|_{op}$$
$$\leq |S|_{op} \cdot |(T - \lambda) \cdot f_n(T)|_{op} \leq |S|_{op}$$

This is impossible, so there is no inverse. This proves that $\tau(T) = \sigma(T)$.   ///

Now we prove the peculiar lemma on polynomials:

[9.D.9] Lemma: Let $f \in \mathbb{R}[x]$ be *nonnegative-valued* on a finite interval $[a, b]$. Then f is expressible as a finite sum of the form

$$f = \sum_i P_i^2 + (x - a) \sum_j Q_j^2 + (b - x) \sum_k R_k^2$$

for polynomials P_i, Q_j, R_k in $\mathbb{R}[x]$.

Proof: It suffices to consider monic f, since positive constants can be absorbed. Factor f into irreducibles over $\mathbb{R}$, show that each of the linear and quadratic factors can be expressed in the given form and then show that a product of such expressions can be rewritten in the same form.

For quadratic irreducibles with complex-conjugate roots $z, \bar{z}$, by completing the square,

$$(x - z)(x - \bar{z}) = x^2 - (z + \bar{z})x + z\bar{z} = (x - \frac{z + \bar{z}}{2})^2 + (z\bar{z} - (\frac{z + \bar{z}}{2})^2)$$

Since

$$z\bar{z} - (\frac{z + \bar{z}}{2})^2 = z\bar{z} - \frac{1}{4}(z^2 + 2z\bar{z} + \bar{z}^2) = -\frac{1}{4}(z - \bar{z})^2$$
$$= (\frac{z - \bar{z}}{2i})^2 = (\text{Im} z)^2 > 0$$

we have the desired expression for $(x - z)(x - \bar{z})$.

A linear factor $x - \alpha$ with $a < \alpha < b$ must occur to an *even* power, since otherwise $f(x)$ would take opposite signs on the two sides of α, contradicting the positivity of f on $[a, b]$.

A linear factor $x - \alpha$ with $\alpha \leq a$ can be rewritten as

$$x - \alpha = (x - a) + (a - \alpha) = (x - a) \cdot 1 + (a - \alpha)$$

Since $a - \alpha \geq 0$, it is a square of an element of $\mathbb{R}$, and this gives the desired expression. Similarly, a linear factor $\alpha - x$ with $\alpha \geq b$ can be rewritten as

$$\alpha - x = (b - x) + (\alpha - b)$$

Thus, all the *factors* of f can be written in the desired form. As for products, we can inductively rewrite them by

$$P^2 \cdot Q^2 = (PQ)^2 \qquad (x - a)P^2 \cdot Q^2 = (x - a) \cdot (PQ)^2$$

$$(x - a)P^2 \cdot (x - a)Q^2 = ((x - a)PQ)^2$$

$$(b - x)P^2 \cdot Q^2 = (b - x) \cdot (PQ)^2$$

$$(b - x)P^2 \cdot (b - x)Q^2 = ((b - x)PQ)^2$$

The only possible issue is the form $(x - a)P^2 \cdot (b - x)Q^2$. By luck,

$$(x - a)(b - x) = (x - a)(b - x) \cdot \frac{(b - x) + (x - a)}{(b - x) + (x - a)}$$

$$= \frac{(x - a) \cdot (b - x)^2 + (b - x) \cdot (x - a)^2}{b - a}$$

which is of the desired form. Iterating these rewritings gives the lemma.   ///

[9.D.10] Corollary: If $\sigma(T) = \{\lambda\}$, then T is the scalar operator λ.

Proof: Because the function $f(x) = x$ restricted to $\{\lambda\}$ is equal to the restriction of the constant function $g(x) = \lambda$,

$$T = f(T) = g(T) = \lambda$$

meaning the scalar operator.   ///

[9.D.11] Remark: Certainly the converse is not true: there easily can be eigenvalues *imbedded* in continuous spectrum.

[9.D.12] Corollary: (*Schur's lemma*) Let R be a set of continuous linear operators on a Hilbert space V, and suppose V is *R-irreducible*, in the sense that there is no R-stable closed subspace of V other than $\{0\}$ and V itself. Let T be a self-adjoint operator commuting with all operators from R. Then T is *scalar*.

Proof: Suppose that $\sigma(T)$ contains at least two distinct points x_1, x_2, and show that V is not R-irreducible. Let f, g be continuous functions with disjoint supports, such that $f(x_1) = 1$ and $g(x_2) = 1$. Thus, $fg = 0$, and $f(T)g(T) = g(T)f(T) = 0$, but neither $f(T)$ nor $g(T)$ is 0, because they are not the zero function on $\sigma(T)$. The image $f(T)(V)$ is not 0 because $f(T) \neq 0$. Also, $f(T)(V)$ is inside the kernel of $g(T)$, because $g(T)f(T) = (gf)(T) = 0$. By continuity of $g(T)$, the closure W of $f(T)(V)$ is also inside the kernel of $g(T)$. Since $g(T) \neq 0$, necessarily $W \neq V$.

Since T commutes with all operators in R, $\mathbb{R}[T]$ commutes with R, and by continuity of operators in R, $\overline{\mathbb{R}[T]}$ commutes with R. Thus, R commutes with $f(T)$ and $g(T)$, so for $S \in R$,

$$S(f(T)(V)) = f(T)(SV) \subset f(T)(V)$$

That is, R stabilizes $f(T)(V)$. By continuity of operators in R, R stabilizes the closure W of $f(T)(V)$. But W is a proper closed subspace of V, so V is not R-irreducible. Since $\sigma(T) \neq \phi$, it is a singleton $\{\lambda\}$. By the previous corollary, T is the scalar operator λ.   ///

Suppose that W is another Hilbert space on which R acts, and let $\mathrm{Hom}_R(V, W)$ be the vector space of $\mathbb{C}$-linear maps $\varphi : V \to W$ such that $\varphi(r \cdot v) = r \cdot \varphi(v)$ for all $r \in R$, $v \in V$. In the situation of the previous corollary, let R act on the orthogonal direct sum $V^n = \underbrace{V \oplus \cdots \oplus W}_{n}$ in the natural fashion, by

$$r \cdot (v_1, \ldots, v_n) = (rv_1, \ldots, rv_n)$$

[9.D.13] Corollary: $\dim_{\mathbb{C}} \mathrm{Hom}_R(V, V^n) = n$ for R-irreducible V.

Proof: Let $p_i : V^n \to V$ be the projection to the ith component. For $\varphi \in \mathrm{Hom}_R(V, V^n)$, each $p_i \circ \varphi : V \to V$ respects the action of R, so by Schur's lemma is scalar. Thus, there are scalars $c_1, \ldots, c_n$ so that

$$\varphi(v) = (c_1 \cdot v, \ c_2 \cdot v, \ldots, c_n \cdot v)$$

as claimed.   ///

For the following, assume that R has an *involution* $r \to r^*$ and that the action of R on all vector spaces respects this involution: we only consider actions of R on Hilbert spaces with the property that the adjoint of $v \to r \cdot v$ is $v \to r^* \cdot v$.[8]

---

[8] When R has a structure of ring or group that is reflected in its action on the vector space, the involution $r \to r^*$ should be an *anti-automorphism*, in the sense that $(r_1 r_2)^* = r_2^* \cdot r_1^*$, since the adjoint map on continuous/bounded endomorphisms of a Hilbert space has that behavior.

Also, now we only consider linear maps $V \to W$ that respect this additional structure on R, still referring to these as R-homomorphisms.

[9.D.14] Corollary: Suppose there is an *injection* in $\mathrm{Hom}_R(V^n, W)$. Then

$$\dim_{\mathbb{C}} \mathrm{Hom}_R(V, W) \geq n$$

Further, if there is *no* injection in $\mathrm{Hom}_R(V^{n+1}, W)$, then

$$\dim_{\mathbb{C}} \mathrm{Hom}_R(V, W) = n$$

Proof: Certainly if there is a copy of V^n inside W, then we can map V to any one of the n summands, respecting the action of R. The converse needs Schur's lemma: suppose $\dim_{\mathbb{C}} \mathrm{Hom}_R(V, W) = n$. Let $\varphi_1, \ldots, \varphi_n$ be n linearly independent homomorphisms. The image $\varphi_1(V) + \varphi_2(V) + \cdots + \varphi_n(V)$ need not be an *orthogonal* direct sum, but we claim that there is another collection of n maps in $\mathrm{Hom}_R(V, W)$ that *does* produce an orthogonal direct sum inside W. In effect, this is a version of a Gram-Schmidt process that refers to copies of the irreducible V rather than to individual vectors.

A key point is that, because of the involution $*$, the orthogonal complement $X^\perp$ to an R-stable subspace X of W is also R-stable. Indeed. For $y \in X^\perp$,

$$\langle r \cdot y, x \rangle = \langle y, r^* \cdot x \rangle \in \langle y, X \rangle = \{0\}$$

This immediately implies that the orthogonal projection $W \to X$ is an R-homomorphism.

Thus, given φ_1 and φ_2, the orthogonal projection p from $\varphi_2(V)$ to $\varphi_1(V)$ is an R-homomorphism. Since φ_1, φ_2 are nonzero, $\varphi_1(V)$ and $\varphi_2(V)$ are R-irreducible, so φ_1 and φ_2 are R-isomorphisms. If the images $\varphi_1(V)$ and $\varphi_2(V)$ are orthogonal, we are done. If not, the map p is not 0, so must be an R-isomorphism by R-irreducibility. Thus, the composition

$$V \xrightarrow{\varphi_1} \varphi_2(V) \xrightarrow{p} \varphi_1(V) \xrightarrow{\varphi_1^{-1}} V$$

is an R-isomorphism $V \to V$. By Schur's lemma, it is a nonzero constant map. That is, there is a uniform constant c such that $p(\varphi_2(v)) = c \cdot \varphi_1(v)$ for all $v \in V$. That is, $c \cdot \varphi_1 - p \circ \varphi_2 = 0$ as element of $\mathrm{Hom}_R(V, W)$. Then

$$p \circ (c \cdot \varphi_1 - p \circ \varphi_2) = c \cdot p \circ \varphi_1 - p^2 \circ \varphi_2 = p \circ \varphi_1 - p \circ \varphi_2 = 0$$

so the image $(c \cdot \varphi_1 - p \circ \varphi_2)(V)$ is orthogonal to $\varphi_1(V)$, as desired. Continue by induction to modify all $\varphi_i(V)$ to be mutually orthogonal. ///

9.E Appendix: Tietze-Urysohn-Brouwer Extension Theorem

Granting Urysohn's lemma [9.E.2], the extension result is not difficult:

[9.E.1] Theorem: For X a *normal* space (meaning that any two disjoint closed sets have disjoint open neighborhoods), closed subset $E \subset X$, every continuous, bounded, real-valued f on E extends to F on X such that $\sup_X |F| = \sup_E |f|$.

Proof: Without loss of generality, the image of f is contained in $[0, 1]$. Urysohn's lemma [9.E.2] will be repeatedly invoked: given disjoint, closed B_n, C_n in X, there is continuous g_n on X taking values in $[0, \frac{1}{2}(2/3)^n]$ such that $g_n = 0$ on B_n and $g_n = \frac{1}{2}(2/3)^n$ on C_n. Specify the subsets B_n, C_n ($n = 1, 2, \ldots$) of E inductively by

$$B_1 = \{x \in E : f(x) \le \frac{1}{3}\} \qquad C_1 = \{x \in E : f(x) \ge \frac{2}{3}\}$$

and

$$B_n = \{x \in E : f(x) - \sum_{i=1}^{n-1} g_i(x) \le \frac{2^{n-1}}{3^n}\}$$

$$C_n = \{x \in E : f(x) - \sum_{i=1}^{n-1} g_i(x) \ge \frac{2^n}{3^n}\}$$

These are disjoint closed subsets of E, so are closed in X. The sum $F = \sum_{i=1}^{\infty} g_i$ converges uniformly, so is continuous. On E, $0 \le f - F \le (2/3)^n$ for all n, so $F = f$ on E.   ///

[9.E.2] Theorem: *(Urysohn)* In a locally compact Hausdorff topological space X, given a compact subset K contained in an open set U, there is a continuous function $0 \le f \le 1$ which is 1 on K and 0 off U.

Proof: First, we prove that there is an open set V such that

$$K \subset V \subset \overline{V} \subset U$$

For each $x \in K$ let V_x be an open neighborhood of x with compact closure. By compactness of K, some finite subcollection $V_{x_1}, \ldots, V_{x_n}$ of these V_x cover K, so K is contained in the open set $W = \bigcup_i V_{x_i}$ which has compact closure $\bigcup_i \overline{V}_{x_i}$ since the union is *finite*.

Using the compactness again in a similar fashion, for each x in the closed set $X - U$, there is an open W_x containing K and a neighborhood U_x of x such that $W_x \cap U_x = \phi$.

Then

$$\bigcap_{x \in X - U} (X - U) \cap \overline{W} \cap \overline{W}_x = \phi$$

These are compact subsets in a Hausdorff space, so (again from compactness) some *finite* subcollection has empty intersection, say

$$(X - U) \cap \left(\overline{W} \cap \overline{W}_{x_1} \cap \ldots \cap \overline{W}_{x_n} \right) = \phi$$

That is,

$$\overline{W} \cap \overline{W}_{x_1} \cap \ldots \cap \overline{W}_{x_n} \subset U$$

Thus, the open set

$$V = W \cap W_{x_1} \cap \ldots \cap W_{x_n}$$

meets the requirements.

Using the possibility of inserting an open subset and its closure between any $K \subset U$ with K compact and U open, we inductively create opens V_r (with compact closures) indexed by rational numbers r in the interval $0 \leq r \leq 1$ such that, for $r > s$,

$$K \subset V_r \subset \overline{V}_r \subset V_s \subset \overline{V}_s \subset U$$

From any such configuration of opens, we construct the desired continuous function f by

$$f(x) = \sup\{r \text{ rational in } [0, 1] : x \in V_r, \}$$
$$= \inf\{r \text{ rational in } [0, 1] : x \in \overline{V}_r, \}$$

It is not immediate that this sup and inf are the same, but if we *grant* their equality, then we can prove the *continuity* of this function $f(x)$. Indeed, the sup description expresses f as the supremum of characteristic functions of open sets, so f is at least *lower semi-continuous*.[9] The inf description expresses f as an infimum of characteristic functions of closed sets so is *upper* semicontinuous. Thus, f would be continuous.

To finish the argument, we must construct the sets V_r and prove equality of the inf and sup descriptions of the function f.

---

[9] A (real-valued) function f is *lower* semicontinuous when for all bounds B the set $\{x : f(x) > B\}$ is open. The function f is *upper* semicontinuous when for all bounds B the set $\{x : f(x) < B\}$ is open. It is easy to show that a sup of lower semicontinuous functions is lower semicontinuous, and an inf of upper semicontinuous functions is upper semicontinuous. As expected, a function both upper and lower semicontinuous is continuous.

To construct the sets V_i, start by finding V_0 and V_1 such that

$$K \subset V_1 \subset \overline{V}_1 \subset V_0 \subset \overline{V}_0 \subset U$$

Fix a well-ordering $r_1, r_2, \ldots$ of the rationals in the open interval $(0, 1)$. Supposing that $V_{r_1}, \ldots, v_{r_n}$ have been chosen. let i, j be indices in the range $1, \ldots, n$ such that

$$r_j > r_{n+1} > r_i$$

and r_j is the *smallest* among $r_1, \ldots, r_n$ *above* r_{n+1}, while r_i is the *largest* among $r_1, \ldots, r_n$ *below* r_{n+1}. Using the first observation of this argument, find $V_{r_{n+1}}$ such that

$$V_{r_j} \subset \overline{V}_{r_j} \subset V_{r_{n+1}} \subset \overline{V}_{r_{n+1}} \subset V_{r_i} \subset \overline{V}_{r_i}$$

This constructs the nested family of opens.

Let $f(x)$ be the sup and $g(x)$ the inf of the characteristic preceding functions. If $f(x) > g(x)$, then there are $r > s$ such that $x \in V_r$ and $x \notin \overline{V}_s$. But $r > s$ implies that $V_r \subset \overline{V}_s$, so this cannot happen. If $g(x) > f(x)$, then there are rationals $r > s$ such that

$$g(x) > r > s > f(x)$$

Then $s > f(x)$ implies that $x \notin V_s$, and $r < g(x)$ implies $x \in \overline{V}_r$. But $V_r \subset \overline{V}_s$, contradiction. Thus, $f(x) = g(x)$.   ///

10

Discrete Decomposition of Pseudo-Cuspforms

Applications of idiosyncracies of Friedrichs self-adjoint extensions of restrictions of Laplace-Beltrami operators are illustrated here, as exploited in [Lax-Phillips 1976] and [Colin de Verdière 1981, 1982/1983], for example, and as illustrated in the next chapter. This device is essentially archimedean, related to differential operators.

On any one of the four simple unicuspidal quotients $\Gamma \backslash G / K$ of Chapter 1, the space of *pseudo-cuspforms* $L_a^2(\Gamma \backslash G / K)$ with cutoff height a is the space of L^2 functions whose constant terms vanish above height $\eta(g) = a$. The case $a = 0$ is the usual space of L^2 cuspforms. We will show that $L_a^2(\Gamma \backslash G / K)$ decomposes discretely for $\widetilde{\Delta}_a$, the Friedrichs extension [9.2] of the restriction Δ_a of Δ to the space

$$D_a = C_c^\infty(\Gamma \backslash G / K) \cap L_a^2(\Gamma \backslash G / K)$$

of test functions inside $L_a^2(\Gamma \backslash G / K)$. The proof proceeds by showing that $\widetilde{\Delta}_a$ has *compact resolvent* $(\widetilde{\Delta}_a - \lambda)^{-1}$ for λ off a discrete set, and then verifying the obvious plausible bijection between the spectrum and eigenfunctions of $\widetilde{\Delta}_a$

and those of $(\widetilde{\Delta}_a - \lambda)^{-1}$ in [10.7]. Then the spectral theorem for self-adjoint compact operators [10.10] yields an orthonormal basis for $L^2_a(\Gamma \backslash G/K)$ consisting of eigenfunctions for $\widetilde{\Delta}_a$, with finite multiplicities.

Further, in those four examples, for $a \gg 1$, the space of pseudo-cuspforms $L^2_a(\Gamma \backslash G/K)$ includes not only cuspforms but infinitely many $\widetilde{\Delta}_a$-eigenfunctions that are (necessarily) *not* Δ-eigenfunctions. *Existence* of further eigenfunctions in $L^2_a(\Gamma \backslash G/K)$ is clear from the spectral decomposition of the orthogonal complement of cuspforms $L^2_o(\Gamma \backslash G/K)$ in $L^2(\Gamma \backslash G/K)$, in terms of integrals of Eisenstein series and residues of Eisenstein series, as in [1.12]. In [11.6], we show that all but finitely many of the the *new* eigenfunctions are the *truncated Eisenstein series* whose constant term *vanishes* at $y = a$.

For the examples $SL_3(\mathbb{Z})$, $SL_4(\mathbb{Z})$, $SL_5(\mathbb{Z})$, . . ., the notion of pseudo-cuspform is more complicated, but the general pattern of the argument is the same. Again, *certain* truncated Eisenstein series comprise most of the new discrete spectrum.

In all examples, the critical point is the estimate on *tails* of pseudo-cuspforms, in [10.3], [10.4], and [10.5]. This is used to prove a Rellich-type compactness lemma, from which the compactness of the resolvent of the Friedrichs extensions follows. The seeming paradox of this *discretization of the continuous spectrum* of Δ on $L^2(\Gamma \backslash G/K)$ is essential to [Colin de Verdière 1981]'s proof of meromorphic continuation of Eisenstein series, which we recapitulate in examples in Chapter 11.

10.1 Compact Resolvents in Simplest Examples

The statements of the theorems are easier for the four simplest examples $\Gamma \backslash G/K$ of Chapter 1. For $a \geq 0$, consider a space of square-integrable *pseudo-cuspforms* including the space of cuspforms: these are functions in $L^2(\Gamma \backslash G/K)$ whose constant terms $c_P f$ vanish above height a:

$$L^2_a(\Gamma \backslash G/K) = \{ f \in L^2(\Gamma \backslash G/K) : c_P f(g) = 0 \text{ for } \eta(g) \geq a \}$$

where the height function is $\eta(nm_y k) = y^r$ with $n \in N$, $m_y = \begin{pmatrix} \sqrt{y} & 0 \\ 0 & 1/\sqrt{y} \end{pmatrix}$, $k \in K$, and $r = 1, 2, 3, 4$ in the respective examples. As for cuspforms, vanishing of the constant term $c_P f$ above height a can be expressed precisely as orthogonality $\langle f, \Psi_\varphi \rangle = 0$ to all pseudo-Eisenstein series Ψ_φ with test-function data $\varphi \in C^\infty_c(0, \infty)$ supported on $[a, +\infty)$. Let

$$D_a = C^\infty_c(\Gamma \backslash G/K) \cap L^2_a(\Gamma \backslash G/K)$$

and Δ_a the restriction of Δ to D_a. In [10.3], we prove that D_a is dense in $L_a^2(\Gamma\backslash G/K)$, so Δ_a has a Friedrichs extension $\widetilde{\Delta}_a$. The main result is:

[10.1.1] Theorem: $\widetilde{\Delta}_a$ has compact resolvent. The space $L_a^2(\Gamma\backslash G/K)$ of square-integrable pseudo-cuspforms with constant term vanishing above height $\eta(g) = a$ has an orthonormal basis of $\widetilde{\Delta}_a$-eigenfunctions, and eigenvalues have finite multiplicities. *(Proof in [10.7].)*

A seeming paradox: Of course, the space $L_a^2(\Gamma\backslash G/K)$ contains the space $L^2(\Gamma\backslash G/K)$ of L^2 cuspforms, for every $a \geq 0$. For $a \gg 1$, the corresponding space of pseudo-cuspforms it is demonstrably properly larger, containing part of the continuous spectrum for Δ, namely, an infinite-dimensional space of pseudo-Eisenstein series Ψ_φ. For example, take $a' < a$, with a' still large enough so that the Siegel set

$$\mathfrak{S} = \mathfrak{S}_{a'} = \{g \in G/K : \eta(g) > a'\}$$

has the property that $\gamma\mathfrak{S} \cap \mathfrak{S} \neq \phi$ implies $\gamma \in \Gamma \cap P$. Then, for any test function φ supported on $[a', a]$, the pseudo-Eisenstein series Ψ_φ is identically 0 in the region $\eta(g) > a$. From the spectral decomposition [1.12], these pseudo-Eisenstein series are integrals of Eisenstein series. Yet the Friedrichs extension $\widetilde{\Delta}_a$ is proven to have entirely discrete spectrum. Evidently, some part of the continuous spectrum of Δ becomes discrete for $\widetilde{\Delta}_a$. That is, some integrals of Eisenstein series E_s become L^2 eigenfunctions for $\widetilde{\Delta}_a$: in the four simple examples:

[10.1.2] Theorem: For $a \gg 1$, truncated Eisenstein series $\wedge^a E_s$ such that $c_P E_s(g) = 0$ for $\eta(g) = a$ become $\widetilde{\Delta}_a$-eigenfunctions. *(Proof in [11.6].)*

From the theory of the constant term [8.1], the truncation $\wedge^a E_s$ is in L^2. However, $\wedge^a E_s$ is not smooth. The possibility that non-smooth functions can be eigenfunctions for $\widetilde{\Delta}_a$ can be understood in terms of the behavior of Friedrichs extensions, and exploited, as in [Colin de Verdière 1981, 1982/1983]. We give the application to meromorphic continuation of Eisenstein series in the next chapter.

Conversely, in these examples, we will show that

[10.1.3] Theorem: *All* noncuspforms with $\widetilde{\Delta}_a$-eigenvalues $\lambda_w < -1/4$ are truncated Eisenstein series $\wedge^a E_s$ such that $c_P E_s(g) = 0$ for $\eta(g) = a$. *(Proof in [11.6].)*

10.2 Compact Resolvents for $SL_3(\mathbb{Z})$, $SL_4(\mathbb{Z})$, $SL_5(\mathbb{Z})$, ...

Now consider $\Gamma = SL_r(\mathbb{Z})$, $G = SL_r(\mathbb{R})$, and $K = SO_r(\mathbb{R})$. Again, we will prove that a certain space of square-integrable functions on $\Gamma\backslash G/K$ with all constant terms vanishing beyond fixed heights has purely discrete spectrum with respect to the Friedrichs extension of the restriction of the invariant Laplacian to (test functions in) this space.

Because we have not discussed a sufficiently general form of *truncations* for automorphic forms on GL_r, we cannot make as strong a statement as we might like. Namely, we will *not* prove that the L^2 closure of a space of test functions D_a, the initial domain for a restriction Δ_a, is as large as we might imagine. Nevertheless, the application to meromorphic continuation of cuspidal-data Eisenstein series in [11.10] and [11.12] does not need the strongest density assertions, so we will have a complete proof of that meromorphic continuation.

As in the simpler examples, the proof will proceed by showing that the resolvent of the Friedrichs extension of a restriction of the invariant Laplacian is *compact*. Specifically, let A be the standard maximal torus consisting of diagonal real matrices, and A^+ its subgroup of positive real diagonal matrices. A standard choice of positive simple roots is

$$\Phi = \{\alpha_i(a) = \frac{a_i}{a_{i+1}} : i = 1, \ldots, r - 1\}$$

with $a = \begin{pmatrix} a_1 \\ & \ddots \\ & & a_r \end{pmatrix}$. Let $N^{\min}$ be the unipotent radical of the standard minimal parabolic $P^{\min}$ consisting of upper-triangular elements of G. For $g \in G$, let $g = n_g m_g k_g$ be the corresponding Iwasawa decomposition with respect to $P^{\min}$, with $m_g \in A^+$. By reduction theory [3.3], there is a sufficiently small $t_o > 0$ such that the standard Siegel set

$$\mathfrak{S} = \mathfrak{S}_{t_o} = \{nmk : n \in N^{\min}, m \in A^+, k \in K, \alpha(m) \geq t_o \text{ for all } \alpha \in \Phi\}$$

satisfies $\Gamma \cdot \mathfrak{S} = G$. Fix such $\mathfrak{S}$ for the following discussion. For real $a \gg 1$, specify a subset of $\mathfrak{S}$ by

$$Y_a = \{nmk \in \mathfrak{S} : \alpha(m) \geq a \text{ for some } \alpha \in \Phi\}$$

where again $n \in N^{\min}$, $m \in A^+$, and $k \in K$. Let Δ be the Casimir operator for G descended to G/K and to $\Gamma\backslash G/K$ as in [4.2], [4.4]. Let Δ_a be the restriction of the invariant Laplace-Beltrami operator Δ to the domain

$$D_a = \{f \in C_c^\infty(\Gamma\backslash G/K) : \text{for } g \in Y_a, c_P f(g) = 0, \text{ for all } P\}$$

where P runs through standard parabolics. Let V_a be the closure of D_a in $L^2(\Gamma\backslash G/K)$. As usual [6.5], integration by parts shows that Δ_a is *symmetric* and nonpositive, in the sense that $\langle \Delta f, f \rangle \leq 0$ for test functions f. Since D_a is dense in V_a, it has a Friedrichs extension $\widetilde{\Delta}_a$, a self-adjoint unbounded operator on V_a.

[10.2.1] Theorem: $\widetilde{\Delta}_a$ has compact resolvent. The space V_a has an orthonormal basis of $\widetilde{\Delta}_a$-eigenfunctions, and eigenvalues occur with finite multiplicities. *(Proof in [10.8] and [10.9].)*

Define the $\mathfrak{B}^1$ norm on D_a by

$$|f|^1_{\mathfrak{B}^1} = \langle (1 - \Delta)f, f \rangle_{L^2(\Gamma \backslash G/K)}$$

and let $\mathfrak{B}^1$ be the completion of D_a with respect to this norm. As in the discussion [9.2] of Friedrichs extensions, we have a natural imbedding $\mathfrak{B}^1 \subset V_a$. As in the simpler examples, for sufficiently high cutoff heights η, we will see that there must be infinitely many eigenfunctions for $\widetilde{\Delta}_a$ that were not eigenfunctions for Δ, by exhibiting some pseudo-Eisenstein series in the $\mathfrak{B}^1$-closure of D_a. Specifically, we consider pseudo-Eisenstein series attached to maximal proper parabolics $P = P^{r,r} \subset SL_{2r}$, with cuspidal data [3.9], with test function data supported just below the cutoff. Via reduction theory [3.3], the P-constant term vanishes for $\alpha(a) \geq a$. From [3.9], all other constant terms along standard parabolics are 0. Similarly, as explicit examples of eigenfunctions for $\widetilde{\Delta}_a$ that are not eigenfunctions for Δ, we again find certain truncated Eisenstein series. The simplest case is the following.

Let $P = P^{r,r} \subset SL_{2r}$, and f cuspidal data on the Levi component $M = M^P$. Let $E_{s,f}$ be the corresponding cuspidal-data Eisenstein series as in [3.11], with constant term $c_P E_{s,f}$ as in [3.11.9]. Let A^P be the center of M, and M^1 the subgroup of M consisting of matrices in r-by-r blocks $\begin{pmatrix} a & 0 \\ 0 & d \end{pmatrix}$ with $\det a = 1 = \det d$.

[10.2.2] Theorem: For $s \in \mathbb{C}$ such that $c_P E_{s,f}(mm_1) = 0$ for $m \in A^P$ with $\alpha_r(m) = a$ and for all $m_1 \in M^1$, the truncation $\wedge^a E_{s,f}$ of $E_{s,f}$ is a $\widetilde{\Delta}_a$-eigenfunction in V_a. *(Proof in [11.11].)*

10.3 Density of Domains of Operators

For an unbounded operator to have a well-defined adjoint, its domain must be *dense* in the ambient Hilbert space. Of course, we could shrink the Hilbert space to be the closure of the domain of the operator, but then there would be the issue of determining that closure, apart from other complications. Test functions are dense in $L^2(\Gamma \backslash G/K)$ for general reasons [6.1], [14.5], and [14.6]: for an *approximate identity* ψ_n in $C_c^\infty(K \backslash G/K)$, the averaged action images $\psi_n \cdot f$ of $f \in L^2_a(\Gamma \backslash G/K)$ are *smooth*. However, each such averaging smears out the support of the constant term of f somewhat, depending on the support of ψ_n. Let X_n be a nested sequence of compact subsets of $\Gamma \backslash G/K$ whose union is $\Gamma \backslash G/K$, and $\alpha_n \in C_c^\infty(\Gamma \backslash G/K)$ identically 1 on X_n, and $0 \le \alpha_n(g) \le 1$. Thus, smoothly cut off $\psi_n \cdot f$ by multiplying by α_n. Thus, $f_n = (\psi_n \cdot f) \cdot \alpha_n$ is a sequence in $C_c^\infty(\Gamma \backslash G/K)$ approaching f in L^2.

However, it is not as trivial to understand the interaction with constant-term vanishing conditions. In these simple examples, density of D_a in $L_a^2(\Gamma \backslash G/K)$ is relatively easily proven for $a \gg 1$, in which case the natural smooth cutting-off of the constant term near the given height a interacts with constant-term vanishing in a controlled manner.

The condition $a \gg 1$ refers to a great-enough height a so that the standard Siegel set $\mathfrak{S}_a$ has the property that $\mathfrak{S}_a \cap \gamma \mathfrak{S}_a \ne \phi$ implies $\gamma \in \Gamma \cap P$. By reduction theory [1.5] there exists such a. For the simplest case of $SL_2(\mathbb{Z}) \subset SL_2(\mathbb{R})$, the explicit bound $a > 1$ suffices, for example. Again, as expected, first approximate $f \in L_a^2(\Gamma \backslash G/K)$ by functions f_n in $C_c^\infty(\Gamma \backslash G/K)$ by general methods, and then use the condition $a \gg 1$ to consider a family of smooth cutoffs of the constant term near height a, with the width of the cutoff region shrinking to 0:

[10.3.1] Lemma: For $a \gg 1$, D_a is dense in $L_a^2(\Gamma \backslash G/K)$.

Proof: As just above, we take $a \gg 1$ so that the Siegel set $\mathfrak{S}_{a-\frac{1}{t}}$ meets its translates $\gamma \mathfrak{S}_{a-\frac{1}{t}}$ only for $\gamma \in \Gamma \cap P$, for all sufficiently large t. This allows separation of variables in $\mathfrak{S}_{a-\frac{1}{t}}$, since the cylinder $C_{a-\frac{1}{t}} = (\Gamma \cap P) \backslash \mathfrak{S}_{a-\frac{1}{t}}$ injects to $\Gamma \backslash G/K$. Let

$$|f|^2_{C_{a-\frac{1}{t}}} = \int_{C_{a-1/t}} |f(z)|^2 \, \frac{dx\,dy}{y^{r+1}} \le \int_{\Gamma \backslash G/K} |f(z)|^2 \, \frac{dx\,dy}{y^{r+1}} = |f|^2_{L^2}$$

Let $f_n \in C_c^\infty(\Gamma \backslash G/K)$ with $f_n \to f$ in L^2. Since $f \in L_a^2(\Gamma \backslash G/K)$, we naturally expect that the constant term is not *too* far from that of f, so that *smooth truncations* of the constant terms of $\psi_n \cdot f$ should produce functions *also* approaching f.

Use the Iwasawa coordinates x, y on G/K with $x \in \mathbb{R}^r$ and $y > 0$ as in [1.3], so the *height* is $\eta(x, y) = y^r$. Let β be a smooth function on $\mathbb{R}$ such that $\beta(y) = 0$ for $y < -1$, $0 \le \beta(y) \le 1$ for $-1 \le y \le 0$, and $\beta(y) = 1$ for $y \ge 0$. For $t > 1$, put $\beta_t(y) = \beta(t(y - a))$, and define a smooth function on $N \backslash G/K$ by

$$\varphi_{n,t}(x, y) = \begin{cases} \beta_t(y^r) \cdot c_P f_n(y) & (\text{for } y^r \ge a - \frac{1}{t}) \\ 0 & (\text{for } y^r < a - \frac{1}{t}) \end{cases}$$

For $t > 0$ large enough so that $\mathfrak{S}_{a-\frac{1}{t}}$ does not meet any of its own translates by $\gamma \in \Gamma$ except $\gamma \in \Gamma \cap P$, let $\Psi_{n,t} = \Psi_{\varphi_{n,t}}$ be the pseudo-Eisenstein series made from $\varphi_{n,t}$. The assumption on t ensures that in the region $y^r > a - \frac{1}{t}$ we have $\Psi_{n,t} = c_P \Psi_{n,t} = \varphi_{n,t}$. Thus, as intended, $c_P(f_n - \Psi_{n,t})$ vanishes in $y \ge a$, so $f_n - \Psi_{n,t} \in L_a^2(\Gamma \backslash G/K)$.

By the triangle inequality,

$$|f - (f_n - \Psi_{n,t})|_{L^2} \le |f - f_n|_{L^2} + |\Psi_{n,t}|_{L^2}$$

and $|f - f_n|_{L^2} \longrightarrow 0$. Thus, it suffices to show that the L^2 norm of the pseudo-Eisenstein series $\Psi_{n,t}$ goes to 0 for large n, t. Since $a \gg 1$,

$$|\Psi_{n,t}|_{L^2} = |\Psi_{n,t}|_{C_{a-\frac{1}{t}}} = |\varphi_{n,t}|_{C_{a-\frac{1}{t}}}$$
$$= |\beta(t(y - a)) \cdot c_P f_n|_{C_{a-\frac{1}{t}}} \le |c_P f_n|_{C_{a-\frac{1}{t}}}$$

The cylinder $C_{a-\frac{1}{t}}$ admits a natural action of the product of circle groups $\mathbb{T}^r = (\Gamma \cap N) \backslash N$, by translation, inherited from the translation of the x-component in coordinates x, y. This induces a continuous action of $\mathbb{T}^r$ on $L^2(C_{a-\frac{1}{t}})$ with the norm $| \cdot |_{C_{a-\frac{1}{t}}}$. Thus, the map $F \to c_P F$, is given by a continuous, compactly supported $L^2(C_{a-\frac{1}{t}})$-valued integrand, so from [14.1] exists as a Gelfand-Pettis integral. Thus, unsurprisingly, the restriction of $c_P f_n$ to $C_{a-\frac{1}{t}}$ goes to $c_P f$ in $L^2(C_{a-\frac{1}{t}})$. Since $c_P f$ is supported in the range $\eta(g) \le a$, and the measure of $C_a - C_{a-\frac{1}{t}}$ goes to 0 as $t \to +\infty$, the $C_{a-\frac{1}{t}}$-norm of $c_P f$ goes to 0 as $t \to +\infty$, since $c_P f$ is locally integrable.

Thus, for example, $\Psi_{n,n}$ goes to 0 in L^2 norm, so the elements $f_n - \Psi_{n,n}$ in L_a^2 go to f in L^2 norm, proving the density of D_a in L_a^2. ///

10.4 Tail Estimates: Simplest Example

The computation for $\Gamma = SL_2(\mathbb{Z})$, $G = SL_2(\mathbb{R})$, and $K = SO_2(\mathbb{R})$ can take advantage of some convenient technical coincidences. Let the $\mathfrak{B}^1$ norm be defined on test functions $C_c^\infty(\Gamma \backslash G/K)$ by

$$|f|_{\mathfrak{B}^1}^2 = \langle (1 - \Delta)f, f \rangle = \langle f, f \rangle + \langle (-\Delta)f, f \rangle$$

Let $\mathfrak{B}^1$ be the completion of $C_c^\infty(\Gamma \backslash G/K)$ with respect to the $\mathfrak{B}^1$ norm. With

$$D_a = C_c^\infty(\Gamma \backslash G/K) \cap L_a^2(\Gamma \backslash G/K)$$

let $\mathfrak{B}_a^1$ be the $\mathfrak{B}^1$-completion of D_a. Note that although it is clear that $\mathfrak{B}_a^1 \subset \mathfrak{B}^1 \cap L_a^2(\Gamma \backslash G/K)$, it is *not* clear that equality holds. We do not need to address this for the moment. As in [10.1], let Δ_a be the restriction of Δ to D_a, and $\widetilde{\Delta}_a$

its Friedrichs extension. The Friedrichs extension $\widetilde{\Delta}_a$ maps from $L_a^2(\Gamma\backslash G/K)$ to $\mathfrak{B}_a^1$. Let B be the unit ball in $\mathfrak{B}_a^1$. As in all cases, the crucial estimate is

[10.4.1] Claim: Given $\varepsilon > 0$, a cutoff $c \geq a$ can be made sufficiently large so that the image of B in $L^2(\Gamma\backslash G/K)$, cut off at height c, lies in a single ε-ball in $L^2(\Gamma\backslash G/K)$. That is, for $f \in \mathfrak{B}_a^1$,

$$\lim_{c\to\infty} \int_{y>c} |f(z)|^2 \, \frac{dx\,dy}{y^2} \longrightarrow 0 \qquad \text{(uniformly for } |f|_{\mathfrak{B}^1} \leq 1)$$

[10.4.2] Remark: To be careful, we note that the inequality of the claim does not directly address the issue of *smooth* truncations of f in $\mathfrak{B}_a^1$ near height c, nor whether a collection of smooth truncations $\varphi_\infty \cdot f$ of all heights $c \gg a$ can be chosen with $\mathfrak{B}^1$-norms *uniformly bounded* for $f \in B$. These somewhat secondary points are addressed just below in [10.4.3]: nothing surprising happens.

Proof: This computation roughly follows [Lax-Phillips 1976], pp. 204–206. To legitimize the following computation, we should be sure that $f \in \mathfrak{B}^1$ has first derivatives in an L^2 sense. This is a *local* fact, and thus follows from the discussion on tori $\mathbb{T}^n$ in [9.5].

Let the Fourier coefficients of $f(x+iy)$ be $\widehat{f}(n)$, functions of y. Take $c > a$ so that the 0th Fourier coefficient $\widehat{f}(0)$ vanishes identically. Use Plancherel for the Fourier expansion in x, and then elementary inequalities: integrating over the part of Y_∞ above $y = c$, letting $\mathcal{F}$ be Fourier transform in x,

$$\int\int_{y>c} |f|^2 \, \frac{dx\,dy}{y^2} \leq \frac{1}{c^2} \int\int_{y>c} |f|^2 \, dx\,dy = \frac{1}{c^2} \sum_{n\neq 0} \int_{y>c} |\widehat{f}(n)|^2 \, dy$$

$$\leq \frac{1}{c^2} \sum_{n\neq 0} (2\pi n)^2 \int_{y>c} |\widehat{f}(n)|^2 \, dy = \frac{1}{c^2} \sum_{n\neq 0} \int_{y>c} \left|\mathcal{F}\frac{\partial f}{\partial x}(n)\right|^2 \, dy$$

$$= \frac{1}{c^2} \int\int_{y>c} \left|\frac{\partial f}{\partial x}\right|^2 \, dx\,dy = \frac{1}{c^2} \int\int_{y>c} -\frac{\partial^2 f}{\partial x^2} \cdot \overline{f}(x) \, dx\,dy$$

$$\leq \frac{1}{c^2} \int\int_{y>c} -\frac{\partial^2 f}{\partial x^2} \cdot \overline{f}(x) - \frac{\partial^2 f}{\partial y^2} \cdot \overline{f}(x) \, dx\,dy$$

$$= \frac{1}{c^2} \int\int_{y>c} -\Delta f \cdot \overline{f} \, \frac{dx\,dy}{y^2} \leq \frac{1}{c^2} \int\int_{\Gamma\backslash G/K} -\Delta f \cdot \overline{f} \, \frac{dx\,dy}{y^2}$$

$$\leq \frac{1}{c^2} \cdot |f|_{\mathfrak{B}^1}^2 \leq \frac{1}{c^2}$$

giving the uniform bound as claimed.   ///

Now we prove the reassuring lemma that the $\mathfrak{B}^1$-norms of systematically specified families of smooth *tails* of functions in $\mathfrak{B}^1$ are uniformly dominated by the $\mathfrak{B}^1$-norms of the original functions. Let φ be a smooth real-valued function on $(0, +\infty)$ with

$$\begin{cases} \varphi(y) = 0 & \text{(for } 0 < y \leq 1) \\ 0 \leq \varphi(y) \leq 1 & \text{(for } 1 < y < 2) \\ 1 \leq \varphi(y) & \text{(for } 1 \leq y) \end{cases}$$

[10.4.3] Claim: For fixed η, for $t \geq 1$, the smoothly cutoff tail $f^{[t]}(x + iy) = \varphi\left(\frac{y}{t}\right) \cdot f(x + iy)$ has $\mathfrak{B}^1$-norm dominated by that of f itself:

$$|f^{[t]}|_{\mathfrak{B}^1} \ll_\varphi |f|_{\mathfrak{B}^1} \qquad \text{(implied constant independent of } f, t \geq 1)$$

Proof: This is essentially elementary. Since $|a + bi|^2 = a^2 + b^2$ and Δ has real coefficients, it suffices to treat real-valued f. Since $0 \leq \varphi \leq 1$, certainly $|\varphi f|_{L^2} \leq |f|_{L^2}$. For the other part of the $\mathfrak{B}^1$-norm, letting $S^1 \approx \mathbb{R}/\mathbb{Z}$ be the circle,

$$\begin{aligned}
\langle -\Delta f^{[t]}, f^{[t]} \rangle &= -\int_{S^1} \int_{y \geq t} \left(\frac{\partial^2}{\partial x^2} + \frac{\partial^2}{\partial y^2} \right) f^{[t]} \cdot f^{[t]} \, dx \, dy \\
&= -\int_{S^1} \int_{y \geq t} \varphi^2\left(\frac{y}{t}\right) f_{xx} f + \frac{1}{t^2} \varphi''\left(\frac{y}{t}\right) \varphi\left(\frac{y}{t}\right) f^2 \\
&\quad + \frac{2}{t} \varphi'\left(\frac{y}{t}\right) \varphi\left(\frac{y}{t}\right) f_y f + \varphi\left(\frac{y}{t}\right)^2 f_{yy} f \, dx \, dy
\end{aligned}$$

Some terms are easy to estimate: using the fact that φ' and φ'' are supported on $[1, 2]$,

$$\begin{aligned}
\int_{S^1} \int_{y \geq t} &-\varphi\left(\frac{y}{t}\right)^2 f_{xx} f + \left| \frac{1}{t^2} \varphi''\left(\frac{y}{t}\right) \varphi\left(\frac{y}{t}\right) f^2 \right| - \varphi\left(\frac{y}{t}\right)^2 f_{yy} f \, dx \, dy \\
&\ll_\varphi \int_{S^1} \int_{t \leq y \leq 2t} \frac{f^2}{t^2} - (f_{xx} f + f_{yy} f) \, dx \, dy \\
&\leq \int_{S^1} \int_{t \leq y \leq 2t} \frac{(2t)^2 f^2}{t^2} - y^2 \left(f_{xx} + f_{yy} \right) f \, \frac{dx \, dy}{y^2} \\
&\leq 4|f|_{L^2}^2 + \int_{\Gamma \backslash G/K} (-\Delta) f \cdot f \, \frac{dx \, dy}{y^2} \ll |f|_{\mathfrak{B}^1}^2
\end{aligned}$$

with uniform implied constants. Transform the remaining term by integration by parts:

$$\int_{S^1} \int_{y \geq t} \frac{2}{t} \varphi'\left(\frac{y}{t}\right) \varphi\left(\frac{y}{t}\right) f_y f \, dx \, dy$$

$$= \int_{S^1} \int_{t \leq y \leq 2t} \frac{1}{t} \varphi'\left(\frac{y}{t}\right) \varphi\left(\frac{y}{t}\right) \cdot \frac{\partial}{\partial y}(f^2) \, dx \, dy$$

$$= \int_{S^1} \int_{t \leq y \leq 2t} \frac{\partial}{\partial y}\left(\frac{1}{t} \varphi'\left(\frac{y}{t}\right) \varphi\left(\frac{y}{t}\right)\right) \cdot f^2 \, dx \, dy$$

This is dominated by

$$\int_{S^1} \int_{t \leq y \leq 2t} \left| \frac{\partial}{\partial y}\left(\frac{1}{t} \varphi'\left(\frac{y}{t}\right) \varphi\left(\frac{y}{t}\right)\right) \right| \cdot f^2 \, dx \, dy$$

$$\leq \int_{S^1} \int_{t \leq y \leq 2t} \left| \frac{\partial}{\partial y}\left(\frac{1}{t} \varphi'\left(\frac{y}{t}\right) \varphi\left(\frac{y}{t}\right)\right) \right| \cdot f^2 \cdot (2t)^2 \frac{dx \, dy}{y^2}$$

$$= 4 \int_{S^1} \int_{t \leq y \leq 2t} \left| \varphi''\left(\frac{y}{t}\right) \varphi\left(\frac{y}{t}\right) + \varphi'\left(\frac{y}{t}\right)^2 \right| \cdot f^2 \frac{dx \, dy}{y^2} \ll_\varphi |f|_{L^2}^2$$

with implied constant independent of f and $t \geq 1$.   ///

10.5 Tail Estimates: Three Additional Small Examples

Now we see how to adapt the previous argument to the other three examples from Chapter 1. Most of the work involves skirting the needless (but convenient) exploitation of coincidences used in that simplest example: the y^2 in the coordinate expression for the invariant Laplacian and in the invariant measure in the $SL_2(\mathbb{R})$ seem to need to cancel to make the computation succeed. Our main point in this section is seeing that that coincidence is irrelevant.

For all four examples from Chapter 1, use the coordinates and conventions there. The coordinates are the Iwasawa coordinates x, y with $x \in \mathbb{R}^r$ for $r = 1, 2, 3, 4$ (with the previous section's example being the case $r = 1$), and $0 < y \in \mathbb{R}$. The invariant Laplacian is

$$\Delta = y^2\left(\frac{\partial^2}{\partial x^2} + \frac{\partial^2}{\partial y^2}\right) - (r - 1)y\frac{\partial}{\partial y}$$

and the invariant measure is $dx \, dy / y^{r+1}$. The [Lax-Phillips 1976] argument as in [10.4] requires not only that $-\Delta$ itself is nonnegative, but that the two natural summands of $-\Delta$ in Iwasawa coordinates are both nonnegative. For example, the seemingly extra term $(r - 1)y\frac{\partial}{\partial y}$ is not only harmless but *necessary*, possibly contrary to a visual appraisal. The $-y^2\frac{\partial^2}{\partial x^2}$ summand is nonnegative because the partial derivative in x does not interact with either the leading coefficient y^2 or the denominator y^n in the measure.

[10.5.1] Claim:

$$\int -\left(y^2 \frac{\partial^2}{\partial y^2} - (r-1)y\frac{\partial}{\partial y}\right)f \cdot \overline{f} \, \frac{dx\,dy}{y^{r+1}} \geq 0$$

Proof: Integrating by parts once in the second-order derivative,

$$\int -y^2 \frac{\partial^2}{\partial y^2}f \cdot \overline{f} \, \frac{dx\,dy}{y^{r+1}} = \int -\frac{\partial^2}{\partial y^2}f \cdot y^{1-r}\overline{f} \, dx\,dy$$

$$= \int \frac{\partial}{\partial y}f \cdot \frac{\partial}{\partial y}(y^{1-r}\overline{f}) \, dx\,dy$$

$$= \int \frac{\partial}{\partial y}f \cdot \left((1-r)y^{-r}\overline{f} + y^{1-r}\frac{\partial}{\partial y}\overline{f}\right) dx\,dy$$

The $\frac{\partial}{\partial y}f \cdot (1-r)y^{-r}\overline{f}$ term cancels the corresponding term in the original expression, so

$$\int -\left(y^2 \frac{\partial^2}{\partial y^2} - (r-1)y\frac{\partial}{\partial y}\right)f \cdot \overline{f} \, \frac{dx\,dy}{y^{r+1}} = \int y\frac{\partial}{\partial y}f \cdot y\frac{\partial}{\partial y}\overline{f} \, \frac{dx\,dy}{y^{r+1}}$$

Thus, for example, with invariant Laplacian Δ,

$$\int -\Delta f \cdot \overline{f}\frac{dx\,dy}{y^{r+1}} = \int \left(y\frac{\partial f}{\partial x}\right)^2 + \left(y\frac{\partial f}{\partial y}\right)^2 \frac{dx\,dy}{y^{r+1}}$$

This is the desired positivity. ///

We grant ourselves that the subordinate issue about uniform estimates on families of smooth cutoffs is resolved, as in [10.4.3]. Let ξ run over characters of $(\Gamma \cap N)\backslash N \approx \mathbb{T}^r$, and take $c \geq c_o \gg 1$. In Iwasawa coordinates x, y, write the Fourier expansion in x as

$$f(x,y) = \sum_{\xi} \widehat{f}(\xi)(y)$$

Toward the compactness of $\mathfrak{B}_a^1 \to L_a^2(\Gamma\backslash G/K)$, the critical point is the tail estimate:

[10.5.2] Claim: For smooth f with support in $y \geq c \geq c_o \gg 1$,

$$\int_{(N\cap\Gamma)\backslash N} \int_{y\geq c} |f|^2 \, \frac{dx\,dy}{y^{r+1}} \ll \frac{1}{c^2} \cdot |f|^2_{\mathfrak{B}^1}$$

with implied constants independent of f and of c.

Proof: By Plancherel in x,

$$\int_{(N\cap\Gamma)\backslash N} \int_{y\geq c} |f|^2 \frac{dx\,dy}{y^{r+1}} = \sum_{\xi} \int_{y\geq c} |\widehat{f}(\xi)(y)|^2 \frac{dy}{y^{r+1}}$$

When $\widehat{f}(0)(y) = 0$ for $y \geq c_o$, since $|\xi| \gg 1$ for $\xi \neq 0$,

$$\sum_{\xi} \int_{y\geq c} |\widehat{f}(\xi)(y)|^2 \frac{dy}{y^{r+1}} \ll \sum_{\xi} \int_{y\geq c} |\xi|^2 \cdot |\widehat{f}(\xi)(y)|^2 \frac{dy}{y^{r+1}}$$

With Δ_x the Euclidean Laplacian in x,

$$|\xi|^2 \cdot \widehat{f}(\xi, y) = \frac{1}{4\pi^2}\left(-\Delta_x f\right)\widehat{}(\xi)(y)$$

Applying this, and going back by Plancherel,

$$\sum_{\xi} \int_{y\geq c} |\xi|^2 \cdot |\widehat{f}(\xi)(y)|^2 \frac{dy}{y^{r+1}}$$

$$\ll \sum_{\xi} \int_{y\geq c} \left(-\Delta_x f\right)\widehat{}(\xi)(y) \cdot \overline{\widehat{f}}(\xi)(y) \frac{dy}{y^{r+1}}$$

$$= \int_{(N\cap\Gamma)\backslash N} \int_{y\geq c} -\Delta_x f \cdot \overline{f} \frac{dx\,dy}{y^{r+1}}$$

Since $y \geq c \geq c_o \gg 1$,

$$\int_{(N\cap\Gamma)\backslash N} \int_{y\geq c} -\Delta_x f \cdot \overline{f} \frac{dx\,dy}{y^{r+1}}$$

$$\leq \frac{1}{c^2} \int_{(N\cap\Gamma)\backslash N} \int_{y\geq c} -y^2 \Delta_x f \cdot \overline{f} \frac{dx\,dy}{y^{r+1}}$$

From the positivity result just above,

$$\int -\left(y^2 \frac{\partial^2}{\partial y^2} - (r-1)y\frac{\partial}{\partial y}\right) f \cdot \overline{f} \frac{dx\,dy}{y^{r+1}} \geq 0$$

so

$$\int_{(N\cap\Gamma)\backslash N} \int_{y\geq c} -y^2 \Delta_x f \cdot \overline{f} \frac{dx\,dy}{y^{r+1}}$$

$$\leq \frac{1}{c^2} \int_{(N\cap\Gamma)\backslash N} \int_{y\geq c} \left(-y^2\Delta_x f - y^2\frac{\partial^2 f}{\partial y^2} + (r-1)y\frac{\partial f}{\partial y}\right) \cdot \overline{f} \frac{dx\,dy}{y^{r+1}}$$

Thus, for smooth f with support in $y \geq c \geq c_o$,

$$\int_{(N \cap \Gamma) \backslash N} \int_{y \geq c} |f|^2 \frac{dx\,dy}{y^{r+1}} \ll \frac{1}{c^2} \int_{(N \cap \Gamma) \backslash N} \int_{y \geq c} -\Delta f \cdot \overline{f} \frac{dx\,dy}{y^{r+1}}$$

$$\leq \frac{1}{c^2} \cdot |f|_{\mathfrak{B}^1}^2$$

as claimed.   ///

10.6 Tail Estimate: $SL_3(\mathbb{Z})$, $SL_4(\mathbb{Z})$, $SL_5(\mathbb{Z})$, ...

As in the smaller examples, the global automorphic Sobolev space $\mathfrak{B}^1$ is the completion of $C_c^\infty(\Gamma \backslash G / K)$ with respect to the $\mathfrak{B}^1$-norm

$$|f|_{\mathfrak{B}^1}^2 = \int_{\Gamma \backslash G / K} (1 - \Delta) f \cdot \overline{f}$$

Let $\mathfrak{S}$ be a sufficiently large standard Siegel set so that it *surjects* to the quotient $\Gamma \backslash G$. For $a > 0$, the set

$$X_a = \{g \in \mathfrak{S} : \alpha(m_g) \leq a \text{ for all } \alpha \in \Phi\}$$

has compact image $\Gamma \backslash (\Gamma \cdot X_a)$. For α in the set Φ of simple roots and $c \geq a$, let

$$Y_c^\alpha = \{g \in \mathfrak{S} : \alpha(m_g) \geq a\}$$

and $Y_a = \bigcup_\alpha Y_a^\alpha$. Certainly

$$\mathfrak{S} = X_a \cup Y_a$$

Let

$$L_a^2(\Gamma \backslash G / K) = \{f \in L^2(\Gamma \backslash G / K) : c_P f(g) = 0, \text{ for all } g \in Y_a, \text{ for all } P\}$$

and

$$D_a = D \cap L_a^2(\Gamma \backslash G / K)$$

It suffices to require the constant-term vanishing just for standard maximal proper parabolics, because $c_{Q \cap P} f = c_Q c_P f$ for two standard parabolics, and every standard parabolic is an intersection of maximal ones. Let $\mathfrak{B}_a^1$ be the $\mathfrak{B}^1$ completion of $C_c^\infty(Z\Gamma \backslash G / K) \cap L_a^2$ in the $\mathfrak{B}^1$ norm.

To eventually show that the injection $\mathfrak{B}_a^1 \to L_a^2$ is *compact*, as in the simpler examples, we will show that the image of the unit ball of $\mathfrak{B}_a^1$ is *totally bounded* in L_a^2. The crucial point is an estimate on the *tails* of functions in the unit ball B in $\mathfrak{B}_a^1$, as follows.

We grant ourselves a suitable analogue of [10.4.3], that we can control the $\mathfrak{B}^1$-norm of smoothly cutoff versions of $f \in B$ when *any* single simple root becomes large:

[10.6.1] Lemma: Fix a positive simple root α. Given $c \geq a + 1$, there are real-valued smooth functions φ_o and φ_1, taking values in $[0, 1]$, summing to 1, such that φ_1 is supported in Y_c^α, and so that there is a bound C *uniform* in $c \geq a + 1$, such that $|f \cdot \varphi_1|_{\mathfrak{B}^1} \leq C \cdot |f|_{\mathfrak{B}^1}$.   ///

The key point is a bound going to 0 when *any* simple root α becomes large:

[10.6.2] Claim: For $\alpha \in \Phi$,

$$\lim_{c \to +\infty} \left(\sup_{f \in \mathfrak{B}_a^1 \text{ and } \mathrm{spt} f \subset Y_c^\alpha} \frac{|f|_{L^2}}{|f|_{\mathfrak{B}^1}} \right) = 0$$

Proof: Fix $\alpha = \alpha_i \in \Phi$, and $f \in \mathfrak{B}_c^1$ with support inside Y_c^α for $c \gg a$. Let $N = N^\alpha$, $P = P^\alpha$, and let $M = M^\alpha$ be the standard Levi component of P. Use exponential coordinates coordinates

$$n_x = \begin{pmatrix} 1_i & x \\ 0 & 1_{r-i} \end{pmatrix}$$

In effect, the coordinate x is in the Lie algebra $\mathfrak{n}$ of N. Let $\Lambda \subset \mathfrak{n}$ be the lattice which exponentiates to $N \cap P$. Give $\mathfrak{n}$ the natural inner product $\langle, \rangle$ invariant under the (Adjoint) action of $M \cap K$ that makes root spaces mutually orthogonal. Fix a nontrivial character ψ on $\mathbb{R}/\mathbb{Z}$. We have the Fourier expansion

$$f(n_x m) = \sum_{\xi \in \Lambda'} \psi \langle x, \xi \rangle \, \widehat{f_\xi}(m) \qquad \text{(with } n \in N \text{ and } m \in M)$$

where Λ' is the dual lattice to Λ in $\mathfrak{n}$ with respect to $\langle, \rangle$, and

$$\widehat{f_\xi}(m) = \int_{\mathfrak{n}/\Lambda} \overline{\psi} \langle x, \xi \rangle \, f(n_x m) \, dx$$

Let $\Delta^\mathfrak{n}$ be the flat Laplacian on $\mathfrak{n}$ associated to the inner product $\langle, \rangle$, normalized so that

$$\Delta^\mathfrak{n} \psi \langle x, \xi \rangle = -\langle \xi, \xi \rangle \cdot \psi \langle x, \xi \rangle$$

Let $U = M \cap N^{\min}$, and $M_\mathfrak{S} = M \cap \mathfrak{S}$. Abbreviating $A_u = \mathrm{Ad} u$,

$$|f|_{L^2}^2 \leq \int_\mathfrak{S} |f|^2 = \int_{M_\mathfrak{S}} \int_{(U \cap \Gamma) \backslash U} \int_{A_u^{-1} \Lambda \backslash \mathfrak{n}} |f(u n_x m)|^2 \, dx \, du \, \frac{dm}{\delta(m)}$$

with Haar measures dx, du, dm, where δ is the modular function of P. Using the Fourier expansion,

$$f(un_x m) = f(un_x u^{-1} \cdot um) = \sum_{\xi \in \Lambda'} \psi \langle A_u x, \xi \rangle \cdot \widehat{f_\xi}(um)$$

$$= \sum_{\xi \in \Lambda'} \psi \langle x, A_u^* \xi \rangle \cdot \widehat{f_\xi}(um)$$

Then

$$-\Delta^n f(un_x m) = \sum_{\xi \in \Lambda'} \langle A_u^* \xi, A_u^* \xi \rangle \cdot \psi \langle x, A_u^* \xi \rangle \cdot \widehat{f_\xi}(um)$$

The compact quotient $(U \cap \Gamma)\backslash U$ has a compact set R of representatives in U, so there is a *uniform* lower bound for $0 \neq \xi \in \Lambda'$:

$$0 < b \leq \inf_{u \in R} \inf_{0 \neq \xi \in \Lambda'} \langle A_u^* \xi, A_u^* \xi \rangle$$

By Plancherel applied to the Fourier expansion in x, using the hypothesis that $\widehat{f_0} = 0$ in X_a^α,

$$\int_{A_u^{-1}\Lambda\backslash\mathfrak{n}} |f(un_x m)|^2 \, dx$$

$$= \int_{A_u^{-1}\Lambda\backslash\mathfrak{n}} |f(un_x u^{-1} \cdot um)|^2 \, dx$$

$$= \sum_{\xi \in \Lambda'} |\widehat{f_\xi}(um)|^2 \leq b^{-1} \sum_{\xi \in \Lambda'} \langle A_u^* \xi, A_u^* \xi \rangle \cdot |\widehat{f_\xi}(um)|^2$$

$$= \sum_{\xi \in \Lambda'} -\widehat{\Delta^n f_\xi}(um) \cdot \overline{\widehat{f}}(um)$$

$$= \int_{u^{-1}\Lambda u\backslash\mathfrak{n}} -\Delta^n f(un_x u^{-1} \cdot um) \cdot \overline{f}(un_x u^{-1} \cdot um) \, dx$$

$$= \int_{A_u^{-1}\Lambda\backslash\mathfrak{n}} \Delta^n f(un_x m) \cdot \overline{f}(un_x m) \, dx$$

Thus, for f with $\widehat{f}(0) = 0$ on Y_a^α,

$$|f|_{L^2}^2 \ll \int_{M_\mathfrak{S}} \int_{(U \cap \Gamma)\backslash U} \int_{A_u^{-1}\Lambda\backslash\mathfrak{n}} -\Delta^n f(un_x m) \cdot \overline{f}(un_x m) \, dx \, du \, \frac{dm}{\delta(m)}$$

Next, we compare Δ^n to the invariant Laplacian Δ. Let $\mathfrak{g}$ be the Lie algebra of $G_\mathbb{R}$, with nondegenerate invariant pairing $\langle u, v \rangle = \text{tr}(uv)$. The Cartan involution $v \to v^\theta = -v^\top$ has $+1$ eigenspace the Lie algebra $\mathfrak{k}$ of K, and -1 eigenspace $\mathfrak{s}$, the space of symmetric matrices.

Let Φ^N be the set of positive roots β whose root-space $\mathfrak{g}_\beta$ appears in $\mathfrak{n}$. For each $\beta \in \Phi^N$, take $x_\beta \in \mathfrak{g}_\beta$ such that $x_\beta + x_\beta^\theta \in \mathfrak{k}$, $x_\beta - x_\beta^\theta \in \mathfrak{s}$, and $\langle x_\beta, x_\beta^\theta \rangle = -1$: for $\beta(a) = a_i/a_j$ with $i < j$, x_β has a single nonzero entry, at the ij^{th} place. Let

$$\Omega' = -\sum_{\beta \in \Phi^N} (x_\beta x_\beta^\theta + x_\beta^\theta x_\beta)$$

in the universal enveloping algebra $U\mathfrak{g}$. Let $\Omega'' \in U\mathfrak{g}$ be the Casimir element for the Lie algebra $\mathfrak{m}$ of $M_\mathbb{R}$, normalized so that Casimir Ω for $\mathfrak{g}$ is the sum $\Omega = \Omega' + \Omega''$. We rewrite Ω' to fit the Iwasawa coordinates: for each β,

$$x_\beta x_\beta^\theta + x_\beta^\theta x_\beta = 2x_\beta x_\beta^\theta + [x_\beta^\theta, x_\beta]$$
$$= -2x_\beta^2 + 2x_\beta(x_\beta + x_\beta^\theta) + [x_\beta^\theta, x_\beta] \in -2x_\beta^2 + [x_\beta^\theta, x_\beta] + \mathfrak{k}$$

Thus,

$$\Omega' = \sum_{\beta \in \Phi^N} 2x_\beta^2 - [x_\beta^\theta, x_\beta] \qquad \text{(modulo } \mathfrak{k})$$

The commutators $[x_\beta^\theta, x_\beta]$ are in $\mathfrak{m}$. In the coordinates $un_x a$ with $U\mathfrak{g}$ acting on the right, $x_\beta \in \mathfrak{n}$ is acted on by a before translating x, by

$$un_x a \cdot e^{tx_\beta} = un_x \cdot e^{t\beta(a) \cdot x_\beta} \cdot a = un_{x+\beta(a)x_\beta} a$$

That is, x_β acts by $\beta(a) \cdot \frac{\partial}{\partial x_\beta}$.

For two symmetric operators S, T on a not necessarily complete inner product space V, write $S \leq T$ when

$$\langle Sv, v \rangle \leq \langle Tv, v \rangle \qquad \text{(for all } v \in V)$$

Say a symmetric operator T is *nonnegative* when $0 \leq T$. Since $m \in M_\mathfrak{S}$, there is an absolute constant so that $\alpha(m) \geq c$ implies $\beta(m) \gg c$. Thus,

$$-\Delta^\mathfrak{n} = -\sum_{\beta \in \Phi^N} \frac{\partial^2}{\partial x_\beta^2} \leq \frac{1}{c^2} \cdot \left(-\sum_{\beta \in \Phi^N} x_\beta^2 \right)$$

as operators on $C_c^\infty(Y_a^\alpha)^K$, where $C_c^\infty(Y_a^\alpha)$ has the L^2 inner product. We claim that

$$\sum_{\beta \in \Phi^N} [x_\beta^\theta, x_\beta] - \Omega'' \geq 0 \qquad \text{(operators on } C_c^\infty(Y_a^\alpha))$$

From this, it would follow that

$$-\Delta^{\mathfrak{n}} \ll \frac{1}{c^2} \cdot \Big(-\sum_{\beta \in \Phi^N} x_\beta^2 \Big) \leq \frac{1}{c^2} \cdot \Big(-\sum_{\beta \in \Phi^N} x_\beta^2 + \sum_{\beta \in \Phi^N} [x_\beta^\theta, x_\beta] - \Omega'' \Big)$$

$$= \frac{1}{c^2} \cdot (-\Delta)$$

Then for $f \in \mathfrak{B}_a^1$ with support in X_a^α we would have

$$|f|_{L^2}^2 \ll \int_{\mathfrak{S}} -\Delta^{\mathfrak{n}} f \cdot \overline{f} \ll \frac{1}{c^2} \int_{\mathfrak{S}} -\Delta f \cdot \overline{f} \ll \frac{1}{c^2} \int_{\Gamma \backslash G} -\Delta f \cdot \overline{f}$$

$$\ll \frac{1}{c^2} \cdot |f|_{\mathfrak{B}^1}^2$$

Taking c large makes this small.

To prove the claimed nonnegativity of $T = \sum_{\beta \in \Phi^N} [x_\beta^\theta, x_\beta] - \Omega''$, exploit the Fourier expansion along N and the fact that $x \in \mathfrak{n}$ does not appear in T: noting that the order of coordinates $n_x u$ differs from the foregoing,

$$\int_{M_{\mathfrak{S}}} \int_{(U \cap \Gamma) \backslash U} \int_{\Lambda \backslash \mathfrak{n}} T f(n_x um) \, \overline{f}(n_x um) \, dx \, du \, \frac{dm}{\delta(m)}$$

$$= \int_{M_{\mathfrak{S}}} \int_{(U \cap \Gamma) \backslash U} \int_{\Lambda \backslash \mathfrak{n}} T \Big(\sum_{\xi} \psi \langle x, \xi \rangle \, \widehat{f}_\xi(um) \Big) \sum_{\xi'} \overline{\psi} \langle x, \xi' \rangle \, \overline{\widehat{f}_\xi}(um) \, dx \, du \, \frac{dm}{\delta(m)}$$

Only the diagonal summands survive the integration in $x \in \mathfrak{n}$, and the exponentials cancel, so this is

$$\int_{M_{\mathfrak{S}}} \int_{(U \cap \Gamma) \backslash U} \sum_{\xi} T \widehat{f}_\xi(um) \cdot \overline{\widehat{f}}_\xi(um) \, du \, \frac{dm}{\delta(m)}$$

Let F_ξ be a left-N-invariant function taking the same values as $\widehat{f}_\xi$ on UA^+K, defined by, for $n_x \in N$, $u \in U$, $m \in M^+$, $k \in K$,

$$F_\xi(n_x umk) = \widehat{f}_\xi(umk)$$

Since T does not involve $\mathfrak{n}$, and since F_ξ is left-N-invariant,

$$T \widehat{f}_\xi(um) = T F_\xi(n_x um) = -\Delta F_\xi(n_x um)$$

and then

$$\int_{M_{\mathfrak{S}}} \int_{(U \cap \Gamma) \backslash U} \sum_{\xi} T \widehat{f}_\xi(um) \cdot \overline{\widehat{f}}_\xi(um) \, du \, \frac{dm}{\delta(m)}$$

$$= \int_{M_{\mathfrak{S}}} \int_{(U \cap \Gamma) \backslash U} \sum_{\xi} -\Delta F_\xi(um) \cdot \overline{F}_\xi(um) \, du \, \frac{dm}{\delta(m)}$$

The individual summands are not left-$U \cap \Gamma$-invariant. Since $\widehat{f_\xi}(\gamma g) = \widehat{f_{A^*_\gamma \xi}}(g)$ for γ normalizing $\mathfrak{n}$, we can group $\xi \in \Lambda'$ by $(U \cap \Gamma)$ orbits to obtain $(U \cap \Gamma)$ subsums, and then *unwind*. Pick a representative ω for each orbit $[\omega]$, and let U_ω be the isotropy subgroup of ω in $(U \cap \Gamma)$, so

$$\int_{(U \cap \Gamma) \backslash U} \sum_\xi -\Delta F_\xi(um) \cdot \overline{F}_\xi(um)\, du$$

$$= \sum_{[\omega]} \int_{(U \cap \Gamma) \backslash U} \sum_{\xi \in [\omega]} -\Delta F_\xi(um) \cdot \overline{F}_\xi(um)\, du$$

$$= \sum_\omega \int_{(U \cap \Gamma) \backslash U} \sum_{\gamma \in U_\omega \backslash (U \cap \Gamma)} -\Delta F_{A^*_\gamma \omega}(um) \cdot \overline{F}_{A^*_\gamma \omega}(um)\, du$$

$$= \sum_\omega \int_{U_\omega \backslash U} -\Delta F_\omega(um) \cdot \overline{F}_\omega(um)\, du$$

Then

$$\int_{M_\Theta} \int_{(U \cap \Gamma) \backslash U} \sum_\xi -\Delta F_\xi(um) \cdot \overline{F}_\xi(um)\, du$$

$$= \sum_\omega \int_{M_\Theta} \int_{U_\omega \backslash U} -\Delta F_\omega(um) \cdot \overline{F}_\omega(um)\, du\, \frac{da}{\delta(a)}$$

Since $-\Delta$ is a nonnegative operator on functions on every quotient $NU_\omega \backslash G/K$ of G/K, each double integral is nonnegative, proving T is non-negative. This completes the estimate of the tails.   ///

10.7   Compact $\mathfrak{B}_a^1 \longrightarrow L_a^2$ in Four Simple Examples

As remarked earlier, the discrete decomposition of $L_a^2(\Gamma \backslash G/K)$, for the Friedrichs extension $\widetilde{\Delta}_a$ of a restriction Δ_a of the differential operator Δ to the dense subspace D_a of $L_a^2(\Gamma \backslash G/K)$, will follow from *compactness* of the resolvent $(1 - \widetilde{\Delta}_a)^{-1}$, which will follow from the compactness of the inclusion $\mathfrak{B}_a^1 \to L_a^2(\Gamma \backslash G/K)$, demonstrated here.

[10.7.1] Theorem: With $a \gg 1$, the inclusion $\mathfrak{B}_a^1(\Gamma \backslash G/K) \to L_a^2(\Gamma \backslash G/K)$ is *compact*.

Proof: Again, we are roughly following [Lax-Phillips 1976], pp. 204–206. The *total boundedness* criterion for precompactness [14.7.1] requires that, given $\varepsilon > 0$, the image of the unit ball B in $\mathfrak{B}_a^1$ in $L_a^2(\Gamma \backslash G/K)$ can be covered by finitely many balls of radius ϵ.

The *idea* is that the usual Rellich compactness lemma, asserting compactness of proper inclusions of *Sobolev spaces* on (multi-)tori as in [9.5.12] and [9.5.15], reduces the issue to estimates [10.4] and [10.5] on the *tails*. In more detail: let $\mathfrak{S}$ be a fixed Siegel set that surjects to the quotient $\Gamma \backslash G$. Given $c \geq a$, let Y_o be the image of $\{g \in \mathfrak{S} : \eta(g) \leq c + 1\}$ in $\Gamma \backslash G/K$, and cover it by opens $U_1, \ldots, U_n$ in $\Gamma \backslash G/K$ with small compact closures, and take one open U_∞ covering the image Y_∞ of $\eta \geq c$. Compactness of Y_o produces a finite subcover. Choose a smooth partition of unity $\{\varphi_i\}$ subordinate to the finite subcover and U_∞, letting φ_∞ be a smooth function that is identically 1 for $y \geq c$. That is, $\varphi_\infty + \sum_i \varphi_i = 1$, and φ_i has compact support inside the open U_i. Note that [10.4.3] showed we can choose a *family* of smooth cutoff functions φ_∞ so that $\varphi_\infty \cdot f$ has a *uniform* $\mathfrak{B}^1$ bound in terms of both f and the family.

Maps among function spaces on the compact part Y_o behave well for more general reasons, as we see now. Let $\mathfrak{B}^1_a(Y_o)$ be the closure of $C^\infty_c(Y_o) \cap L^2_a(\Gamma \backslash G/K)$ in $\mathfrak{B}^1_a$, and $L^2_a(Y_o)$ the closure of $C^\infty_c(Y_o) \cap L^2_a(\Gamma \backslash G/K)$ in $L^2_a(\Gamma \backslash G/K)$.

[10.7.2] Theorem: For $\Gamma = SL_2(\mathbb{Z})$ and $G = SL_2(\mathbb{R})$, $\mathfrak{B}^1_a(Y_o) \to L^2_a(Y_o)$ is compact.

Proof: To take advantage of some fortunate, simplifying (but not strictly necessary) coincidences, we first carry out this part of the argument just for $\Gamma = SL_2(\mathbb{Z})$ and $G = SL_2(\mathbb{R})$.

For finite j, without loss of generality take the opens U_j to be small rectangles in the upper half-plane, and the coordinate maps ψ_j simply the inclusions. Fix j, and let $U_j = \{z = x + iy : x_1 < x < x_2, y_1 < y < y_2\}$. On U_j, the measure $dx\,dy/y^2$ and the coefficients of the differential operator $\Delta = y^2(\frac{\partial^2}{\partial x^2} + \frac{\partial^2}{\partial y^2})$ differ by bounded amounts from the Euclidean $dx\,dy$ and $\Delta^E = \frac{\partial^2}{\partial x^2} + \frac{\partial^2}{\partial y^2}$. Thus, the corresponding $\mathfrak{B}^1$ and L^2 norms are comparable, as follows. Let

$$L^2_a(U_j) = \text{closure of } C^\infty_c(U_j) \cap L^2_a(\Gamma \backslash G/K) \text{ in } L^2_a(\Gamma \backslash G/K)$$

and

$$\mathfrak{B}^1_a(U_j) = \text{closure of } C^\infty_c(U_j) \cap L^2_a(\Gamma \backslash G/K) \text{ in } \mathfrak{B}^1_a(\Gamma \backslash G/K)$$

Letting

$$|f|^2_{L^2(\psi_j U_j)} = \int_{\psi_j U_j} |f|^2 \, dx\,dy$$

$$|f|^2_{\mathfrak{B}^1(\psi_j U_j)} = \int_{\psi_j U_j} \left(|f|^2 - \Delta^E f \cdot \overline{f} \right) dx\,dy$$

there are easy comparisons: for $f \in C_c^\infty(U_j) \cap L_a^2(\Gamma \backslash G/K)$,

$$\frac{1}{y_2} \cdot |f|_{L^2(\psi_j U_j)} \le |f|_{L^2(U_j)} \le \frac{1}{y_1} \cdot |f|_{L^2(\psi_j U_j)}$$

and for $f \in C_c^\infty(U_j) \cap L_a^2(\Gamma \backslash G/K)$,

$$\frac{1}{y_2} \cdot |f|_{\mathfrak{B}^1(\psi_j U_j)} \le |f|_{\mathfrak{B}^1(U_j)} \le \frac{1}{y_1} \cdot |f|_{L^2(\psi_j U_j)}$$

Identification of opposite edges of the rectangles $\psi_i U_i$ produces a two-torus T_j, with $L^2(T_j)$ and $\mathfrak{B}^1(T_j)$ defined from the Euclidean measure and Euclidean Laplacian. The usual Rellich Lemma asserts the compactness of the inclusion $\mathfrak{B}^1(T_j) \to L^2(T_j)$. We will repeatedly use

[10.7.3] Lemma: Let A, B, C, D be Hilbert spaces, with a commutative diagram of continuous linear maps

$$
\begin{array}{ccc}
A & \longrightarrow & B \\
\downarrow & & \downarrow \\
C & \xrightarrow{\ \ s\ \ } & D
\end{array}
$$

with $B \to D$ *compact*, and $S : C \to D$ with constant $m > 0$ such that $|v|_C \le m \cdot |Sv|_D$ for all $v \in C$. Then $A \to C$ is also compact.

Proof: (of lemma) Let X be the closed unit ball in A, with image Y in C. By continuity, the image of X in B is inside a finite-radius ball Z. By compactness of $B \to D$, given $\varepsilon > 0$, the image of Z in D is covered by finitely many $\frac{\varepsilon}{m}$-balls $V_1, \dots, V_n$. The condition on S ensures that the inverse images $S^{-1}(SY \cap V_j)$ are contained in ε-balls in C. Thus, Y is covered by finitely many ε-balls in C. This holds for every $\varepsilon > 0$, so the image Y is precompact, and $A \to C$ is compact.   ///

The lemma applies to the situation

$$
\begin{array}{ccc}
\mathfrak{B}_a^1(\psi_j U_j) & \longrightarrow & \mathfrak{B}^1(T_j) \\
\downarrow & & \downarrow \\
L_a^2(\psi_j U_j) & \longrightarrow & L^2(T_j)
\end{array}
$$

The standard Rellich lemma is that $\mathfrak{B}^1(T_j) \to L^2(T_j)$ is compact, and the inclusion $L_a^2(U_j) \to L^2(T_j)$ satisfies the hypothesis of the lemma with $m = 1$, so $\mathfrak{B}_a^1(U_j) \to L_a^2(U_j)$ is *compact*.

Map $\mathfrak{B}_a^1(Y_o)$ to $\bigoplus_{j=1}^n \mathfrak{B}_a^1(U_j)$ by $f \to \bigoplus_j \varphi_j \cdot f$, and similarly for L^2.
Applying the lemma to

$$
\begin{array}{ccccc}
\mathfrak{B}_a^1(Y_o) & \longrightarrow & \bigoplus_{j=1}^n \mathfrak{B}_a^1(U_j) & \longrightarrow & \bigoplus_{j=1}^n \mathfrak{B}^1(\psi_j U_j) \\
\downarrow & & & & \downarrow \\
L_a^2(Y_o) & \longrightarrow & \bigoplus_{j=1}^n L_a^2(U_j) & \longrightarrow & \bigoplus_{j=1}^n L^2(\psi_j U_j)
\end{array}
$$

yields the compactness of $\mathfrak{B}_a^1(Y_o) \to L_a^2(Y_o)$.   ///

Returning to the proof of the theorem: let B be the unit ball in $\mathfrak{B}_a^1(\Gamma \backslash G/K)$.
Given $\varepsilon > 0$, take $c \geq a$ sufficiently large and smooth cutoff function φ_∞
such that $\varphi_\infty \cdot B$ lies in a single $\varepsilon/2$-ball in $L_a^2(\Gamma \backslash G/K)$. By compactness of
$\mathfrak{B}_a^1(Y_o) \to L_a^2(Y_o)$, the image of B in

$$
B \longrightarrow (1 - \varphi_\infty) \cdot B \subset \mathfrak{B}_a^1(Y_o) \longrightarrow L_a^2(Y_o)
$$

is precompact, so can be covered by finitely many $\varepsilon/2$-balls in $L_a^2(\Gamma \backslash G/K)$.
Thus, $B = \varphi_\infty \cdot B + (1 - \varphi_\infty) \cdot B$ can be covered by finitely many ε-balls in
$L_a^2(\Gamma \backslash G/K)$, so is precompact there. This proves the compactness of the inclusion $\mathfrak{B}_a^1(\Gamma \backslash G/K) \to L_a^2(\Gamma \backslash G/K)$.   ///

Now we give an argument for compactness applicable more generally. To
cope more sanely with comparisons of norms, we want a sort of *gradient* oper-
ator ∇ on functions on $\Gamma \backslash G/K$ such that there is an integration by parts property

$$
\int_{\Gamma \backslash G/K} -\Delta f \cdot \overline{f} = \int_{\Gamma \backslash G/K} \langle \nabla f, \nabla f \rangle_{\mathfrak{s}}
$$

with an inner product on the vector space $\mathfrak{s}$ in which ∇ takes values. This can
be accomplished quite generally as follows.

Let $\gamma \to \gamma^\theta$ be an involutive automorphism[1] on the Lie algebra of G as $\mathfrak{g} = \mathfrak{s} + \mathfrak{k}$ such that the Lie algebra of K is the $+1$ eigenspace, and let $\mathfrak{s}$ be the -1
eigenspace. For example, for $G = SL_2(\mathbb{R})$ we can take $\gamma^\theta = -\gamma^\top$. Let $\langle , \rangle_{\mathfrak{s}}$
be a positive-definite real-valued inner product on $\mathfrak{s}$, invariant under the action
of K:

$$
\langle k\alpha k^{-1}, k\beta k^{-1} \rangle_{\mathfrak{s}} = \langle \alpha, \beta \rangle_{\mathfrak{s}} \qquad (\text{for all } \alpha, \beta \in \mathfrak{s} \text{ and } k \in K)
$$

In all these examples, $\langle , \rangle_s$ can be obtained by restricting the *trace form*[2]

$$
\langle \alpha, \beta \rangle_{\text{trace}} = \operatorname{Re}\operatorname{tr}(\alpha \cdot \beta) \qquad (\text{for } \alpha, \beta \in \mathfrak{g})
$$

---

[1] This is a *Cartan involution*.
[2] This trace form is a concrete instantiation of the *Cartan-Killing* form.

where tr is matrix trace. Let $x \in \mathfrak{g}$ act on functions on G or $\Gamma \backslash G$ by differentiating right translation X_x, as in Chapters 4, 6, and subsequently. We need a name for this map, so let $\rho(x) = X_x$. It is K-equivariant. To describe ∇ independently of coordinates, consider a sequence of K-equivariant maps, reminiscent of the analogue in the coordinate-independent description [4.2] of the Casimir operator:

$$\text{End}_{\mathbb{R}}(\mathfrak{s}) \longrightarrow \mathfrak{s} \otimes_{\mathbb{R}} \mathfrak{s}^* \xrightarrow{\langle,\rangle_{\mathfrak{s}}} \mathfrak{s} \otimes_{\mathbb{R}} \mathfrak{s} \xrightarrow{\rho \otimes 1_{\mathfrak{s}}} \rho(\mathfrak{s}) \otimes_{\mathbb{R}} \mathfrak{s}$$

$$1_{\mathfrak{s}} - - - - - - - - - - - - - - - - - - \to \nabla$$

where $\mathfrak{s} \otimes \mathfrak{s}^* \to \text{End}_{\mathbb{R}}(\mathfrak{s})$ is the natural map $(x \otimes \lambda)(y) = \lambda(y) \cdot x$, and where $\mathfrak{s}^*$ is identified with $\mathfrak{s}$ via $x \to \langle -, x \rangle_{\mathfrak{s}}$. Thus,

[10.7.4] Lemma: The image ∇ of the identity automorphism $1_{\mathfrak{s}}$ of $\mathfrak{s}$ is K-equivariant. ///

Thus, for any orthonormal basis $\{x_j\}$ of $\mathfrak{s}$, in coordinates

$$\nabla = \sum_j X_{x_j} \cdot x_j \qquad \nabla f = \sum_j X_{x_j} f \cdot x_j \quad (\text{for } f \in C^\infty(G))$$

Since ∇ is right K-equivariant, it descends to an operator on functions on G/K and $\Gamma \backslash G/K$.

[10.7.5] Lemma: For $f \in C_c^\infty(\Gamma \backslash G/K) = C_c^\infty(\Gamma \backslash G)^K$,

$$\int_{\Gamma \backslash G} -\Delta f \cdot \overline{f} = \int_{\Gamma \backslash G} \langle \nabla f, \nabla f \rangle_{\mathfrak{s}}$$

Proof: Now write simply x for the operator $X_x = \rho(x)$. Let $\{\theta_j\}$ be a basis for $\mathfrak{k}$ such that $\langle \theta_i, \theta_j \rangle_{\text{trace}} = -\delta_{ij}$ with Kronecker δ and the trace pairing. As in [4.2], the Casimir operator is (the image of) $\sum_j x_j^2 - \sum_i \theta_i^2$ in the universal enveloping algebra. Thus, on right K-invariant functions it is $\sum_j x_j^2$. Thus, integrating by parts,

$$\int_{\Gamma \backslash G} -\Delta f \cdot \overline{f} = \int_{\Gamma \backslash G} -\sum_j x_j^2 f \cdot \overline{f} = \int_{\Gamma \backslash G} \sum_j x_j f \cdot x_j \overline{f}$$

$$= \int_{\Gamma \backslash G} \sum_j \left\langle x_j f \cdot x_j, \, x_j \overline{f} \cdot x_j \right\rangle_{\mathfrak{s}} = \int_{\Gamma \backslash G} \langle \nabla f, \nabla f \rangle_{\mathfrak{s}}$$

as desired. ///

Using ∇, now we can give a more general proof of compactness, as follows. As earlier, let $\mathfrak{S}$ be a fixed Siegel set that surjects to the quotient $\Gamma \backslash G / K$. Given $c \geq a$, let Y_o be the image of $\{g \in \mathfrak{S} : \eta(g) \leq c + 1\}$ in $\Gamma \backslash G / K$, and cover it by opens $U_1, \ldots, U_n$ in $\Gamma \backslash G / K$ with small compact closures, and take one open U_∞ covering the image Y_∞ of $\eta \geq c$. Compactness of Y_o produces a finite subcover. Choose a smooth partition of unity $\{\varphi_i\}$ subordinate to the finite subcover and U_∞, letting φ_∞ be a smooth function identically 1 for $y \geq c$. That is, $\varphi_\infty + \sum_i \varphi_i = 1$, and φ_i has compact support inside the open U_i. A general version of [10.4.3] shows that we can choose a *family* of smooth cutoff functions φ_∞ so that $\varphi_\infty \cdot f$ has a *uniform* $\mathfrak{B}^1$ bound in terms of both f and the family. Now we have the somewhat more general version of [10.7.2]: as earlier, let $\mathfrak{B}_a^1(Y_o)$ be the closure of $C_c^\infty(Y_o) \cap L_a^2(\Gamma \backslash G / K)$ in $\mathfrak{B}_a^1$, and $L_a^2(Y_o)$ the closure of $C_c^\infty(Y_o) \cap L_a^2(\Gamma \backslash G / K)$ in $L_a^2(\Gamma \backslash G / K)$.

[10.7.6] Theorem: $\mathfrak{B}_a^1(Y_o) \to L_a^2(Y_o)$ is compact.

Proof: As earlier, let $r = 1, 2, 3, 4$ in the respective cases. For finite j, without loss of generality take the opens U_j to be small rectangles in Iwasaw a coordinates x, y on G / K with $x \in \mathbb{R}^r$ and $y > 0$, and the coordinate maps ψ_j the inclusions. As in the simplest case, on U_j, the measure $dx \, dy / y^{r+1}$ differs by a bounded amount from from the Euclidean invariant measure $dx \, dy$.

Identifying $\mathfrak{s}$ with the tangent space at every point of G / K, on a subset U with compact closure the inner product $\langle , \rangle_{\mathfrak{s}}$ differs from the Euclidean inner product $\langle , \rangle_E$ by bounded amounts, simply because continuous functions on compacts are uniformly bounded. Similarly, the coefficients of ∇ differ from those of the Euclidean gradient ∇^E by bounded amounts, for the same reason. Thus, the $\mathfrak{B}^1$ and L^2 norms are comparable to the Euclidean ones on each of the finitely many U_j. Specifically, let

$$L_a^2(U_j) = \text{closure of } C_c^\infty(U_j) \cap L_a^2(\Gamma \backslash G / K) \text{ in } L_a^2(\Gamma \backslash G / K)$$

and

$$\mathfrak{B}_a^1(U_j) = \text{closure of } C_c^\infty(U_j) \cap L_a^2(\Gamma \backslash G / K) \text{ in } \mathfrak{B}_a^1(\Gamma \backslash G / K)$$

As earlier in the simplest case, denote the Euclidean versions by

$$|f|_{L^2(\psi_j U_j)}^2 = \int_{\psi_j U_j} |f|^2 \, dx \, dy$$

$$|f|_{\mathfrak{B}^1(\psi_j U_j)}^2 = \int_{\psi_j U_j} \left(|f|^2 + \langle \nabla^E f, \nabla^E \overline{f} \rangle_E \right) dx \, dy$$

Then the comparisons, less explicit than in the proof of [10.7.2], are, for $f \in C_c^\infty(U_j) \cap L_a^2(\Gamma \backslash G / K)$,

$$|f|_{L^2(\psi_j U_j)} \ll |f|_{L^2(U_j)} \ll |f|_{L^2(\psi_j U_j)}$$

and, for $f \in C_c^\infty(U_j) \cap L_a^2(\Gamma \backslash G / K)$,

$$|f|_{\mathfrak{B}^1(\psi_j U_j)} \ll |f|_{\mathfrak{B}^1(U_j)} \ll |f|_{L^2(\psi_j U_j)}$$

with implied constants uniform in f. The rest of the argument proceeds as in [10.7.2]: first, identification of opposite edges of the rectangles $\psi_i U_i$ produces an $(r+1)$-torus T_j, with $L^2(T_j)$ and $\mathfrak{B}^1(T_j)$ defined from the Euclidean measure and Euclidean Laplacian. The usual Rellich Lemma asserts the compactness of the inclusion $\mathfrak{B}^1(T_j) \to L^2(T_j)$. Use the lemma [10.7.3] in the situation

$$
\begin{array}{ccc}
\mathfrak{B}_a^1(\psi_j U_j) & \longrightarrow & \mathfrak{B}^1(T_j) \\
\downarrow & & \downarrow \\
L_a^2(\psi_j U_j) & \longrightarrow & L^2(T_j)
\end{array}
$$

As in the proof of [10.7.2], the standard Rellich lemma asserts that $\mathfrak{B}^1(T_j) \to L^2(T_j)$ is compact, and the inclusion $L_a^2(U_j) \to L^2(T_j)$ satisfies the hypothesis of the lemma, so $\mathfrak{B}_a^1(U_j) \to L_a^2(U_j)$ is *compact*.

As in the proof of [10.7.2], map $\mathfrak{B}_a^1(Y_o)$ to $\bigoplus_{j=1}^n \mathfrak{B}_a^1(U_j)$ by $f \to \bigoplus_j \varphi_j \cdot f$, and similarly for L^2. Applying the lemma [10.7.3] to

$$
\begin{array}{ccccc}
\mathfrak{B}_a^1(Y_o) & \longrightarrow & \bigoplus_{j=1}^n \mathfrak{B}_a^1(U_j) & \longrightarrow & \bigoplus_{j=1}^n \mathfrak{B}^1(\psi_j U_j) \\
\downarrow & & & & \downarrow \\
L_a^2(Y_o) & \longrightarrow & \bigoplus_{j=1}^n L_a^2(U_j) & \longrightarrow & \bigoplus_{j=1}^n L^2(\psi_j U_j)
\end{array}
$$

yields the compactness of $\mathfrak{B}_a^1(Y_o) \to L_a^2(Y_o)$. ///

10.8 Compact $\mathfrak{B}_a^1 \longrightarrow L_a^2$ for $SL_3(\mathbb{Z})$, $SL_4(\mathbb{Z})$, $SL_5(\mathbb{Z})$, ...

Now let $\Gamma = SL_r(\mathbb{Z})$, $G = SL_r(\mathbb{R})$, and $K = SO_r(\mathbb{R})$. As in the smaller examples, the global automorphic Sobolev space $\mathfrak{B}^1$ is the completion of $C_c^\infty(\Gamma \backslash G / K)$ with respect to the $\mathfrak{B}^1$-norm

$$|f|_{\mathfrak{B}^1}^2 = \int_{\Gamma \backslash G / K} (1 - \Delta) f \cdot \overline{f}$$

For a cutoff height $a \gg 1$, let $\mathfrak{B}^1_a$ be the $\mathfrak{B}^1$ completion of $C^\infty_c(Z\Gamma\backslash G/K) \cap L^2_a$ in the $\mathfrak{B}^1$ norm. The resolvent of the Friedrichs extension maps continuously from L^2_a to an automorphic Sobolev space $\mathfrak{B}^1_a$ with its finer topology. Thus, it suffices to show that the injection $\mathfrak{B}^1_a \to L^2_a$ is compact. As in the simpler examples, to prove this compactness, we will show that the image of the unit ball of $\mathfrak{B}^1_a$ is totally bounded in L^2_a.

[10.8.1] Theorem: The Friedrichs self-adjoint extension $\widetilde{\Delta}_a$ of the restriction of the symmetric operator Δ to test functions D_a in L^2_a has *compact resolvent*, thus has purely *discrete spectrum*.

Proof: First, we grant that we can control smooth cutoff functions:

[10.8.2] Lemma: Fix a positive simple root α. Given $\mu \geq \eta(\alpha) + 1$, there are smooth functions φ^α_a for $\alpha \in \Phi$ and φ^o_a such that: all these functions are real-valued, taking values between 0 and 1, φ^o is supported in $C_{\mu+1}$ and $\varphi^\alpha \mu$ is supported in X^α_a, and $\varphi^o_a + \sum_\alpha \varphi^\alpha_a = 1$. Further, there is a bound C *uniform* in $\mu \geq \eta(\alpha) + 1$, such that $|f \cdot \varphi^o_a|_{\mathfrak{B}^1} \leq C \cdot |f|_{\mathfrak{B}^1}$ and

$$|f \cdot \varphi^\alpha_a|_{\mathfrak{B}^1} \leq C \cdot |f|_{\mathfrak{B}^1} \qquad \text{(for all } \mu \geq \eta(\alpha) + 1)$$

(Proof almost identical to [10.4.3].)

The key point is the estimation of tails as in [10.3] and [10.4]. To prove total boundedness of $\mathfrak{B}^1_a \to L^2_a$, given $\varepsilon > 0$, take $\mu \geq \eta(\alpha) + 1$ for all $\alpha \in \Phi$, large enough so that $|f \cdot \varphi^\alpha_a|_{L^2} < \varepsilon$ for all $\alpha \in \Phi$, for all $f \in \mathfrak{B}^1_a$ with $|f|_{\mathfrak{B}^1} \leq 1$. This covers the images $\{f \cdot \varphi^\alpha_a : f \in \mathfrak{B}^1_a\}$ with $\alpha \in \Phi$ with card(Φ) open balls in L^2 of radius ε.

The remaining part $\{f \cdot \varphi^o_a : f \in \mathfrak{B}^1_a\}$ consists of smooth functions supported on the compact C_a. The latter can be covered by finitely many coordinate patches $\psi_i : U_i \to \mathbb{R}^d$. Take smooth cutoff functions φ_i for this covering. The functions $(f \cdot \varphi_i) \circ \psi^{-1}_i$ on $\mathbb{R}^d$ have support strictly inside a Euclidean box, whose opposite faces can be identified to form a flat d-torus $\mathbb{T}^d$. As in the proof of [10.7.6], because continuous functions on compacts are *uniformly* continuous, the flat gradient and the gradient inherited from G admit uniform comparison on each $\psi(U_i)$, as do the measures, so the $\mathfrak{B}^1(\mathbb{T}^d)$-norm of $(f \cdot \varphi_i) \circ \psi^{-1}_i$ is uniformly bounded by the $\mathfrak{B}^1$-norm. The classical Rellich lemma asserts compactness of $\mathfrak{B}^1(\mathbb{T}^d) \to L^2(\mathbb{T}^d)$. By restriction, this gives the compactness of each $\mathfrak{B}^1 \cdot \varphi_i \to L^2$. A finite sum of compact maps is compact, so $\mathfrak{B}^1 \cdot \varphi^o_a \to L^2$ is compact. In particular, the image of the unit ball from $\mathfrak{B}^1$ admits a cover by finitely many ε-balls for any $\varepsilon > 0$.

Combining these finitely many ε-balls with the card(Φ) balls covers the image of $\mathcal{B}_a^1$ in L^2 by finitely many ε-balls, proving that $\mathcal{B}_a^1 \to L^2$ is compact.   ///

10.9 Compact Resolvents and Discrete Spectrum

The overall corollary in all these examples:

[10.9.1] Corollary: For λ off a *discrete* set X of points in $\mathbb{C}$, the inverse $(\widetilde{\Delta}_a - \lambda)^{-1}$ *exists*, is a *compact* operator, and

$$\lambda \longrightarrow \left((\widetilde{\Delta}_a - \lambda)^{-1} : L_a^2(\Gamma \backslash G/K) \longrightarrow L_a^2(\Gamma \backslash G/K) \right)$$

is *meromorphic* in $\lambda \in \mathbb{C} - X$. The decomposition of $L_a^2(\Gamma \backslash G/K)$ with respect to $\widetilde{\Delta}_a$ is *discrete*: there is an orthogonal basis of $L_a^2(\Gamma \backslash G/K)$ consisting of $\widetilde{\Delta}_a$-eigenvectors. The eigenvectors of $\widetilde{\Delta}_a$ are eigenvectors of $(1 - \widetilde{\Delta}_a)^{-1}$, and eigenvalues λ of $\widetilde{\Delta}_a$ are in bijection with nonzero eigenvalues of $(1 - \widetilde{\Delta}_a)^{-1}$ by $\lambda \longleftrightarrow (1 - \lambda)^{-1}$.

Proof: The previous preparations and [9.4.1].   ///

11

Meromorphic Continuation of Eisenstein Series

This proof of meromorphic continuation of various Eisenstein series is in part an elaboration of [Colin de Verdière 1981], and parts of [Colin de Verdière 1982/1983]. The less-simple examples [11.7-11.12] of cuspidal-data Eisenstein series for maximal proper parabolics in GL_r constitute a natural extension.

In the four simplest examples, the compactness [10.7] of the inclusion map of $\mathcal{B}_a^1 \to L_a^2(\Gamma \backslash G/K)$ of pseudo-cuspforms yields [10.9] the compactness of the resolvent of the Friedrichs self-adjoint extension [9.2] $\widetilde{\Delta}_a$ of the restriction of the invariant Laplacian to (a dense subspace of) that subspace, giving its *meromorphy*. Eisenstein series differ by elementary functions from Eisenstein-series-like functions in the domain of $\widetilde{\Delta}_a$, giving the meromorphic continuation of the Eisenstein series.

A noteworthy preliminary result, reminiscent of [Avakumović 1956], [Roelcke 1956b], and [Selberg 1956], immediately extends Eisenstein series E_s up to the critical line $\mathrm{Re}\,(s) = \frac{1}{2}$. Analytic continuation of the zeta function $\zeta(s)$ to $\mathrm{Re}\,(s) > 0$ is a corollary of this preliminary result, the simplest example of the arguments of [Langlands 1967/1976], [Langlands 1971], and [Shahidi 1978] about meromorphic continuation of automorphic L-functions.

11.1 Up to the Critical Line: Four Simple Examples

In this section, we consider the four simplest cases of Chapter 1, with Iwasawa coordinates x, y with $x \in \mathbb{R}^r$ with $r = 1, 2, 3, 4$ respectively, and $y > 0$.

Precise discussion of an unbounded operator and its resolvent requires a specified *domain* [9.1]. Let $\widetilde{\Delta}$ be the Friedrichs extension [9.2] of the restriction of Δ to $C_c^\infty(\Gamma\backslash G/K)$. The Friedrichs construction shows that the domain of $\widetilde{\Delta}$ is *contained in* a Sobolev space

$$\mathfrak{B}^1 = \left(\text{completion of } C_c^\infty(\Gamma\backslash G/K) \text{ under} \langle v, w\rangle_{\mathfrak{B}^1} = \langle (1 - \Delta)v, w\rangle \right)$$

The domain of $\widetilde{\Delta}$ *contains* the smaller Sobolev space

$$\mathfrak{B}^2 = \left(\text{completion of } C_c^\infty(\Gamma\backslash G/K) \text{ under} \langle v, w\rangle_{\mathfrak{B}^2} = \langle (1 - \Delta)^2 v, w\rangle \right)$$

As in the previous chapter, the quotient $\Gamma\backslash G/K$ is a union of a *compact* part X_{cpt}, whose (conceivably complicated) geometry does not matter, and a geometrically simpler *noncompact* part:

$$\Gamma\backslash G/K = X_{\mathrm{cpt}} \cup X_\infty$$

with compact X_{cpt}, cusp neighborhood X_∞, where, with $a \gg 1$, with normalized height function $\eta(nm_y k) = y^r$ as in [1.9],

$$X_\infty = \text{image of } \{g \in G/K : \eta(g) \geq a\}$$
$$= \Gamma_\infty\backslash\{g \in G/K : \eta(g) \geq a\} \approx \mathbb{Z}^r\backslash\mathbb{R}^r \times [a, +\infty)$$

Define a smooth cutoff function τ as usual: fix $a'' < a'$ large enough so that the image of $\{(x, y) \in G/K : y > a''\}$ in the quotient is in X_∞, and let

$$\tau(g) = \begin{cases} 1 & (\text{for } \eta(g) > a') \\ 0 & (\text{for } \eta(g) < a'') \end{cases}$$

Form a pseudo-Eisenstein series h_s by winding up the smoothly cutoff function $\tau(g) \cdot \eta(g)^s$:

$$h_s(g) = \sum_{\gamma \in \Gamma_\infty\backslash\Gamma} \tau(\gamma g) \cdot \eta(\gamma g)^s$$

Since τ is supported on $\eta \geq a''$ for large a'', for any $g \in G/K$, there is at most one nonvanishing summand in the expression for h_s, and convergence is not an issue. Thus, the pseudo-Eisenstein series h_s is *entire* as a function-valued function of s. Let

$$\widetilde{E}_s = h_s - (\widetilde{\Delta} - \lambda_s)^{-1} (\Delta - \lambda_s) h_s$$

where $\lambda = r^2 \cdot s(s-1)$ with $r = 1, 2, 3, 4$.

[11.1.1] Claim: $\widetilde{E}_s - h_s$ is a holomorphic $\mathfrak{B}^1$-valued function of s for $\mathrm{Re}\,(s) > \frac{1}{2}$ and $\mathrm{Im}\,(s) \neq 0$.

Proof: From Friedrichs' construction [9.2], the resolvent $(\widetilde{\Delta} - \lambda_s)^{-1}$ *exists* as an everywhere-defined, continuous operator for $s \in \mathbb{C}$ for λ_s not a nonpositive real number, because of the nonpositiveness of Δ. Further, for λ_s not a nonpositive real, this resolvent is a *holomorphic* operator-valued function. In fact, for such λ_s, the resolvent $(\widetilde{\Delta} - \lambda_s)^{-1}$ injects from $L^2(\Gamma\backslash G/K)$ to $\mathfrak{B}^1$.   ///

[11.1.2] Remark: The smooth function $(\Delta - \lambda_s)h_s$ is supported on the image of $b \leq y \leq b'$ in $\Gamma\backslash G/K$, which is *compact*. Thus, it is in $L^2(\Gamma\backslash G/K)$. It might seem $\widetilde{E}_s$ *vanishes*, if it is forgotten that the indicated resolvent maps to the domain of $\widetilde{\Delta}$ *inside* $L^2(\Gamma\backslash G/K)$, and that h_s is not in $L^2(\Gamma\backslash G/K)$ for $\mathrm{Re}\,(s) > \frac{1}{2}$. Indeed, since h_s is not in $L^2(\Gamma\backslash G/K)$ and $(\widetilde{\Delta} - \lambda_s)^{-1}(\Delta - \lambda_s)h_s$ is in $L^2(\Gamma\backslash G/K)$, the difference cannot vanish.

[11.1.3] Theorem: With $\lambda_s = r^2 \cdot s(s-1)$ not nonpositive real, $u = \widetilde{E}_s - h_s$ is the unique element of the domain of $\widetilde{\Delta}$ such that

$$(\widetilde{\Delta} - \lambda_s)\,u = -(\Delta - \lambda_s)h_s$$

Thus, $\widetilde{E}_s$ is the usual Eisenstein series E_s of [1.9] for $\mathrm{Re}\,(s) > 1$, and gives an analytic continuation of $E_s - h_s$ as $\mathfrak{B}^1$-valued function to $\mathrm{Re}\,(s) > \frac{1}{2}$ with $s \notin (\frac{1}{2}, 1]$.

Proof: Uniqueness follows from Friedrichs' construction [9.2] and construction of resolvents because $\widetilde{\Delta} - \lambda_s$ is a *bijection* of its domain to $L^2(\Gamma\backslash G/K)$.

On the other hand, for $\mathrm{Re}\,(s) > \frac{1}{2}$ and $s \notin (\frac{1}{2}, 1]$, $\widetilde{E}_s - h_s$ is in $L^2(\Gamma\backslash G/K)$, is *smooth*, and

$$\Delta(\widetilde{E}_s - h_s) = (\Delta - \lambda_s)(\widetilde{E}_s - h_s) + \lambda_s \cdot (\widetilde{E}_s - h_s)$$
$$= (\Delta - \lambda_s)h_s + \lambda_s \cdot (\widetilde{E}_s - h_s)$$
$$= (\text{smooth, compactly supported}) + \lambda_s \cdot (\widetilde{E}_s - h_s)$$

so is in $\mathfrak{B}^2$, so certainly in the domain of $\widetilde{\Delta}$. Abbreviating $H_s = (\Delta - \lambda_s)h_s$, it is legitimate to compute

$$(\widetilde{\Delta} - \lambda_s)(\widetilde{E}_s - h_s) = (\widetilde{\Delta} - \lambda_s)\Big((h_s - (\widetilde{\Delta} - \lambda_s)^{-1}H_s) - h_s\Big)$$
$$= (\widetilde{\Delta} - \lambda_s)\Big(-(\widetilde{\Delta} - \lambda_s)^{-1}H_s\Big) = -H_s$$

Thus, $\widetilde{E}_s - h_s$ is a solution. Also, $E_s - h_s$ is a solution:

$$(\Delta - \lambda_s)(E_s - h_s) = (\Delta - \lambda_s)E_s - (\Delta - \lambda_s)h_s = 0 - (\Delta - \lambda_s)h_s$$

By uniqueness, we are done.   ///

[11.1.4] Remark: Thus, the Eisenstein series E_s has an analytic continuation to $\mathrm{Re}(s) > \frac{1}{2}$ and $s \notin (\frac{1}{2}, 1]$ as an $h_s + \mathfrak{B}^1$-valued function. Further, the Friedrichs construction gives a bound for the L^2-norm of $E_s - h_s$ via an estimate on the operator norm of $(\widetilde{\Delta} - \lambda_s)^{-1}$. The L^2-norm of $(\Delta - \lambda_s)h_s$ is not difficult to estimate, since its support is $b \le y \le b'$:

$$|(\Delta - \lambda_s)h_s|_{L^2}^2 \le \int_0^1 \int_b^{b'} (|\Delta h_s| + |\lambda_s h_s|)^2 \, \frac{dx\,dy}{y^2} \ll_{b,b'} |\lambda_s|^2$$

Since $\widetilde{\Delta}$ is negative-definite, as in the proof of [9.17], with $\lambda_s = a + bi$

$$|(\widetilde{\Delta} - \lambda_s)v|^2 = |(\widetilde{\Delta} - a)v|^2 - ib\langle(T - a)v, v\rangle + ib\langle v, (T - a)v\rangle + b^2|v|^2$$
$$\ge b^2|v|^2$$

Thus, the operator norm of the resolvent is estimated by

$$\|(\widetilde{\Delta} - \lambda_s)^{-1}\| \le \frac{1}{\mathrm{Im}(\lambda_s)^2} = \frac{1}{2r^2(\mathrm{Re}(s) - \frac{1}{2}) \cdot \mathrm{Im}(s)}$$

for $\mathrm{Re}(s)\sigma > \frac{1}{2}$, $\mathrm{Im}(s) \ne 0$. Thus,

$$|E_s - h_s|_{L^2} \le \|(\widetilde{\Delta} - \lambda_s)^{-1}\| \cdot |(\Delta - \lambda_s)h_s|_{L^2}$$
$$\ll_{b,b'} \frac{1}{(\mathrm{Re}(s) - \frac{1}{2}) \cdot \mathrm{Im}(s)} \cdot |\lambda_s|$$
$$= \frac{1}{(\mathrm{Re}(s) - \frac{1}{2}) \cdot \mathrm{Im}(s)} \cdot \left((\mathrm{Re}(s) - \tfrac{1}{2})^2 + \mathrm{Im}(s)^2\right)^{\frac{1}{2}}$$

[11.1.5] Remark: From [1.9.4], the Eisenstein series E_s has *constant term* of the form $\eta^s + c_s\eta^{1-s}$. Thus, the analytic continuation of E_s to $\mathrm{Re}(s) > \frac{1}{2}$ analytically continues c_s to $\mathrm{Re}(s) > \frac{1}{2}$. In the case $\Gamma = SL_2(\mathbb{R})$, since $c_s = \xi(2s - 1)/\xi(2s)$ with $\xi(s)$ the completed zeta-function $\xi(s) = \pi^{-s/2}\,\Gamma(s/2)\,\zeta(s)$ this yields the analytic continuation of $\zeta(s)$ to $\mathrm{Re}(s) > 0$, off the interval $[0, 1]$. A similar conclusion holds for $\Gamma = SL_2(\mathbb{Z}[i])$ and the zeta function of $\mathbb{Z}[i] \subset \mathbb{Q}(i)$.

11.2 Recharacterization of Friedrichs Extensions

Friedrichs extensions of restrictions of Δ admit simple alternative descriptions facilitating finer analysis in terms of distributions. Up to a point, this can be done abstractly, in the same context as the construction of the Friedrichs extension [9.2].

Let V be a Hilbert space with a complex conjugation map $v \to \bar{v}$, with expected behavior with respect to the Hermitian inner product. This gives a complex-linear isomorphism $c : V \to V^*$ of V to its dual V^* via Riesz-Fréchet composed with complex conjugation, by $c : v \to \langle -, \bar{v} \rangle$. Let S be a symmetric operator on V with dense domain D, with $\langle Sv, v \rangle \geq \langle v, v \rangle$ for $v \in D$. Suppose that S commutes with the conjugation map. Put $\langle x, y \rangle_{V^1} = \langle Sx, y \rangle$ for $x, y \in D$, and let V^1 be the completion of D with respect to this norm. The identity map $D \to D$ induces a continuous injection $j : V^1 \to V$ with dense image. This much is the same as in [9.2].

Write V^{-1} for the Hilbert-space dual V^* of V^{-1}, with Hermitian inner product $\langle, \rangle_{V^{-1}}$. Let j^* be the adjoint map $j^* : V^* \to (V^{-1})^*$ of j, so composition with complex conjugation c gives

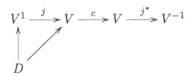

There is a continuous linear map $S^{\#} : V^1 \longrightarrow V^{-1}$, with the respective topologies, given by

$$S^{\#}(x)(y) = \langle x, \bar{y} \rangle_{V^1} \qquad (\text{for } x, y \in V^1)$$

By Riesz-Frechet, this map is a topological isomorphism.

[11.2.1] Claim: The restriction of $S^{\#}$ to the domain of $\tilde{S}$ is $j^* \circ c \circ \tilde{S}$. The domain of $\tilde{S}$ is

$$\text{domain } \tilde{S} = \tilde{D} = \{x \in V^1 \ : \ S^{\#}x \in (j^* \circ c)V\}$$

Proof: By construction of the Friedrichs extension [9.2], its domain is exactly $\tilde{D} = \tilde{S}^{-1}V$. Thus, for $x = \tilde{S}^{-1}x'$ with $x' \in V$, for all $y \in V^1$

$$(S^{\#}x)(y) = (S^{\#}\tilde{S}^{-1}x')(y) = \langle \tilde{S}^{-1}x', \bar{y} \rangle_{V^{-1}} = \langle x, \bar{y} \rangle$$
$$= ((j^* \circ c)x')(y) = ((j^* \circ c \circ \tilde{S})x)(y)$$

Thus, the restriction of $S^{\#}$ to the domain $\tilde{D}$ of $\tilde{S}$ is essentially $\tilde{S}$, namely,

$$S^{\#}\Big|_{\tilde{D}} = (j^* \circ c \circ \tilde{S})\Big|_{\tilde{D}}$$

Thus, $S^{\#} : V^1 \to V^{-1}$ extends $\widetilde{S}$. On the other hand, for $S^{\#}x = (j^* \circ c)y$ with $y \in V$, for all $z \in V^1$

$$\langle z, \bar{x} \rangle_{V^1} = (S^{\#}x)(z) = ((j^* \circ c)y)(z)$$
$$= (\lambda y)(jz) = \langle jz, \bar{y} \rangle = \langle z, \widetilde{S}^{-1}\bar{y} \rangle_{V^1}$$

Thus, $\bar{x} = \widetilde{S}^{-1}\bar{y}$. Thus, the domain of $\widetilde{S}$ is as claimed.   ///

Let $\Theta \subset D$ be stable under conjugation, and stable under S. For subsequent application, in the simplest examples we are thinking of the collection of pseudo-Eisenstein series Ψ_φ with $\varphi \in C_c^\infty(a, \infty)$. Let V^Θ be the orthogonal complement to Θ in V. Let S_Θ be S restricted to $D_\Theta = D \cap V_\Theta$. The S-stability assumption on Θ gives $S(D_\Theta) \subset V_\Theta$. Certainly $D_\Theta \subset V^1 \cap V_\Theta$, so the V^1 closure of D_Θ is a subset of $V^1 \cap V_\Theta$. However, V^1-density of D_Θ in $V^1 \cap V_\Theta$ equality is not clear in general: we must *assume* that D_Θ is V^1-dense in $V^1 \cap V_\Theta$. In the cases of interest, we have proven this under mild hypotheses [10.3]. This density assumption legitimizes the natural sequel: S_Θ with domain D_Θ is densely defined and symmetric on V_Θ, so has Friedrichs extension $\widetilde{S}_\Theta$, with domain $\widetilde{D}_\Theta$.

The extension

$$(S_\Theta)^{\#} : V^1 \cap V_\Theta \longrightarrow (V^1 \cap V^\Theta)^*$$

is described by

$$(S_\Theta)^{\#}(x)(y) = \langle x, y \rangle_{V^1} \qquad (\text{for } x, y \in V^1 \cap V^\Theta)$$

Let

$$i_\Theta : V^1 \cap V^\Theta \longrightarrow V^1 \qquad i_\Theta^* : V^{-1} = (V^1)^* \longrightarrow (V^1 \cap V^\Theta)^*$$

be the inclusion and its adjoint, fitting into a diagram

$$
\begin{array}{ccccc}
V^1 & \xrightarrow{\;j\;} & V & \xrightarrow{\;j^* \circ c\;} & V^{-1} \\
{\scriptstyle i_\Theta}\big\uparrow & & \big\uparrow & & \big\downarrow{\scriptstyle i_\Theta^*} \\
V^1 \cap V_\Theta & \longrightarrow & V_\Theta & & (V^1 \cap V_\Theta)^*
\end{array}
$$

[11.2.2] Claim: $(S_\Theta)^{\#} = i_\Theta^* \circ S^{\#} \circ i_\Theta$, and the domain of $\widetilde{S}_\Theta$ is

$$\widetilde{D}_\Theta = \{x \in V^1 \cap V_\Theta : (S^{\#} \circ i_\Theta)x \in (j^* \circ c)V + \Theta\}$$
$$= \{x \in V^1 \cap V_\Theta : S_\Theta^{\#}x \in (i_\Theta^* \circ j^* \circ c)V\}.$$

and $\widetilde{S}_\Theta x = y$, with $x \in V^1 \cap V_\Theta$ and $y \in V$, if and only if $(S^{\#} \circ i_\Theta)x = (j^* \circ c)y + \theta$ for some θ in the V^{-1}-closure of $(j^* \circ c)\Theta$.

Proof: The assumption of denseness of D_Θ in $V^1 \cap V^\Theta$ legitimizes formation of the Friedrichs extension as an unbounded self-adjoint operator (densely defined) on V. For $x, y \in V^1 \cap V^\Theta$

$$(i_\Theta^* \circ S^\# \circ i_\Theta)(x)(y) = S^\#(x)(y) = \langle i_\Theta x, i_\Theta \overline{y}\rangle_{V^1}$$
$$= \langle x, \overline{y}\rangle_{V^1} = (S_\Theta)^\#(x)(y),$$

which is the first statement of the claim.

From the preceding, the Friedrichs extension $\widetilde{S}_\Theta$ is characterized by

$$\langle z, \widetilde{S}_\Theta^{-1} y\rangle_{V^1} = \langle z, y\rangle \qquad \text{(for } z \in D_\Theta \text{ and } y \in V_\Theta)$$

Given $S^\# x = (j^* \circ c)y + \theta$ with $x \in V^1 \cap V_\Theta$, $y \in V$, and θ in the V^{-1} closure of $(j^* \circ c)\Theta$, take $z \in D_\Theta$ and compute

$$\langle x, \overline{z}\rangle_{V^1} = (S^\# x)(z) = ((j^* \circ c)y + \theta)(z) = (j^* \overline{y})(z) + \theta(z)$$
$$= \langle z, \overline{y}\rangle + 0 = \langle y, \widetilde{S}_\Theta^{-1} S\overline{z}\rangle = \langle \widetilde{S}_\Theta^{-1} y, S\overline{z}\rangle = \langle \widetilde{S}_\Theta^{-1} y, \overline{z}\rangle_{V^1}$$

thus showing that $\widetilde{S}_\Theta^{-1} x = y$. On the other hand, $(S_\Theta)^\# x = (i_\Theta^* \circ j^* \circ c)y$ if and only if $(S^\# \circ i_\Theta)x = y + \theta$ for some $\theta \in \ker i_\Theta^*$, and $\ker i_\Theta^*$ is the closure of Θ in V^{-1}.   ///

11.3 Distributional Characterization of Pseudo-Laplacians

The previous section applies to the pseudo-Laplacians $\widetilde{\Delta}_a$ of Chapter 10 for $a \gg 1$ large enough so that the density result [10.3.1] legitimizes the discussion. This recharacterization is needed for meromorphic continuation of Eisenstein series beyond the critical line.

Referring to the notation of the previous section, take $V = L^2(\Gamma\backslash G/K)$, use the pointwise conjugation map $c : L^2(\Gamma\backslash G/K) \to L^2(\Gamma\backslash G/K)$, let $D = C_c^\infty(\Gamma\backslash G/K)$, put $S = 1 - \Delta|_D$, and let $\Theta = \Theta_a$ be the space of pseudo-Eisenstein series Ψ_φ with $\varphi \in C_c^D(a, +\infty)$ with $a \gg 1$ large enough so that the density lemma [10.3.1] holds. Let $V^1 = \mathfrak{B}^1$ be the completion of D with respect to the norm given by

$$|f|_{\mathfrak{B}^1}^2 = \int_{\Gamma\backslash G/K} (1 - \Delta)f \cdot \overline{f} = \langle(1 - \Delta)f, f\rangle$$

Let $\mathfrak{B}^{-1}$ be the Hilbert space dual of $\mathfrak{B}^1$. With inclusion $j : \mathfrak{B}^1 \to V$, let j^* be its adjoint, and we have a picture

$$\mathfrak{B}^1 \xrightarrow{\ j\ } V \xrightarrow{\ j^* \circ c\ } \mathfrak{B}^{-1}$$

Let η_a be the functional on D which evaluates constant terms at height a.

[11.3.1] Remark: In the present context, we have to prove the following lemma without using any *spectral* description of $\mathfrak{B}^1$ or $\mathfrak{B}^{-1}$ because we are in the process of proving meromorphic continuation of Eisenstein series, which must be done (logically) before spectral decompositions. In the following chapter, we can revisit spaces $\mathfrak{B}^s$ for $s \in \mathbb{R}$ in a more congenial context, with spectral theory available.

Indeed, the proof of the following lemma uses the already-available spectral theory on multitori:

[11.3.2] Lemma: For $a \gg 1$ sufficiently large, $\eta_a \in \mathfrak{B}^{-1}$.

Proof: As expected, take $b' \gg 1$ large enough so that the standard Siegel set $\mathfrak{S}_{b'}$ meets no translate $\gamma \mathfrak{S}_{b'}$ with $\gamma \in \Gamma$ unless $\gamma \in N \cap \Gamma$, so that the cylinder $C_{b'} = (P \cap \Gamma)\backslash\mathfrak{S}_{b'}$ injects to $\Gamma\backslash G/K$. Take $a > b'$. Since the support of η_a is compact and properly inside $\mathfrak{S}_{b'}$, there is a test function ψ identically 1 on the support of η_a, and supported inside $\mathfrak{S}_{b'}$. Then $\psi \cdot \eta_a = \eta_a$, in the sense that $\eta_a(f) = \eta_a(\psi f)$ for all test functions f. Thus, it suffices to consider test functions with support in a subset $X = (N \cap \Gamma)\backslash N \times (b', b'')$ of the cylinder $C_{b'} = (N \cap \Gamma)\backslash N \times (b', +\infty) \approx (\mathbb{Z}\backslash\mathbb{R})^r \times (b', b'')$, with $b'' < +\infty$.

Identifying the endpoints of the finite interval $(b', b'') \subset [b', b'']$ identifies it with another circle, thus imbedding $X \subset \mathbb{T}^{r+1}$. As in the proofs of [10.7.2] and [10.7.6], the $\mathfrak{B}^1$ and L^2 norms on X are uniformly comparable to those on $\mathbb{T}^{r+1}$ descended from the Euclidean versions. Thus, to prove $\eta_a \in \mathfrak{B}^{-1}$, it suffices to prove that the functional θ given by integration along $\mathbb{T}^r \times \{0\}$ inside $\mathbb{T}^{r+1}$ is in the corresponding $\mathfrak{B}^{-1}$ space there. The advantage is that we can use Fourier series, since the spectral theory of $\mathbb{T}$ and $\mathbb{T}^n$ is already available, as in [9.5], especially [9.5.9]. That is, parametrizing $\mathbb{T}^{r+1}$ as $\mathbb{Z}^{r+1}\backslash\mathbb{R}^{r+1}$, let ψ_ξ be $\psi(x) = e^{2\pi i \xi \cdot x}$ for $\xi, x \in \mathbb{R}$ and $\xi \cdot x$ the usual inner product on $\mathbb{R}^{r+1}$. Letting $\xi = (\xi_1, \ldots, \xi_{r+1})$, the Fourier coefficients of θ are

$$\widehat{\theta}(\xi) = \theta(\overline{\psi}_\xi) = \int_{\mathbb{T}^r \times \{0\}} \psi_\xi(x)dx = \begin{cases} 0 & (\text{for } \xi \neq 0) \\ 1 & (\text{for } \xi = 0) \end{cases}$$

Thus, the sth Sobolev norm of θ is

$$\sum_{\xi \in \mathbb{Z}^{r+1}} |\widehat{\theta}(\xi)|^2 \cdot (1 + |\xi|^2)^s = \sum_{\xi_{r+1} \in \mathbb{Z}} 1 \cdot (1 + |\xi_{r+1}|^2)^s$$

which is finite for $\text{Re}(s) < -\frac{1}{2}$. Certainly it is finite for $s = -1$, giving the desired conclusion. ///

In the previous lemma, on $\mathbb{T}^{r+1}$, θ is certainly the suitable Sobolev space limit of its finite subsums, which are smooth. This pulls back to an assertion that η_a is in the $\mathfrak{B}^1$ closure of test functions. We need a stronger assertion to use the recharacterization of the previous section:

[11.3.3] Lemma: η_a is in the $\mathfrak{B}^{-1}$-closure of Θ.

Proof: Again, by the previous lemma, η_a is a $\mathfrak{B}^{-1}$-limit of a sequence $\{f_n\}$ of test functions on $\Gamma\backslash G/K$ or on the cylinder $C_{b'}$. Much as in [10.3], we want to show that suitable smooth truncations of the f_n, to put them into Θ, still converge to η_a in $\mathfrak{B}^{-1}$. As in the previous proof, using $a \gg 1$, we can convert the question to one on $\mathbb{T}^{r+1}$ or on $\mathbb{T}^r \times \mathbb{R}$. Further, since nothing is happening in the first r coordinates, it suffices to prove the following claim on $\mathbb{R}$.

That is, in the standard Sobolev spaces H^s on $\mathbb{R}$ [9.7], we claim that the standard Dirac δ on $\mathbb{R}$ is an H^{-1} limit of a sequence of test functions supported in $[0, +\infty)$. Let u be a test function on $\mathbb{R}$ which is 0 in $(-\infty, 0]$, is nonnegative with integral 1 on $[0, +\infty)$. For $n = 1, 2, 3, \ldots$, let $u_n(t) = n \cdot u(nt)$. We claim that $u_n \to \delta$ in H^{-1}. Taking Fourier transforms,

$$\widehat{u_n}(\xi) = \int_{\mathbb{R}} e^{-2\pi i \xi t} n \cdot u(nt)\,dt = \int_{\mathbb{R}} e^{-2\pi i \xi t/n} u(t)\,dt = \widehat{u}(\xi/n)$$

The Fourier transform of δ is 1, since $\delta(t \to e^2 \pi i \xi t) = 1$ for all $\xi \in \mathbb{R}$. The function $\widehat{u}$ is still a Schwartz function. We want to show that, as $n \to +\infty$,

$$\int_{\mathbb{R}} \left|\widehat{u}(\xi/n) - 1\right|^2 \cdot (1 + \xi^2)^{-1} d\xi \longrightarrow 0$$

Certainly $\widehat{u}$ is bounded, so, given $\varepsilon > 0$, there is $N \gg 1$ such that for *all* n

$$\int_{|\xi|\geq N} \left|\widehat{u}(\xi/n) - 1\right|^2 \cdot (1 + \xi^2)^{-1} d\xi < \varepsilon$$

By the differentiability of $\widehat{u}$, for some t_o between 0 and ξ/n

$$\widehat{u}(\xi/n) = \widehat{u}(0) + (\xi/n) \cdot \widehat{u}'(t_o)$$

Since the integral of u is 1, $\widehat{u}(0) = 1$. The derivative $\widehat{u}'$ is continuous, so has a bound B on $[-1, 1]$. For $|\xi| \leq N$, take n large enough so that $|\xi/n| < \varepsilon \leq 1$. Then

$$\int_{|\xi|\leq N} \left|\widehat{u}(\xi/n) - 1\right|^2 \cdot (1 + \xi^2)^{-1} d\xi$$

$$= \int_{|\xi|\leq N} \left|(\xi/n) \cdot \widehat{u}'(t_o)\right|^2 \cdot (1 + \xi^2)^{-1} d\xi$$

$$\leq \int_{|\xi|\leq N} \varepsilon^2 \cdot B^2 \cdot (1 + \xi^2)^{-1} d\xi \leq \varepsilon^2 \cdot B^2 \int_{\mathbb{R}} (1 + \xi^2)^{-1} d\xi \ll \varepsilon$$

Thus, in the spectral-side description of the topology on H^{-1}, we have the desired convergence.   ///

In the four simplest cases, we have

[11.3.4] Corollary: $\widetilde{\Delta}u = f$ for $f \in L^2_a(\Gamma\backslash G/K)$ if and only if $u \in \mathfrak{B}^1 \cap L^2_a(\Gamma\backslash G/K)$, and $\Delta u = f + c \cdot \eta_a$ for some constant c.

[11.3.5] Remark: In particular, the preceding proof mechanisms show that $u \in \mathfrak{B}^1 \cap L^2_a(\Gamma\backslash G/K)$ implies that the constant term is in the Euclidean Sobolev space $H^1(\mathbb{R})$ as a function of the coordinate y. By Sobolev imbedding [9.5.4], [9.5.11], and [9.5.14], this implies *continuity* of the constant term, so vanishing in $\eta > a$ implies $\eta_a u = 0$. Conversely, if $u \in \mathfrak{B}^1$ and $\eta_a u = 0$, we could *truncate* u at height a without disturbing the condition $u \in \mathfrak{B}^1$, to put $\wedge^a u$ in $\mathfrak{B}^1 \cap L^2_a(\Gamma\backslash G/K)$. In fact, after we have the meromorphic continuation of Eisenstein series in hand, and once we have a spectral form of global automorphic Sobolev spaces $\mathfrak{B}^s$, one can easily prove that the conditions $(\Delta - \lambda)u = \eta_a$, $u \in \mathfrak{B}^1$, and $\eta_a u = 0$ *imply* $\eta_{b'} u = 0$ for all $b' \geq a$.

[11.3.6] Remark: For λ_w not the eigenvalue of a cuspform, the homogeneous equation $(\Delta - \lambda_w)u = 0$ has no nonzero solution, so the constant c must be nonzero for nonzero u.

Proof: Use the characterization [11.2.2]. The previous lemma shows that η_a is in the $\mathfrak{B}^{-1}$ closure Θ_{-1} of $\Theta = \Theta_a$. Using $a \gg 1$, we must show that the intersection of that closure with the image $\Delta\mathfrak{B}^1$ is *at most* $\mathbb{C} \cdot \eta_a$.

On one hand, because $a \gg 1$, Θ_{-1} consists of distributions which, on a Siegel set $\mathfrak{S}_{b'}$ with b' just slightly less than a, have support inside $\mathfrak{S}_a \subset \mathfrak{S}_{b'}$. On the cylinder $C_{b'} = \Gamma_\infty\backslash\mathfrak{S}_{b'}$, the product of circles $(N \cap \Gamma)\backslash N \approx \mathbb{T}^r$ acts by translations, descending to the quotient from G/K. By reduction theory, the restrictions to $C_{a'}$ of every pseudo-Eisenstein series Ψ_φ with $\varphi \in C^\infty_c[a, \infty)$ are invariant under $(N \cap \Gamma)\backslash N$, so anything in the $\mathfrak{B}^{-1}$ closure is likewise invariant.

On the other hand, consider the possible images of $\mathfrak{B}^1 \cap L^2_a(\Gamma\backslash G/K)$ by Δ. Certainly $D \cap V^\Theta$ consists of functions with constant term vanishing in $\eta \geq a$, and taking $\mathfrak{B}^1$ completion preserves this property. Since Θ_{-1} is $(N \cap \Gamma)\backslash N$-invariant and the Laplacian commutes with the group action, it suffices to look at $(N \cap \Gamma)\backslash N$-integral averages restricted to the cylinder $C_{b'}$. Such an integral is a restriction of the constant term $c_P v$ to $C_{b'}$, and vanishes in $\eta > a$.

Thus, the intersection of possible images by $\widetilde{\Delta}_a$ with Θ_{-1} consists of $(N \cap \Gamma)\backslash N$-invariant distributions in $\mathfrak{B}^{-1}$ supported on $Z = \{\eta \leq a\} \cap \{\eta \geq a\} \approx (N \cap \Gamma)\backslash N$. By [11.A], such distributions are obtained as compositions of derivatives transverse to Z composed with a distribution supported on Z. By uniqueness of invariant distributions [14.4], the only $(N \cap \Gamma)\backslash N$-invariant distribution on $Z \approx (N \cap \Gamma)\backslash N$ is (a scalar multiple of) integration on $(N \cap \Gamma)\backslash N$.

Certainly η_a itself is among these functionals. No higher-order derivative (composed with η_a) gives a functional in $\mathfrak{B}^{-1}$, as is visible already on $\mathbb{R}$: computing the sth Sobolev norm of the nth derivative $\delta^{(n)}$ of the Euclidean Dirac δ,

$$|\delta^{(n)}|^2_{H^s} = \int_{\mathbb{R}} |\widehat{\delta^{(n)}}(\xi)|^2 \cdot (1 + \xi^2)^s d\xi = \int_{\mathbb{R}} |(-2\pi i\xi)^n|^2 \cdot (1 + \xi^2)^s d\xi$$

This is finite only for $s < -(\frac{1}{2} + n)$.   ///

11.4 Key Density Lemma: Simple Cases

Similar to the description of E_s as $\widetilde{E}_s$ in [11.1], but with $\widetilde{\Delta}_a$ in place of $\widetilde{\Delta}$, with the pseudo-Eisenstein series h_s formed from the smooth cutoff $\tau \cdot \eta^s$ of η^s as in [11.1], put

$$\widetilde{E}_{a,s} = h_s - (\widetilde{\Delta}_a - \lambda_s)^{-1} (\Delta - \lambda_s) h_s$$

Since $(\Delta - \lambda_s)h_s$ is compactly supported, it is in $L^2_a(\Gamma\backslash G/K)$. For λ_s *not* a nonpositive real, $(\widetilde{\Delta}_a - \lambda_s)^{-1}$ is a bijection of $L^2(\Gamma\backslash G/K)_a$ to the domain of $\widetilde{\Delta}_a$, so $u = \widetilde{E}_{s,a} - h_s$ is the *unique* element of the domain of $\widetilde{\Delta}_a$ satisfying

$$(\widetilde{\Delta}_a - \lambda_s) u = -(\Delta - \lambda_s) h_s$$

Since the pseudo-Eisenstein series h_s is entire, the meromorphy of the resolvent $(\widetilde{\Delta}_a - \lambda_s)^{-1}$ [10.9] yields the meromorphy of $\widetilde{E}_{a,s} - h_s$ as $\mathfrak{B}^1 \cap L^2_a(\Gamma\backslash G/K)$-valued function.

Recall that, with $D = C^\infty_c(\Gamma\backslash G/K)$ and $D_a = D \cap L^2(\Gamma\backslash G/K)$, $\mathfrak{B}^1$ is the $\mathfrak{B}^1$-norm completion of D, while $\mathfrak{B}^1_a$ is the $\mathfrak{B}^1$-completion of D_a. We are counting on $\widetilde{E}_{a,s} - h_s$ to be in $\mathfrak{B}^1_a$. This depends on the assumption $a \gg 1$, and it is critical to verify the following:

[11.4.1] Lemma: For $a \gg 1$, $\mathfrak{B}^1_a = \mathfrak{B}^1 \cap L^2_a(\Gamma\backslash G/K)$. That is, for $a \gg 1$, D_a is $\mathfrak{B}^1$-dense in $\mathfrak{B}^1 \cap L^2_a(\Gamma\backslash G/K)$.

Proof: The containment $\mathfrak{B}^1_a \subset \mathfrak{B}^1 \cap L^2_a(\Gamma\backslash G/K)$ is immediate. For the opposite containment, given a sequence $\{\Psi_{\varphi_i} \in D\}$ of pseudo-Eisenstein series converging to $f \in \mathfrak{B}^1 \cap L^2_a(\Gamma\backslash G/K)$ in the $\mathfrak{B}^1$-topology, we must produce a sequence of pseudo-Eisenstein series in D_a converging to f in the topology of $\mathfrak{B}^1$. We will do so by smooth cut-offs of the constant terms of the Ψ_{φ_i}. Since the limit f of the Ψ_{φ_i} has constant term vanishing above height $y = a$ and is in $L_a^2(\Gamma\backslash G/K)$, that part of the constant terms of the Ψ_{φ_i} becomes small. More precisely, we proceed as follows.

Let F be a smooth real-valued function on $\mathbb{R}$ with $F(t) = 0$ for $t < -1$, $0 \leq F(t) \leq 1$ for $-1 \leq t \leq 0$, and $F(t) = 1$ for $t \geq 0$. For $\varepsilon > 0$, let $F_\varepsilon(t) = F((t - a)/\varepsilon)$. Fix real b with $a > b > 1$. Given $\Psi_{\varphi_i} \to f \in L_a^2(\Gamma\backslash G/K)$, the *b-tail* of the constant term of Ψ_{φ_i} is $\tau_i(g) = c_P \Psi_{\varphi_i}(g)$ for $\eta(g) \geq a'$, and $\tau_i(g) = 0$ for $0 < \eta(g) \leq a''$. By design, $\Psi_{\varphi_i} - \Psi_{F_\varepsilon \cdot \tau_i} \in D_a$ for small ε. We will show that, as $i \to +\infty$, for ε_i sufficiently small depending on i, the $\mathfrak{B}^1$-norms of $\Psi_{F_{\varepsilon_i} \cdot \tau_i}$ go to 0, and $\Psi_{\varphi_i} - \Psi_{F_{\varepsilon_i} \cdot \tau_i} \to f$ in the $\mathfrak{B}^1$-norm.

Let $\mathfrak{S} = \mathfrak{S}_b$ with $b \gg 1$, and put $C_b = (N \cap \Gamma)\backslash\mathfrak{S}$. The cylinders C_b admit natural actions of the compact abelian group $(N \cap \Gamma)\backslash N$, by translation. For $b \gg 1$, by reduction theory [1.5], the further quotient $(\Gamma \cap M)\backslash C_b$ injects to its image in $\Gamma\backslash G/K$. Conveniently, $\Gamma \cap M$ is *finite* in these examples, so, for $f \in C_c^\infty(\Gamma\backslash G/K)$, letting

$$|f|_{\mathfrak{B}^1(C_b)}^2 = \int_{C_b} |f(z)|^2 - \Delta f \cdot \overline{f}$$

we have

$$|f|_{\mathfrak{B}^1(C_b)}^2 \ll \int_{\Gamma\backslash G/K} |f(z)|^2 - \Delta f \cdot \overline{f}$$

For each $b > 1$, let $\mathfrak{B}^1(C_b)$ be the completion of $C_c^\infty(\Gamma\backslash G/K)$ with respect to the seminorm $|\cdot|_{\mathfrak{B}^1(C_b)}$ (with collapsing since the $\mathfrak{B}^1$-norm ignores function values outside C_b).

As usual, we have a continuous action of $(N \cap \Gamma)\backslash N$ on $\mathfrak{B}^1(C_b)$. Thus, the map $u \to c_P u$ gives continuous maps of the spaces $\mathfrak{B}^1(C_b)$ to themselves and $c_P \Psi_{\varphi_i}$ goes to $c_P f$ in $\mathfrak{B}^1(C_b)$, and $c_P \Psi_{\varphi_i} \to c_P f = 0$ in $\mathfrak{B}^1(C_a)$.

To have a Leibniz rule, write the norms as *energy norms* by integrating by parts: for $f \in C_c^\infty(\Gamma\backslash G/K)$, put

$$|f|_{\mathfrak{B}^1}^2 = |f|_{L^2(\Gamma\backslash G/K)}^2 + ||\nabla f|_{\mathfrak{s}}^2|_{L^2(\Gamma\backslash G/K)}$$

where ∇ is the left G-invariant, right K-equivariant tangent-space-valued gradient on G, as in [10.7]. Thus, ∇ descends to G/K and to $\Gamma\backslash G/K$, and $|\cdot|_{\mathfrak{s}}$ is a K-invariant norm on the tangent space(s). Explicitly, as in [10.7], for an involutive automorphism θ of the Lie algebra $\mathfrak{g}$ with the Lie algebra $\mathfrak{k}$ of K the $+1$-eigenspace, the -1-eigenspace $\mathfrak{s}$ can be identified with the tangent space at every point of G/K, via left translation of the exponential map: for $\beta \in \mathfrak{s}$, the associated left G-invariant differential operator X_β is

$$(X_\beta f)(g) = \frac{\partial}{\partial t}\Big|_{t=0} f(ge^{t\cdot\beta})$$

It is easy to describe ∇ in coordinates, even though it is independent of coordinates: for an orthonormal basis $\{\sigma_i\}$ for $\mathfrak{s}$,

$$\nabla f(g) = \sum_i X_{\sigma_i} f(g) \cdot \sigma_i \in \mathfrak{s} \otimes_{\mathbb{R}} \mathbb{C}$$

Let $|\cdot|_{\mathfrak{s}}$ be the K-invariant norm on $\mathfrak{s}$. The essential property is the integration by parts identity

$$\int_{\Gamma \backslash G/K} \langle \nabla F_1, \nabla F_2 \rangle_{\mathfrak{s}} = \int_{\Gamma \backslash G/K} -\Delta F_1 \cdot \overline{F}_2$$

for $F_1, F_2 \in C_c^{\infty}(\Gamma \backslash G/K)$. Thus, extending ∇ by continuity in the $\mathfrak{B}^1$ topology, ∇F exists (in an L^2 sense) for $F \in \mathfrak{B}^1(C_b)$. Likewise,

$$|f|_{\mathfrak{B}^1(C_b)}^2 = |f|_{L^2(C_b)}^2 + \||\nabla f|_{\mathfrak{s}}^2|_{L^2(C_b)}$$
$$\ll |f|_{L^2(\Gamma \backslash G/K)}^2 + \||\nabla f|_{\mathfrak{s}}^2|_{L^2(\Gamma \backslash G/K)}$$

Since $a \gg 1$, $\Psi_{F_\varepsilon \cdot \tau_i}$ is just $F_\varepsilon \cdot \tau_i$ on $\mathfrak{S}_b$, and the support of $F_\varepsilon \cdot \tau_i$ is inside the image of the cylinder $C_{a-\varepsilon}$. The map $C_{a-\varepsilon} \to \Gamma \backslash G/K$ is (uniformly) finite-to-one, so

$$|\Psi_{F_\varepsilon \cdot \tau_i}|_{\mathfrak{B}^1} \ll |F_\varepsilon \cdot \tau_i|_{\mathfrak{B}^1(C_{a-\varepsilon})}$$
$$\leq |(F_\varepsilon - 1) \cdot \tau_i|_{\mathfrak{B}^1(C_{a-\varepsilon})} + |\tau_i - c_P f|_{\mathfrak{B}^1(C_{a-\varepsilon})} + |c_P f|_{\mathfrak{B}^1(C_{a-\varepsilon})}$$

by the triangle inequality. The middle summand goes to 0: from earlier, by design,

$$|\tau_i - c_P f|_{\mathfrak{B}^1(C_{a-\varepsilon})} \ll |c_P \Psi_{\varphi_i} - c_P f|_{\mathfrak{B}^1} \ll |\Psi_{\varphi_i} - f|_{\mathfrak{B}^1} \longrightarrow 0$$

The first and third summands require somewhat more care. Estimate

$$|(F_\varepsilon - 1) \cdot \tau_i|_{\mathfrak{B}^1(C_{a-\varepsilon})}^2 = \int_{C_{a-\varepsilon}} |(F_\varepsilon - 1)\tau_i|^2 + |\nabla(F_\varepsilon - 1)\tau_i|_{\mathfrak{s}}^2$$
$$\leq \int_{C_{a-\varepsilon}} |F_\varepsilon - 1|^2 \cdot (|\tau_i|^2 + |\nabla \tau_i|_{\mathfrak{s}}^2) + \int_{C_{a-\varepsilon}} |\nabla F_\varepsilon|_{\mathfrak{s}}^2 \cdot |\tau_i|^2$$
$$+ \int_{C_{a-\varepsilon}} 2|F_\varepsilon| \cdot |\nabla F_\varepsilon|_{\mathfrak{s}} \cdot |\tau_i| \cdot |\nabla \tau_i|_{\mathfrak{s}}$$

The first summand in the latter expression goes to 0 as $\varepsilon \to 0^+$ because $F_\varepsilon - 1 = 0$ when $y \geq a$, and τ_i and $|\nabla \tau_i|_{\mathfrak{s}}$ are continuous.

We can take the orthonormal basis $\{\sigma_i\}$ for $\mathfrak{s}$ to have $\sigma_1 = \begin{pmatrix} 1 & 0 \\ 0 & 1 \end{pmatrix}$, and $\sigma_i = \begin{pmatrix} 0 & * \\ * & 0 \end{pmatrix}$ for $i \geq 2$. Thus, $\sigma_i \in \mathfrak{n} + \mathfrak{k}$ for $i \geq 2$, and in terms of the Iwasawa coordinates $x = (x_1, \ldots, x_r) \in \mathbb{R}^r$ and $0 < y \in \mathbb{R}$, for a smooth function φ on $N\backslash G/K$, only the σ_1 component is nonzero:

$$\nabla\varphi = y\frac{\partial\varphi}{\partial y} \cdot \sigma_1$$

Thus,

$$|\nabla F_\varepsilon(x + iy)|_\mathfrak{s} = |\frac{1}{\varepsilon} \cdot y\, g'((y - a)/\varepsilon) \cdot h|_\mathfrak{s} = \frac{1}{\varepsilon} \cdot |y\, g'((y - a)/\varepsilon)| \ll_F \frac{1}{\varepsilon}$$

Similarly, since τ_i is a function of y independent of x, $\nabla\tau_i = y\tau_i'(y) \cdot h$. The fundamental theorem of calculus and Cauchy-Schwarz-Bunyakowsky recover an easy instance of a Sobolev inequality:

$$|\tau_i(a - v)| = \left|0 - \int_0^v \tau_i'(a - v)dv\right|$$

$$\leq \left(\int_0^v |\tau_i'(a - v)|^2 dv\right)^{\frac{1}{2}} \cdot \left(\int_0^v 1^2 dv\right)^{\frac{1}{2}} \leq o(1) \cdot \sqrt{v}$$

with Landau's little-o notation, since τ_i' is locally L^2. Thus,

$$\int_{C_{a-\varepsilon}} |F_\varepsilon| \cdot |\nabla F_\varepsilon|_\mathfrak{s} \cdot |\tau_i| \cdot |\nabla\tau_i|_\mathfrak{s} \leq \frac{1}{\varepsilon} \cdot o(1) \cdot \sqrt{\varepsilon} \cdot \int_0^\varepsilon |\nabla\tau_i|_\mathfrak{s}$$

$$\leq \frac{1}{\varepsilon} \cdot o(1) \cdot \sqrt{\varepsilon} \cdot \left(\int_0^\varepsilon |\tau_i'|^2\right)^{\frac{1}{2}} \cdot \left(\int_0^\varepsilon 1^2\right)^{\frac{1}{2}} \ll_{\tau_i} \frac{1}{\varepsilon} \cdot o(1) \cdot \sqrt{\varepsilon} \cdot \sqrt{\varepsilon} = o(1)$$

That is, the summand $\int_{C_{a-\varepsilon}} |F_\varepsilon| \cdot |\nabla F_\varepsilon|_\mathfrak{s} \cdot |\tau_i| \cdot |\nabla\tau_i|_\mathfrak{s}$ goes to 0. By the same estimates,

$$\int_{C_{a-\varepsilon}} |\nabla F_\varepsilon|_\mathfrak{s}^2 \cdot |\tau_i|^2 \ll \frac{1}{\varepsilon^2}\int_0^\varepsilon \left(o(1) \cdot \sqrt{v}\right)^2 dv = \frac{1}{\varepsilon^2} \cdot o(1) \cdot \frac{\varepsilon^2}{2} \longrightarrow 0$$

Thus, taking the ε_i sufficiently small, the smooth truncations $\Psi_{\phi_i} - \Psi_{F_{\varepsilon_i} \cdot \tau_i}$ are in $D \cap L_a^2(\Gamma\backslash G/K)$, and still converge to f in $\mathfrak{B}^1$.   ///

11.5 Beyond the Critical Line: Four Simple Examples

We return to the continuation argument. Since $(\widetilde{\Delta}_a - \lambda_s)^{-1}$ maps $(\Delta - \lambda_s)h_s$ to a function with constant term vanishing above $\eta = a$, above $\eta = a$ the constant term of $\widetilde{E}_{a,s}$ is that of h_s, namely, η^s. More generally, evaluate $\widetilde{\Delta}_a - \lambda_s$ distributionally by application of $\Delta - \lambda_s$: for some constant C_s, as distributions,

$$-(\Delta - \lambda_s)h_s = (\widetilde{\Delta}_a - \lambda_s)(\widetilde{E}_{a,s} - h_s) = (\Delta - \lambda_s)(\widetilde{E}_{a,s} - h_s) + C_s \cdot \eta_a$$

Everything else in the latter equation is meromorphic in s, so C_s must be, as well. Thus, rearranging,

$$(\Delta - \lambda_s)\widetilde{E}_{a,s} = -C_s \cdot \eta_a \qquad \text{(as distributions)}$$

Since Δ is G-invariant, it commutes with the constant-term map, and the distribution $(\Delta - \lambda_s)c_P\widetilde{E}_{a,s}$ is 0 away from $\eta = a$. The distributional differential equation

$$\left(y^2 \frac{\partial^2}{\partial y^2} - (r-1)y\frac{\partial}{\partial y} - \lambda_s\right) u = 0 \qquad \text{(on } 0 < y^r = \eta < a)$$

has solutions exactly of the form $A_s\eta^s + B_s\eta^{1-s}$ for constants A_s, B_s, so $c_P\widetilde{E}_{a,s}$ must be of this form in $0 < \eta < a$. Since $\widetilde{E}_{a,s}$ is meromorphic in s, so are A_s, B_s. In summary,

$$c_P\widetilde{E}_{a,s} = \begin{cases} \eta^s & \text{(for } \eta > a) \\ A_s\eta^s + B_s\eta^{1-s} & \text{(for } 0 < \eta < a) \end{cases}$$

By construction, h_s is smooth, and $(\widetilde{\Delta} - \lambda_s)^{-1} f \in \mathfrak{B}^1$ for all $f \in L^2_a(\Gamma\backslash G/K)$. Thus, $\widetilde{E}_{a,s}$ is *locally* in $\mathfrak{B}^1$ in the sense that $\psi \cdot \widetilde{E}_{a,s}$ is in $\mathfrak{B}^1$ for any smooth cutoff $\psi \in C_c^\infty(\Gamma\backslash G/K)$. In particular, taking ψ with support near $\eta = a$ and identically 1 on a neighborhood of the set where $\eta = a$, since $(N \cap \Gamma)\backslash N$ acts continuously on $\mathfrak{B}^1$, the constant term $c_P(\psi \cdot \widetilde{E}_{a,s})$ is in $\mathfrak{B}^1$. Since that constant term is a function on the one-dimensional $N\backslash G/K \approx A^+ \approx (0, +\infty)$, as in the previous section we can conclude that this constant term as a function of $t = \log y$ is in the Euclidean Sobolev space $\mathfrak{B}^1$ on $\mathbb{R}$. By Sobolev's imbedding on $\mathbb{R}$ [9.7], the constant term is *continuous*. Since ψ was identically 1 near $\eta = a$, we conclude that $c_P\widetilde{E}_{a,s}$ itself is continuous at $\eta = a$, and

$$A_s \cdot a^s + B_s \cdot a^{1-s} = a^s \qquad \text{(for all } s)$$

Let $\mathrm{ch}_{[a,\infty)}$ be the characteristic function of $[a, \infty)$, and

$$\beta_{a,s} = \mathrm{ch}_{[a,\infty)}(\eta) \cdot \left(A_s\eta^s + B_s\eta^{1-s} - \eta^s\right)$$

and form a pseudo-Eisenstein series

$$\Phi_{a,s}(g) = \sum_{\gamma \in \Gamma_\infty\backslash\Gamma} \beta_{a,s}(\gamma g)$$

The support of $\beta_{a,s}$ is inside the set where $\eta \geq a$, and $a \gg 1$, so for each $g \in G/K$ the series has at most one nonzero summand, so converges for all $s \in \mathbb{C}$.

[11.5.1] Theorem: $A_s \cdot E_s = \widetilde{E}_{a,s} + \Phi_{a,s}$ and $\widetilde{E}_{a,s} + \Phi_{a,s} = B_s \cdot E_{1-s}$. Thus, E_s has a meromorphic continuation and $E_s - h_s$ is a meromorphic $\mathfrak{B}^1$-valued function.

Proof: With $\widetilde{\Delta}$ as in [11.1], we have shown that $u = E_s - h_s$ is the unique solution in $\mathfrak{B}^1$ to

$$(\widetilde{\Delta} - \lambda_s)\, u = -(\Delta - \lambda_s)\, h_s$$

Thus, multiplying through by A_s, it suffices to prove that $\widetilde{E}_{a,s} + \Phi_{a,s} - A_s \cdot h_s$ is in $\mathfrak{B}^1$ and satisfies

$$(\widetilde{\Delta} - \lambda_s)\,(\widetilde{E}_{a,s} + \Phi_{a,s} - A_s \cdot h_s) = -(\Delta - \lambda_s)\,(A_s \cdot h_s)$$

The fact that $\widetilde{E}_{a,s} - h_s$ is in $\mathfrak{B}^1 \cap L_a^2(\Gamma\backslash G/K)$ motivates the rearrangement

$$\widetilde{E}_{a,s} + \Phi_{a,s} - A_s \cdot h_s = (\widetilde{E}_{a,s} - h_s) + (\Phi_{a,s} - A_s\, h_s + h_s)$$

Thus, we must show that the pseudo-Eisenstein series $F = \Phi_{a,s} - A_s h_s + h_s$ is in $\mathfrak{B}^1$.

For integrability, by reduction theory, $\Phi_{a,s}$ is just $\varphi_s = \mathrm{ch}_{[a,\infty)}(\eta) \cdot \left(A_s \eta^s + B_s \eta^{1-s} - \eta^s\right)$ on $\eta > a$, so on $\eta > a$,

$$F = \Phi_{a,s} - A_s h_s + h_s = (A_s \eta^s + B_s \eta^{1-s} - \eta^s) - A_s \eta^s + \eta^s = B_s \eta^{1-s}$$

For $\mathrm{Re}\,(s) > 1$, η^{1-s} is square-integrable on $\eta > a$, so F is in $L^2(\Gamma\backslash G/K)$.

To demonstrate the additional smoothness required for F to be in $\mathfrak{B}^1$, from the rewriting of Sobolev norms in [10.7], especially [10.7.5], it suffices to show that the right-translation derivatives αF are in $L^2(\Gamma\backslash G)$ for $\alpha \in \mathfrak{g}$. By the left invariance of the right action of $\mathfrak{g}$, it suffices to prove square-integrability, on standard Siegel sets, of the derivatives of the data $\beta_{a,s} - A_s \tau \cdot \eta^s + \tau \cdot \eta^s$ used to form the pseudo-Eisenstein series. This data is smooth everywhere but at $\eta = a$, where it is *continuous*, since $A_s a^s + B_s a^s - a^s = 0$. Further, it possesses continuous left and right derivatives at $\eta = a$, so is *locally* in a $+1$-index Sobolev space at $\eta = a$. The data is left N-invariant and right K-invariant, and A^+ normalizes N, so we need only consider the differential operator $y\frac{\partial}{\partial y}$ coming from the Lie algebra of A^+: the derivative of F is discontinuous at $\eta = a$, and as a distribution it is, recalling that $\eta = y^r$, so $\eta' = r y^{r-1}$, and $\varphi_s = \mathrm{ch}_{[a,\infty)} \cdot (A_s \eta^s + B_s \eta^{1-s} - \eta^s)$,

$$y\frac{\partial}{\partial y} F = y\frac{\partial}{\partial y}\Big(\Phi_{a,s} - A_s h_s + h_s\Big) = y\frac{\partial}{\partial y}\Big(\beta_{a,s} - A_s \cdot \tau \cdot \eta^s + \tau \cdot \eta^s\Big)$$

$$= y\frac{\partial}{\partial y}\begin{cases} B_s \eta^{1-s} & \text{(for } \eta > a) \\ -A_s \cdot \eta^s + \eta^s & \text{(for } a' \leq \eta < a) \\ -A_s \cdot \tau \cdot \eta^s + \tau \cdot \eta^s & \text{(for } a'' \leq \eta \leq a') \\ 0 & \text{(for } \eta \leq a'') \end{cases}$$

$$
= \begin{cases}
B_s \cdot (1-s) \cdot \dfrac{\partial \eta}{\partial y} \cdot \eta^{1-s} & \text{(for } \eta > a) \\[2mm]
(1-A_s) \cdot s\eta^s & \text{(for } a' \le \eta < a) \\[2mm]
(1-A_s)(\dfrac{\partial \tau}{\partial y} \cdot \eta^s + \tau \cdot \dfrac{\partial \eta}{\partial y} \cdot s\eta^s) & \text{(for } a'' \le \eta \le a') \\[2mm]
0 & \text{(for } \eta \le a')
\end{cases}
$$

On $a'' \le \eta \le a$, this derivative is bounded, so the truly relevant behavior is in $\eta > a$: for $\mathrm{Re}\,(s) > 1$ this derivative is square-integrable on standard Siegel sets. Thus, $\Phi_{a,s} - A_s h_s + h_s$ is in $\mathfrak{B}^1$, proving that $\widetilde{E}_{a,s} + \Phi_{a,s} - A_s h_s$ is in $\mathfrak{B}^1$.

To show that $\widetilde{E}_{a,s} + \Phi_{a,s} - A_s h_s$ satisfies the expected equation, we justify computing the effect of differential operators on $\widetilde{E}_{a,s} + \Phi_{a,s} - A_s h_s$ distributionally, as follows. For $f \in C_c^\infty(\Gamma \backslash G / K)$, with $\widetilde{\Delta}$ the Friedrichs extension of the restriction of Δ to $C_c^\infty(\Gamma \backslash G / K)$ as in [11.1],

$$
\left\langle (\widetilde{\Delta} - \lambda_s)(\widetilde{E}_{a,s} + \Phi_{a,s} - A_s h_s),\, f \right\rangle = \left\langle \widetilde{E}_{a,s} + \Phi_{a,s} - A_s h_s,\, (\Delta - \overline{\lambda}_s)f \right\rangle
$$
$$
= \left\langle (\Delta - \lambda_s)(\widetilde{E}_{a,s} + \Phi_{s,f} - A_s h_s),\, f \right\rangle
$$

By design, using the invariance of Δ and the local finiteness of the sum for Φ_s, it is legitimate to compute

$$
(\Delta - \lambda_s)(\widetilde{E}_{a,s} + \Phi_{a,s}) = (\Delta - \lambda_s)\widetilde{E}_{a,s} + \sum_{\gamma \in \Gamma_\infty \backslash \Gamma} (\Delta - \lambda_s)\beta_{a,s} \circ \gamma
$$
$$
= -C_s \cdot \eta_a + C_s \cdot \eta_a = 0 \qquad \text{(as distributions)}
$$

Thus,

$$
(\widetilde{\Delta} - \lambda_s)(\widetilde{E}_{a,s} + \Phi_{a,s} - A_s h_s) = (\Delta - \lambda_s)(\widetilde{E}_{a,s} + \Phi_{a,s} - A_s h_s)
$$
$$
= 0 - A_s(\Delta - \lambda_s)h_s
$$

as desired, proving $\widetilde{E}_{a,s} + \Phi_{a,s} = A_s \cdot E_s$ for $\mathrm{Re}\,(s) > 1$. For $\mathrm{Re}\,(1-s) > 1$, the same argument shows that $\widetilde{E}_{a,s} + \Phi_{a,s} = B_s \cdot E_{1-s}$. This proves the formulas in the claim. Since not both A_s and B_s can be identically 0, we obtain the meromorphic continuation of E_s. ///

[11.5.2] Corollary: $A_s \cdot E_s = B_s \cdot E_{1-s}$. ///

In particular, *neither* A_s nor B_s is identically 0, and with $a(s) = B_s / A_s$, $E_{1-s} = a(s) \cdot E_s$. The relation $c_P E_s = \eta^s + c_s \eta^{1-s}$ gives the meromorphic continuation of c_s. Since $c_P E_{1-s} = \eta^{1-s} + c_{1-s}\eta^s$, apparently $c_s = a(s) = B_s / A_s$. Since $1 - (1-s) = s$, we obtain $c_s \cdot c_{1-s} = 1$:

[11.5.3] Corollary: c_s has a meromorphic continuation, and $c_s \cdot c_{1-s} = 1$. ///

On $\text{Im}(s) = 0$ and $\text{Re}(s) > 1$, E_s and $c_P E_s$ are real-valued. Thus, the two holomorphic functions E_s and $\overline{E_{\bar{s}}}$ agree on $(1, +\infty)$, so agree everywhere. That is, $\overline{E_s} = E_{\bar{s}}$. In particular, on $\text{Re}(s) = \frac{1}{2}$, where $\bar{s} = 1 - s$,

$$|c_s|^2 = c_s \cdot \overline{c_s} = c_s \cdot c_{\bar{s}} = c_s \cdot c_{1-s} = 1 \qquad (\text{on } \text{Re}(s) = \tfrac{1}{2})$$

proving

[11.5.4] Corollary: $|c_s| = 1$ on $\text{Re}(s) = \frac{1}{2}$, and c_s has no pole on $\text{Re}(s) = \frac{1}{2}$.   ///

Further, we have

[11.5.5] Corollary: E_s has no pole on $\text{Re}(s) = \frac{1}{2}$.

Proof: Suppose E_s had a pole of order $N > 0$ at s_o on the critical line $\text{Re}(s) = \frac{1}{2}$. Then $(s - s_o)^N \cdot E_s$ is holomorphic at $s = s_o$, gives a not identically automorphic form and has vanishing constant term there. From

$$\wedge^a (s - s_o)^N E_s = (s - s_o)^N \wedge^a E_s$$

and using the Maass-Selberg relations [1.11] with $s = s_o + \varepsilon$ and $r = \bar{s}_o + \varepsilon = 1 - s_o + \varepsilon$ with $0 < \varepsilon \in \mathbb{R}$, since $(s - s_o) \cdot c_s \to 0$ at $s = s_o$, suppressing measure-normalizations,

$$|(s - s_o)^N E_s|^2 = \varepsilon^{2N} \cdot \left(\frac{a^{s+r-1}}{s + r - 1} + c_s \frac{a^{(1-s)+r-1}}{(1 - s) + r - 1} \right.$$
$$\left. + c_r \frac{a^{s+(1-r)-1}}{s + (1 - r) - 1} + c_s c_r \frac{a^{(1-s)+(1-r)-1}}{(1 - s) + (1 - r) - 1} \right)$$
$$= \varepsilon^{2N} \cdot \left(\frac{a^{2\varepsilon}}{2\varepsilon} + c_{s_o + \varepsilon} \frac{a^{1 - 2s_o - 2\varepsilon}}{1 - 2s_o - 2\varepsilon} \right.$$
$$\left. + c_{1 - s_o + \varepsilon} \frac{a^{2s_o - 1 + 2\varepsilon}}{2s_o - 1 + 2\varepsilon} + c_{s_o + \varepsilon} c_{1 - s_o + \varepsilon} \frac{a^{-2\varepsilon}}{-2\varepsilon} \right) \longrightarrow 0$$

contradiction. Thus, E_s has no pole on the critical line.   ///

Toward proving *moderate growth* of the meromorphic continuation of E_s:

[11.5.6] Claim: E_s meromorphically continues as a $C^\infty(\Gamma \backslash G / K)$-valued function.

Proof: As earlier, let

$$\chi_s \begin{pmatrix} a & 0 \\ 0 & a^{-1} \end{pmatrix} = |a|^{2s}$$

and put

$$\varphi_s(nmk) = \chi_s(m) \qquad (\text{for } n \in N, k \in K, \text{ and } m \in M)$$

Up to scalar multiples, φ_s is the unique function on G that is right K-invariant, left N-invariant, and $\varphi_s(mg) = \chi_s(m) \cdot \varphi_s(g)$. The function $s \to \varphi_s$ is a holomorphic $C^o(G/K)$-valued function on $\mathbb{C}$. For $\psi \in C_c^\infty(K \backslash G/K)$, the image

$$(\psi \cdot \varphi_s)(g) = \int_G \varphi(h)\varphi_s(gh)dh$$

is again left N-invariant, left M, χ_s-equivariant, and right K-invariant. Thus, $\psi \cdot \varphi_s = \mu_s(\psi) \cdot \varphi_s$ for scalar $\mu_s(\psi)$. Since $s \to \varphi_s$ is holomorphic $C^o(G/K)$-valued, $s \to \mu_s(\psi)$ is holomorphic $\mathbb{C}$-valued for each such ψ. By nondegeneracy [14.1.5], there exists ψ such that the function $s \to \mu_s(\psi)$ is not identically 0. In the region of convergence $\text{Re}(s) > 1$, from $E_s = \sum_{\gamma \in (\Gamma \cap P) \backslash \Gamma} \varphi_s \circ \gamma$, also $\psi \cdot E_s = \mu_s(\psi) \cdot E_s$. Exactly what we are missing at this point is knowledge of what topological vector space of functions (or distributions) the meromorphically continued Eisenstein series may lie in, so we cannot directly assert much about $\psi \cdot E_s$ outside the region of convergence. (Otherwise we could apply the identity principle from complex analysis to the latter identity.) Rather, we approach this a little indirectly, as follows.

Since Δ commutes with G, $\mathfrak{B}^1$ is stable under the action of $\psi \in C_c^\infty(K \backslash G/K)$. From the meromorphic continuation of $E_s - h_s$ as $\mathfrak{B}^1$-valued function, we have the meromorphic continuation of

$$\psi \cdot (E_s - h_s) = \mu_s(\psi) \cdot E_s - \psi \cdot h_s$$

as $\mathfrak{B}^1$-valued function. In fact, for $F \in \mathfrak{B}^1$, by [14.5], $\psi \cdot F$ is in $C^\infty(\Gamma \backslash G/K)$. By construction, $h_s \in C^\infty(\Gamma \backslash G/K)$. Rearranging,

$$\mu_s(\psi) \cdot E_s = \psi \cdot (E_s - h_s) + \psi \cdot h_s$$

Dividing through by $\mu_s(\eta)$ for some η with $\mu_s(\eta) \neq 0$ exhibits the meromorphically continued E_s as a smooth-function-valued function. ///

[11.5.7] Corollary: E_s has a meromorphic continuation as a $C^o(\Gamma \backslash G/K)$-valued function, so it makes sense to address the issue of its moderate growth. ///

Finally, we have

[11.5.8] Theorem: Away from poles, the meromorphically continued E_s is of moderate growth.

Proof: By [11.5.1] and [11.5.7], (at least) the *pointwise* values of the meromorphic continuation are given by

$$E_s = A_s^{-1} \cdot (\widetilde{E}_{a,s} + \Phi_{a,s})$$

where $\widetilde{E}_{a,s} = h_s - (\widetilde{\Delta}_a - \lambda_s)^{-1}(\Delta - \lambda_s)h_s$ and $\Phi_{a,s}$ is the pseudo-Eisenstein series formed from $\beta_{a,s} = \text{ch}_{[a,\infty)} \cdot (A_s\eta^s + B_s\eta^{1-s} - \eta^s)$. Since $a \gg 1$, in the region $\eta \geq a$ the function $\Phi_{a,s}$ is just $\beta_{a,s}$ itself, which is $A_s\eta^s + B_s\eta^{1-s} - \eta^s$, which is of moderate growth in standard Siegel sets. The preceding computation shows continuity at $\eta = a$. The pseudo-Eisenstein series h_s of [11.1] made from $\tau \cdot \eta^s$ with smooth cutoff τ is a locally finite sum, so is smooth, so certainly continuous. For $\eta \geq a$, its value is just η^s, which is of moderate growth for all s. Thus, to show that $\widetilde{E}_{a,s}$ is of moderate growth even after meromorphic continuation, it suffices to show that $(\widetilde{\Delta}_a - \lambda_s)^{-1}(\Delta - \lambda_s)h_s$ is of moderate growth.

Again, the pseudo-Eisenstein series $h_s = \Psi_{\tau \cdot \eta^s}$ is a locally finite sum, so it is legitimate to compute

$$(\Delta - \lambda_s)h_s = (\Delta - \lambda_s)\Psi_{\tau \cdot \eta^s} = \Psi_{(\Delta - \lambda_s)(\tau \cdot \eta^s)}$$

Since differential operators do not increase support, $f_s = (\Delta - \lambda_s)(\tau \cdot \eta^s)$ is smooth and supported in $[a'', a']$. It is visibly a holomorphic $C^\infty(0, +\infty)$-valued function of $s \in \mathbb{C}$. Its uniform compact support in $[a'', a']$ implies that $s \to f_s$ is certainly a holomorphic $C_c^\infty(\Gamma\backslash G/K)$-valued function of s, in fact taking values in the Fréchet subspace of functions supported in $[a'', a']$.

Given a uniformly compactly supported holomorphic family $f_s \in C_c^\infty(N\backslash G/K) \approx C_c^\infty(0, +\infty)$, in light of [11.3.4] we solve equations $(\Delta - \lambda_s)u = f_s + c \cdot \eta_a$ with $c \in \mathbb{C}$ (depending on s) for u on $N\backslash G/K \approx (0, +\infty)$ with sufficient decay at 0^+ to form a pseudo-Eisenstein series Ψ_u, giving $(\Delta - \lambda_s)\Psi_{u_s} = \Psi_{f_s}$. In Iwasawa coordinates, the equation is

$$y^2 \frac{\partial^2}{\partial y^2}u - (r-2)\frac{\partial}{\partial y}u - \lambda_s u = f_s + c \cdot \delta_a$$

Letting $x = \log y$, with $F_s(x) = f_s(e^x)$ and $v(x) = u(e^x)$, this becomes

$$v'' - (r-1)v' - \lambda_s v = F + c \cdot \delta_{\log a}$$

Taking Fourier transform in a normalization that suppresses some factors of 2π,

$$(-i\xi)^2\widehat{v} - (r-2)(-i\xi)\widehat{v} - \lambda_s\widehat{v} = \widehat{F}_s + c \cdot a^{-i\xi}$$

or

$$\widehat{v}(\xi) = -\frac{\widehat{F}(\xi) + c \cdot a^{-i\xi}}{\xi^2 - (r-2)i\xi + \lambda_s}$$

Since F is a test function, $\widehat{F}$ is an *entire* function such that $\widehat{F}(x + iy_o)$ is (uniformly) in the Schwartz space for each fixed y_o. (We need little more about Paley-Wiener spaces than this idea.) Division by a quadratic polynomial produces a function holomorphic in a *strip* along $\mathbb{R}$ not including either of the two poles. The two poles occur at the zeros of the denominator:

$$\frac{(r-2)i \pm \sqrt{-(r-2)^2 - 4\lambda_s}}{2}$$

Fix $\varepsilon > 0$. Given a bound $|\mathrm{Re}(s)| \le B$, for $\mathrm{Im}(s) \gg_B 1$, those poles are outside the strip $S = \{z \in \mathbb{C} : |\mathrm{Im}(z)| \le 1 + \varepsilon\}$. Thus, $\widehat{v}$ is holomorphic on an open set containing S and has decay like $1/\xi^2$ on horizontal lines inside that strip. Thus, in the Fourier inversion integral

$$v(x) = \frac{1}{2\pi} \int_{\mathbb{R}} e^{i\xi x} \widehat{v}(\xi) d\xi$$

we can move the contour up to $\mathbb{R} + i(1 + \varepsilon)$, giving

$$v(x) = \frac{1}{2\pi} \int_{\mathbb{R}} e^{i(\xi + i\varepsilon)x} \widehat{v}(\xi + i\varepsilon) \, d\xi = e^{-\varepsilon x} \frac{1}{2\pi} \int_{\mathbb{R}} e^{i\xi x} \widehat{v}(\xi + i\varepsilon) d\xi$$

Thus, $v(x) \ll e^{-(1+\varepsilon)x}$, giving genuine exponential decrease for $x \to +\infty$. Similarly, moving the contour down gives exponential decrease $v(x) \ll e^{-(1+\varepsilon)|x|}$ for $x \to -\infty$. Then $u(y) = u_s(y) = v(\log y)$ satisfies $u_s(y) \ll y^{1+\varepsilon}$ as $y \to 0^+$, and $u(y) \ll y^{-(1+\varepsilon)}$ as $y \to +\infty$. Thus, the pseudo-Eisenstein series Ψ_{u_s} converges absolutely since the sum for Ψ_{u_s} is dominated termwise by the sum for an absolutely convergent Eisenstein series [1.9.1]. Further, as it is termwise dominated by an absolutely convergent Eisenstein series, by [1.9.1] Ψ_{u_s} is continuous and of moderate growth.

Having available a choice of the constant c is necessary, since we must adjust Ψ_u to have constant term vanishing above height $\eta = a$. Choose it so that $c_P \Psi_u$ vanishes *at* $\eta = a$. Since $a \gg 1$, by reduction theory the truncation $\wedge^a \Psi_u$ has constant term vanishing at and above height a. Since $a \gg 1$, this truncation is itself a pseudo-Eisenstein series, and still $(\Delta - \lambda_s) \wedge^a \Psi_u$ differs from Ψ_f only by a multiple of η_a. By [11.3.4], $(\widetilde{\Delta}_u - \lambda_s) \wedge^a \Psi_u - \Psi_f$.

Thus, for a given bound $|\mathrm{Re}(s)| \le B$, there is C sufficiently large so that for $|\mathrm{Im}(s)| \ge C$ we have meromorphic continuation of E_s as a (continuous) moderate-growth function.

For $|\mathrm{Im}(s)| < C$, we can express E_s as a vector-valued Cauchy integral along a circular path γ that lies inside the union U of regions $\mathrm{Re}(s) \ge B, \mathrm{Re}(s) \le 1 - B$, and $|\mathrm{Im}(s)| \ge C$, and does not run through any poles of E_s. In $\mathrm{Re}(s) \le 1 - B$ the Eisenstein series is (continuous) of moderate growth, via the functional equation. Thus, E_s is of moderate growth throughout U, and in particular along γ. Let Z be the collection of poles of E_s (as meromorphic $C^o(\Gamma \backslash G / K)$-valued function) inside γ, and $P(z) = \prod_{z_j \in Z}(z - z_j)$. For each $g \in G$

$$P(s) \cdot E_s(g) = \frac{1}{2\pi i} \int_\gamma \frac{P(z) \cdot E_z(g)}{z - s} dz$$

In fact, on γ, $z \to (s \to P(z)E_s/(z - s)$ is a compactly supported, continuous, moderate-growth-function-valued function of z, so the vector-valued Cauchy integral

$$P(s) \cdot E_s = \frac{1}{2\pi i} \int_\gamma \frac{P(z) \cdot E_z}{z - s} dz$$

as in [15.2] exists as a Gelfand-Pettis integral [14.1] lying in that same space of functions.   ///

11.6 Exotic Eigenfunctions: Four Simple Examples

In addition to cuspforms, there must be new, *exotic* eigenfunctions for the operators $\widetilde{\Delta}_a$, which are *not* eigenfunctions for Δ.

[11.6.1] Claim: Take $a \gg 1$. If $a^w + c_w a^{1-w} = 0$, then the truncation $\wedge^a E_w$ is an eigenfunction for $\widetilde{\Delta}_a$. Conversely, if $\wedge^a E_w$ is an eigenfunction for $\widetilde{\Delta}_a$, then $a^w + c_w a^{1-w} = 0$. In particular, for $a^w + c_w a^{1-w} = 0$, we have $(\Delta - \lambda_w) \wedge^a E_w = 2(1 - 2w)a^{w+\frac{1}{r}} \cdot \eta_a$. *(Proof just below.)*

Since $\widetilde{\Delta}_a$ is a nonpositive self-adjoint operator, any eigenvalues are nonpositive real, giving

[11.6.2] Corollary: If $a^w + c_w a^{1-w} = 0$, then either $\mathrm{Re}(w) = \frac{1}{2}$ or $w \in [0, 1]$.   ///

[11.6.3] Remark: An argument-principle discussion shows that there are infinitely many values w on $\mathrm{Re}(w) = \frac{1}{2}$ such that $a^w + c_w a^{1-w} = 0$.

[11.6.4] Remark: Thus, zeros w of $a^w + c_w a^{1-w}$ give eigenvalues $\lambda_w = w(w - 1)$ of $\widetilde{\Delta}_a$ for $a \gg 1$. A spectral characterization of the global automorphic Sobolev spaces $\mathfrak{B}^s$ will prove a converse, that for $\lambda_w < -1/4$, the only eigenvalues arise from zeros of $a^w + c_w a^{1-w}$, and the corresponding exotic eigenfunctions are corresponding trucated Eisenstein series.

Proof: With $a \gg 1$, in a fundamental domain, away from $\eta = a$ we have $(\Delta - \lambda_w) \wedge^a E_w = 0$ *locally*. In $\eta \gg 1$, the differential operator annihilates all Fourier components of E_w but the constant term, and in the lower part of a Siegel set the operator does also annihilate the constant term.

We first do the slightly simpler version of this computation for $SL_2(\mathbb{Z})$, in which case $\eta = y$. To compute near $y = a$, let H be the Heaviside function

$H(y) = 0$ for $y < 0$ and $H(y) = 1$ for $y > 0$. Thus, near $y = a$, as functions of y independent of x,

$$(\Delta - \lambda_w) \wedge^a E_w$$

$$= (\Delta - \lambda_w)\Big(H(a - y) \cdot (y^w + c_w y^{1-w})\Big)$$

$$= (y^2 \frac{\partial^2}{\partial y^2} - w(w - 1))\Big(H(a - y) \cdot (y^w + c_w y^{1-w})\Big)$$

$$= y^2 \Big(H''(a - y)(y^w + c_w y^{1-w}) + 2H'(a - y)(y^w + c_w y^{1-w})'$$

$$+ H(a - y)(y^w + c_w y^{1-w})''\Big)$$

$$- w(w - 1)H(a - y)(y^w + c_w y^{1-w})$$

$$= y^2 \Big(-\delta'_a \cdot (y^w + c_w y^{1-w}) - 2\delta_a \cdot (wy^{w-1} + (1 - w)c_w y^{-w})\Big)$$

For $a^w + c_w a^{1-w} = 0$, the term with δ'_a vanishes, and the rest simplifies to

$$(\Delta - \lambda_w) \wedge^a E_w = -2a\delta_a \cdot (wa^w + (1 - w)c_w a^{1-w}) = -2\delta_a \cdot (2w - 1)a^{w+1}$$

on functions of y independent of x. Thus, this is $2(2w - 1)a^{w+1} \cdot \eta_a$. If $a^w + c_w a^{1-w} \neq 0$, the term with δ'_a remains, and is *not* inside $\mathfrak{B}^{-1}$, so in that case $\wedge^a E_w$ is *not* an eigenfunction.

More generally, with height $\eta = y^r$ with $r = 1, 2, 3, 4$, $\lambda_w = r^2 \cdot w(w - 1)$, and $\Delta = y^2(\Delta_x + \frac{\partial^2}{\partial y^2}) - (r - 1)y\frac{\partial}{\partial y}$, the truncated Eisenstein series $\wedge^a E_w$ is annihilated by Δ except near (images of) $\eta = a$, at which a messier computation gives

$$(\Delta - \lambda_w) \wedge^a E_w$$

$$= (\Delta - \lambda_w)\Big(H(a - y) \cdot (y^{rw} + c_w y^{r(1-w)})\Big)$$

$$= \Big(y^2 \frac{\partial^2}{\partial y^2} - r^2 w(w - 1) - (r - 1)y\frac{\partial}{\partial y}\Big)\Big(H(a - y) \cdot (y^{rw} + c_w y^{r(1-w)})\Big)$$

$$= y^2 \Big(H''(a - y)(y^{rw} + c_w y^{r(1-w)}) + 2H'(a - y)(y^{rw} + c_w y^{r(1-w)})'$$

$$+ H(a - y)(y^{rw} + c_w y^{r(1-w)})''\Big)$$

$$- (r - 1)y\Big(H'(a - y)(y^{rw} + c_w y^{r(1-w)}) + H(a - y)(y^{rw} + c_w y^{r(1-w)})'\Big)$$

$$- r^2 w(w - 1)H(a - y)(y^{rw} + c_w y^{r(1-w)})$$

$$= y^2 \Big(-\delta'_a \cdot (y^{rw} + c_w y^{r(1-w)}) - 2\delta_a \cdot (rwy^{rw-1} + r(1 - w)c_w y^{r(1-w)-1})\Big)$$

$$- (r - 1)y\Big(-\delta_a(y^{rw} + c_w y^{r(1-w)})\Big)$$

At $y^r = \eta = a$, for $a^w + c_w a^{1-w} = 0$, the term with δ_a' vanishes, as does the $(r-1)y\big(-\delta_a(y^{rw} + c_w y^{r(1-w)})\big)$ term, and the rest simplifies to the indicated expression.   ///

11.7 Up to the Critical Line: $SL_r(\mathbb{Z})$

Now take $G = SL_r(\mathbb{R})$, $\Gamma = SL_r(\mathbb{Z})$, and $K = SO_r(\mathbb{R})$. At various moments, it is convenient to consider $G = Z \backslash GL_r$ instead, where Z is the center, but nothing we do depends on the distinction. We only consider Eisenstein series for *maximal* proper parabolics, and with cuspidal data on the Levi components, as in [3.9] and [3.11]. As in the four simplest cases, it is relatively easy to meromorphically continue these Eisenstein series up to the critical line.

As in Chapter 3, there are two qualitatively different types of maximal proper parabolics, namely, the *self-associate* $P = P^{r,r} \subset GL_{2r}$, and *non-self-associate* $P = P^{r_1,r_2} \subset GL_{r_1+r_2}$ with $r_1 \neq r_2$, with associate $Q = P^{r_2,r_1}$.

Fix a maximal proper parabolic $P = P^{r_1,r_2}$ with Levi decomposition $P = NM$, and fix cuspidal data $f = f_1 \otimes f_2$ on the Levi component $M \approx GL_{r_1} \times GL_{r_2}$. We assume that f_1 and f_2 are cuspforms in a strong sense: they are eigenfunctions for the corresponding invariant Laplacians, are of rapid decay in Siegel sets (in particular, are bounded on the respective groups), and there exist test functions β_1, β_2 on the respective groups such that $\beta_j \cdot f_j = f_j$. In particular, all derivatives (whether right invariant under maximal compacts or not) have similarly good decay and smoothness. Recall [3.9] that pseudo-Eisenstein series with cuspidal data f are formed from test functions $\psi \in C_c^\infty(0, +\infty)$ as follows. Let

$$\varphi(znmk) = \varphi_{\psi,f}(znmk) = \psi\left(\frac{|\det m_1|^{r_2}}{|\det m_2|^{r_1}}\right) \cdot f_1(m_1) \cdot f_2(m_2)$$

and the corresponding pseudo-Eisenstein series is

$$\Psi_\varphi = \Psi_{\psi,f} = \sum_{\gamma \in (P\cap\Gamma)\backslash\Gamma} \varphi_{\psi,f} \circ \gamma$$

By [3.11.1], because of the cuspidal data, the only nonvanishing constant terms $c_Q \Psi_\varphi^P$ are for $Q = P$, or $Q = P^{r_2,r_1}$ when P is not self-associate. As in [3.16], these pseudo-Eisenstein series admit spectral decompositions in terms of the (genuine) Eisenstein series $E_{s,f} = E_{s,f}^P$ for P with the same cuspidal data f: with

$$\varphi_{s,f}(nmk) = \left|\frac{(\det m_1)^{r_2}}{(\det m_2)^{r_1}}\right|^s \cdot f_1(m_1) \cdot f_2(m_2)$$

with $m = \begin{pmatrix} m_1 & 0 \\ 0 & m_2 \end{pmatrix} \in M^P$, $n \in N$, $k \in K$, the corresponding Eisenstein series is

$$E_{s,f} = \sum_{\gamma \in (P \cap \Gamma) \backslash \Gamma} \varphi_{s,f} \circ \gamma$$

Again from [3.11.3], $c_Q E^P_{s,f} = 0$ unless $Q = P$ or Q is the *associate* of P. For f_1 and f_2 eigenfunctions of the respective Laplacians, by [3.11.11] the function $\varphi^P_{s,f}$ is an eigenfunction for the invariant Laplacian on G, and $E^P_{s,f}$ is an eigenfunction. In particular, letting μ_j be the eigenvalue of f_j for the Laplacian on GL_{r_j} for $j = 1, 2$, letting

$$\lambda_{s,f} = r_1 r_2 (r_1 + r_2)(s^2 - s) + \mu_1 + \mu_2$$

we have

$$\Delta \cdot E^P_{s,f} = \lambda_{s,f} \cdot E^P_{s,f}$$

With this normalization, the eigenvalue is invariant under $s \longrightarrow 1 - s$.

Let $\mathfrak{E}(P, f)$ be the space of pseudo-Eisenstein series for P formed with the given cuspidal data f. An analogue of [3.11.11] for pseudo-Eisenstein series with cuspidal data:

[11.7.1] Claim: $\mathfrak{E}(P, f)$ is stable under Δ. Explicitly, using coordinate $y > 0$ on the ray $(0, +\infty)$, $\Delta \Psi_{\psi, f} = \Psi_{\beta, f}$ with test function β given in terms of ψ by

$$\beta = \left(r_1 r_2 (r_1 + r_2) y \frac{\partial}{\partial y} \left(y \frac{\partial}{\partial y} - 1 \right) + \mu_1 + \mu_2 \right) \psi$$

Proof: This reduces to [3.11.11] via a primitive initial form of the spectral decomposition of $\Psi_{\psi, f}$ in terms of $E_{s,f}$ in the proof of [3.16.1], not requiring any meromorphic continuation of $E_{s,f}$:

$$\Psi_{\psi, f} = \frac{1}{2\pi i} \int_{\sigma - i\infty}^{\sigma + i\infty} \mathcal{M} \psi(s) \cdot E_{s,f} \, ds \qquad \text{(for } \sigma \gg 1\text{)}$$

Applying Δ multiplies $E_{s,f}$ by $r_1 r_2 (r_1 + r_2) s(s - 1) + \lambda_1 + \lambda_2$ by [3.11.11]. At the same time, $y \frac{\partial}{\partial y} y^s = s \cdot y^s$, so from Mellin inversion

$$s \cdot \mathcal{M} \psi = \mathcal{M} \left(y \frac{\partial}{\partial y} \psi \right)$$

This gives the assertion. ///

Let $\mathfrak{E}^0$ be the completion of $\mathfrak{E}(P, f)$ in $L^2(\Gamma \backslash G / K)$. Let T_f be the restriction of Δ to $\mathfrak{E}(P, f)$, and $\widetilde{S}_f$ its Friedrichs extension. Let $\mathfrak{E}^1$ be the completion of

$\mathcal{E}(P, f)$ in the $\mathfrak{B}^1$-norm given by

$$|f|^2_{\mathfrak{B}^1} = \langle (1 - \Delta)f, f \rangle_{L^2(\Gamma \backslash G / K)}$$

and $\mathcal{E}^2$ the completion of $\mathcal{E}(P, f)$ in the $\mathfrak{B}^2$ norm

$$|f|^2_{\mathfrak{B}^2} = \langle (1 - \Delta)^2 f, f \rangle_{L^2(\Gamma \backslash G / K)}$$

The domain of any self-adjoint extension of T_f necessarily contains $\mathcal{E}^2$, and the domain of $\widetilde{S}_f$ is contained in $\mathcal{E}^1$. More generally, for non-negative integer k, let $\mathcal{E}^k$ be the completion of $\mathcal{E}(P, f)$ in the $\mathfrak{B}^k$ norm

$$|f|^2_{\mathfrak{B}^k} = \langle (1 - \Delta)^k f, f \rangle_{L^2(\Gamma \backslash G / K)}$$

Let $\mathfrak{B}^\infty = \bigcap_k \mathfrak{B}^k = \lim_k \mathfrak{B}^k$.

[11.7.2] Corollary: For positive integer k, the $\mathcal{E}^k$-norm of $\Psi_{\psi, f}$ is the L^2 norm of $\Psi_{\beta, f}$, where

$$\beta = \left(1 - \left(r_1 r_2 (r_1 + r_2) y \frac{\partial}{\partial y} \left(y \frac{\partial}{\partial y} - 1 \right) + \mu_1 + \mu_2 \right) \right)^k \psi$$

(Immediate from [11.7.1].) ///

We grant the general form of the constant term along P (see [3.11.9]): this requires an assumption that in the cuspidal data $f = f_1 \otimes f_2$, both f_1 and f_2 are the unique cuspforms on GL_{r_i} with their respective eigenvalues (and right invariant under compacts, and left invariant under the respective groups $SL_{r_i}(\mathbb{Z})$, as opposed to other subgroups). Then, with meromorphic $c_{s,f}$ and $c^Q_{s,f}$, with $Q = P^{r_2, r_1}$,

$$\begin{cases} c_P E^P_{s,f} = \varphi^P_{s,f} & \text{(for } r_1 \neq r_2) \\ c_P E^P_{s,f} = \varphi^P_{s,f} + c_{s,f} \varphi^P_{1-s,f^w} & \text{(for } r_1 = r_2) \\ c_Q E^P_{s,f} = c^Q_{s,f} \cdot \varphi^Q_{1-s,f^w} & \text{(for } r_1 \neq r_2) \end{cases}$$

In fact, we do not use the precise nature of $c_{s,f}$.

Let

$$\alpha(m) = \left| \frac{(\det m_1)^{r_2}}{(\det m_2)^{r_1}} \right| \qquad \text{(for } m = \begin{pmatrix} m_1 & 0 \\ 0 & m_2 \end{pmatrix} \in M)$$

For $1 \ll_b a'' < a'$, define a real-valued smooth cutoff function by

$$\tau(y) = \begin{cases} 1 & \text{(for } y > a') \\ 0 & \text{(for } y < a'') \end{cases} \qquad \text{(for } y \in (0, +\infty))$$

Form a cuspidal-data pseudo-Eisenstein series $h_{s,f}$ by winding up a smoothly cutoff version of $\varphi_{s,f}$: with $\psi(y) = \tau(y) \cdot y^s$, put

$$h_{s,f} = \Psi_{\psi,f}$$

[11.7.3] Lemma: The sum for $h_{s,f}$ is absolutely convergent, uniformly on compacts, for all $s \in \mathbb{C}$. Further, $h_{s,f} \in \mathfrak{B}^\infty$.

Proof: By reduction theory [3.3], for $a \gg b$ large enough so that, for $\gamma \in \Gamma$, if $\gamma \mathfrak{S}_a \cap \mathfrak{S}_b \neq \phi$ then $\gamma \in B \cap \Gamma$ with minimal parabolic B. We increase a'' and a' in the definition of τ, if necessary, so that this property holds for them. Of course, $\mathfrak{S}_a$ is not $(M^P \cap \Gamma)$-stable, so no (strong-sense) cuspform f on M could be supported on any single copy of $\mathfrak{S}_a$. Thus, we need a type of Siegel set adapted to P: with B the standard minimal parabolic, let

$$\mathfrak{S}_a^P = \bigcup_{\gamma \in (\Gamma \cap P)/(\Gamma \cap B)} \gamma \mathfrak{S}_a$$

Then $\mathfrak{S}_a^P$ is $P \cap \Gamma$-stable, and for $\gamma \in \Gamma$, if $\gamma \mathfrak{S}_a^P \cap \mathfrak{S}_b^P \neq \phi$ then $\gamma \in P \cap \Gamma$: Indeed, suppose $\gamma \mathfrak{S}_a^P \cap \mathfrak{S}_b^P \neq \phi$. Then $\gamma \gamma_1 \mathfrak{S}_a \cap \gamma_2 \mathfrak{S}_b \neq \phi$ for some $\gamma_1, \gamma_2 \in \Gamma \cap P$. By the choice of $a \gg b$, this implies that $\gamma_2^{-1} \gamma \gamma_1 \in B \cap \Gamma$, or

$$\gamma \in \gamma_2 (B \cap \Gamma) \gamma_1^{-1} = \gamma_2 B \gamma_1^{-1} \cap \Gamma \subset P \cap \Gamma$$

Thus, for each $g \in G$ there is at most one nonzero summand in the sum defining $h_{s,f}$. The same is true of its image under $(1 - \Delta)^k$ for every k, so the sum converges in $\mathfrak{B}^k$ for every k.   ///

Thus, the pseudo-Eisenstein series $h_{s,f}$ is *entire* as a function-valued function of s. Let

$$\widetilde{E}_{s,f} = h_{s,f} - (\widetilde{S}_f - \lambda_{s,f})^{-1} (\Delta - \lambda_{s,f}) h_{s,f}$$

[11.7.4] Claim: $\widetilde{E}_{s,f} - h_{s,f}$ is a holomorphic $\mathfrak{C}^1$-valued function of s for $\mathrm{Re}(s) > \frac{1}{2}$ and $\mathrm{Im}(s) \neq 0$.

Proof: From Friedrichs' construction [9.2], the resolvent $(\widetilde{\Delta} - \lambda_{s,f})^{-1}$ *exists* as an everywhere-defined, continuous operator for $s \in \mathbb{C}$ for $\lambda_{s,f}$ not a nonpositive real number, because of the nonpositive-ness of Δ. Further, for $\lambda_{s,f}$ not a nonpositive real, this resolvent is a *holomorphic* operator-valued function. In fact, for such $\lambda_{s,f}$, the resolvent $(\widetilde{S}_f - \lambda_{s,f})^{-1}$ injects from $L^2(\Gamma \backslash G/K)$ to $\mathfrak{C}^1$.   ///

[11.7.5] Theorem: With $\lambda_{s,f}$ not nonpositive real, $u = \widetilde{E}_{s,f} - h_{s,f}$ is the unique element of the domain of $\widetilde{S}_f$ such that

$$(\widetilde{S}_f - \lambda_{s,f}) u = -(\Delta - \lambda_{s,f}) h_{s,f}$$

Thus, $\widetilde{E}_{s,f}$ is the usual Eisenstein series $E_{s,f}$ of [3.11] for $\mathrm{Re}\,(s) > 1$, and gives an analytic continuation of $E_{s,f} - h_{s,f}$ as $\mathfrak{C}^1$-valued function to $\mathrm{Re}\,(s) > \frac{1}{2}$ with $s \notin (\frac{1}{2}, 1]$.

Proof: The proof is very similar to that of [11.1.3]. *Uniqueness* follows from Friedrichs' construction [9.2] and construction of resolvents, because $\widetilde{S} - \lambda_s$ is a *bijection* of its domain to $L^2(\Gamma \backslash G / K)$.

On the other hand, for $\mathrm{Re}\,(s) > \frac{1}{2}$ and $s \notin (\frac{1}{2}, 1]$, $\widetilde{E}_{s,f} - h_{s,f}$ is in $L^2(\Gamma \backslash G / K)$, is *smooth*, and

$$\Delta(\widetilde{E}_{s,f} - h_{s,f}) = (\Delta - \lambda_{s,f})(\widetilde{E}_{s,f} - h_{s,f}) + \lambda_{s,f} \cdot (\widetilde{E}_{s,f} - h_{s,f})$$
$$= (\Delta - \lambda_{s,f})h_{s,f} + \lambda_{s,f} \cdot (\widetilde{E}_{s,f} - h_{s,f})$$
$$= (\text{element of } \mathfrak{B}^\infty) + \lambda_{s,f} \cdot (\widetilde{E}_{s,f} - h_{s,f})$$

so is in $\mathfrak{B}^2$, so certainly in the domain of $\widetilde{\Delta}$. Abbreviating $H_{s,f} = (\Delta - \lambda_s)h_{s,f}$, it is legitimate to compute

$$(\widetilde{S}_f - \lambda_{s,f})(\widetilde{E}_{s,f} - h_{s,f}) = (\widetilde{S}_f - \lambda_{s,f})\Big((h_{s,f} - (\widetilde{S}_f - \lambda_{s,f})^{-1}H_{s,f}) - h_{s,f}\Big)$$
$$= (\widetilde{S}_f - \lambda_{s,f})\Big(-(\widetilde{S}_f - \lambda_{s,f})^{-1}H_{s,f}\Big) = -H_{s,f}$$

Thus, $\widetilde{E}_{s,f} - h_{s,f}$ is a solution. Also, $E_{s,f} - h_{s,f}$ is a solution:

$$(\Delta - \lambda_{s,f})(E_{s,f} - h_{s,f}) = (\Delta - \lambda_{s,f})E_{s,f} - (\Delta - \lambda_{s,f})h_{s,f}$$
$$= 0 - (\Delta - \lambda_{s,f})h_{s,f}$$

By uniqueness, we are done.   ///

[11.7.6] Corollary: $E_{s,f}$ has an analytic continuation to $\mathrm{Re}\,(s) > \frac{1}{2}$ and $s \notin (\frac{1}{2}, 1]$ as an $h_{s,f} + \mathfrak{C}^1$-valued function.   ///

[11.7.7] Corollary: The function $s \to c_{s,f}$ has a meromorphic continuation to $\Re(s) > \frac{1}{2}$ (off $(\frac{1}{2}, 1]$).   ///

11.8 Distributional Characterization of Pseudo-Laplacians

First, we consider the self-associate maximal proper parabolic $P = P^{r,r} \subset G = SL_{2r}$, and cuspidal data of the symmetrical form $f = f_1 \otimes f_1$, so that $f^w = f$. Further, without loss of generality $\overline{f} = f$. Then the argument is nearly identical to that for the simple examples in [11.3]. However, since strong-sense cusp-forms f_1 are not likely to be compactly supported, the simple *local* argument for [11.3.2] requires some adaptation. What we do have, from [7.3.19] (and from [7.2.20] and [7.1.20] for simpler situations), is that strong-sense cusp-forms are smooth, of rapid decay, that there are test functions φ_j such that $\varphi_j \cdot f_j = f_j$, and that the f_j are eigenfunctions for the Laplacians on the factors of the Levi component. From the theory of the constant term [8.2.5], relations $\varphi_j \cdot f_j = f_j$ imply that all *derivatives* of such cuspforms (with respect to the universal enveloping algebra of the Lie algebra) are also of rapid decay.

Take symmetrical cuspidal data $f = f_1 \otimes f_1$ on $M = M^P$, with f_1 a cuspform in a strong sense, and with L^2 norm 1. Put

$$\mathfrak{E}(P, f) = \{\Psi^P_{\psi, f} : \psi \in C_c^\infty(0, +\infty)\}$$

We recall some context from [10.6]. Let B be the standard minimal parabolic, with unipotent radical N^B and standard Levi component M^B. Write Iwasawa decompositions $g = nm_g k$ with $n \in N^B$, $m \in M^B$. We let $\mathfrak{S}$ be a standard Siegel set stable under the (left) action of N^B:

$$\mathfrak{S} = \mathfrak{S}_b = \{g \in G : |\alpha_j(m_g)| \geq b, \text{ for all simple roots } \alpha_j\}$$

Take $0 < b \ll 1$ such that $\mathfrak{S}_b \to \Gamma \backslash G$ is a surjection. For $a > b$, let X_a be the subset of $\mathfrak{S}$ where $\beta(m_g) \leq a$ for all simple roots β. The quotient $(\Gamma \cap B) \backslash X_a$ is compact, since $(N^B \cap \Gamma) \backslash N$ is compact. For each simple root β, let

$$Y_a^\beta = \{g \in \mathfrak{S} : \beta(m_g) \geq a\}$$

and $Y_a = \bigcup_\beta Y_a^\beta$. Thus, $\mathfrak{S} = X_a \cup Y_a$. Parallel to [10.6], let

$$\mathfrak{E}(P, f)_a = \{F \in \mathfrak{E}(P, f) : c_{P'} F(g) = 0, \text{ for all } g \in Y_a, \text{ for all } P'\}$$

where P runs through standard parabolics, and

$$\mathfrak{E}_a^0 = \mathfrak{B}^0\text{-closure of } \mathfrak{E}(P, f)_a \qquad \mathfrak{E}_a^1 = \mathfrak{B}^1\text{-closure of } \mathfrak{E}(P, f)_a$$
$$\mathfrak{E}_a^2 = \mathfrak{B}^2\text{-closure of } \mathfrak{E}(P, f)_a$$

It suffices to require vanishing of constant terms for maximal proper parabolics P'. Further, from [3.11.1], since all pseudo-Eisenstein series in $\mathfrak{E}(P, f)$ have cuspidal data, the vanishing condition is automatically satisfied for all parabolics P' except P.

To be careful, since our unbounded operators should be densely defined, we note

[11.8.1] Lemma: For $a \gg 1$, $\mathfrak{E}(P, f)_a = \mathfrak{E}(P, f) \cap \mathfrak{E}_a^0$ is dense in $\mathfrak{E}_a^0$.

Proof: Having restricted our attention to the relatively small space $\mathfrak{E}(P, f)$, with $a \gg 1$, the observation [11.6.2] essentially reduces the issue to a generic, local, one-dimensional issue of smooth cutoffs, much as addressed in the proof of [10.3.1]. ///

Let $S_{a,f}$ be Δ restricted to $\mathfrak{E}(P, f)_a$. Since $\Delta\Psi^P_{\psi,f} = \Psi^P_{\beta,f}$ from [11.6.1], and differential operators do not enlarge supports, Δ does stabilize $\mathfrak{E}(P, f)_a$. Let $\widetilde{S}_{a,f}$ be the Friedrichs extension of $S_{a,f}$ to an unbounded self-adjoint operator on $\mathfrak{E}^0_a$, with domain contained in $\mathfrak{E}^1_a$ and containing $\mathfrak{E}^2_a$.

[11.8.2] Corollary: *(of [10.8])* $\widetilde{S}_{a,f}$ has compact resolvent $(\widetilde{S}_{a,f} - \lambda_{s,f})^{-1}$ (away from poles).

Proof: As usual, the crucial point is that the inclusion $\mathfrak{E}^1_a \to \mathfrak{E}^0_a$ is a restriction of the inclusion $\mathfrak{B}^1_a \to L^2_a$, the latter shown to be compact in [10.8]. The restrictions of compact operators are compact. The resolvents of the Friedrichs extension are continuous maps $\mathfrak{E}^0_a \to \mathfrak{E}^1_a$ composed with the inclusion $\mathfrak{E}^1_a \to \mathfrak{E}^0_a$. Continuous maps composed with compact maps are compact.   ///

Let M^1 be the copy of $SL_r \times SL_r$ inside $M = M^P$, and Z^M the center of M. We take representatives

$$z_a = \begin{pmatrix} a^{\frac{1}{r_1}} \cdot 1_r & 0 \\ 0 & 1_r \end{pmatrix} \qquad (\text{for } 0 < a \in \mathbb{R}^\times)$$

for the connected component $Z\backslash\mathbb{Z}^M$ containing 1_r, and let η_a be the functional on $\mathfrak{E}(P, f)$ defined by

$$\eta_a(F) = \int_{Z(\Gamma\cap M^P_1) \cdot \mathcal{M}_1} c_P F(m' \cdot z_a)\overline{f}(m')dm'$$

for $F \in \mathfrak{E}(P, f)$. Then $F \in \mathfrak{E}(P, f)_a$ if and only if $\eta_{b'}(F) = 0$ for all $b' \geq a$.

As in [3.8.2], pointwise vanishing conditions for constant terms can be rewritten as L^2 orthogonality to corresponding pseudo-Eisenstein series. With

$$\Theta = \{\Psi^P_{\psi,f} : \psi \in C^\infty_c(0, +\infty) \text{ with support inside } [a, +\infty)\}$$

$\mathfrak{E}(P, f)_a$ is the intersection of $\mathfrak{E}(P, f)$ with the orthogonal complement to Θ in $L^2(\Gamma\backslash G/K)$.

Let c be the pointwise conjugation map $c : \mathfrak{E}^0 \to \mathfrak{E}^0$. Let $\mathfrak{E}^{-1}$ be the Hilbert-space dual of $\mathfrak{E}^1$. Let j^* be the adjoint of the inclusion $j : \mathfrak{E}^1 \to \mathfrak{E}^0$, let j^* be its adjoint. Let $\mathfrak{E}^{-1}_a$ be the Hilbert-space dual of $\mathfrak{E}^1_a$, let $t : \mathfrak{E}^1_a \to \mathfrak{E}^1$ be the inclusion, with adjoint $t^* : \mathfrak{E}^{-1} \to \mathfrak{E}^{-1}_a$, giving a picture

$$\begin{array}{ccccc} \mathfrak{E}^1 & \xrightarrow{\ j\ } & \mathfrak{E}^0 & \xrightarrow{\ j^* \circ c\ } & \mathfrak{E}^{-1} \\ {\scriptstyle t}\uparrow & & & & \downarrow{\scriptstyle t^*} \\ \mathfrak{E}^1_a & & & & \mathfrak{E}^{-1}_a \end{array}$$

Let Δ_a be Δ restricted to $C_c^\infty(\Gamma \backslash G/K) \cap L_a^2$, with Friedrichs extension $\widetilde{\Delta}_a$. Let $S^\# : \mathfrak{E}_a^1 \to \mathfrak{E}_a^{-1}$ be as in [11.2], namely, $S^\#(x)(y) = \langle x, y \rangle_{\mathfrak{E}^1}$. Recall the recharacterization of Friedrichs extensions in [11.2]: $\widetilde{S}_{a,f}x = y$ for $x \in \mathfrak{E}_a^1$ and $y \in \mathfrak{E}_a^0$ if and only if $S^\# x = (t^* \circ (j^* \circ c))y$. Thus, we have

[11.8.3] Corollary: *(of [11.2.2])* $\widetilde{S}_{a,f}x = y$ for $x \in \mathfrak{E}_a^1$ and $y \in \mathfrak{E}_a^0$ if and only $(\Delta \circ t)x = y + \theta$ for some θ in the $\mathfrak{B}^{-1}$-closure of Θ.   ///

Thus, as in the simpler cases, the critical fact is

[11.8.4] Claim: For $a \gg_b 1$, the intersection of the $(\Delta \circ t)\mathfrak{E}_a^1$ and the $\mathfrak{E}^{-1}$-closure of Θ is at most $\mathbb{C} \cdot \eta_a$.

Proof: Use Siegel sets $\mathfrak{S}_{b'}^P$ adapted to P, as in the proof of [11.7.3]. Take $b' < a$ but still $b' \gg_b 1$ so that $\mathfrak{S}_{b'}^P$ has the same features as a. The compact abelian group $A = (N^P \cap \Gamma) \backslash N^P$ acts on $C_{a'} = (N^P \cap \Gamma) \backslash \mathfrak{S}_{a'}^P$, and $C_{a'}$ contains the image C_a of $\mathfrak{S}_a^P$ in the quotient.

On one hand, by the choice of $a \gg_b 1$, for a test function ψ supported in $[a, +\infty)$, on $\mathfrak{S}_b^P$ the pseudo-Eisenstein series $\Psi_{\psi,f} \in \Theta$ is just $\varphi_{\psi,f}$. These distributions are N^P-invariant on $\mathfrak{S}_a^P$. Taking $\mathfrak{E}^{-1}$ closure does not increase support and does not harm the N^P-invariance. Thus, the $\mathfrak{E}^{-1}$-closure of Θ consists of A-invariant distributions supported in $C_a \subset C_{b'}$.

On the other hand, A-invariants in $\mathfrak{E}_a^1$ are obtained as constant-term integrals, which are averaging integrals over the compact A, which exist as Gelfand-Pettis integrals with values at least as distributions. For each small $\varepsilon > 0$, there is a sequence $F_i \in \mathfrak{E}^1$ supported in $C_{a-\varepsilon}$ approaching $\theta \in \mathfrak{E}^{-1}$, with $F_i = \Delta u - f$ with $u \in \mathfrak{E}_a^1$ and $f \in \mathfrak{E}_a^0$, since $a - \varepsilon \gg 1$, in $C_{b'}$

$$F_i = c_P\theta = c_P(\Delta u - f) = \Delta(c_P u) - c_P f$$

and the intersection of $C_{a-\varepsilon}$ with the supports of $c_P u$ and $c_P f$ is contained in the complement $C_{a-\varepsilon} - C_a$. The differential operator Δ does not enlarge supports. Thus, the support of θ is contained in the boundary ∂C_a.

Thus, for each $\varepsilon > 0$, we can approximate in $\mathfrak{E}^{-1}$ such θ by a sequence $\Psi_{\psi_i,f}$ with $\psi_i \in C_c^\infty(0, +\infty)$ supported in $[a - \varepsilon, a + \varepsilon]$. For $\beta \in C_c^\infty(0, +\infty)$, without loss of generality suppose the support of β is similarly restricted. Since $a - \varepsilon \gg 1$, by reduction theory

$$\int_{\Gamma \backslash G} \Psi_{\psi_i,f} \cdot \Psi_{\beta,f} = \int_{C_{a-\varepsilon}} \varphi_{\psi_i,f} \cdot \varphi_{\beta,f}$$

and this integral simplifies to $\int_{a-\varepsilon}^{a+\varepsilon} \psi_i \cdot \beta$ with suitable measure on $(0, +\infty)$. Thus, such θ is specified by a distribution θ_o on $\mathbb{R}$ supported at the point a. By the classification of distributions supported at a point [13.14.3], θ_o must be a finite linear combination of Dirac delta δ_a and its derivatives. As in the local computations in the proof of [11.3.2], the condition $\theta \in \mathfrak{E}^{-1}$ requires that θ_o be at worst in $\mathfrak{B}^{-1}(\mathbb{R})$, which then requires that θ_o be a constant multiple of δ_a. Thus, the $\mathfrak{E}^{-1}$-limit is a constant multiple of η_a.   ///

Thus, the relatively simple characterization of the Friedrichs extension for $a \gg 1$:

[11.8.5] Corollary: With $a \gg 1$, $\widetilde{S}_{a,f} x = y$ for $x \in \mathfrak{E}_a^1$ and $y \in \mathfrak{E}_a^0$ if and only $(\Delta \circ t)x = y + c \cdot \eta_a$ for some constant c.   ///

11.9 Density Lemma for $P^{r,r} \subset SL_{2r}$

Similar to the description of $E_{s,f}$ as $\widetilde{E}_{s,f}$ above in [11.7], but with $\widetilde{S}_{a,f}$ in place of $\widetilde{S}_f$, with the pseudo-Eisenstein series $h_{s,f}$ formed from the smooth cutoff $\tau \cdot \varphi_{s,f}$ of $\varphi_{s,f}$ as in [11.7], put

$$\widetilde{E}_{a,s,f} = h_{s,f} - (\widetilde{S}_{a,f} - \lambda_{s,f})^{-1} (\Delta - \lambda_{s,f}) h_{s,f}$$

We already noted in [11.7] that $h_{s,f}$ is an entire, $\mathfrak{E}^1$-valued function, for simple reasons, given reduction theory.

For $\lambda_{s,f}$ *not* a nonpositive real, $(\widetilde{S}_{a,f} - \lambda_{s,f})^{-1}$ is a bijection of $\mathfrak{E}_a^0$ to the domain of $\widetilde{S}_{a,f}$, so $u = \widetilde{E}_{a,s,f} - h_{s,f}$ is the *unique* element of the domain of $\widetilde{S}_{a,f}$ satisfying

$$(\widetilde{S}_{a,f} - \lambda_{s,f}) u = -(\Delta - \lambda_{s,f}) h_{s,f}$$

Since $s \to h_{s,f}$ is entire, the meromorphy of the resolvent $(\widetilde{S}_{a,f} - \lambda_{s,f})^{-1}$ [10.9] yields the meromorphy of $\widetilde{E}_{a,s} - h_{s,f}$ as $\mathfrak{E}_a^1$-valued function, assuming that we have the following lemma, parallel to [11.4.1]:

[11.9.1] Lemma: For $a \gg 1$, $\mathfrak{E}_a^1 = \mathfrak{E}^1 \cap L_a^2(\Gamma \backslash G / K)$. That is, for $a \gg 1$, $\mathfrak{E}(P, f)_a$ is $\mathfrak{E}^1$-dense in $\mathfrak{E}^1 \cap L_a^2(\Gamma \backslash G / K)$.

Proof: The proof is also parallel to that of [11.4.1], with a few minor complications. Indeed, after suitable setup observations, the necessary estimates reduce to essentially one-dimensional estimates as in the proof of [11.4.1].

Since $\mathfrak{E}_a^1$ is the $\mathfrak{E}^1$-closure of $\mathfrak{E}(P, f)_a$, the containment is $\mathfrak{E}_a^1 \subset \mathfrak{E}^1 \cap L_a^2(\Gamma \backslash G / K)$ is immediate.

For the opposite containment, given a sequence $\{\Psi_{\varphi_i,f} \in \mathfrak{E}^1\}$ of pseudo-Eisenstein series converging to $\Psi \in \mathfrak{E}^1 \cap L_a^2(\Gamma \backslash G / K)$ in the $\mathfrak{E}^1$-topology, we produce a sequence of pseudo-Eisenstein series in $\mathfrak{E}(P, f)_a$ converging to Ψ in the $\mathfrak{E}^1$-topology, by smooth cutoffs of the constant terms of the $\Psi_{\varphi_i, f}$. Again, by [3.11], as noted in [10.6], all constant terms along parabolics other than P vanish entirely because of the cuspidal data f. As in [11.8], let

$$\alpha(nmk) = \left| \frac{\det m_1}{\det m_2} \right|^r$$

with $n \in N^P, k \in K, m = \begin{pmatrix} m_1 & 0 \\ 0 & m_2 \end{pmatrix}$. The constant term along P of the limit Ψ of the $\Psi_{\varphi_i, f}$ vanishes above $\alpha(g) = a$ by definition, and Ψ is in $L^2_a(\Gamma \backslash G / K)$. We will show that this entails that the part of the constant terms of the $\Psi_{\varphi_i, f}$ above $\alpha(g) = a$ must become small. Thus, smoothly cutting off the part of the constant terms above $\alpha(g) = a$ has a vanishingly small effect on the $\Psi_{\varphi_i, f}$. More precisely, proceed as follows.

Let F be a smooth real-valued function on $\mathbb{R}$ with $F(t) = 0$ for $t < -1, 0 \le F(t) \le 1$ for $-1 \le t \le 0$, and $F(t) = 1$ for $t \ge 0$. For $\varepsilon > 0$, let $F_\varepsilon(t) = F((t - a)/\varepsilon)$. Fix real b with $a > b > 1$. Given $\Psi_{\varphi_i, f} \to \Psi \in L^2_a(\Gamma \backslash G / K)$, the *b-tail* of the P-constant term of $\Psi_{\varphi_i, f}$ is $\tau_i(g) = c_P \Psi_{\varphi_i, f}(g)$ for $\alpha(g) \ge a'$, and $\tau_i(g) = 0$ for $0 < \alpha(g) \le a''$. By design, $\Psi_{\varphi_i, f} - \Psi_{F_\varepsilon \cdot \tau_i, f} \in \mathfrak{E}(P, f)_a$ for small ε. We will show that, as $i \to +\infty$, for ε_i sufficiently small depending on i, the $\mathfrak{E}^1$-norms of $\Psi_{F_{\varepsilon_i} \cdot \tau_i, f}$ go to 0, while still $\Psi_{\varphi_i, f} - \Psi_{F_{\varepsilon_i} \cdot \tau_i, f} \to f$ in the $\mathfrak{E}^1$-norm.

Let $\mathfrak{S}^P_a$ be Siegel sets adapted to P, as in [11.8], and put $C_a = (P \cap \Gamma) \backslash \mathfrak{S}^P_a$. As in [11.8], by reduction theory, for $a \gg 1$, C_a injects to $\Gamma \backslash G / K$. For $\Phi \in C^\infty_c(\Gamma \backslash G / K)$, let

$$|\Phi|^2_{\mathfrak{E}^1(C_a)} = \int_{C_a} |\Phi|^2 - \Delta \Phi \cdot \overline{\Phi}$$

We have

$$|\Phi|^2_{\mathfrak{E}^1(C_a)} \ll \int_{\Gamma \backslash G / K} |\Phi|^2 - \Delta \Phi \cdot \overline{\Phi}$$

For each $a > 0$, let $\mathfrak{E}^1(C_a)$ be the completion of $C^\infty_c(\Gamma \backslash G / K)$ with respect to the seminorm $|\cdot|_{\mathfrak{E}^1(C_b)}$ (with collapsing). In contrast to the four simpler examples, $\Gamma \cap M^P$ is not finite, so we need a slightly different argument for estimates on tails of constant terms than in [11.4.1].

By reduction theory, there is $0 < b \ll 1$ such that $\mathfrak{S}^P_b$ surjects to $\Gamma \backslash G / K$. Then take $a \gg_b 1$ such that, for $\eta \in C^\infty(0, +\infty)$ with support in $[a, +\infty)$, on $\mathfrak{S}^P_b$

$$\Psi_{\eta, f}(g) = \varphi_{\eta, f}(g) \qquad \text{(for } g \in \mathfrak{S}^P_b)$$

Since $\varphi_{\eta,f}$, for such η, in $\mathfrak{S}_b^P$ the P-constant term is $c_P\Psi_{\eta,f} = \varphi_{\eta,f}$. Thus, for $a \gg_b 1$,

$$\int_{C_a} |c_P\Psi_{\eta,f}|^2 = \int_{C_b} |\varphi_{\eta,f}|^2 = \int_{C_a} |\varphi_{\eta,f}|^2 = |\Psi_{\eta,f}|^2_{L^2(\Gamma\backslash G/K)}$$

That is, for such η, the P-constant term has L^2 norm dominated by (in fact, equal to) that of $\Psi_{\eta,f}$. By [11.7.1],

$$\Delta\Psi_{\eta,f} = \Psi_{T\eta,f} \qquad (\text{with } T = 2r^3\, y^2 \frac{\partial^2}{\partial y^2} + \mu_1 + \mu_2)$$

and similarly for $\varphi_{\eta,f}$. Thus, we have similar inequalities for the $\mathfrak{C}^1(C_a)$ norms for such η:

$$|c_P\Psi_{\eta,f}|^2_{\mathfrak{C}^1(C_a)} = |\varphi_{\eta,f}|^2_{\mathfrak{C}^1(C_a)} = |\Psi_{\eta,f}|^2_{\mathfrak{C}^1(C_a)} \le |\Psi_{\eta,f}|^2_{\mathfrak{C}^1}$$

That is, for fixed cuspidal data, for η supported in $[a, +\infty)$ with $a \gg_b 1$, the $\mathfrak{C}^1$ norm of the constant term of $\Psi_{\eta,f}$ is dominated by that of $\Psi_{\eta,f}$, as desired.

To an extent, we can dodge entirely rewriting the norms as energy norms, instead using the earlier computations [3.11.11] and [11.7.1] which give $\Delta\Psi_{\psi,f} = \Psi_{T\psi,f}$ with

$$T = 2r^3\, y\frac{\partial}{\partial y}\Big(y\frac{\partial}{\partial y} - 1\Big) + \mu_1 + \mu_2 = 2r^3\, y^2 \frac{\partial^2}{\partial y^2} + \mu_1 + \mu_2$$

to reduce to a one-dimensional computation more directly. As in the simplest case of $G = SL_2(\mathbb{R})$, the coefficient y^2 on $\frac{\partial^2}{\partial y^2}$ exactly cancels a denominator in an invariant measure in that coordinate, as we see in the following.

Turning to the main argument: since $a \gg_b 1$, $\Psi_{F_\varepsilon \cdot \tau_i, f}$ is just $\varphi_{F_\varepsilon \cdot \tau_i, f}$ on $\mathfrak{S}_b$, and the support of $\varphi_{F_\varepsilon \cdot \tau_i, f}$ is inside the image of the cylinder $C_{a-\varepsilon} \subset C_b$, by the triangle inequality,

$$|\Psi_{F_\varepsilon \cdot \tau_i, f}|_{\mathfrak{C}^1} \le |\Psi_{F_\varepsilon \cdot \tau_i, f}|_{\mathfrak{C}^1(C_b)} = |\varphi_{F_\varepsilon \cdot \tau_i, f}|_{\mathfrak{C}^1(C_{a-\varepsilon})}$$
$$\le |\varphi_{(F_\varepsilon - 1)\cdot\tau_i, f}|_{\mathfrak{C}^1(C_{a-\varepsilon})} + |\varphi_{\tau_i, f} - c_P\Psi|_{\mathfrak{C}^1(C_{a-\varepsilon})} + |c_P\Psi|_{\mathfrak{C}^1(C_{a-\varepsilon})}$$

From above, by design,

$$|\tau_i - c_P\Psi|_{\mathfrak{C}^1(C_{a-\varepsilon})} \ll |c_P\Psi_{\varphi_i, f} - c_P\Psi|_{\mathfrak{C}^1} \ll |\Psi_{\varphi_i, f} - \Psi|_{\mathfrak{C}^1} \longrightarrow 0$$

so the middle summand goes to 0. The first and third summands require somewhat more care.

Up to measure normalization constants, integrating away the cuspidal data,

$$|\varphi_{(F_\varepsilon - 1)\cdot\tau_i, f}|^2_{\mathfrak{C}^1(C_{a-\varepsilon})} = \int_{a-\varepsilon}^\infty |(F_\varepsilon - 1)\tau_i|^2 - T(F_\varepsilon - 1)\tau_i \cdot \overline{(F_\varepsilon - 1)\tau_i}\frac{dy}{y^2}$$

Since T is of the form $Ay^2 \frac{\partial^2}{\partial y^2} + B$ for real constants $A > 0$ and $B \le 0$, it is a symmetric operator with respect to the measure dy/y^2, and the previous expression is dominated by

$$\int_{a-\varepsilon}^{\infty} |(F_\varepsilon - 1)\tau_i|^2 \frac{dy}{y^2} + \int_{a-\varepsilon}^{\infty} \frac{\partial}{\partial y}(F_\varepsilon - 1)\tau_i \cdot \overline{\frac{\partial}{\partial y}(F_\varepsilon - 1)\tau_i} dy$$

$$\le \int_{a-\varepsilon}^{\infty} |F_\varepsilon - 1|^2 \cdot (|\tau_i|^2 + |y\tau_i'|^2)\frac{dy}{y^2} + \int_{C_{a-\varepsilon}} |F_\varepsilon'|^2 \cdot |\tau_i|^2 dy$$

$$+ \int_{C_{a-\varepsilon}} 2|F_\varepsilon| \cdot |F_\varepsilon'| \cdot |\tau_i| \cdot |\tau_i'| dy$$

by Leibniz's rule for derivatives. The first summand in the latter expression goes to 0 as $\varepsilon \to 0^+$ because $F_\varepsilon - 1 = 0$ when $y \ge a$, and τ_i and τ_i' are continuous. Thus,

$$|F_\varepsilon'(y)| = |\frac{1}{\varepsilon} \cdot F'((y-a)/\varepsilon)| = \frac{1}{\varepsilon} \cdot |F'((y-a)/\varepsilon)| \ll_F \frac{1}{\varepsilon}$$

The fundamental theorem of calculus and the Cauchy-Schwarz-Bunyakowsky inequality recover an easy instance of a Sobolev inequality:

$$|\tau_i(a-v)| = \left| 0 - \int_0^v \tau_i'(a-v)dv \right|$$

$$\le \left(\int_0^v |\tau_i'(a-v)|^2 dv \right)^{\frac{1}{2}} \cdot \left(\int_0^v 1^2 dv \right)^{\frac{1}{2}} \le o(1) \cdot \sqrt{v}$$

with Landau's little-o notation, since τ_i' is locally L^2. Thus,

$$\int_{C_{a-\varepsilon}} |F_\varepsilon| \cdot |F_\varepsilon'| \cdot |\tau_i| \cdot |\tau_i'| \le \frac{1}{\varepsilon} \cdot o(1) \cdot \sqrt{\varepsilon} \cdot \int_0^\varepsilon |\tau_i'|$$

$$\le \frac{1}{\varepsilon} \cdot o(1) \cdot \sqrt{\varepsilon} \cdot \left(\int_0^\varepsilon |\tau_i'|^2 \right)^{\frac{1}{2}} \cdot \left(\int_0^\varepsilon 1^2 \right)^{\frac{1}{2}} \ll_{\tau_i} \frac{1}{\varepsilon} \cdot o(1) \cdot \sqrt{\varepsilon} \cdot \sqrt{\varepsilon} = o(1)$$

That is, the summand $\int_{C_{a-\varepsilon}} |F_\varepsilon| \cdot |F_\varepsilon'|_s \cdot |\tau_i| \cdot |\tau_i'|_s$ goes to 0 as $\varepsilon \to 0^+$. By the same estimates,

$$\int_{C_{a-\varepsilon}} |F_\varepsilon'|_s^2 \cdot |\tau_i|^2 \ll \frac{1}{\varepsilon^2} \int_0^\varepsilon \left(o(1) \cdot \sqrt{v} \right)^2 dv = \frac{1}{\varepsilon^2} \cdot o(1) \cdot \frac{\varepsilon^2}{2} \longrightarrow 0$$

Thus, taking the ε_i sufficiently small, the smooth truncations $\Psi_{\varphi_i,f} - \Psi_{F_{\varepsilon_i},\tau_i,f}$ are in $D \cap L_a^2(\Gamma \backslash G/K)$ and still converge to Ψ in the $\mathfrak{E}^1$-topology.   ////

11.10 Beyond the Critical Line: $P^{r,r} \subset SL_{2r}$

Returning to the meromorphic continuation, we continue to consider symmetrical cuspidal data, of the symmetrical form $f = f_1 \otimes f_1$. The discussion continues to resemble that for the four simple cases [11.5].

Since $(\widetilde{S}_{a,f} - \lambda_{s,f})^{-1}$ maps $(\Delta - \lambda_{s,f})h_{s,f}$ to a function with P-constant term vanishing above height a, above that height the constant term of $\widetilde{E}_{a,s,f}$ is that of $h_{s,f}$. More generally, evaluate $\widetilde{S}_{a,f} - \lambda_{s,f}$ distributionally by application of $\Delta - \lambda_{s,f}$: for some constant C_s, by [11.8.5],

$$-(\Delta - \lambda_{s,f})h_{s,f} = (\widetilde{S}_{a,f} - \lambda_{s,f})(\widetilde{E}_{a,s,f} - h_{s,f})$$
$$= (\Delta - \lambda_{s,f})(\widetilde{E}_{a,s,f} - h_{s,f}) + C_s \cdot \eta_a \qquad \text{(as distributions)}$$

Everything else in the latter equation is meromorphic in s, so C_s must be, as well. Rearranging,

$$(\Delta - \lambda_{s,f})\widetilde{E}_{a,s,f} = -C_s \cdot \eta_a \qquad \text{(as distributions)}$$

Since Δ is G-invariant, it commutes with the constant-term map, so the distribution $(\Delta - \lambda_{s,f})c_P\widetilde{E}_{a,s,f}$ is 0 away from height a.

That constant term is of the form $\varphi_{\psi,f}$ for some function ψ on $(0, +\infty)$. By [11.7.1], in coordinates $m' \cdot z_y$, the distributional differential equation $(\Delta - \lambda_{s,f})\varphi_{\psi,f} = 0$ has solutions with ψ exactly of the form $A_s y^s f(m') + B_s y^{1-s} f(m')$ with constants A_s, B_s, so $c_P\widetilde{E}_{a,s}$ must be of this form in $0 < y < a$. Since $\widetilde{E}_{a,s,f}$ is meromorphic in s, so are A_s, B_s. Thus,

$$c_P\widetilde{E}_{a,s,f}(m'z_y) = \begin{cases} y^s \cdot f(m') & \text{(for } y > a) \\ A_s y^s f(m') + B_s y^{1-s} f(m') & \text{(for } 0 < y < a) \end{cases}$$

By construction, $h_{s,f}$ is smooth, and $(\widetilde{S}_{s,f} - \lambda_{s,f})^{-1}\Psi \in \mathfrak{E}_a^1$ for all $\Psi \in \mathfrak{E}_a^0$.

Next, we claim that the constant term of $\widetilde{E}_{a,s,f}$ is *continuous*, in particular along the set of values $nm'z_a$ with $m' \in M^1, n \in N^P$. To this end, for $F \in C_c^\infty((N^P \cap \Gamma)\backslash G/K)$, define a norm via

$$|F|^2_{\mathfrak{B}^1((N^P\cap\Gamma)\backslash G/K)} = \int_{(N^P\cap\Gamma)\backslash G/K} (1 - \Delta)F \cdot \overline{F}$$

and form the corresponding Hilbert space of functions by completion:

$$\mathfrak{B}^1((N^P \cap \Gamma)\backslash G/K) = \mathfrak{B}^1\text{-completion of} C_c^\infty((N^P \cap \Gamma)\backslash G/K)$$

Since the topological space $(N^P \cap \Gamma)\backslash G/K$ is a bit too large, we define a *localized* version of $\mathfrak{B}^1((N^P \cap \Gamma)\backslash G/K)$ via seminorms using smooth cutoffs β:

$$\nu_\beta(F) = |\beta \cdot F|_{\mathfrak{B}^1((N^P\cap\Gamma)\backslash G/K)}$$

for $\beta \in C_c^\infty((N^P \cap \Gamma)\backslash G/K)$. We can take larger-and-larger smooth cutoffs, so there is a countable cofinal subset of these seminorms so they give a locally convex invariant-metric topology T, with completion a Fréchet space:

$$\mathfrak{B}_{\mathrm{loc}}^1 = \mathfrak{B}_{\mathrm{loc}}^1((N^P \cap \Gamma)\backslash G/K)$$
$$= T\text{-completion of } C_c^\infty((N^P \cap \Gamma)\backslash G/K)$$

As usual, the compact group $(N^P \cap \Gamma)\backslash N$ acts continuously on $\mathfrak{B}_{\mathrm{loc}}^1$, for general reasons. The smoothly cutoff tail $h_{s,f}$ is an entire smooth-function-valued function, so is in $\mathfrak{B}_{\mathrm{loc}}^1$. Since $\widetilde{E}_{a,s,f} - h_{s,f}$ is in $\mathfrak{E}_a^1 \subset \mathfrak{B}^1((N^P \cap \Gamma)\backslash G/K) \subset \mathfrak{B}_{\mathrm{loc}}^1$, certainly $\widetilde{E}_{a,s,f} \in \mathfrak{B}_{\mathrm{loc}}^1$. Thus, the constant term $c_P \widetilde{E}_{a,s,f}$ exists as a Gelfand-Pettis integral, so is in $\mathfrak{B}_{\mathrm{loc}}^1$.

Continuity is a local property, so to prove that the constant term is continuous, it suffices to show that smooth cutoffs are continuous. For sufficiently small support of the smooth cutoffs, we can reduce the local problem to functions on a multitorus $\mathbb{T}^n$. The dimension $n = \dim_\mathbb{R}(N^P \cap \Gamma)\backslash G/K$ is too high for local Sobolev imbedding theorems to promise that every element of $\mathfrak{B}_{\mathrm{loc}}^1$ is continuous. Indeed, $\mathfrak{B}_{\mathrm{loc}}^k \subset C^o((N^P \cap \Gamma)\backslash G/K)$ for $k > \frac{n}{2}$ [9.5.14]. Fortunately, with strong-sense cuspidal data f, the constant term $c_P \widetilde{E}_{a,s,f}$ is in a better situation, namely, it is *smooth* (or even constant) in all but one coordinate. Indeed, it allows separation of variables, being of the form $\varphi_{\psi,f}(nm'z_y) = \psi(y) \cdot f(m')$. Thus, we smoothly truncate in a fashion that preserves the separation of variables, giving a function

$$F(x_1, \ldots, x_n) = F_1(x_1) \cdot F_2(x_2, \ldots, x_n)$$

on $\mathbb{T}^n$, with F_2 known to be C^∞ in $x' = x_2, \ldots, x_n$. We claim that $F \in \mathfrak{B}^1(\mathbb{T}^n)$ implies $F_1 \in \mathfrak{B}^1(\mathbb{T})$. Since F is a product, its Fourier coefficients likewise are products: letting $\xi' = (\xi_2, \ldots, \xi_n)$,

$$\widehat{F}(\xi_1, \xi_2, \ldots, \xi_n) = \int_{\mathbb{T}^n} F_1(x_1) F_2(x') \cdot e^{-2\pi i(x_1 \xi_1 + x' \cdot \xi')} dx_1 \, dx'$$
$$= \int_{\mathbb{T}} F_1(x_1) e^{-2\pi i x_1 \xi_1} dx_1 \cdot \int_{\mathbb{T}^{n-1}} F_2(x') \cdot e^{-2\pi i x' \cdot \xi'} dx'$$
$$= \widehat{F_1}(\xi_1) \cdot \widehat{F_2}(\xi')$$

The smoothness condition on F_2 gives *rapid decrease* of $\widehat{F_2}$, that is, for all $N \gg 1$

$$|\widehat{F_2}(\xi')| \ll_N |\xi'|^{-N}$$

Then the $\mathfrak{B}^1(\mathbb{T}^n)$ condition gives

$$\sum_{\xi_1} |\widehat{F_1}(\xi_1)|^2 \cdot (1 + |\xi_1|^2) \leq \sum_{\xi \in \mathbb{Z}^n} |\widehat{F}(\xi)|^2 \cdot (1 + |\xi|^2)$$

$$= \sum_{\xi_1, \xi'} |\widehat{F_1}(\xi_1)|^2 \cdot |\widehat{F_2}(\xi')|^2 \cdot (1 + |\xi|^2)$$

$$= \sum_{\xi_1} |\widehat{F_1}(\xi_1)|^2 \cdot \left(\sum_{\xi'} |\widehat{F_2}(\xi')|^2 \right) \cdot (1 + |\xi|^2)$$

$$\ll \sum_{\xi_1} |\widehat{F_1}(\xi_1)|^2 \cdot \left(\sum_{\xi'} |\xi'|^{-N} \right) \cdot (1 + |\xi|^2)$$

Taking $N > n - 1$ gives convergence of the inner sum, so F_1 is in the $+1$ Sobolev space on $\mathbb{T}$. Then Sobolev imbedding [9.5.4] implies that F_1 is *continuous*. Thus, $c_P \widetilde{E}_{a,s,f}$ is continuous at $\eta = a$, as claimed.

Thus, the values above and below $y = a$ must match:

$$A_s \cdot a^s \cdot f(m') + B_s \cdot a^{1-s} \cdot f(m') = a^s \cdot f(m') \qquad \text{(for all } s\text{)}$$

and since $f(m')$ is not identically 0,

$$A_s \cdot a^s + B_s \cdot a^{1-s} = a^s \qquad \text{(for all } s\text{)}$$

As in the proof of [11.5] (with somewhat different notation), let $\text{ch}_{[a,\infty)}$ be the characteristic function of $[a, \infty)$, and

$$\beta_{a,s,f}(nm'z_y) = \text{ch}_{[a,\infty)}(y) \cdot \left(A_s y^s + B_s y^{1-s} - y^s \right) \cdot f(m')$$

and form a pseudo-Eisenstein series

$$\Phi_{a,s,f} = \sum_{\gamma \in \Gamma_\infty \backslash \Gamma} \beta_{a,s,f} \circ \gamma$$

The support of $\beta_{a,s,f}$ is inside $\mathfrak{S}_a^P$, and $a \gg 1$, so by reduction theory for each $g \in G$ the series has at most one nonzero summand, so converges for all $s \in \mathbb{C}$.

[11.10.1] Theorem: $A_s \cdot E_{s,f} = \widetilde{E}_{a,s,f} + \Phi_{a,s,f}$ and $\widetilde{E}_{a,s,f} + \Phi_{a,s,f} = B_s \cdot E_{1-s,f}$. Thus, $E_{s,f}$ has a meromorphic continuation and $E_{s,f} - h_{s,f}$ is a meromorphic $\mathfrak{C}^1$-valued function.

Proof: With $\widetilde{S}$ as in [11.7], we have shown that $u = E_{s,f} - h_{s,f}$ is the unique solution $u \in \mathfrak{C}^1$ to

$$(\widetilde{S} - \lambda_{s,f}) u = -(\Delta - \lambda_{s,f}) h_{s,f}$$

Thus, multiplying through by A_s, it suffices to prove that $\widetilde{E}_{a,s,f} + \Phi_{a,s,f} - A_s \cdot h_{s,f}$ is in $\mathfrak{C}^1$ and satisfies

$$(\widetilde{S} - \lambda_{s,f})(\widetilde{E}_{a,s,f} + \Phi_{a,s,f} - A_s \cdot h_{s,f}) = -(\Delta - \lambda_{s,f})(A_s \cdot h_{s,f})$$

That $\widetilde{E}_{a,s,f} - h_{s,f}$ is in $\mathfrak{C}^1_a$ motivates the rearrangement

$$\widetilde{E}_{a,s,f} + \Phi_{a,s,f} - A_s \cdot h_{s,f} = (\widetilde{E}_{a,s,f} - h_{s,f}) + (\Phi_{a,s,f} - A_s h_{s,f} + h_{s,f})$$

We claim that the pseudo-Eisenstein series $F = \Phi_{a,s,f} - A_s h_{s,f} + h_{s,f}$ is in $\mathfrak{C}^1$.

Regarding integrability, by reduction theory, in $\mathfrak{S}^P_a$, $\Phi_{a,s,f}$ is just

$$\beta_{a,s,f}(nm'z_y) = \mathrm{ch}_{[a,\infty)}(y) \cdot (A_s y^s + B_s y^{1-s} - y^s) \cdot f(m')$$

so in $\mathfrak{S}^P_a$

$$F = \Phi_{a,s,f} - A_s h_{s,f} + h_{s,f} = \left((A_s y^s + B_s y^{1-s} - \eta^s) - A_s y^s + y^s\right) \cdot f(m')$$
$$= B_s y^{1-s} \cdot f(m') \qquad (\text{in } \mathfrak{S}^P_a)$$

For $\mathrm{Re}(s) > 1$, $y^{1-s} \cdot f(m')$ is square-integrable on $(P \cap \Gamma)\backslash \mathfrak{S}^P_a$, so F is in $L^2(\Gamma\backslash G/K)$.

To demonstrate the additional smoothness required for F to be in $\mathfrak{C}^1$, from the rewriting of Sobolev norms in [10.7], especially [10.7.5], it suffices to show that the right-translation derivatives xF are in $L^2(\Gamma\backslash G)$ for $x \in \mathfrak{g}$. By the left invariance of the right action of $\mathfrak{g}$, it suffices to prove square-integrability on the adapted standard Siegel sets $\mathfrak{S}^P_t$ of the derivatives of the data

$$\beta_{a,s,f} - A_s \tau(y) y^s \cdot f(m') + \tau(y) y^s \cdot f(m')$$

wound up to form $F = \Phi_{a,s,f} - A_s h_{s,f} + h_{s,f}$. This data is smooth everywhere but along $y = a$, where it is at least *continuous*, since $A_s a^s + B_s a^s - a^s = 0$. Further, it possesses continuous left and right derivatives in y at $y = a$, and is smooth in all other directions on $y = a$, so is *locally* in a $+1$-index Sobolev space near $y = a$.

Derivatives in the directions coming from the unipotent radical $\mathfrak{n}^P$ give 0. Since the cuspidal data f is strong-sense, f is smooth, and all derivatives are still of uniform rapid decay in Siegel sets, by [7.3.19] and [7.3.15]. Thus, for such derivatives, the L^2 estimate above is sufficient.

In coordinates $nm'z_y$ as earlier, the remaining differential operator is $y\frac{\partial}{\partial y}$. The derivative of F is discontinuous at $y = a$, although it has left and right limits.

As a distribution, it is

$$
y\frac{\partial}{\partial y}F = y\frac{\partial}{\partial y}\Big(\Phi_{s,f} - A_s h_{s,f} + h_{s,f}\Big)
$$

$$
= y\frac{\partial}{\partial y}\Big(\varphi_{s,f} - A_s \cdot \tau(y) \cdot y^s \cdot f(m') + \tau(y) \cdot y^s \cdot f(m')\Big)
$$

$$
= y\frac{\partial}{\partial y}
\begin{cases}
B_s y^{1-s} \cdot f(m') & (y > a) \\
-A_s \cdot y^s \cdot f(m') + y^s \cdot f(m') & (a' \le y < a) \\
-A_s \cdot \tau(y) \cdot y^s \cdot f(m') + \tau(y) \cdot y^s \cdot f(m') & (a'' \le y \le a') \\
0 & (y \le a'')
\end{cases}
$$

$$
=
\begin{cases}
B_s \cdot (1 - s) \cdot y^{1-s} \cdot f(m') & (y > a) \\
(1 - A_s) \cdot s y^s \cdot f(m') & (a' \le y < a) \\
(1 - A_s)(\frac{\partial \tau}{\partial y} \cdot y^s + \tau(y) \cdot s y^s) \cdot f(m') & (a'' \le y \le a') \\
0 & (y \le a'')
\end{cases}
$$

On $a' \le y \le a$, this derivative is bounded, so the truly relevant behavior is in $y > a$: for $\mathrm{Re}(s) > 1$ this derivative is square-integrable on quotients $(P \cap \Gamma)\backslash \mathfrak{S}_b^P$. Thus, $\Phi_{s,f} - A_s h_{s,f} + h_{s,f}$ is in $\mathfrak{E}^1$, proving that $\widetilde{E}_{a,s,f} + \Phi_{s,f} - A_s h_{s,f}$ is in $\mathfrak{E}^1$.

To show that $\widetilde{E}_{a,s,f} + \Phi_{s,f} - A_s h_{s,f}$ satisfies the expected equation, we justify computing the effect of differential operators on $\widetilde{E}_{a,s,f} + \Phi_{a,s,f} - A_s h_{s,f}$ distributionally, as follows. For $F \in C_c^\infty(\Gamma\backslash G/K)$, with $\widetilde{S}$ the Friedrichs extension of the restriction of Δ to $C_c^\infty(\Gamma\backslash G/K)$ as in [11.7],

$$
\Big\langle (\widetilde{S} - \lambda_{s,f})(\widetilde{E}_{a,s,f} + \Phi_{a,s,f} - A_s h_{s,f}),\ F \Big\rangle
$$

$$
= \Big\langle \widetilde{E}_{a,s,f} + \Phi_{a,s,f} - A_s h_{s,f},\ (\Delta - \overline{\lambda}_{s,f})F \Big\rangle
$$

$$
= \Big\langle (\Delta - \lambda_{s,f})(\widetilde{E}_{a,s,f} + \Phi_{a,s,f} - A_s h_{s,f}),\ F \Big\rangle
$$

By design, using the invariance of Δ and the local finiteness of the sum for $\Phi_{s,f}$, it is legitimate to compute

$$
(\Delta - \lambda_{s,f})(\widetilde{E}_{a,s,f} + \Phi_{a,s,f}) = (\Delta - \lambda_{s,f})\widetilde{E}_{a,s,f} + \sum_{\gamma \in \Gamma_\infty\backslash\Gamma} (\Delta - \lambda_{s,f})\beta_{a,s,f} \circ \gamma
$$

$$
= -C_s \cdot \eta_a + C_s \cdot \eta_a = 0 \qquad \text{(as distributions)}
$$

Thus,

$$
(\widetilde{S} - \lambda_{s,f})(\widetilde{E}_{a,s,f} + \Phi_{a,s,f} - A_s h_{s,f})
$$

$$
= (\Delta - \lambda_{s,f})(\widetilde{E}_{a,s,f} + \Phi_{a,s,f} - A_s h_{s,f}) = 0 - A_s(\Delta - \lambda_{s,f})h_{s,f}
$$

as desired, proving $\widetilde{E}_{a,s,f} + \Phi_{a,s,f} = A_s \cdot E_{s,f}$ for $\mathrm{Re}(s) > 1$. For $\mathrm{Re}(1 - s) > 1$, the same argument shows that $\widetilde{E}_{a,s,f} + \Phi_{a,s,f} = B_s \cdot E_{1-s,f}$. This proves the formulas in the claim. Since not both A_s and B_s can be identically 0, we obtain the meromorphic continuation of $E_{s,f}$. ///

[11.10.2] Corollary: $A_s \cdot E_{s,f} = B_s \cdot E_{1-s,f}$.   ///

In particular, *neither* A_s nor B_s is identically 0, and with $a(s) = B_s/A_s$, $E_{1-s,f} = a(s) \cdot E_{s,f}$. The relation $c_P E_{s,f} = (y^s + c_{s,f} y^{1-s}) \cdot f(m')$ gives the meromorphic continuation of $c_{s,f}$. Since $c_P E_{1-s,f} = (y^{1-s} + c_{1-s,f} y^s) \cdot f(m')$, apparently $c_{s,f} = a(s) = B_s/A_s$. Since $1 - (1 - s) = s$, we obtain $c_{s,f} \cdot c_{1-s,f} = 1$:

[11.10.3] Corollary: $c_{s,f}$ has a meromorphic continuation, and $c_{s,f} \cdot c_{1-s,f} = 1$.   ///

On $\mathrm{Im}(s) = 0$ and $\mathrm{Re}(s) > 1$, $E_{s,f}$ and $c_P E_{s,f}$ are real-valued. We assume without loss of generality that f is real-valued. Thus, the two holomorphic functions $E_{s,f}$ and $\overline{E_{\bar{s},f}}$ agree on $(1, +\infty)$, so agree everywhere. That is, $\overline{E_{s,f}} = E_{\bar{s},f}$. In particular, on $\mathrm{Re}(s) = \frac{1}{2}$, where $\bar{s} = 1 - s$,

$$|c_{s,f}|^2 = c_{s,f} \cdot \overline{c_{s,f}} = c_{s,f} \cdot c_{\bar{s},f} = c_{s,f} \cdot c_{1-s,f} = 1$$

proving

[11.10.4] Corollary: $|c_{s,f}| = 1$ on $\mathrm{Re}(s) = \frac{1}{2}$, and $c_{s,f}$ has no pole on $\mathrm{Re}(s) = \frac{1}{2}$.   ///

Further, we have

[11.10.5] Corollary: $E_{s,f}$ has no pole on $\mathrm{Re}(s) = \frac{1}{2}$.

Proof: Suppose $E_{s,f}$ had a pole of order $N > 0$ at s_o on the critical line $\mathrm{Re}(s) = \frac{1}{2}$. Then $(s - s_o)^N \cdot E_{s,f}$ is holomorphic at $s = s_o$, gives a not identically automorphic form, and has vanishing constant term there. From

$$\wedge^a (s - s_o)^N E_{s,f} = (s - s_o)^N \wedge^a E_{s,f}$$

and using the Maass-Selberg relations [3.14.2] with $s = s_o + \varepsilon$ and $r = \bar{s}_o + \varepsilon = 1 - s_o + \varepsilon$ with $0 < \varepsilon \in \mathbb{R}$, since $(s - s_o) \cdot c_s \to 0$ at $s = s_o$, suppressing measure-normalizations,

$$
\begin{aligned}
|(s - s_o)^N E_s|^2 &= \varepsilon^{2N} \cdot \Big(\frac{a^{s+r-1}}{s + r - 1} + c_{s,f} \frac{a^{(1-s)+r-1}}{(1 - s) + r - 1} \\
&\quad + c_{r,f} \frac{a^{s+(1-r)-1}}{s + (1 - r) - 1} + c_{s,f} c_{r,f} \frac{a^{(1-s)+(1-r)-1}}{(1 - s) + (1 - r) - 1} \Big) \\
&= \varepsilon^{2N} \cdot \Big(\frac{a^{2\varepsilon}}{2\varepsilon} + c_{s_o+\varepsilon,f} \frac{a^{1-2s_o-2\varepsilon}}{1 - 2s_o - 2\varepsilon} \\
&\quad + c_{1-s_o+\varepsilon,f} \frac{a^{2s_o-1+2\varepsilon}}{2s_o - 1 + 2\varepsilon} + c_{s_o+\varepsilon,f} c_{1-s_o+\varepsilon,f} \frac{a^{-2\varepsilon}}{-2\varepsilon} \Big) \longrightarrow 0
\end{aligned}
$$

contradiction. Thus, $E_{s,f}$ has no pole on the critical line.   ///

Toward proving *moderate growth* of the meromorphic continuation of $E_{s,f}$:

[11.10.6] Claim: $E_{s,f}$ meromorphically continues as a $C^\infty(\Gamma\backslash G/K)$-valued function.

Proof: As we have assumed throughout, to know the form of the constant term of $E_{s,f}$ with $f = f_1 \otimes f_1$, as in [3.11.9] we need to assume that f_1 is a Δ-eigenfunction on $SL_r(\mathbb{Z})\backslash SL_r(\mathbb{R})/SO(n, \mathbb{R})$, with eigenvalue λ_1, and up to scalar multiples is the only cuspform there with Δ-eigenvalue λ_1. From the computation in [3.11.11], it follows that $E_{s,f}$ is a Δ-eigenfunction, with eigenvalue $\lambda_{s,f} = 2r^3 s(s-1) + 2\lambda_1$. Thus, in the region $\mathrm{Re}(s) > 1$, there is at most a single s making $\lambda_{s,f}$ assume a given value. As above, let

$$\varphi_{s,f}(nm'z_yk) = y^s \cdot f(m') \qquad (\text{for } n \in N^P, \ k \in K, \text{ and } m \in M')$$

The computation in the proof of [3.11.11] also shows that $\varphi_{s,f}$ is a Δ-eigenfunction with eigenvalue $\lambda_{s,f}$. Let $\eta \in C_c^\infty(K\backslash G/K)$ act on spaces of right K-invariant functions on G as usual, by integral operators. From [8.4.1], for every $\eta \in C_c^\infty(K\backslash G/K)$, there is $\mu_{s,f}(\eta) \in \mathbb{C}$ such that $\eta \cdot \varphi_{s,f} = \mu_{s,f}(\eta) \cdot \varphi_{s,f}$, and there exists η giving $\mu_{s,f}(\eta) \neq 0$. In the region of convergence $\mathrm{Re}(s) > 1$, from $E_{s,f} = \sum_{\gamma \in (\Gamma\cap P)\backslash\Gamma} \varphi_{s,f} \circ \gamma$, also $\eta \cdot E_{s,f} = \mu_{s,f}(\eta) \cdot E_{s,f}$. Exactly what we are missing at this point is knowledge of in what topological vector space of functions (or distributions) the meromorphically continued Eisenstein series may lie, so we cannot directly assert much about $\eta \cdot E_{s,f}$ outside the region of convergence. (Otherwise we could apply the identity principle from complex analysis to the latter identity.) Rather, we approach this a little indirectly, as follows.

Since Δ commutes with G, $\mathfrak{E}^1$ is stable under the action of $\eta \in C_c^\infty(K\backslash G/K)$. From the meromorphic continuation of $E_{s,f} - h_{s,f}$ as $\mathfrak{E}^1$-valued function, we have the meromorphic continuation of

$$\eta \cdot (E_{s,f} - h_{s,f}) = \mu_{s,f}(\eta) \cdot E_{s,f} - \eta \cdot h_{s,f}$$

as $\mathfrak{E}^1$-valued function. In fact, for $F \in \mathfrak{E}^1$, by [14.5], $\eta \cdot F$ is in $C^\infty(\Gamma\backslash G/K)$. By construction, $h_{s,f} \in C^\infty(\Gamma\backslash G/K)$. Rearranging,

$$\mu_{s,f}(\eta) \cdot E_{s,f} = \eta \cdot (E_{s,f} - h_{s,f}) + \eta \cdot h_{s,f}$$

Dividing through by $\mu_{s,f}(\eta)$ for some η with $\mu_{s,f}(\eta) \neq 0$ exhibits the meromorphically continued $E_{s,f}$ as a smooth-function-valued function.   ///

[11.10.7] Corollary: E_s has a meromorphic continuation as $C^\infty(\Gamma\backslash G/K)$-valued function, so it makes sense to address the issue of its moderate growth.   ///

Finally, we have

[11.10.8] Theorem: Away from poles, the meromorphically continued $E_{s,f}$ is of moderate growth.

Proof: By [11.10.1] and [11.10.7], (at least) the *pointwise* values of the meromorphic continuation are given by

$$E_{s,f} = A_s^{-1} \cdot (\widetilde{E}_{a,s,f} + \Phi_{a,s,f})$$

Since $a \gg 1$, in $\mathfrak{S}_a^P$ the function $\Phi_{a,s,f}$ is just $\varphi_{a,s,f}$ itself, which is $(A_s y^s + B_s y^{1-s} - y^s) \cdot f(m')$, which is of moderate growth in standard Siegel sets. The preceding computation shows continuity at $y = a$. The pseudo-Eisenstein series $h_{s,f}$ of [11.7] made from $\tau \cdot y^s \cdot f(m')$ with smooth cutoff τ is a locally finite sum, so is smooth, so certainly continuous. For $\eta \geq a$, its value is just η^s, which is of moderate growth for all s. Thus, to show that $\widetilde{E}_{a,s,f}$ is of moderate growth even after meromorphic continuation, it suffices to show that $(\widetilde{S}_{a,f} - \lambda_{s,f})^{-1}(\Delta - \lambda_{s,f})h_{s,f}$ is of moderate growth.

Let T be the operator determined in [11.7.1], such that $\Delta \Psi_{\psi,f} = \Psi_{T\psi,f}$, namely, with μ_1 the eigenvalue of the cuspform f_1,

$$T = 2r^3 y \frac{\partial}{\partial y}\left(y\frac{\partial}{\partial y} - 1\right) + 2\mu_1$$

Thus,

$$T - \lambda_{s,f} = 2r^3 y \frac{\partial}{\partial y}\left(y\frac{\partial}{\partial y} - 1\right) + 2\mu_1 - 2r^3 s(s-1) - 2\mu_1$$

$$= 2r^3 \cdot \left(y\frac{\partial}{\partial y}\left(y\frac{\partial}{\partial y} - 1\right) - s(s-1)\right)$$

The constant $2r^3$ can be dropped without changing anything.

Again, the pseudo-Eisenstein series $h_{s,f} = \Psi_{\tau \cdot y^s, f}$ is a locally finite sum, so it is legitimate to compute

$$(\Delta - \lambda_{s,f})h_{s,f} = (\Delta - \lambda_{s,f})\Psi_{\tau \cdot y^s, f} = \Psi_{(T - \lambda_{s,f})(\tau \cdot y^s), f}$$

and $H_s = (T - \lambda_{s,f})(\tau \cdot y^s)$ is smooth and uniformly compactly supported. It suffices to demonstrate solvability of the differential equation $(T - \lambda_{s,f})u = H_s$ for a function u of sufficient decay at both 0^+ and $+\infty$. Then hope to form a pseudo-Eisenstein series $\Psi_{u,f}$ giving $(\widetilde{S}_{a,f} - \lambda_{s,f})\Psi_{u,f} = \Psi_{H_s,f}$. From the distributional characterization [11.8] of $\widetilde{S}_{a,f}$, this equation is equivalent to

$$(\widetilde{S}_{a,f} - \lambda_{s,f})\Psi_{u,f} = \Psi_{H_s,f} + c \cdot \eta_a \qquad \text{(for some } c \in \mathbb{C}\text{)}$$

Thus, in the y-coordinate, given $f \in C_c^\infty(0, +\infty)$, we solve equations $(T - \lambda_{s,f})u = H_s + c \cdot \delta_a$ with $c \in \mathbb{C}$ for u in $C^\infty(0, +\infty)$, with behavior at 0^+ and $+\infty$ to be adjusted suitably.

From [11.7.1], the differential equation is

$$(y\frac{\partial}{\partial y}(y\frac{\partial}{\partial y} - 1) - s(s - 1))u = H_s + c \cdot \delta_a$$

We can divide through by A to suppose without loss of generality that it is 1, and still the renormalized is $B < 0$. Letting $x = \log y$, with $F(x) = H_s(e^x)$ and $v(x) = u(e^x)$, this becomes

$$v'' - v' - s(s - 1)v = F + c \cdot \delta_{\log a}$$

Taking Fourier transform in a normalization that suppresses some factors of 2π,

$$(-i\xi)^2\widehat{v} - (-i\xi)\widehat{v} - s(s - 1)\widehat{v} = \widehat{F} + c \cdot a^{-i\xi}$$

or

$$\widehat{v}(\xi) = -\frac{\widehat{F}(\xi) + c \cdot a^{-i\xi}}{\xi^2 - i\xi + s(s - 1)}$$

Since F is a test function, $\widehat{F}$ is an *entire* function such that $\widehat{F}(x + iy_o)$ is (uniformly) in the Schwartz space for each fixed y_o. Division by a quadratic polynomial produces a function holomorphic in a *strip* along $\mathbb{R}$ not including either of the two poles at the zeros of the denominator:

$$\frac{i \pm \sqrt{(-i)^2 - 4s(s - 1)}}{2}$$

Fix $\varepsilon > 0$. Given a bound $|\text{Re}(s)| \leq B$, for $\text{Im}(s) \gg_B 1$, those poles are outside the strip $S = \{z \in \mathbb{C} : |\text{Im}(z)| \leq 1 + \varepsilon\}$. Thus, $\widehat{v}$ is holomorphic on an open set containing S and has decay like $1/\xi^2$ on horizontal lines inside that strip. Thus, in the Fourier inversion integral

$$v(x) = \frac{1}{2\pi} \int_{\mathbb{R}} e^{i\xi x}\widehat{v}(\xi)d\xi$$

we can move the contour up to $\mathbb{R} + i(1 + \varepsilon)$, giving

$$v(x) = \frac{1}{2\pi} \int_{\mathbb{R}} e^{i(\xi + i\varepsilon)x}\widehat{v}(\xi + i\varepsilon)\,d\xi = e^{-\varepsilon x}\frac{1}{2\pi} \int_{\mathbb{R}} e^{i\xi x}\widehat{v}(\xi + i\varepsilon)d\xi$$

Thus, $v(x) \ll e^{-(1+\varepsilon)x}$, giving genuine exponential decrease for $x \to +\infty$. Similarly, moving the contour down gives exponential decrease $v(x) \ll e^{-(1+\varepsilon)|x|}$ for $x \to -\infty$. Then $u(y) = v(\log y)$ satisfies $u(y) \ll y^{1+\varepsilon}$ as $y \to 0^+$, and $u(y) \ll y^{-(1+\varepsilon)}$ as $y \to +\infty$. Thus, the pseudo-Eisenstein series $\Psi_{u,f}$ converges absolutely, being dominated termwise by the sum expressing an absolutely convergent Eisenstein series [3.9] and [3.11]. Further, being termwise dominated by an absolutely convergent Eisenstein series, $\Psi_{u,f}$ is continuous and of moderate growth.

Having available a choice of the constant c is necessary, to adjust $\Psi_{u,f}$ to have P-constant term vanishing above height $y = a$. Choose the constant so that $c_P \Psi_u$ vanishes at $y = a$. Since $a \gg 1$, by reduction theory the truncation $\wedge^a \Psi_{u,f}$ has P-constant term vanishing at and above height a. Since $a \gg 1$, this truncation is itself a pseudo-Eisenstein series, and still $(\Delta - \lambda_{s,f}) \wedge^a \Psi_{u,f}$ differs from $\Psi_{u,f}$ only by a multiple of η_a. Again by the distributional characterization of $\widetilde{S}_{a,f}$, we have $(\widetilde{S}_{a,f} - \lambda_{s,f}) \wedge^a \Psi_{u,f} = \Psi_{H_s,f}$.

Thus, for a given bound $|\mathrm{Re}(s)| \leq B$, there is C sufficiently large so that for $|\mathrm{Im}(s)| \geq C$ we have meromorphic continuation of $E_{s,f}$ as a (continuous) moderate-growth function.

For $|\mathrm{Im}(s)| < C$, we can express $E_{s,f}$ as a vector-valued Cauchy integral along a circular path γ that lies inside the union U of regions $\mathrm{Re}(s) \geq B$, $\mathrm{Re}(s) \leq 1 - B$, and $|\mathrm{Im}(s)| \geq C$, and does not run through any poles of $E_{s,f}$. In $\mathrm{Re}(s) \leq 1 - B$ the Eisenstein series is (continuous) of moderate growth, via the functional equation. Thus, $E_{s,f}$ is of moderate growth throughout U, and in particular along γ. Let Z be the collection of poles of $E_{s,f}$ (as meromorphic $C^o(\Gamma \backslash G / K)$-valued function) inside γ, and $P(z) = \prod_{z_j \in Z}(z - z_j)$. For each $g \in G$

$$P(s) \cdot E_{s,f}(g) = \frac{1}{2\pi i} \int_\gamma \frac{P(z) \cdot E_z(g)}{z - s} dz$$

In fact, on γ, $z \to (s \to P(z)E_{s,f}/(z - s)$ is a compactly supported, continuous, moderate-growth-function-valued function of z, so the vector-valued Cauchy integral

$$P(s) \cdot E_{s,f} = \frac{1}{2\pi i} \int_\gamma \frac{P(z) \cdot E_{z,f}}{z - s} dz$$

as in [15.2] exists as a Gelfand-Pettis integral [14.1] lying in that same space of functions.   ///

11.11 Exotic Eigenfunctions: $P^{r,r} \subset SL_{2r}$

Since $\mathfrak{E}(P, f)$ contained no eigenfunctions for Δ except the finitely many possible residues of $E^P_{s,f}$ in $\mathrm{Re}(s) > \frac{1}{2}$ (see [3.14]), the eigenfunctions for $\widetilde{S}_{a,f}$ cannot be eigenfunctions for Δ, so must be *exotic*.

Continue to consider *symmetrical* cuspidal data $f = f_1 \otimes f_1$, so that $f^w = f$. On genuine Eisenstein series $E_{s,f}$ the functional η_a makes sense: unwinding, and using the explicit form of the constant terms, we have absolutely convergent integrals

$$
\eta_a(E_{s,f}) = \int_{Z(\Gamma \cap M^1)\backslash \mathcal{M}^1} c_P E_{s,f}(m' \cdot z_a)\overline{f}(m')dm'
$$

$$
= \int_{Z(\Gamma \cap M^1)\backslash \mathcal{M}^1} \Big(a^s \cdot f(m') + c_{s,f} a^{1-s} \cdot f^w(m') \Big) \cdot \overline{f}(m')dm'
$$

$$
= a^s + c_{s,f} a^{1-s}
$$

[11.11.1] Claim: Take $a \gg 1$. For values of s such that $a^s + c_{s,f} a^{1-s} = 0$ the truncation $\wedge^a E_{s,f}$ is an eigenfunction for $\widetilde{S}_{a,f}$, and $(\Delta - \lambda_{s,f}) \wedge^a E_{s,f}$ is a constant multiple of η_a.

Proof: Let H be the usual Heaviside function on $\mathbb{R}$: identically 0 on $(-\infty, 0)$ and identically 1 on $(0, +\infty)$. The truncation [3.14] (along P) of $E^P_{s,f}$ is a pseudo-Eisenstein series:

$$
\wedge^a E^P_{s,f} = \Psi_{H(a-y)\beta, f}
$$

with $\beta = c_P E^P_{s,f}(z_y)$, with $z_y = \begin{pmatrix} y^{\frac{1}{r}} \cdot 1_r & 0 \\ 0 & 1_r \end{pmatrix}$. The identity [11.5.1] shows the effect of applying $\Delta - \lambda_{s,f}$ to pseudo-Eisenstein series: at first for test functions ψ,

$$
(\Delta - \lambda_{s,f})\Psi^P_{\psi, f} = \Psi_{D_s\psi, f}
$$

where D_s is the differential operator

$$
D_s = \Big(2r^3 y \frac{\partial}{\partial y} \Big(y\frac{\partial}{\partial y} - 1 \Big) + \mu_1 + \mu_2 \Big) - \Big(2r^3 s(s-1) + \mu_1 + \mu_2 \Big)
$$

$$
= 2r^3 \Big(\frac{\partial}{\partial y} \Big(y\frac{\partial}{\partial y} - 1 \Big) - s(s-1) \Big)
$$

Extend the identity by continuity. Then

$$
(\Delta - \lambda_{s,f}) \wedge^a E^P_{s,f} = (\Delta - \lambda_{s,f})\Psi_{H(a-y)\beta, f} = \Psi_{D_s H(a-y)\beta, f}
$$

Thus, this computation reduces to an elementary computation on $(0, +\infty)$:

$$
D_s\big(H(a-y)\beta\big)
$$

$$
= 2r^3 \Big(\frac{\partial}{\partial y} \Big(y\frac{\partial}{\partial y} - 1 \Big) - s(s-1) \Big)\Big(H(a-y) \cdot (y^s + c_{s,f} y^{1-s}) \Big)
$$

Apart from the leading coefficient $2r^3$, this is the same expression appearing in the proofs in [11.6] for $SL_2(\mathbb{Z})$ and for the other three simple cases. That is, a *derivative* of a Dirac δ appears unless $y^s + c_{s,f} y^{1-s} = 0$, in which case the truncated Eisenstein series is indeed in $\mathfrak{C}^1$, and is in the domain of $\widetilde{S}$.   ///

[11.11.2] Corollary: Let f_1 have eigenvalue μ_1 for the Laplacian on $SL_r(\mathbb{Z})\backslash SL_r(\mathbb{R})/SO_r(\mathbb{R})$, rewritten as $\mu_1 = \frac{1}{4} - \tau^2$ with $\tau \geq 0$. For $a \gg 1$, if $a^s + c_{s,f}a^{1-s} = 0$, then either $\operatorname{Re}(s) = \frac{1}{2}$ or $s \in [\frac{1}{2} - \tau, \frac{1}{2} + \tau]$.

Proof: If $a^s + c_{s,f}a^{1-s} = 0$, then the corresponding truncated Eisenstein series is an eigenfunction for the the nonpositive self-adjoint differential operator $\widetilde{S}_{a,f}$, so the corresponding eigenvalue computed in [3.11.11] must be real and nonpositive:

$$\lambda_{s,f} = 2r^3(s^2 - s) + \mu_1 + \mu_1 \leq 0$$

Thus, $s(s-1) + \frac{1}{4} - \tau^2 \leq 0$. ///

11.12 Non-Self-Associate Cases

The general argument for cuspidal-data Eisenstein series for maximal proper parabolics in SL_r has the same shape as in the four simple cases, and as for $P^{r,r} \subset SL_{2r}$ with cuspidal data $f = f_1 \otimes f_1$, treated in [11.8], [11.9], and [11.10]. For self-associate maximal proper parabolic $P = P^{r,r} \subset G = SL_{2r}$, the case opposite to that already considered is $f_2 \perp f_1$, with strong-sense cuspforms f_1, f_2. We recapitulate the argument in outline for $r_1 \neq r_2$, highlighting only the complications and differences from the previous examples.

For non-self-associate $P = P^{r_1, r_2}$ with $r_1 \neq r_2$, let $Q = P^{r_2, r_1}$, and only consider cuspidal data $f = f_1 \otimes f_2$ with strong-sense cuspforms f_1, f_2, and put

$$\mathfrak{E}(P, Q, f) = \{\Psi^P_{\psi, f} : \psi \in C^\infty_c(0, +\infty)\} + \{\Psi^Q_{\psi, f^w} : \psi \in C^\infty_c(0, +\infty)\}$$

Again, recall context from [10.6]. Let B be the standard minimal parabolic, with unipotent radical N^B and standard Levi component M^B. Write Iwasawa decompositions $g = nm_gk$ with $n \in N^B, m \in M^B$. We let $\mathfrak{S}$ be a standard Siegel set stable under the (left) action of N^B:

$$\mathfrak{S} = \mathfrak{S}_b = \{g \in G : |\alpha_j(m_g)| \geq b, \text{ for all simple roots } \alpha_j\}$$

Take $0 < b \ll 1$ such that $\mathfrak{S}_b \to \Gamma\backslash G$ is a surjection. For $a > b$, let X_a be the subset of $\mathfrak{S}$ where $\beta(m_g) \leq a$ for all simple roots β. The quotient $(\Gamma \cap B)\backslash X_a$ is compact, since $(N^B \cap \Gamma)\backslash N$ is compact. For each simple root β, let

$$Y^\beta_a = \{g \in \mathfrak{S} : \beta(m_g) \geq a\}$$

and $Y_a = \bigcup_\beta Y^\beta_a$. Thus, $\mathfrak{S} = X_a \cup Y_a$. Parallel to [10.6], let

$$\mathfrak{E}(P, Q, f)_a$$
$$= \{F \in \mathfrak{E}(P, Q, f) : c_{P'}F(g) = 0, \text{ for all } g \in Y_a, \text{ for all } P'\}$$

as P' runs through standard parabolics, and

$$\mathfrak{E}_a^0 = \mathfrak{B}^0\text{-closure of } \mathfrak{E}(P, Q, f)_a \qquad \mathfrak{E}_a^1 = \mathfrak{B}^1\text{-closure of } \mathfrak{E}(P, Q, f)_a$$

$$\mathfrak{E}_a^2 = \mathfrak{B}^2\text{-closure of } \mathfrak{E}(P, Q, f)_a$$

It suffices to require vanishing of constant terms for maximal proper parabolics P'. Further, from [3.11.1], since all pseudo-Eisenstein series in $\mathfrak{E}(P, Q, f)$ have cuspidal data, the vanishing condition is automatically satisfied for all parabolics P' except P (and Q, in case $Q \neq P$).

As earlier, to be careful, since unbounded operators should be densely defined, we need

[11.12.1] Lemma: For $a \gg 1$, $\mathfrak{E}(P, Q, f)_a = \mathfrak{E}(P, Q, f) \cap \mathfrak{E}_a^0$ is dense in $\mathfrak{E}_a^0$.

Proof: On the relatively small space $\mathfrak{E}(P, Q, f)$, with $a \gg 1$, the observation [11.6.2] again reduces the issue to a generic, local, one-dimensional issue of smooth cutoffs, as addressed in the proof of [10.3.1] but now admitting the minor complication that constant terms along P and along Q are related.   ///

Let $S_{a,f}$ be Δ restricted to $\mathfrak{E}(P, Q, f)_a$. Since $\Delta \Psi_{\psi,f}^P = \Psi_{\beta,f}^P$ from [11.6.1], and similarly for Ψ_{ψ,f^w}^Q, and differential operators do not enlarge supports, Δ does stabilize $\mathfrak{E}(P, Q, f)_a$. Let $\widetilde{S}_{a,f}$ be the Friedrichs extension of $S_{a,f}$ to an unbounded self-adjoint operator on $\mathfrak{E}_a^0$, with domain contained in $\mathfrak{E}_a^1$ and containing $\mathfrak{E}_a^2$.

[11.12.2] Corollary: $\widetilde{S}_{a,f}$ has compact resolvents $(\widetilde{S}_{a,f} - \lambda_{s,f})^{-1}$ (away from poles).

Proof: As earlier, the point is that the inclusion $\mathfrak{E}_a^1 \to \mathfrak{E}_a^0$ is a restriction of the inclusion $\mathfrak{B}_a^1 \to L_a^2$, the latter compact from [10.8]. Restrictions of compact operators are compact. The resolvents of the Friedrichs extension are continuous maps $\mathfrak{E}_a^0 \to \mathfrak{E}_a^1$ composed with the inclusion $\mathfrak{E}_a^1 \to \mathfrak{E}_a^0$. Continuous maps composed with compact maps are compact.   ///

Let M^1 be the copy of $SL_{r_1} \times SL_{r_2}$ inside $M = M^P$, and Z^M the center of M. We take representatives

$$z_a = \begin{pmatrix} a^{\frac{1}{r_1 r_2}} \cdot 1_{r_1} & 0 \\ 0 & 1_{r_2} \end{pmatrix} \qquad \text{(for } 0 < a \in \mathbb{R}^\times)$$

for the connected component $Z \backslash \mathbb{Z}^M$ containing 1_r, and let η_a be the functional on $\mathfrak{E}(P, Q, f)$ defined by

$$\eta_a(F) = \int_{Z(\Gamma \cap M_1^P) \backslash \mathcal{M}_1} c_P F(m' \cdot z_a) \overline{f}(m') dm'$$

for $F \in \mathfrak{E}(P, Q, f)$. Similarly, both for P self-associate and not, let M^1 be the copy of $SL_{r_2} \times SL_{r_1}$ inside M^Q, let

$$z'_a = \begin{pmatrix} a^{\frac{1}{r_1 r_2}} \cdot 1_{r_2} & 0 \\ 0 & 1_{r_1} \end{pmatrix} \qquad (\text{for } 0 < a \in \mathbb{R}^\times)$$

and

$$\eta_a^w(F) = \int_{Z(\Gamma \cap M^1) \backslash \mathcal{M}_1} c_P F(m' \cdot z_a) \overline{f^w}(m') dm'$$

for $F \in \mathfrak{E}(P, Q, f)$. Then $F \in \mathfrak{E}(P, Q, f)_a$ if and only if $\eta_{b'}(F) = 0 = \eta_{b'}^w(F)$ for all $b' \geq a$. Certainly η_a and η_a^w do also depend on the cuspidal data.

On genuine Eisenstein series $E_{s,f}^P$ and E_{s,f^w}^Q the functionals η_a and η_a^w also make sense: unwinding, and using the explicit form of the constant terms, we have absolutely convergent integrals

$$\eta_a(E_{s,f}^P) = \int_{Z(\Gamma \cap M^1) \backslash \mathcal{M}^1} c_P E_{s,f}(m' \cdot z_a) \overline{f}(m') dm'$$

$$= \int_{Z(\Gamma \cap M^1) \backslash \mathcal{M}^1} (a^s \cdot f(m')) \cdot \overline{f}(m') dm' = a^s \qquad (\text{for } r_1 \neq r_2)$$

Similarly,

$$\eta_a^w(E_{s,f}^P) = c_{s,f}^Q a^{1-s} \qquad (\text{for } r_1 \neq r_2)$$

As earlier in [11.3.4] and [11.8.5], the Friedrichs extension can be usefully recharacterized:

[11.12.3] Claim: $\widetilde{S}_{a,f} x = y$ for $x \in \mathfrak{E}_a^1$ and $y \in \mathfrak{E}_a^0$ if and only if $\Delta x = y + A \cdot \eta_a + B \cdot \eta_a^w$ for some constants A, B. ///

As earlier, but now with two different tails to accommodate, form two smooth pseudo-Eisenstein series: let $h_{s,f}$ be the smooth pseudo-Eisenstein series formed from a smooth tail of $c_P E_{s,f}^P$, and $h_{s,f}^w$ a smooth pseudo-Eisenstein series formed from a smooth tail of $c_Q E_{s,f}^P$. To subtract multiples of $h_{s,f}$ and $h_{s,f}^w$ from $E_{s,f}^P$ to obtain an element of $\mathfrak{E}_a^1$, use a linear combination of $h_{s,f}$ and $h_{s,f}^w$ whose constant terms along both P and Q are both *eventually* (that is, sufficiently high up in the corresponding Siegel sets) exactly those of $E_{s,f}^P$.

From the computation of constant terms in [3.11.3] and [3.11.5], there is a tight relationship between the constant terms of $E_{s,f}^P$ and E_{s,f^w}^Q, which after meromorphic continuation gives the functional equation $E_{1-s,f} = (c_{s,f^w}^P)^{-1} \cdot E_{s,f^w}^Q$ and $c_{1-s,f}^Q \cdot c_{s,f^w}^P = 1$. ///

11.A Appendix: Distributions Supported on Submanifolds

The fact that distributions supported at a single point are finite linear combinations of Dirac delta and its derivatives is the simplest special case of the following, which reduces questions about distributions supported on smooth submanifolds to the *local* situation of Euclidean spaces.

[11.A.1] Theorem: A distribution u on $\mathbb{R}^{m+n} \approx \mathbb{R}^m \times \mathbb{R}^n$ supported on $\mathbb{R}^m \times \{0\}$, is uniquely expressible as a locally finite sum of transverse differentiations followed by restriction and evaluations, namely, a locally finite sum

$$u = \sum_\alpha u_\alpha \circ \mathrm{Res}_{\mathbb{R}^m \times \{0\}}^{\mathbb{R}^m \times \mathbb{R}^n} \circ D^\alpha$$

where α is summed over multi-indices $(\alpha_1, \ldots, \alpha_n)$, D^α is the corresponding differential operator on $\{0\} \times \mathbb{R}^n$, and u_α are distributions on $\mathbb{R}^m \times \{0\}$. Further,

$$\mathrm{spt}\, u_\alpha \times \{0\} \subset \mathrm{spt}\, u \qquad \text{(for all multi-indices } \alpha\text{)}$$

Proof: For brevity, let

$$\rho = \mathrm{Res}_{\mathbb{R}^m \times \{0\}}^{R^m \times R^n} : C_c^\infty(\mathbb{R}^m \times \mathbb{R}^n) \longrightarrow C_c^\infty(\mathbb{R}^m)$$

be the natural restriction map of test functions on $\mathbb{R}^m \times \mathbb{R}^n$ to $\mathbb{R}^m \times \{0\}$, by

$$(\rho f)(x) = f(x, 0) \qquad \text{(for } x \in \mathbb{R}^m\text{)}$$

The adjoint $\rho^* : C_c^\infty(\mathbb{R}^m)^* \to C_c^\infty(\mathbb{R}^{m+n})^*$ is a continuous map of distributions on $\mathbb{R}^m$ to distributions on $\mathbb{R}^m \times \mathbb{R}^n$, defined by

$$(\rho^* u)(f) = u(\rho(f))$$

First, *if* we could apply u to functions of the form $F(x, y) = f(x) \cdot y^\beta$, and *if* u had an expression as a sum as in the statement of the theorem, then

$$u\left(f(x) \cdot \frac{y^\alpha}{\alpha!}\right) = (-1)^{|\beta|} \cdot u_\beta(f) \cdot \beta!$$

since most of the transverse derivatives evaluated at 0 vanish. This is not quite legitimate, since y^α is not a test function. However, we can take a test function ψ on $\mathbb{R}^n$ that is identically 1 near 0, and consider $\psi(y) \cdot y^\alpha$ instead of y^α, and reach the same conclusion.

Thus, if there *exists* such an expression for u, it is unique. Further, this computation suggests how to specify the u_α, namely,

$$u_\beta(f) = u\left(f(x) \otimes \frac{y^\beta}{\beta!} \cdot \psi(y) \cdot (-1)^{|\beta|}\right)$$

This would also show the containment of the supports.

Show that the sum of these u_β's does give u. Given an open U in $\mathbb{R}^{m+n}$ with compact closure, u on $C^\infty_c{}_U$ has some finite order k. As a slight generalization of the fact that distributions supported on $\{0\}$ are finite linear combinations of Dirac delta and its derivatives, we have

[11.A.2] Lemma: Let v be a distribution of finite order k supported on a compact set K. For a test function φ whose derivatives up through order k vanish on K, $v(\varphi) = 0$.   ///

For any test function $F(x, y)$,

$$\Phi(x, y) = F(x, y) - \sum_{|\alpha| \leq k} (-1)^{|\alpha|} \frac{y^\alpha}{\alpha!} \, \psi(y) \, (D^\alpha F)(x, 0)$$

has all derivatives vanishing to order k on the closure of U. Thus, by the lemma, $u(\Phi) = 0$, which proves that u is equal to that sum, and also proves the local finiteness.   ///

12

Global Automorphic Sobolev Spaces, Green's Functions

The pretrace formulas in this chapter depend on estimates of eigenfunctions of integral operators on automorphic forms, in terms of the eigenvalues for the invariant Laplacian. This is accomplished by two observations. First, the eigenvalues depend only on the *isomorphism class* of the space generated by the group acting by right translations on the given automorphic form, so that these eigenvalues can be computed on any isomorphic copy of such a space, as representation space for G. Second, it happens that there are much simpler isomorphic copies of the relevant representations, parametrized essentially by one or more complex numbers, namely, *principal series* representations. We emphasize the archimedean aspect in this chapter, for which the general result is the subrepresentation theorem of [Casselman 1978/1980], [Casselman-Miličić 1982], improving the subquotient theorem of [Harish-Chandra 1954]. A simple argument sufficient for the four simple examples follows from older results on asymptotics for solutions of second-order ordinary differential equations, recalled in an appendix (Chapter 16).

The pretrace formulas ground the discussion of *global automorphic Sobolev spaces*. Among other goals, an important one is interpretation and legitimization of termwise differentiation of L^2 automorphic spectral expansions, especially by the invariant Laplacian. Of course, Plancherel theorems refer to L^2 expansions. Significantly, Plancherel theorems do *not* refer to sup norms of the eigenfunctions (such as cuspforms) entering in a spectral decomposition, nor sup norms of non-L^2 eigenfunctions (such as Eisenstein series) entering in the L^2 decomposition. This is already manifest in Plancherel for Fourier inversion on $L^2(\mathbb{R}^n)$. Typically, L^2 expansions do not produce continuous functions and are not continuously differentiable, so the goal cannot possibly be *proving* classical differentiability, since it does not hold. Especially with respect to *invariant* operators such as Casimir operators, and in delicate situations such as *automorphic forms*, Plancherel theorems most naturally yield corollaries about an *extension by continuity* of the classical limit-of-difference-quotient notion of differentiation. This L^2-differentiation is a usefully refined *distributional* differentiation. Termwise differentiability of L^2 spectral expansions in a *distributional* sense is of course correct, but needlessly very weak, and specifically too weak for many applications, since it is difficult to *return* from the larger world of distributions to the smaller world of L^2 functions.

Further, already for Fourier transforms on $\mathbb{R}^n$, the apparent integral expressing Fourier inversion is *not* a superposition of L^2 functions, since the exponential functions are not in $L^2(\mathbb{R}^n)$. Similarly, the spectral decomposition of pseudo-Eisenstein series involves *integrals* of the corresponding non-L^2 Eisenstein series.

The *global-ness* of the automorphic Sobolev spaces first refers to the expression of the norms (on automorphic test functions) as integrals over the whole space $\Gamma \backslash G / K$, rather than as a collection of seminorms given by integrals over smaller sets. Equivalently, the norms have expressions in terms of L^2 spectral expansions in terms of eigenfunctions for Δ. Global L^2 Sobolev spaces balance the simplicity of Hilbert space structures with extensions of notions of differentiability, insofar as solving more-or-less elliptic partial differential equations of sufficiently high degree can move back to L^2 from nearby Sobolev spaces of distributions. That is, in terms of the basic processes of analysis, Sobolev spaces are within *finite distance* of L^2.

Among other applications of global automorphic Sobolev spaces, we can immediately write a spectral expansion of an automorphic Green's function.

12.1 A Simple Pretrace Formula

Let $G, \Gamma, K, P, M, N, A^+$ be as in the four examples from Chapter 1, with Iwasawa coordinates x, y [1.3] with $x \in \mathbb{R}^{r-1}$ and $0 < y \in \mathbb{R}$, with $r = 2, 3, 4, 5$ the dimension of G/K. From [4.5], [4.6], [4.7], and [4.8], the invariant Laplacian is

$$\Delta = y^2 \Big(\frac{\partial^2}{\partial x_1^2} + \cdots + \frac{\partial^2}{\partial x_r^2} \Big) - (r - 2) y \frac{\partial}{\partial y}$$

For complex s, let $\lambda_s = (r - 1)^2 \cdot s(s - 1)$. For cuspforms F in an orthonormal basis, let $s_F \in \mathbb{C}$ be such that the Δ-eigenvalue of F is λ_{s_F}, and let t_F be the imaginary part of s_F.

[12.1.1] Theorem: Fix $z_o = (x, y) \in G/K$. Then, as $T \to +\infty$,

$$\sum_{F : |t_F| \leq T} |F(z_o)|^2 + \int_{|t| \leq T} |E_{\frac{1}{2} + it}(z_o)|^2 \, dt \ll_C T^r$$

with implied constant uniform for z_o in a fixed compact C.

Proof: We consider *integral* operators attached to compactly supported (regular Borel) measures η on the group G, and their operation on any reasonable representation space V for G, for example, Hilbert, Banach, Fréchet, and LF (strict colimits of Fréchet), or, generally, quasi-complete, locally convex spaces. For a continuous action $G \times V \to V$ of G on such a space V, and compactly supported measure η, the action is

$$\eta \cdot v = \int_G g \cdot v \, d\eta(g) \qquad \text{(for } v \in V\text{)}$$

as Gelfand-Pettis integral. The further nontrivial fact used in the proof[1] is that the eigenvalues of these integral operators on automorphic forms on $\Gamma\backslash G/K$ depend only on their eigenvalues for the Laplacian. This itself will follow from the fact that for $\Delta f = \lambda_s \cdot f$, a suitable topological vector space of functions on $\Gamma\backslash G$ generated by right translates of f is isomorphic to a subquotient of the *principal series* representation I_s (below), a relatively elementary object. The same is true of Eisenstein series E_s more immediately, since the Eisenstein series is (the meromorphic continuation of) a wound-up function from I_s (below).

Spaces V, W with continuous actions $G \times V \to V$ and $G \times W \to W$ are *representations of* G. Continuous $\mathbb{C}$-linear maps $T : V \to W$ among such spaces, respecting the action of G: $T(g \cdot v) = g \cdot T(v)$ for $v \in V$ and $g \in G$, are *G-homomorphisms* or *G-intertwinings*. The eigenvalues and eigenvectors of integral operators are preserved by G-homomorphisms:

---

[1] This is a very small instance of a *subquotient theorem* from [Harish-Chandra 1954]. This was strengthened to the *subrepresentation* theorem [Casselman 1978/1980], and [Casselman-Miličić 1982].

[12.1.2] Lemma: Let $T : V \to W$ be a homomorphism of G-spaces. For η a compactly supported measure on G, the action of η commutes with T:

$$\eta \cdot T(v) = T(\eta \cdot v) \qquad (\text{for all } v \in V)$$

In particular, for $\eta \cdot v = \lambda \cdot v$ with $v \in V$, the image Tv is also an eigenvector, with the same eigenvalue λ.

Proof: By properties of Gelfand-Pettis integrals and the fact that T commutes with the action of g,

$$T(\eta \cdot v) = T \int_G g \cdot v \, d\eta(g) = \int_G T(g \cdot v) \, d\eta(g)$$

$$= \int_G g \cdot T(v) \, d\eta(g) = \eta \cdot T(v)$$

as claimed. ///

For v in a G-representation V, the subrepresentation *generated by* v is the topological closure of the span of finite linear combinations of images $g \cdot v$ of v by $g \in G$. Among K-invariant vectors in a G-representation, eigenvectors for the spherical Hecke-algebra $\mathcal{H}$ occupy a privileged position:

[12.1.3] Claim: A strong-sense cuspform f, or Eisenstein series E_s, is the unique K-invariant vector in the representation it generates under right translation by G, up to a constant. More generally, for $v \neq 0$ a K-fixed vector in a G-representation V with V quasi-complete and locally convex, for v also an $\mathcal{H}$-eigenvector, the subrepresentation generated by v has K-fixed vectors exactly $\mathbb{C} \cdot v$.

Proof: Let α be the average-over-K map $v \to \int_K k \cdot v \, dk$, giving K total mass 1. This Gelfand-Pettis integral maps V to the K-fixed vectors V^K, and is the identity map on V^K.

First, consider $w = \sum_{i=1}^{n} c_i g_i \cdot v$ a K-fixed vector in the (algebraic) span of images $g_i \cdot v$ of v by $g_i \in G$. From a basic property [14.1.4] of Gelfand-Pettis integrals, $\varphi_j \cdot w \to w$, for any approximate identity $\{\varphi_j\}$. Since $\alpha \cdot w = w$ and $\alpha \cdot v = v$,

$$w = \alpha \cdot w = \alpha \cdot \lim_j \varphi_j \cdot w = \alpha \lim_j \varphi_j \cdot \sum_i c_i g_i \cdot \alpha \cdot v$$

By basic properties of Gelfand-Pettis integrals, the operator α commutes with the limit:

$$w = \lim_j \left(\sum_i c_i \alpha \circ \varphi_j \circ g_i \circ \alpha \right) \cdot v$$

The function

$$\eta_j = \sum_i c_i \alpha \circ \varphi_j \circ g_i \circ \alpha$$

is in $C_c^o(G)$ and is left and right K-invariant, so is in the spherical Hecke algebra $\mathcal{H}$. Since v is an $\mathcal{H}$-eigenvector, $\eta_j \cdot v = \lambda_j \cdot v$ for some scalar λ_j. That is, $\lambda_j \cdot v \to w$, so $w \in \mathbb{C} \cdot v$. Next, for $w = \lim_j w_j$ a limit of w_j of w_j in the span of images $g_i \cdot v$, $\alpha w = \alpha \lim w_j = \lim \alpha w_j$, and by the previous paragraph every αw_j is a scalar multiple of v, so w must be, as well.

Since Eisenstein series and strong-sense cuspforms are $\mathcal{H}$-eigenfunctions, the conclusion applies to them, as well.   ///

Thus, for any left-and-right K-invariant compactly supported measure η the integral operator action

$$(\eta \cdot f)(x) = \int_G f(xy) \cdot d\eta(y)$$

produces another right K-invariant vector in the representation space V_f generated by f. By the claim, $\eta \cdot f$ is a scalar multiple of f. Let $\chi_f(\eta)$ denote the eigenvalue:

$$\eta \cdot f = \chi_f(\eta) \cdot f \qquad (\text{with } \chi_f(\eta) \in \mathbb{C})$$

By the lemma, this is an intrinsic representation-theoretic relation, meaning that the scalar $\chi_f(\eta)$ can be computed in any image of V_f. As demonstrated subsequently, for $\Delta f = (r-1)^2 \cdot s(s-1)$, the representation generated by f has a common image[2] with an *unramified principal series* for $s \in \mathbb{C}$,

$$I_s = \left\{ \varphi \in C^\infty(G) : \varphi\left(\begin{pmatrix} a & * \\ 0 & d \end{pmatrix} \cdot g \right) = \left| \frac{a}{d} \right|^{(r-1)s} \cdot \varphi(g) \right\}$$

under right translation action by G. The Iwasawa decomposition $G = P \cdot K$ shows that the space of K-fixed vectors is one-dimensional. Thus, by the lemma, we can compute eigenvalues of elements of $\mathcal{H}$ on Eisenstein series or strong-sense cuspforms by computing eigenvalues on I_s.

Choice of integral operators: Let $\|g\|$ be the square of the *operator norm* on G for a standard representation of G on $\mathbb{C}^2$ or $\mathbb{C}^4$ (depending on cases) by matrix multiplication. In a Cartan decomposition,

$$\left\| k_1 \cdot \begin{pmatrix} e^{\rho/2} & 0 \\ 0 & e^{-\rho/2} \end{pmatrix} \cdot k_2 \right\| = e^r \qquad (\text{with } k_1, k_2 \in K, \rho \geq 0)$$

---

[2] As will be visible, this common image is inside a *Whittaker space* of smooth functions on G with suitable left equivariance under N.

This norm gives a left G-invariant metric $d(,)$ on G/K by

$$d(gK, hK) = \log \|g^{-1}h\| = \log \|h^{-1}g\|$$

The triangle inequality follows from the submultiplicativity of the norm. Take η to be the characteristic function of the left and right K-invariant set of group elements of operator norm at most e^δ, with small $\delta > 0$. That is,

$$\eta(g) = \begin{cases} 1 & (\text{for } \|g\| \le e^\delta) \\ \\ 0 & (\text{for } \|g\| > e^\delta) \end{cases}$$

or

$$\eta\left(k_1 \cdot \begin{pmatrix} e^{\rho/2} & 0 \\ 0 & e^{-\rho/2} \end{pmatrix} \cdot k_2\right) = \begin{cases} 1 & (\text{for } \rho \le \delta) \\ \\ 0 & (\text{for } \rho > \delta) \end{cases} \quad (\text{with } \rho \ge 0)$$

Upper bound on a kernel: The map $f \to (\eta \cdot f)(x)$ on automorphic forms f can be expressed as integration of f against a sort of automorphic form q_x by *winding up* the integral, as follows.

$$(\eta \cdot f)(x) = \int_G f(xy)\,\eta(y)\,dy = \int_G f(y)\,\eta(x^{-1}y)\,dy$$

$$= \int_{\Gamma\backslash G} \left(\sum_{\gamma\in\Gamma} f(\gamma y)\,\eta(x^{-1}\gamma y)\right) dy = \int_{\Gamma\backslash G} f(y) \cdot \left(\sum_{\gamma\in\Gamma} \eta(x^{-1}\gamma y)\right) dy$$

Thus, for $x, y \in G$ put

$$q_x(y) = \sum_{\gamma\in\Gamma} \eta(x^{-1}\gamma y)$$

The norm-squared of q_x, as a function of y alone, is

$$|q_x|^2_{L^2(\Gamma\backslash G)} = \int_{\Gamma\backslash G} \sum_{\gamma\in\Gamma}\sum_{\gamma'\in\Gamma} \eta(x^{-1}\gamma\gamma y)\,\overline{\eta}(x^{-1}\gamma'y)\,dy$$

$$= \int_G \sum_{\gamma\in\Gamma} \eta(x^{-1}\gamma y)\,\overline{\eta}(x^{-1}y)\,dy$$

after unwinding. For both $\eta(x^{-1}\gamma y)$ and $\eta(x^{-1}y)$ to be nonzero, the distance from x to both y and γy must be at most δ. By the triangle inequality, the distance from y to γy must be at most 2δ. For x in a fixed compact C, this requires that y be in ball of radius δ and that $\gamma y = y$. Since K is compact and Γ is discrete, the isotropy groups of all points in G/K are finite. Thus,

$$|q_x|^2_{L^2(\Gamma\backslash G)} \ll \int_{d(x,y)\le\delta} 1\,dy \asymp_C \delta^r \quad (\text{as } \delta \to 0^+)$$

Lower bound on eigenvalues: Let $\Delta f = (r-1)^2 \cdot s_f(s_f - 1)$, with $s_f = \frac{1}{2} + it_f$. A nontrivial *lower* bound for $\chi_f(\eta)$ can be given for $\delta \ll 1/t_f$, as follows. With *spherical function*

$$\varphi_s^o(\begin{pmatrix} a & * \\ 0 & d \end{pmatrix} \cdot k) = \left|\frac{a}{d}\right|^{(r-1)s}$$

in the sth principal series, the corresponding eigenvalue is

$$\chi_s(\eta) = \int_G \eta(g)\,\varphi_s^o(g)\,dg = \int_{r \le \delta} \varphi_s^o(k \cdot \begin{pmatrix} e^{r/2} & 0 \\ 0 & e^{-r/2} \end{pmatrix})\,dg$$

In fact, a qualitative argument clearly indicates the outcome, although we will also carry out a more explicit computation. For the qualitative argument, we need qualitative metrical properties of the Iwasawa decomposition. Let $g \to n_g a_g^+ k_g$ be the Iwasawa decomposition. We claim that $\|g\| \le \delta$ implies $\|n_g a_g^+\| \ll \delta$ for small $\delta > 0$. This is immediate, since the Jacobian of the map $N \times A^+ \to G/K$ near $e \in NA^+$ is *invertible*.

But, also, the Iwasawa decomposition is easily computed here for $G = SL_2(\mathbb{R})$ and $SL_2(\mathbb{C})$, and the integral expressing the eigenvalue can be estimated explicitly: elements of K can be parametrized as

$$k = \begin{pmatrix} \overline{\alpha} & \overline{\beta} \\ -\beta & \alpha \end{pmatrix} \qquad \text{(where } |\alpha|^2 + |\beta|^2 = 1)$$

and let $a = e^{r/2}$. Then

$$k \cdot \begin{pmatrix} a & 0 \\ 0 & a^{-1} \end{pmatrix} = \begin{pmatrix} * & * \\ -a\beta & \alpha/a \end{pmatrix}$$

Right multiplication by a suitable element k_2 of $SU(2)$ rotates the bottom row to put the matrix into NA^+:

$$k \cdot \begin{pmatrix} a & 0 \\ 0 & a^{-1} \end{pmatrix} \cdot k_2 = \begin{pmatrix} * & * \\ 0 & \sqrt{(-a|\beta|)^2 + (|\alpha|/a)^2} \end{pmatrix}$$

Thus,

$$\chi_s(\eta) = \int_{r \le \delta} \left((-a|\beta|)^2 + (|\alpha|/a)^2\right)^{-s}\,dg$$

Rather than compute the integral exactly, make δ small enough to give a lower bound on the integrand, such as would arise from

$$\left| \left((-a|\beta|)^2 + (|\alpha|/a)^2\right)^{-s} - 1 \right| < \frac{1}{2} \qquad \text{(for all elements of } K)$$

Since $|\alpha|^2 + |\beta|^2 = 1$, for small ρ,

$$(-e^{\rho/2}|\beta|)^2 + (|\alpha|/e^{\rho/2})^2 = e^\rho |\beta|^2 + |\alpha|^2/e^\rho$$
$$\asymp (1+\rho)|\beta|^2 + (1-\rho)|\alpha|^2 \ll 1 + \rho$$

Thus, for small $0 \leq \rho \leq \delta$,

$$\left| \left(e^{\rho} |\beta|^2 + |\alpha|^2 / e^{\rho} \right)^{-s} - 1 \right| \ll |s| \cdot \rho$$

Thus, $0 \leq \rho \leq \delta \ll \frac{1}{|s|}$ suffices to make this less than $\frac{1}{2}$.

From either a qualitative or quantitative approach, we see that with η the characteristic function of the δ-ball, we have the lower bound

$$|\chi_s(\eta)| = \int_G \eta(g) \varphi_s^o(g) \, dg \gg \int_{\rho \leq \delta} 1 = \mathrm{vol}(\delta\text{-ball}) \asymp \delta^r$$

for $|s| \ll 1/\delta$. Taking δ as large as possible compatible with $\delta \ll 1/|s|$ gives the lower bound

$$\chi_s(\eta) \gg \delta^r \qquad (\text{for } |s| \ll 1/\delta)$$

Combining the estimates: From the L^2 automorphic spectral expansion of q_x, apply Plancherel, dropping the finitely many terms from residues of Eisenstein series, and dropping normalization constants,

$$\sum_F |\langle q_x, F \rangle|^2 + \int_{-\infty}^{+\infty} |\langle q_x, E_s \rangle|^2 \, dt \leq |q_x|_{L^2(\Gamma \backslash G / K)}^2 \ll \delta^r$$

Truncating this to Bessel's inequality

$$\sum_{|t_F| \leq T} |\langle q_x, F \rangle|^2 + \int_{-T}^{+T} |\langle q_x, E_s \rangle|^2 \, dt \ll \delta^r$$

From the preceeding eigenvalue computation,

$$\langle q_x, f \rangle = \chi_s(\eta) \cdot f$$

and use the earlier inequality $\chi_s(\eta) \gg \delta^r$ from for this restricted parameter range, obtaining

$$\sum_{|t_F| \leq T} \left(\delta^r \cdot |F(x)| \right)^2 + \int_{-T}^{+T} \left(\delta^r \cdot |E_s(x)| \right)^2 \, dt \ll \delta^r$$

Multiply through by $T^{2r} \asymp 1/\delta^{2r}$ to obtain the *standard estimate* or *pretrace formula*

$$\sum_{|t_F| \leq T} |F(x)|^2 + \int_{-T}^{+T} |E_s(x)|^2 \, dt \ll T^r$$

as claimed earlier. Since the argument succeeds for both s and $1 - s$, the ambiguity in determining s from $s(s-1)$ is irrelevant. ///

12.2 Pretrace Formula for Compact Periods

The argument of the preceding section is a prototype. Now consider somewhat more general G, including not only $SL_2(\mathbb{R})$ but also $G = SL_n(\mathbb{R})$ or $SL_n(\mathbb{C})$, with $\Gamma = SL_n(\mathbb{Z})$ or $\Gamma = SL_n(\mathbb{Z}[i])$, respectively. We will not prove the corresponding subquotient or subrepresentation theorems: see [Harish-Chandra 1954], [Casselman 1978/1980], and [Casselman-Miličić 1982].

For closed subgroup H of G, let $\Theta = H \cap \Gamma$, and suppose that $\Theta \backslash H$ is *compact*. The $\Theta \backslash Hx$-*period* of f is

$$\Theta \backslash Hx\text{-period of } f = f_{\Theta \backslash Hx} = \int_{\Theta \backslash H} f(hx)\, dh$$

Similarly, with ψ an automorphic form on $\Theta \backslash H$, the period of $\psi \otimes f$ is

$$\langle f, \psi \rangle_{\Theta \backslash Hx} = \int_{\Theta \backslash H} \psi(h) \cdot f(hx)\, dh$$

[12.2.1] Theorem: Using abbreviated notation for the spectral expansions to implicitly include the appropriate integrals of cuspidal-data Eisenstein series, as well as their residues,

$$\sum_{\mathrm{cfm}F : |t_F| \ll T} |F_{\Theta \backslash Hx}|^2 + \dots \ll_{x, H} T^{\dim X - \dim Y}$$

and, similarly

$$\sum_{\mathrm{cfm}F : |t_F| \ll T} |\langle \eta \cdot F, \psi \rangle|^2 + \dots \ll_{x, H, \psi} T^{\dim X - \dim Y}$$

Proof: The usual action of compactly supported measures η on suitable f on $\Gamma \backslash G$ is $(\eta \cdot f)(x) = \int_G \eta(g) f(xg)\, dg$. The $\Theta \backslash Hx$-period of $\eta \cdot f$ admits a useful rearrangement

$$(\eta \cdot f)_{\Theta \backslash Hx} = \int_{\Theta \backslash H} (\eta \cdot f)(hx)\, dh = \int_{\Theta \backslash H} \int_G \eta(g) f(hxg)\, dg\, dh$$

$$= \int_{\Theta \backslash H} \int_G \eta(x^{-1} h^{-1} g) f(g)\, dg\, dh$$

$$= \int_{\Theta \backslash H} \int_{\Gamma \backslash G} \sum_{\gamma \in \Gamma} \eta(x^{-1} h^{-1} \gamma g) f(g)\, dg\, dh$$

$$= \int_{\Gamma \backslash G} f(g) \left(\int_{\Theta \backslash H} \sum_{\gamma \in \Gamma} \eta(x^{-1} h^{-1} \gamma g)\, dh \right) dg$$

Denote the inner sum-and-integral by $q(g) = q_{H,x}(g)$.

Similarly, with ψ an automorphic form on $\Theta\backslash H$, the period of $\psi \otimes f$ rearranges to

$$\langle \eta \cdot f, \psi \rangle_{\Theta\backslash Hx} = \int_{\Gamma\backslash G} f(g) \left(\int_{\Theta\backslash H} \psi(h) \cdot \sum_{\gamma \in \Gamma} \eta(x^{-1}h^{-1}\gamma g) \, dh \right) dg$$

For η left-and-right K-invariant, for f the spherical vector in a copy of a principal series representation of G, necessarily $\eta \cdot f = \lambda_f(\eta) \cdot f$ for some constant $\lambda_f(\eta)$. Thus, the action of such η changes the period by the eigenvalue:

$$(\eta \cdot f)_{\Theta\backslash Hx} = \lambda_f(\eta) \cdot f_{\Theta\backslash Hx}$$

An *upper* bound for the $L^2(\Gamma\backslash G)$ norm of q, and a *lower* bound for $\lambda_f(\eta)$ contingent on restrictions on the spectral parameter of f, yield, by Bessel's inequality, an *upper* bound for a sum-and-integral of periods $\langle f, \psi \rangle_{\Theta\backslash Hx}$:

Estimating the L^2 norm,

$$\int_{\Gamma\backslash G} |q(g)|^2 \, dg$$

$$= \int_{\Gamma\backslash G} \int_{\Theta\backslash H} \int_{\Theta\backslash H} \sum_{\gamma \in \Gamma} \sum_{\gamma_2 \in \Gamma} \eta(x^{-1}h^{-1}\gamma g) \, \overline{\eta}(x^{-1}h_2^{-1}\gamma_2 g) \, dh \, dh_2 \, dg$$

$$= \int_G \int_{\Theta\backslash H} \int_{\Theta\backslash H} \sum_{\gamma \in \Gamma} \eta(x^{-1}h^{-1}\gamma g) \, \overline{\eta}(x^{-1}h_2^{-1}g) \, dh \, dh_2 \, dg$$

With $C \subset H$ a large-enough compact to surject to $\Theta\backslash H$,

$$\int_{\Gamma\backslash G} |q(g)|^2 \, dg \leq \int_G \int_C \int_C \sum_{\gamma \in \Gamma} |\eta|(x^{-1}h^{-1}\gamma g) \, |\eta|(x^{-1}h_2^{-1}g) \, dh \, dh_2 \, dg$$

Let η be the characteristic function of a small ball B_ε in G/K, of geodesic radius $\varepsilon > 0$, for a G-invariant metric $d(x, y) = \nu(x^{-1}y)$ on G/K, where $\nu(g) = \log \sup(|g|, |g^{-1}|)$, where $|\cdot|$ is operator norm on G. The triangle inequality follows from submultiplicativity of operator norm.

Identify B_ε with its preimage $B_\varepsilon \cdot K$ in G. The set

$$\Phi = \Phi_{H,x,\eta} = \{\gamma \in \Gamma : \eta(x^{-1}h^{-1}\gamma g) \, \eta(x^{-1}h_2^{-1}g) \neq 0$$

$$\text{for some } h, h_2 \in C \text{ and } g \in G\}$$

$$= \{\gamma \in \Gamma : \gamma \in CxB_\varepsilon g^{-1}, \ g \in CxB_\varepsilon\}$$

$$\subset \Gamma \cap CxB_\varepsilon \cdot (CxB_\varepsilon)^{-1} = \text{discrete subgroup} \cap \text{compact}$$

is *finite* and can only *shrink* as $\varepsilon \to 0^+$.

For each $\gamma \in \Phi$, for each $h \in C$, $\eta(x^{-1}h^{-1}\gamma g) \neq 0$ only for g in a ball in $X = G/K$ of radius ε, with volume dominated by $\varepsilon^{\dim X}$. For each h and g,

$\eta(x^{-1}h_2^{-1}g) \neq 0$ only for h_2x in a ball in $Y = HxK/K$ of radius ε, with volume dominated by $\varepsilon^{\dim Y}$. Thus,

$$\int_{\Gamma\backslash G} |q(g)|^2 \, dg \ll \int_C \varepsilon^{\dim X + \dim Y} \, dh \ll_{x,H} \varepsilon^{\dim X + \dim Y}$$

By Plancherel for $L^2(\Gamma\backslash X)$, with η the characteristic function of the ε-ball,

$$\sum_{\text{cfm}F} |\lambda_F(\eta)|^2 \cdot |F_{\Theta\backslash Hx}|^2 + \cdots = \sum_{\text{cfm}F} |(\eta \cdot F)_{\Theta\backslash Hx}|^2 + \cdots = |q_{H,x,\eta}|^2$$

$$\ll_{x,H} \varepsilon^{\dim X + \dim Y}$$

Similarly,

$$\sum_{\text{cfm}F} |\lambda_F(\eta)|^2 \cdot |\langle \eta \cdot F, \psi \rangle|^2 + \cdots \ll_{x,H,\psi} \varepsilon^{\dim X + \dim Y}$$

Next, a bound on the spectral data is determined to give a nontrivial lower bound for $|\lambda_f(\eta)|$. For f the spherical vector in a copy of a principal series representation of G, left-and-right K-invariant η necessarily gives $\eta \cdot f = \lambda_f(\eta) \cdot f$, since up to scalars f is the unique spherical vector in the irreducible representation it generates.

The eigenvalues $\lambda_f(\eta)$ can be computed in the usual model of principal series, as $\eta \cdot \varphi_s^o = \lambda_f(\eta) \cdot \varphi_s^o$ for φ_s^o the normalized spherical vector for $s \in \mathfrak{a}^* \otimes_{\mathbb{R}} \mathbb{C}$, and $\varphi^o(1) = 1$. Thus,

$$\lambda_f(\eta) = (\eta \cdot \varphi_s^o)(1) = \int_G \eta(g) \cdot \varphi_s^o(g) \, dg = \int_{B_\varepsilon} \varphi_s^o(g) \, dg$$

Let P^+ be the connected component of the identity in the minimal parabolic. The Jacobian of the map $P^+ \times K \to G$ is nonvanishing at 1, and $\varphi^o(1) = 1$, so a suitable bound in terms of ε on the spectral parameter $s \in \mathfrak{a}^* \otimes_{\mathbb{R}} \mathbb{C}$ will keep $\varphi_s^o(g)$ near 1 on B_ε. In the example of $SL_n(\mathbb{R})$, with

$$\varphi_s^o \begin{pmatrix} a_1 & \cdots & * \\ 0 & \ddots & \vdots \\ 0 & 0 & a_n \end{pmatrix} = |a_1|^{s_1 + \rho_1} \cdots |a_n|^{s_n + \rho_n}$$

for whatever normalizing constants ρ_j, bounds of the form $|s_j| \ll 1/\varepsilon$ assure that $\operatorname{Re}\varphi_s^o(g) \geq \frac{1}{2}$ on B_ε, which prevents cancellation in the *real part* of $\varphi_s^o(g)$ for $g \in B_\varepsilon$, so

$$|\lambda_f(\eta)| = \left| \int_{B_\varepsilon} \varphi_s^o(g) \, dg \right| \gg \int_{B_\varepsilon} \operatorname{Re}\varphi_s^o(g) \, dg \gg \int_{B_\varepsilon} \frac{1}{2} \, dg \gg \varepsilon^{\dim X}$$

Combining the upper bound on $|q|^2_{L^2}$ with this lower bound on eigenvalues, letting $T \sim 1/\varepsilon$,

$$(\varepsilon^{\dim X})^2 \times \Big(\sum_{\mathrm{cfm} F : |t_F| \ll T} |F_{\Theta \backslash Hx}|^2 + \cdots \Big) \ll_{x,H} \varepsilon^{\dim X + \dim Y}$$

proving the first assertion of the theorem. The proof of the second is essentially identical.   ///

12.3 Global Automorphic Sobolev Spaces H^ℓ

Again, let G, Γ, K be any one of the archimedean examples, such as $SL_2(\mathbb{R})$, $SL_2(\mathbb{Z}), SO_2(\mathbb{R})$ or $SL_n(\mathbb{R}), SL_n(\mathbb{Z}), SO_n(\mathbb{R})$ or $SL_n(\mathbb{C}), SL_n(\mathbb{Z}[i]), SU_n(\mathbb{C})$. Let Δ be the invariant Laplacian on $\Gamma \backslash G/K$. Functions f in $L^2(\Gamma \backslash G/K)$ decompose in an L^2 sense [1.14] and [3.18]. To write the spectral expansion succinctly, let Ξ be a locally compact, Hausdorff, σ-compact topological space parametrizing cuspforms, Eisenstein series appearing in the spectral decomposition and Plancherel, as well as their residues, with corresponding Δ-eigenfunction $\Phi_\xi \in C^\infty(\Gamma \backslash G/K)$ for $\xi \in \Xi$, and a positive regular Borel measure $d\xi$ on Ξ, to write the expansions of [1.14] and [3.18] uniformly as

$$f = \int_\Xi \langle f, \Phi_\xi \rangle \cdot \Phi_\xi \, d\xi \qquad (\text{for } f \in C_c^\infty(\Gamma \backslash G/K))$$

and Plancherel as

$$|f|^2_{L^2} = \int_\Xi |\langle f, \Phi_\xi \rangle|^2 \, d\xi \qquad (\text{for } f \in C_c^\infty(\Gamma \backslash G/K))$$

For example, for $SL_2(\mathbb{Z})$, and similarly for the other three simplest examples, the explicit spectral expansion

$$f = \sum_{\mathrm{cfm} \, F} \langle f, F \rangle \cdot F + \frac{\langle f, 1 \rangle \cdot 1}{\langle 1, 1 \rangle} + \frac{1}{4\pi i} \int_{(\frac{1}{2})} \langle f, E_s \rangle \cdot E_s \, ds$$

for test functions f would parametrize the cuspform components by an infinite discrete set, the constant-function component by a further point, and the integrals-of-Eisenstein series component by $\mathbb{R}$ or by $[0, +\infty)$. For test functions f, the implied integrals $\langle f, E_s \rangle$ against Eisenstein series do converge absolutely.

Although many of the eigenfunctions Φ_ξ are not in $L^2(\Gamma \backslash G/K)$, they are all in $C^\infty(\Gamma \backslash G/K)$, and we can easily arrange that $\xi \to \Phi_\xi$ is a *continuous* $C^\infty(\Gamma \backslash G/K)$-valued function on Ξ. Integration of elements of $C^\infty(\Gamma \backslash G/K)$ against a fixed $f \in C_c^\infty(\Gamma \backslash G/K)$ is a continuous linear functional on $C^\infty(\Gamma \backslash G/K)$, so

$$\xi \; \longrightarrow \; \Phi_\xi \; \longrightarrow \; \langle f, \Phi_\xi \rangle \qquad \text{(for fixed } f \in C_c^\infty(\Gamma \backslash G/K)\text{)}$$

is a *continuous* $\mathbb{C}$-valued function on Ξ and thus has unambiguous pointwise values.

[12.3.1] Remark: For $\Xi_1 \subset \Xi_2 \subset$ a sequence of compact subsets of Ξ whose union is Ξ, the integrals

$$\int_{\Xi_n} \langle f, \Phi_\xi \rangle \cdot \Phi_\xi \qquad \text{(for test functions } f\text{)}$$

do exist as $C^\infty(\Gamma \backslash G/K)$-valued Gelfand-Pettis integrals. However, as is already the case with ordinary Fourier series on the circle, the fact that finite partial sums are invariably smooth tells little about the nature of the limit. On the basis of other examples and of folklore, we imagine that for test functions f the compactly supported integrals should converge to f in some topology finer than L^2, but this requires proof, as will follow.

The implied literal integrals $\langle f, \Phi_\xi \rangle$ against Eisenstein series do not necessarily converge for all f in L^2, and certainly L^2 expansions do not reliably converge *pointwise*. Nevertheless, the Plancherel theorem asserts that the literal integrals $f \to (\xi \to \langle f, \Phi_\xi \rangle)$ on *test functions* do extend to an isometry

$$\mathcal{F} \; : \; L^2(\Gamma \backslash G/K) \; \longrightarrow \; L^2(\Xi)$$

Similarly, the spectral synthesis integrals $f = \int_\Xi c(\xi) \cdot \Phi_\xi \, d\xi$ do make literal sense (in fact, as $C^\infty(\Gamma \backslash G/K)$-valued integrals) for test functions f. Then the integrals and pairings for $f \in L^2$ are understood as *extensions by continuity* of the literal integrals. The (extensions of the) integrals $\langle f, \Phi_\xi \rangle$ are the *spectral coefficients* of f.

[12.3.2] Remark: Having said all that, just as is done with the Plancherel extension of the Fourier transform and Fourier inversion on $\mathbb{R}^n$, *eventually* we will write integrals and pairings that do not literally converge but do exist as extensions by continuity of those integrals and pairings.

[12.3.3] Remark: Notably, Plancherel neither needs nor asserts anything directly about pointwise values of cuspforms or Eisenstein series or residues of Eisenstein series. This is fortunate, since already in the simplest case various pointwise values of Eisenstein series E_s for $SL_2(\mathbb{Z})$ are $\zeta_k(s)/\zeta(2s)$ for complex quadratic extensions k of $\mathbb{Q}$ [2.C], and sharp pointwise bounds on the critical line presumably include the Lindelöf Hypothesis.

[12.3.4] Claim: The eigenvalues λ_ξ of Δ on Φ_ξ are *real* and nonpositive.

Proof: For square-integrable eigenfunctions (such as cuspforms or square-integrable residues of Eisenstein series), these eigenvalues are real because a suitable restriction of Δ to a dense subspace of L^2 is symmetric and nonpositive, so has a self-adjoint, nonpositive Friedrichs extension.

In the four simple cases, the non-L^2 eigenfunctions entering the spectral expansion and Plancherel are Eisenstein series E_s with eigenvalues $s(s-1)$ (up to real constants) and Re$(s) = \frac{1}{2}$, and residues of E_s with $s \in (0, 1]$, with eigenvalue $s(s-1)$ (up to real constants). Somewhat more generally, [3.11.11] shows that cuspidal-data Eisenstein series $E^P_{s,f}$ for maximal proper parabolics P in GL_n, with the cuspidal data $f = f_1 \otimes f_2$ eigenfunctions for Casimir operators on the Levi component, are eigenfunctions for Casimir on GL_n. The explicit formula for eigenvalues shows that they are real and nonpositive for Re$(s) = \mathfrak{H}$, using the corresponding fact for the cuspidal data on the Levi components. The part of the spectral decomposition [3.16.1] and Plancherel [3.17.1], [3.17.4] corresponding to a given maximal proper parabolic uses Re$(s) = \frac{1}{2}$ and residues (if any) on $(\frac{1}{2}, 1]$.   ///

[12.3.5] Remark: Another proof that these eigenvalues are real would follow from a suitable description of the spectral decomposition of L^2 and Plancherel in terms of *Hilbert integrals* of representations of G. This would show that all representations appearing must be *unitary*, allowing various continuations proving the previous claim, using some form of a subquotient theorem.

[12.3.6] Claim: The invariant Laplacian Δ commutes with pointwise complex conjugation on functions on G/K or on $\Gamma\backslash G/K$.

Proof: Any expression $\Omega = \sum_i x_i x_i^*$, with basis $\{x_i\}$ for $\mathfrak{g}$ and dual basis $\{x_i^*\}$, expresses Casimir Ω as a real-linear combination of compositions of operators of the form

$$(x_i f)(g) = \left.\frac{\partial}{\partial t}\right|_{t=0} f(g \cdot e^{tx_i})$$

These visibly commute with pointwise complex conjugation.   ///

[12.3.7] Claim: For $f \in C_c^\infty(\Gamma\backslash G/K)$, the spectral coefficients of Δf are

$$\langle \Delta f, \Phi_\xi \rangle = \lambda_\xi \cdot \langle f, \Phi_\xi \rangle \qquad (\text{for } f \in C_c^\infty(X))$$

Thus, for test functions,

$$\Delta f = \int_\Xi \langle \Delta f, \Phi_\xi \rangle \, \Phi_\xi \, d\xi = \int_\Xi \lambda_\xi \cdot \langle f, \Phi_\xi \rangle \, \Phi_\xi \, d\xi$$

That is, succinctly,

$$\Delta f = \mathcal{F}^{-1}\mathcal{F}\Delta f = \mathcal{F}^{-1}\lambda_\xi \mathcal{F}f \qquad \text{(for test functions } f\text{)}$$

Proof: For test functions f, integration by parts is legitimate:

$$\langle \Delta f, \Phi_\xi \rangle = \int_X f \, \overline{\Delta \Phi_\xi} = \overline{\lambda_\xi} \cdot \int_X f \, \overline{\Phi_\xi} = \lambda_\xi \cdot \int_X f \, \overline{\Phi_\xi}$$

as asserted.   ///

[12.3.8] Corollary: The differential operator Δ differentiates spectral expansions of test functions *termwise*, in the sense of moving inside the integration and summation over Ξ giving the *spectral synthesis*.   ///

For $0 \le \ell \in \mathbb{Z}$, the ℓth Sobolev norm on $C_c^\infty(\Gamma\backslash G/K)$ is given by

$$|f|_{H^\ell}^2 = \int_{\Gamma\backslash G/K} \overline{f} \cdot (1 - \Delta)^\ell f$$

and

$$H^\ell = H^\ell(\Gamma\backslash G/K)$$
$$= \text{completion of } C_c^\infty(\Gamma\backslash G/K) \text{ with respect to } |\cdot|_{H^\ell}$$

[12.3.9] Claim: $|f|_{H^{k+1}} \ge |f|_{H^k}$ for test functions f, for $0 \le k \in \mathbb{Z}$, and there is a canonical continuous injection $H^{k+1}(\Gamma\backslash G/K) \to H^k(\Gamma\backslash G/K)$ with dense image.

Proof: For all test functions f, $\langle -\Delta f, f \rangle \ge 0$. For a polynomial P with nonnegative real coefficients, we claim that $P(-\Delta)$ is nonnegative on test functions, in the sense that for all test functions f

$$\langle P(-\Delta)f, f \rangle \ge 0$$

It suffices to prove this for monomials $(-\Delta)^n$. For even $n = 2m$,

$$\langle (-\Delta)^{2m}f, f \rangle = \langle (-\Delta)^m f, (-\Delta)^m f \rangle \ge 0$$

For odd $n = 2m + 1$,

$$\langle (-\Delta)^{2m+1}f, f \rangle = \langle (-\Delta)((-\Delta)^m f), ((-\Delta)^m f) \rangle \ge 0$$

For test functions f, the desired comparison is

$$\begin{aligned}
|f|_{H^{k+1}}^2 &= \langle (1 - \Delta)^{k+1} f, f \rangle_{L^2} \\
&= \langle (1 + (-\Delta))^k f, f \rangle_{L^2} + \langle (1 + (-\Delta))^k (-\Delta)f, f \rangle_{L^2} \\
&\ge \langle (1 + (-\Delta))^k f, f \rangle_{L^2} + 0 = |f|_{H^k}^2
\end{aligned}$$

Thus, the identity map $C_c^\infty(\Gamma\backslash G/K) \to C_c^\infty(\Gamma\backslash G/K)$ extends to a continuous injection $H^{k+1} \to H^k$. Since C_c^∞ is dense in both, necessarily the image is dense.   ///

The following result is true *by design*.

[12.3.10] Claim: The differential operator $\Delta : C_c^\infty(\Gamma\backslash G/K) \longrightarrow C_c^\infty(\Gamma\backslash G/K)$ is continuous when the source is given the $H^{\ell+2}$ topology and the target is given the H^ℓ topology, for $0 \geq \ell \in \mathbb{Z}$.

Proof: Using the latter nonnegativity property of the previous proof,

$$
\begin{aligned}
|\Delta f|_{H^\ell}^2 &= \langle (1-\Delta)^\ell(\Delta f),\, (\Delta f)\rangle = \langle (-\Delta)^2(1+(-\Delta))^\ell f,\, f\rangle \\
&\leq \langle (-\Delta)^2(1+(-\Delta))^\ell f,\, f\rangle + \langle (2(-\Delta)+1)f,\, f\rangle \\
&= \langle (1+(-\Delta))^{\ell+2} f,\, f\rangle = |f|_{H^{\ell+2}}^2
\end{aligned}
$$

as asserted.   ///

[12.3.11] Corollary: Δ extends by continuity from test functions to a continuous linear map

$$
\Delta :\ H^{\ell+2}(\Gamma\backslash G/K) \ \longrightarrow\ H^\ell(\Gamma\backslash G/K) \qquad \text{(for } 0 \leq \ell \in \mathbb{Z})
$$

Proof: That is, for test functions $\{f_n\}$ forming a Cauchy sequence in the $H^{\ell+2}$ topology, the continuity in the respective topologies on *test functions* means that the extension-by-continuity definition

$$
\Delta\Big(H^{\ell+2}\text{-}\lim_n f_n\Big) = H^\ell\text{-}\lim_n \Delta f_n
$$

is well defined and gives a continuous map in those topologies.   ////

[12.3.12] Remark: This extension of Δ is L^2-*differentiation* for non-negative index Sobolev spaces. This extension is a refined version of *distributional* differentiation. Nevertheless, to examine *global* automorphic Sobolev spaces, the present discussion of L^2-differentiation does not directly depend on distributional notions.

[12.3.13] Corollary: For f in H^ℓ with $\ell \geq 2$,

$$
\mathcal{F}(\Delta f)(\xi) = \lambda_\xi \cdot \mathcal{F}f
$$

Proof: Since $\mathcal{F} : L^2(\Gamma\backslash G/K) \to L^2(\Xi)$ is an isometric isomorphism obtained by extension by continuity from $\mathcal{F}$ on $C_c^\infty(\Gamma\backslash G/K)$, the literal integral computation for test functions

$$(\mathcal{F}\Delta f)(\xi) = \int_X \Delta f \, \Phi_\xi = \int_X f \, \Delta\overline{\Phi}_\xi$$

$$= \int_X f \, \lambda_\xi \cdot \overline{\Phi}_\xi = \lambda_\xi \int_X f \cdot \overline{\Phi}_\xi = \lambda_\xi \cdot (\mathcal{F}f)(\xi)$$

extends by continuity to give the result.   ///

[12.3.14] Corollary: Termwise differentiation is valid:

$$\Delta f = \mathcal{F}^{-1}\mathcal{F}\Delta f = \mathcal{F}^{-1}\lambda_\xi \mathcal{F}f \qquad (\text{for } f \in H^\ell \text{ with } \ell \geq 2)$$

This differentiation is in the extended, nonclassical sense.   ///

Negative-index Sobolev spaces are not easily described via differential operators. Instead, characterize negative-index Sobolev spaces as Hilbert-space duals

$$H^{-\ell} = H^{-\ell}(\Gamma\backslash G/K) = \text{Hilbert-space dual to } H^\ell$$

for $0 \leq \ell \in \mathbb{Z}$. To identify $H^0 = L^2$ with its own dual $\mathbb{C}$-linearly, combine the $\mathbb{C}$-conjugate-linear Riesz-Fréchet map $\Lambda : f \to \langle -, f \rangle$ with pointwise conjugation $c : f \to \overline{f}$, so that $\Lambda \circ c : H^0 \to H^0$ is $\mathbb{C}$-linear. Mapping $(H^0)^* \to H^{-1}$ by the adjoint of the inclusion $H^1 \to H^0$, and generally $H^{-k} \to H^{-k-1}$ by the adjoint of the inclusion $H^{k+1} \to H^k$, we have a chain of continuous linear maps

$$\cdots \xrightarrow{} H^1 \xrightarrow{\text{inc}} H^0 \xrightarrow[\approx]{\Lambda \circ c} (H^0)^* \xrightarrow{\text{inc}^*} H^{-1} \xrightarrow{\text{inc}^*} \cdots$$

[12.3.15] Claim: For $0 \leq k \in \mathbb{Z}$, the maps $H^{-k} \to H^{-k-1}$ adjoint to inclusions $H^{k+1} \to H^k$ are themselves inclusions with dense images. Thus, we have a chain of continuous injections with dense images:

$$\cdots \subset H^2 \subset H^1 \subset H^0 \approx (H^0)^* \subset H^{-1} \subset H^{-2} \subset \cdots$$

Proof: For $0 \leq k \in \mathbb{Z}$, since $H^{k+1} \to H^k$ has dense image, the adjoint $H^{-k} \to H^{-k-1}$ is injective. If $H^{-k} \to H^{-k-1}$ did not have dense image, then its adjoint would not be injective. By the reflexivity of Hilbert spaces, its adjoint is the original $H^{k+1} \to H^k$, which is injective.   ////

In the sequel, we identify H^0 with its dual via $\Lambda \circ c$.

The continuous L^2-differentiation $\Delta : H^{2\ell} \to H^{2\ell-2}$ for $2\ell \geq 2$ on positive-index Sobolev spaces gives an adjoint, still denoted Δ, on negative-index spaces:

$$\Delta : H^{-2\ell} \longrightarrow H^{-2\ell-2} \qquad (\text{for } 0 \leq \ell \in \mathbb{Z})$$

The extension of Δ to $\Delta : H^1 \to H^{-1}$ can be characterized by

$$((1 - \Delta)f)(F) = \langle f, \overline{F} \rangle_{H^1} \qquad (\text{for } f, F \in H^1)$$

The compatibility of this extension with the others is best clarified by the spectral characterizations of the next section. Anticipating that clarification, let

$$H^\infty(\Gamma\backslash G/K) = \bigcap_{k\in\mathbb{Z}} H^k(\Gamma\backslash G/K) = \lim_{k\in\mathbb{Z}} H^k(\Gamma\backslash G/K)$$

and

$$H^{-\infty}(\Gamma\backslash G/K) = \bigcup_{k\in\mathbb{Z}} H^k(\Gamma\backslash G/K) = \mathrm{colim}_{k\in\mathbb{Z}} H^k(\Gamma\backslash G/K)$$

Pending the spectral characterization, for the following corollary we can temporarily take

$$H^\infty = \lim_k H^{2k} \qquad\qquad H^{-\infty} = \mathrm{colim}_k H^{2k}$$

[12.3.16] Corollary: Both $H^\infty(\Gamma\backslash G/K)$ and $H^{-\infty}(\Gamma\backslash G/K)$ are stable under the extension of Δ (from test functions to global Sobolev spaces), and Δ gives a continuous linear operator on both.

Proof: For $H^\infty = \bigcap_k H^{2k}$, the extended Δ maps $H^{2k+2} \to H^{2k}$, so $\Delta H^\infty \subset H^{2k}$ for every k, and the intersection is H^∞. More precisely, by the characterization of (projective) limits, the family of compatible maps

$$H^\infty \xrightarrow{\mathrm{inc}} H^{2k+2} \xrightarrow{\Delta} H^{2k}$$

induces a unique compatible *continuous* (linear) map $H^\infty \to H^\infty$.

Oppositely, the extension-by-adjoint Δ maps $H^{-2k} \to H^{-2k-2}$, so the image of the ascending union is contained in the ascending union. More precisely, by the characterization of colimits, the compatible family of maps

$$H^{-2k} \xrightarrow{\Delta} H^{-2k-2} \xrightarrow{\mathrm{inc}} H^{-\infty}$$

gives a unique compatible continuous (linear) map $H^{-\infty} \to H^{-\infty}$.   ///

[12.3.17] Claim: $H^{-\infty}(\Gamma\backslash G/K)$ is a subset of the space $C_c^\infty(\Gamma\backslash G/K)$ of *distributions* on $\Gamma\backslash G/K$.

Proof: First, check that the inclusion $C_c^\infty(\Gamma\backslash G/K) \subset H^k(\Gamma\backslash G/K)$ is continuous for every k. Recall from [13.9] and [6.3] that the space of test functions is the colimit of spaces

$$C_E^\infty = \{f \in C_c^\infty : \mathrm{spt} f \subset E\}$$

for compact $E \subset \Gamma \backslash G/K$. By the characterization of colimit, it suffices to prove continuity of each $C_E^\infty \to H^k$. Among the seminorms defining the topology on C_E^∞ (see [6.4], [13.9]) are

$$\nu_k(f) = \sup_{x \in E} |(1 - \Delta)^k f(x)|$$

Note that the volume of $\Gamma \backslash G/K$ is finite. Given $f \in C_E^\infty$, by Cauchy-Schwarz-Bunyakowsky,

$$\begin{aligned}
|f|_{H^k}^2 &= \int_{\Gamma \backslash G/K} (1 - \Delta)^k f \cdot \overline{f} = \int_E (1 - \Delta)^k f \cdot \overline{f} \\
&\leq \left(\int_E |(1 - \Delta)f|^2 \right)^{\frac{1}{2}} \cdot \left(\int_E |f|_{L^2} \right)^{\frac{1}{2}} \\
&\leq \nu_k(f) \cdot \operatorname{meas}(E)^{\frac{1}{2}} \cdot \nu_0(f) \cdot \operatorname{meas}(E)^{\frac{1}{2}} \\
&\leq \nu_k(f) \cdot \nu_o(f) \cdot \operatorname{meas}(\Gamma \backslash G/K)
\end{aligned}$$

Thus, $|f|_{H^k} \ll_E \nu_k(f) + \nu_0(f)$, giving the desired continuity. Thus, $H^{-k} = (H^k)^*$ gives continuous linear functionals on C_c^∞. Thus, the ascending union does so as well.   ///

[12.3.18] Claim: $H^{-\infty}(\Gamma \backslash G/K)$ is the dual of $H^\infty(\Gamma \backslash G/K)$.

Proof: This is an instance of the general fact that every continuous linear functional on a limit of Banach spaces factors through some limitand [13.14.4].
///

Although we will prove a stronger result in terms of *global* spectral expansions in the following section, we can reduce to *local* considerations to prove the following:

[12.3.19] Claim: For $n = \dim_{\mathbb{R}} \Gamma \backslash G/K$, for all $\ell > k + \frac{n}{2}$ and $\ell \in \mathbb{Z}$,

$$H^\ell(\Gamma \backslash G/K) \subset C^k(\Gamma \backslash G/K)$$

Proof: Smoothness is a local property, which allows reduction to a *local* version of Sobolev spaces. Namely, to show that $f \in H^\ell$ is C^k at g_o, it suffices to show that ηf is in C^k, for η a smooth cutoff function near g_o. We can take η with sufficiently small support so that its compact support E lies inside a small open subset diffeomorphic to a cube of the dimension of $\Gamma \backslash G/K$. Identifying opposite faces of the cube imbeds E into a multitorus $\mathbb{T}^n$. Further, it suffices to show a suitable Sobolev inequality for *test* functions f. Thus, we can apply the literal differential operator Δ to f.

For fixed η and E, the Laplacian on $\Gamma \backslash G/K$ restricted to E and the Laplacian on $\mathbb{T}^n$ are *comparable* on functions ηf for test functions f, giving constants

$0 < A_\ell, B_\ell < \infty$ such that

$$A_\ell \cdot |\eta f|_{H^k(\mathbb{T}^n)} \leq |\eta f|_{H^\ell(\Gamma \backslash G/K)} \leq B_k \cdot |\eta f|_{H^\ell(\mathbb{T}^n)}$$

for all test functions f, for $0 \leq \ell \in \mathbb{Z}$. By continuity, the same inequalities hold for $f \in H^\ell(\Gamma \backslash G/K)$, for every k. This reduces the problem to $H^\ell(\mathbb{T}^n) \subset C^k(\mathbb{T}^n)$, which we know from [9.5]. ///

[12.3.20] Remark: The quotient $\Gamma \backslash G/K$ can fail to be a smooth manifold at various points, due to the possibility that the isotropy group G_x of $x \in G/K$ can have nontrivial intersection with Γ. However, this is surmountable in various ways. For example, G_x is compact, so $G_x \cap \Gamma$ is *finite*, and to examine smoothness of functions at x, we can harmlessly shrink Γ by finite index to shrink $G_x \cap \Gamma$ to act trivially on $\Gamma \backslash G/K$, so that $\Gamma \backslash G/K$ is smooth near x. In fact, as earlier, we can identify $C^\infty(\Gamma \backslash G/K)$ with the right K-fixed vectors $C^\infty(\Gamma \backslash G)^K$ in $C^\infty(\Gamma \backslash G)$.

[12.3.21] Remark: The smooth cutoff device allows elementary local comparison of nonnegative integer index Sobolev spaces on $\Gamma \backslash G/K$ with those on $\mathbb{T}^n$, but for noninteger index, as in the following section, such a comparison is less elementary.

[12.3.22] Corollary: $C^\infty(\Gamma \backslash G/K)^* \subset H^{-\infty}(\Gamma \backslash G/K) \subset C_c^\infty(\Gamma \backslash G/K)^*$. ///

12.4 Spectral Characterization of Sobolev Spaces H^s

By expressing Sobolev norm and differentiation via spectral transforms $\mathcal{F}$, for $0 \leq \ell \in \mathbb{Z}$ certainly $\mathcal{F}H^\ell$ is *contained in* V^ℓ, where for $s \in \mathbb{R}$

$$V^s = \{\text{measurable } v \text{ on } \Xi \ : \ (1 - \lambda_\xi)^{s/2} \cdot v \in L^2(\Xi)\}$$

For any $s \in \mathbb{R}$, give V^s the Hilbert-space structure from the expected norm

$$|v|_{V^s}^2 = \int_\Xi (1 - \lambda_\xi)^s |v|^2$$

For $s > t$, certainly there is a continuous inclusion $V^s \to V^t$ with dense image. The space V^{-s} is naturally the Hilbert-space dual $(V^s)^*$ of V^s, with $\mathbb{C}$-bilinear pairing given by integration

$$\int_\Xi v(\xi) \, w(\xi) \, d\xi \qquad (\text{complex-bilinear, for } v \in V^s \text{ and } w \in V^{-s})$$

The asymmetrical extension of the *Hermitian* pairing $V^0 \times V^0 \to \mathbb{C}$ by $v \times w \to \langle v, w \rangle_{V^0}$ to a *Hermitian* pairing on $V^s \times V^{-s}$ is

$$\langle v, w \rangle_{V^s \times V^{-s}} = \int_\Xi v(\xi) \overline{w(\xi)} \, d\xi$$

[12.4.1] Claim: The spectral transform $\mathcal{F} : C_c^\infty(\Gamma\backslash G/K) \to L^2(\Xi)$ induces a Hilbert-space isomorphism $\mathcal{F} : H^{2\ell} \xrightarrow{\approx} V^{2\ell}$ for all $0 \le \ell \in \mathbb{Z}$, and we have commuting rectangles

$$
\begin{array}{ccc}
H^{2k} & \xrightarrow[\approx]{1-\Delta} & H^{2k-2} \\
\mathcal{F} \downarrow \approx & & \mathcal{F} \downarrow \approx \\
V^{2k} & \xrightarrow[\approx]{\times(1-\lambda_\xi)} & V^{2k-2}
\end{array}
\qquad \text{(for all } 1 \le k \in \mathbb{Z})
$$

Proof: Plancherel asserts that $\mathcal{F} : H^0 \to V^0$ is an isomorphism. By design, $1 - \Delta$ gives an isomorphism $H^{2k} \to H^{2k-2}$ for all $k \in \mathbb{Z}$. Even more directly, multiplication by $1 - \lambda_\xi$ gives an isomorphism $V^s \to V^{s-2}$ for all $s \in \mathbb{R}$. Corollary [12.3.12] shows that $\mathcal{F}$ intertwines $1 - \Delta$ and multiplication by $1 - \lambda_\xi$ on Sobolev spaces with positive index, and dualization gives the same result on negative-index spaces.   ///

This allows us to define an *isomorphism* $\mathcal{F} : H^{-2k} \to V^{-2k}$ for $0 > -k \in \mathbb{Z}$ as the adjoint to the *isomorphism* $\mathcal{F}^{-1} : V^{2k} \to H^{2k}$. Proof of the analogous assertion for *odd*-index H^k and V^k is slightly complicated by the fact that the Sobolev space H^k does not have an elementary isomorphism to H^0, so Plancherel does not immediately resolve the issue.

[12.4.2] Claim: The spectral transform $\mathcal{F} : C_c^\infty(\Gamma\backslash G/K) \to L^2(\Xi)$ induces Hilbert-space isomorphisms $\mathcal{F} : H^{2k+1} \to V^{2k+1}$ for all $0 \le k \in \mathbb{Z}$, and we have commuting rectangles

$$
\begin{array}{ccc}
H^{2k+1} & \xrightarrow[\approx]{1-\Delta} & H^{2k-1} \\
\mathcal{F} \downarrow \approx & & \mathcal{F} \downarrow \approx \\
V^{2k+1} & \xrightarrow[\approx]{\times(1-\lambda_\xi)} & V^{2k-1}
\end{array}
\qquad \text{(for all } 1 \le k \in \mathbb{Z})
$$

Proof: In the commuting rectangle

$$
\begin{array}{ccc}
H^2 & \xrightarrow{\text{inc}} & H^1 \\
\mathcal{F} \downarrow \approx & & \downarrow \mathcal{F} \\
V^2 & \xrightarrow{\text{inc}} & V^1
\end{array}
$$

the horizontal maps are injections with dense images, and the left vertical map is an isomorphism, from the previous. Thus, $\mathcal{F}(H^1)$ is dense in V^1. Since $\mathcal{F}$ intertwines $1 - \Delta$ with multiplication by $1 - \lambda_\xi$, the spectral map $\mathcal{F} : H^1 \to V^1$

is an isometry to its image. Since H^1 is complete, the image is *closed*. A dense, closed subspace of V^1 is the whole V^1.

As in the even-index case, by design, $(1 - \Delta) : H^{2k+1} \to H^{2k-1}$ is an isomorphism, and multiplication by $1 - \lambda_\xi$ is an isomorphism $V^{2k+1} \to V^{2k-1}$. The intertwining of these two operators by $\mathcal{F}$, by [12.3.12], gives the commutativity.   ///

Thus, we can define isomorphisms $\mathcal{F} : H^{-1-2k} \to V^{-1-2k}$ as the inverses of the (isomorphism) adjoints $\mathcal{F}^* : V^{-1-2k} \to H^{-1-2k}$ to (the isomorphism) $\mathcal{F} : H^{2k+1} \to V^{2k+1}$. Unlike the even-index case, we need

[12.4.3] Lemma: We have a commutative rectangle with all maps isomorphisms:

$$
\begin{array}{ccc}
H^1 & \xrightarrow[\approx]{\ 1-\Delta\ } & H^{-1} \\[2pt]
{\scriptstyle\mathcal{F}}\Big\downarrow{\scriptstyle\approx} & & {\scriptstyle\approx}\Big\downarrow{\scriptstyle\mathcal{F}} \\[2pt]
V^1 & \xrightarrow[\approx]{\ \times(1-\lambda_\xi)\ } & V^{-1}
\end{array}
$$

Proof: Again, the horizontal maps are (isometric) isomorphisms by design, the left vertical map is an (isometric) isomorphism from earlier, and the right vertical map is the inverse of the adjoint of the left vertical map. Then the commutativity is immediate: composing $\mathcal{F}^* \circ (1 - \lambda_\xi) \circ \mathcal{F}$ around three sides, for $f, F \in C_c^\infty(\Gamma \backslash G / K)$, the characterization of adjoints and Plancherel,

$$
\langle (\mathcal{F}^* \circ (1 - \lambda_\xi) \circ \mathcal{F})f, \ F \rangle_{H^{-1} \times H^1}
$$
$$
= \langle ((1 - \lambda_\xi) \circ \mathcal{F})f, \ \mathcal{F}F \rangle_{V^{-1} \times V^1}
$$
$$
= \langle \mathcal{F}(1 - \Delta)f, \ \mathcal{F}F \rangle_{V^{-1} \times V^1} = \langle (1 - \Delta)f, \ F \rangle_{H^{-1} \times H^1}
$$
$$
= \langle (1 - \Delta)f, \ F \rangle_{H^{-1} \times H^1}
$$

since the asymmetrical pairings are the extensions by continuity of the L^2 pairing restricted to test functions. Since $H^{-1} = (H^1)^*$, this gives the assertion.///

Thus, as in the even-index case, we have

[12.4.4] Corollary: For all $k \in \mathbb{Z}$, we have commuting rectangles

$$
\begin{array}{ccc}
H^{2k+1} & \xrightarrow[\approx]{\ 1-\Delta\ } & H^{2k-1} \\[2pt]
{\scriptstyle\mathcal{F}}\Big\downarrow{\scriptstyle\approx} & & {\scriptstyle\mathcal{F}}\Big\downarrow{\scriptstyle\approx} \\[2pt]
V^1 & \xrightarrow[\approx]{\ \times(1-\lambda_\xi)\ } & V^{2k-1}
\end{array}
$$

That is, for $k \in \mathbb{Z}$, the spaces V^k and the multiplication operator $1 - \lambda_\xi$ are a faithful spectral-side mirror of the spaces H^k and operator $1 - \Delta$. In fact, on the spectral side, greater flexibility is afforded by the spaces V^s for $s \in \mathbb{R}$. Given the compatibility just proven, we can define Sobolev norms

$$|f|_{H^s} = |\mathcal{F}f|_{V^s} \qquad \text{(for } s \in \mathbb{R} \text{ and } f \in C_c^\infty(\Gamma\backslash G/K))$$

and for $s \in \mathbb{R}$

$$H^s = H^s(\Gamma\backslash G/K)$$
$$= \text{completion of } C_c^\infty(\Gamma\backslash G/K) \text{ with respect to } |\cdot|_{H^s}$$

As an application of the pretrace formulas of [12.1] and [12.2]:

[12.4.5] Claim: With n the dimension of $\Gamma\backslash G/K$, we have $\int_\Xi |\Phi_\xi(z_o)|^2 \cdot (1 - \lambda_\xi)^{-\frac{n}{2}-\varepsilon} < \infty$, with a bound depending uniformly on z_o in compacts in $\Gamma\backslash G/K$.

Proof: In current notation, the pretrace formulas assert that, for z_o in a fixed compact $C \subset \Gamma\backslash G/K$,

$$\int_{\xi : |\lambda_\xi| \leq T^2} |\Phi_\xi(z_o)|^2 \ll_C T^n \qquad \text{(as } T \to +\infty)$$

or

$$\int_{\xi : |\lambda_\xi| \leq T} |\Phi_\xi(z_o)|^2 \ll_C T^{\frac{n}{2}} \qquad \text{(as } T \to +\infty)$$

Thus, summing by parts,

$$\int_\Xi |\Phi_\xi(z_o)|^2 \cdot (1 - \lambda_\xi)^{-\frac{n}{2}-\varepsilon}$$

$$= \sum_{\ell=1}^\infty \int_{\xi : \ell - 1 \leq |\lambda_\xi| < \ell} |\Phi_\xi(z_o)|^2 \cdot (1 - \lambda_\xi)^{-\frac{n}{2}-\varepsilon}$$

$$\ll_C \sum_{\ell=1}^\infty \int_{\xi : |\lambda_\xi| < \ell} |\Phi_\xi(z_o)|^2 \cdot \left((1 + \ell)^{-\frac{n}{2}-\varepsilon} - (1 + (\ell+1))^{-\frac{n}{2}-\varepsilon} \right)$$

$$\ll_C \sum_{\ell=1}^\infty \int_{\xi : |\lambda_\xi| < \ell} |\Phi_\xi(z_o)|^2 \cdot (1 + \ell)^{-\frac{n}{2}-\varepsilon-1}$$

$$\ll_C \sum_{\ell=1}^\infty \ell^{\frac{n}{2}} \cdot (1 + \ell)^{-\frac{n}{2}-\varepsilon-1} < \infty$$

as claimed.   ///

Still n is the dimension of $\Gamma\backslash G/K$. Now we can prove

[12.4.6] Theorem: For $s > \frac{n}{2}$, for $f \in C_c^\infty(\Gamma\backslash G/K)$, and for compact $C \subset \Gamma\backslash G/K$, we have a global Sobolev inequality

$$\sup_{z \in C} |f(z)| \ll_{C,s} |f|_{H^s}$$

Thus, $H^s(\Gamma\backslash G/K) \subset C^o(\Gamma\backslash G/K)$. Further, with $\Xi_\ell = \{\xi \in \Xi : |\lambda_\xi| \leq \ell\}$,

$$\lim_\ell \int_{\Xi_\ell} \langle f, \Phi_\xi \rangle \cdot \Phi_\xi \, d\xi \; = \; f \qquad \text{(in the } C^o \text{ topology)}$$

[12.4.7] Remark: That is, for $s > \frac{n}{2}$, $H^s(\Gamma\backslash G/K)$ is an improved version of $C^o(\Gamma\backslash G/K)$, in the sense that this H^s not only consists of continuous functions but also the spectral expansion of every $f \in H^s$ converges uniformly pointwise to f on compacts. In contrast, already on the circle $\mathbb{T}$, spectral expansions (Fourier series) of continuous functions need not converge pointwise, much less uniformly so.

Proof: For $0 \leq \ell < +\infty$, Ξ_ℓ is compact. Every Φ_ξ is smooth, and $\xi \to \Phi_\xi$ is a continuous $C^\infty(\Gamma\backslash G/K)$-valued function on Ξ. Thus, for a test function f,

$$f_\ell \; = \; \int_{\Xi_\ell} \langle f, \Phi_\xi \rangle \cdot \Phi_\xi(z) \, d\xi$$

exists as a $C^\infty(\Gamma\backslash G/K)$-valued Gelfand-Pettis integral, so is certainly continuous. By the spectral characterization of H^s, the sequence $\{f_\ell\}$ approaches f in the H^s topology.

For $f \in H^s$, by Cauchy-Schwarz-Bunyakowsky, for $z_o \in C$,

$$\left| \int_\Xi \mathcal{F}f(\xi) \cdot \Phi(z_o) \, d\xi \right|$$

$$= \left| \int_\Xi \mathcal{F}f(\xi)(1 - \lambda_\xi)^{s/2} \cdot (1 - \lambda_\xi)^{-s/2} \Phi(z_o) \, d\xi \right|$$

$$\leq \left(\int_\Xi |\mathcal{F}f(\xi)|^2 (1 - \lambda_\xi)^s \, d\xi \right)^{\frac{1}{2}} \cdot \left(\int_\Xi (1 - \lambda_\xi)^{-s} |\Phi(z_o)|^2 \, d\xi \right)^{\frac{1}{2}}$$

$$= |f|_{H^s} \cdot \left(\int_\Xi (1 - \lambda_\xi)^{-s} |\Phi(z_o)|^2 \, d\xi \right)^{\frac{1}{2}}$$

The previous claim shows that the latter integral is finite for $s > \frac{n}{2}$, with a bound uniform in $z_o \in C$. That is, the sup norm on C of the pointwise function $z \to \int_\Xi \mathcal{F}f(\xi) \cdot \Phi_\xi(z) \, d\xi$ *exists*, and is dominated by $|f|_{H^s}$.

Thus, for test function f, the H^s convergence of the continuous function f_ℓ to f implies C^o convergence $f_\ell \to f$. That is, the spectral expansion of f converges pointwise to f, uniformly on compacts. Extending by continuity, the same result follows for $f \in H^s$. ▢

Since test functions are dense in H^s and H^s convergence implies C^o convergence, $H^s \subset C^o$, and the spectral expansion of every function f in H^s converges pointwise to f, uniformly on compacts.   ///

[12.4.8] Corollary: The automorphic Dirac δ_{z_o} at $z_o \in \Gamma\backslash G/K$ is in H^{-s} for every $s > \frac{n}{2}$, and has spectral expansion

$$\delta_{z_o} = \int_\Xi \overline{\Phi_\xi(z_o)} \cdot \Phi_\xi \, d\xi \qquad \text{(convergent in } H^{-s}\text{)}$$

[12.4.9] Remark: Unsurprisingly, the indicated integral does not converge pointwise, but there is no claim that it does so.

Proof: Fix $s > \frac{n}{2}, z_o \in \Gamma\backslash G/K$, and take $f \in H^s$. By the theorem, $H^s \subset C^o$, so $f \to f(z_o) = \delta_{z_o}(f)$ is a continuous linear functional on H^s, so is in H^{-s}. To determine its spectral coefficients, consider

$$\delta_{z_o}(f) = f(z_o) = \int_\Xi \mathcal{F}f(\xi) \cdot \Phi_\xi(z_o) \, d\xi$$

The claim shows that $\varphi(\xi) = \Phi_\xi(z_o)$ is in V^{-s}, so this integral is the complex bilinear pairing on $V^s \times V^{-s}$ applied to $\mathcal{F}f$ and φ. By uniqueness of spectral expansions, $\mathcal{F}\delta_{z_o} = \varphi$.   ///

12.5 Continuation of Solutions of Differential Equations

Given $f \in H^{-\infty}(\Gamma\backslash G/K)$ and $\lambda \in \mathbb{C}$, we want to solve

$$(\Delta - \lambda)u = f$$

for $u \in H^{-\infty}(\Gamma\backslash G/K)$, when possible. Here Δ is the extension of the invariant Laplacian from test functions to $H^{-\infty}$. Applying $\mathcal{F}$ to both sides gives

$$\mathcal{F}f = \mathcal{F}(\Delta - \lambda)u = (\lambda_\xi - \lambda)\mathcal{F}u$$

and both $\mathcal{F}f$ and $\mathcal{F}u$ are in weighted L^2 spaces on Ξ. For $\lambda \notin (-\infty, 0]$, the function $\lambda_\xi - \lambda$ is bounded away from 0 on Ξ, so we can simply *divide* to obtain

$$\mathcal{F}u = \frac{\mathcal{F}f}{\lambda_\xi - \lambda}$$

That is,

$$u = \int_\Xi \frac{\mathcal{F}f(\xi)}{\lambda_\xi - \lambda} \cdot \Phi_\xi \, d\xi \qquad \text{(converging in } H^{-\infty}\text{)}$$

[12.5.1] Claim: For $\lambda \notin (-\infty, 0]$, the previous solution u to the differential equation $(\Delta - \lambda)u = f$ is the *unique* solution in $H^{-\infty}$. That is, the corresponding homogeneous equation has no solutions in $H^{-\infty}$.

Proof: The difference $v \in H^{-\infty}$ between the previous solution and any other would be a solution to the homogeneous equation $(\Delta - \lambda)v = 0$. Since $v \in H^{-\infty}$, it has a spectral expansion $v = \int_{\Xi} \mathcal{F}v(\xi) \cdot \Phi_\xi \, d\xi$, and the differential equation gives $(\lambda_\xi - \lambda)\mathcal{F}v = 0$. Since $\lambda_\xi - \lambda \neq 0$ on Ξ, this requires that $\mathcal{F}v = 0$ almost everywhere, so v is 0 in $H^{-\infty}$. ///

For $\lambda \in (-\infty, 0]$ there is potential interaction with the eigenvalues λ_ξ and eigenfunctions Φ_ξ. For simplicity, we consider only the simplest example of $SL_2(\mathbb{Z})$. The other three simple examples admit nearly identical treatment, with uninteresting minor complications due to normalizations of constants. The spectral expansion, convergent in $H^{-\infty}$, is

$$f = \sum_{\text{cfm } F} \langle f, F \rangle \cdot F + \frac{\langle f, 1 \rangle \cdot 1}{\langle 1, 1 \rangle} + \frac{1}{4\pi i} \int_{(\frac{1}{2})} \langle f, E_s \rangle \cdot E_s \, ds$$

where the indicated pairings and integrals are extensions by continuity of the literal pairings and integrals, as earlier, and F runs through a orthonormal basis for cuspforms, consisting of strong-sense cuspforms.

[12.5.2] Claim: $f \in H^r$ if and only if the discrete-spectrum part

$$\sum_{\text{cfm } F} \langle f, F \rangle \cdot F + \frac{\langle f, 1 \rangle \cdot 1}{\langle 1, 1 \rangle}$$

and the continuous-spectrum part

$$\frac{1}{4\pi i} \int_{(\frac{1}{2})} \langle f, E_s \rangle \cdot E_s \, ds$$

are both in H^r, individually.

Proof: Use the spectral characterization. ///

Use notation $\lambda_s = s(s-1)$, and for cuspform eigenfunction F let $s_F \in \mathbb{C}$ be such that $\Delta F = \lambda_{s_F} \cdot F$. For $\text{Re}(w) > \frac{1}{2}$ and $w \neq 1$, by division, the equation $(\Delta - \lambda_w)u = f$ has solution

$$u = \sum_{\text{cfm } F} \frac{\langle f, F \rangle \cdot F}{\lambda_{s_F} - \lambda_w} + \frac{\langle f, 1 \rangle \cdot 1}{(\lambda_1 - \lambda_w) \cdot \langle 1, 1 \rangle} + \frac{1}{4\pi i} \int_{(\frac{1}{2})} \frac{\langle f, E_s \rangle \cdot E_s}{\lambda_s - \lambda_w} \, ds$$

This spectral expansion converges at least in $H^{-\infty}$. For $f \in H^r$ with $r \in \mathbb{R}$, the spectral characterization shows that $u \in H^{r+2}$, and the spectral expansion converges in H^{r+2}.

To examine possible solutions for all $w \in \mathbb{C}$, it is useful to consider a solution $u = u_w$ to $(\Delta - \lambda_w)u = f$ as a holomorphic or meromorphic function-valued function of w. By the spectral characterization of H^r and H^{r+2}, the cuspidal component

$$\sum_{\text{cfm } F} \frac{\langle f, F \rangle \cdot F}{\lambda_{s_F} - \lambda_w}$$

is visibly a meromorphic H^{r+2}-valued function of $w \in \mathbb{C}$, with poles at most at $w = s_F$: the decomposition [7.1] of cuspforms shows that the multiplicity of s_F is finite and that the points s_F are *discrete* in C. The constant component is similar.

The continuous spectrum component

$$\frac{1}{4\pi i} \int_{(\frac{1}{2})} \frac{\langle f, E_s \rangle \cdot E_s}{\lambda_s - \lambda_w} \, ds$$

is subtler and does *not* generally meromorphically continue as an H^{r+2}-valued function, but only in a broader sense, as follows.

[12.5.3] Claim: For $\lambda_w \leq -1/4$, if $(\Delta - \lambda_w)u = f$ has a solution u in $H^{-\infty}$, then $\langle f, E_w \rangle = 0$, in the strong sense that $\langle f, E_s \rangle / (\lambda_s - \lambda_w)$ is locally integrable near $s = w$.

Proof: The continuous-spectrum part of the spectral transform of the differential equation gives

$$(\lambda_s - \lambda_w)\langle u, E_s \rangle = \langle f, E_s \rangle$$

almost everywhere in s with $\text{Re}(s) = \frac{1}{2}$, where the pairings are extensions by continuity of the literal integrals and are at least locally integrable functions. Since $\lambda_s - \lambda_w = (s - w)(s - 1 + w)$, we have the indicated vanishing.   ///

[12.5.4] Theorem: Let X be a quasi-complete, locally convex topological vector space containing both H^{r+2} and Eisenstein series. The function

$$u_w = \frac{1}{4\pi i} \int_{(\frac{1}{2})} \frac{\langle f, E_s \rangle \cdot E_s}{\lambda_s - \lambda_w} \, ds \qquad \text{(convergent in } H^{r+2}\text{)}$$

has a meromorphic continuation as X-valued function of w, with functional equation

$$u_w = u_{1-w} - \frac{\langle f, E_w \rangle \cdot E_w}{2w - 1}$$

[12.5.5] Remark: Thus, although u_{1-w} is in H^{r+2} for $\text{Re}(w) < \frac{1}{2}$, the extra term is not in $H^{-\infty}$ unless it is 0, that is, unless $\langle f, E_w \rangle = 0$.

[12.5.6] Remark: Despite the seeming symmetry of the spectral integral for u_w under $w \to 1 - w$, there is no such symmetry.

Proof: This begins with a natural regularization:

$$
u_w = \frac{1}{4\pi i} \int_{(\frac{1}{2})} \frac{\langle f, E_s \rangle \cdot E_s - \langle f, E_w \rangle \cdot E_w}{\lambda_s - \lambda_w} \, ds
$$

$$
+ \frac{1}{4\pi i} \int_{(\frac{1}{2})} \frac{\langle f, E_w \rangle \cdot E_w}{\lambda_s - \lambda_w} \, ds
$$

$$
= \frac{1}{4\pi i} \int_{(\frac{1}{2})} \frac{\langle f, E_s \rangle \cdot E_s - \langle f, E_w \rangle \cdot E_w}{\lambda_s - \lambda_w} \, ds
$$

$$
+ \langle f, E_w \rangle \, E_w \frac{1}{4\pi i} \int_{(\frac{1}{2})} \frac{ds}{\lambda_s - \lambda_w}
$$

$$
= \frac{1}{4\pi i} \int_{(\frac{1}{2})} \frac{\langle f, E_s \rangle \cdot E_s - \langle f, E_w \rangle \cdot E_w}{\lambda_s - \lambda_w} \, ds - \frac{\langle f, E_w \rangle \, E_w}{2(2w - 1)}
$$

by residues.

The integral appears to be better behaved near $s = w$, but since it is not necessarily a *literal* integral, the appearance is potentially misleading. With $t = \operatorname{Im}(s)$, rewrite

$$
\int_{(\frac{1}{2})} \frac{\langle f, E_s \rangle \cdot E_s - \langle f, E_w \rangle \cdot E_w}{\lambda_s - \lambda_w} \, ds
$$

$$
= \int_{|t| > T} \frac{\langle f, E_s \rangle \cdot E_s - \langle f, E_w \rangle \cdot E_w}{\lambda_s - \lambda_w} \, ds
$$

$$
+ \int_{|t| \leq T} \frac{\langle f, E_s \rangle \cdot E_s - \langle f, E_w \rangle \cdot E_w}{\lambda_s - \lambda_w} \, ds
$$

The first integral is

$$
\int_{|t| > T} \frac{\langle f, E_s \rangle \cdot E_s}{\lambda_s - \lambda_w} \, ds - \langle f, E_w \rangle \, E_w \cdot \int_{|t| > T} \frac{ds}{\lambda_s - \lambda_w} \, ds
$$

The latter integral on $|t| > T$ is H^{r+2}-valued. The extra term is meromorphic in w but takes values in some function space adequate to contain Eisenstein series. Of course, if $\langle f, E_w \rangle = 0$, then that extra term disappears.

The second integral

$$
\int_{|t| \leq T} \frac{\langle f, E_s \rangle \cdot E_s - \langle f, E_w \rangle \cdot E_w}{\lambda_s - \lambda_w} \, ds
$$

is compactly supported and is a holomorphic X-valued function of two complex variables s, w away from the diagonal $s = w$. By design, there is cancellation on the diagonal, as is visible from a vector-valued power series expansion [15.8]. Thus, the integrand is a holomorphic X-valued function of s, w.

Let $\mathrm{Hol}(\Omega, X)$ be the space of holomorphic X-valued functions on a region Ω, with seminorms

$$\nu_\mu(f) = \sup_{w \in K} \mu(f(w))$$

where $K \subset \Omega$ is compact, and the topology on X is given by seminorms μ. With this topology, $\mathrm{Hol}(\Omega, X)$ is quasi-complete and locally convex [15.3.2]. By [15.3.3] and [15.3.4], for a complex-analytic X-valued function $f(s, w)$ in two variables, on a domain $\Omega_1 \times \Omega_2 \subset \mathbb{C}^2$, function $s \to (w \to f(z, w)))$ is a holomorphic $\mathrm{Hol}(\Omega_1, X)$-valued function on Ω_2.

Thus, letting Ω be an appropriate bounded open containing the set where $|t| \leq T$, the integrand in the integral over $|t| \leq T$ is a compactly supported, continuous, $\mathrm{Hol}(\Omega, X)$-valued function of s and has a Gelfand-Pettis integral in $\mathrm{Hol}(\Omega, X)$. That is, it has a meromorphic continuation as an X-valued function of w.

To obtain the functional equation of u_w, from the first part of the proof, at first for $\mathrm{Re}\,(w) > \frac{1}{2}$,

$$u_w = \frac{1}{4\pi i} \int_{(\frac{1}{2})} \frac{\langle f, E_s \rangle \cdot E_s - \langle f, E_w \rangle \cdot E_w}{\lambda_s - \lambda_w} \, ds - \frac{\langle f, E_w \rangle E_w}{2(2w - 1)}$$

and then this holds by meromorphic continuation. Now take $\mathrm{Re}\,(w) < \frac{1}{2}$, and bring the regularizing term back out, producing

$$u_w = \frac{1}{4\pi i} \int_{(\frac{1}{2})} \frac{\langle f, E_s \rangle \cdot E_s}{\lambda_s - \lambda_w} \, ds$$

$$- \langle f, E_w \rangle \cdot E_w \cdot \frac{1}{4\pi i} \int_{(\frac{1}{2})} \frac{ds}{\lambda_s - \lambda_w} - \frac{\langle f, E_w \rangle E_w}{2(2w - 1)}$$

$$= \frac{1}{4\pi i} \int_{(\frac{1}{2})} \frac{\langle f, E_s \rangle \cdot E_s}{\lambda_s - \lambda_w} \, ds - \frac{\langle f, E_w \rangle E_w}{(2w - 1)} \qquad (\text{for } \mathrm{Re}\,(w) < \tfrac{1}{2})$$

That is, the two extra terms do not cancel but reinforce. For $\mathrm{Re}\,(w) < \frac{1}{2}$, $\mathrm{Re}\,(1 - w) > \frac{1}{2}$, so the latter integral is u_{1-w}. That is,

$$u_w = u_{1-w} - \frac{\langle f, E_w \rangle E_w}{(2w - 1)}$$

at first for $\mathrm{Re}\,(w) < \frac{1}{2}$, but then for all w (away from poles), by the identity principle.   ///

[12.5.7] Remark: One simple choice for a topological vector space containing both suitable global automorphic Sobolev spaces and Eisenstein series is as follows. A preliminary choice of topological vector space $\mathfrak{E}$ containing Eisenstein series is needed. An easy choice is the Fréchet space $\mathfrak{E} = C^\infty(\Gamma\backslash G/K)$. Others are spaces of moderate-growth continuous functions or moderate-growth smooth functions. The essential point is that $\mathcal{D} = C_c^\infty(\Gamma\backslash G/K)$ should be dense in $\mathfrak{E}$. Since $\mathcal{D}$ is dense in H^r, we could hope for a topological vector space X fitting into a *pushout diagram*[3]

$$
\begin{array}{ccc}
\mathcal{D} & \xrightarrow{\text{inc}} & H^r \\
\text{\scriptsize inc} \downarrow & & \downarrow \\
\mathfrak{E} & \longrightarrow & X
\end{array}
$$

meaning that, for every topological vector space Y fitting into a commutative diagram

$$
\begin{array}{ccc}
\mathcal{D} & \xrightarrow{\text{inc}} & H^r \\
\text{\scriptsize inc} \downarrow & & \downarrow \\
\mathfrak{E} & \dashrightarrow & Y
\end{array}
$$

there is a unique $X \to Y$ making a commutative diagram

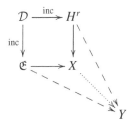

For the usual diagrammatic reasons, there is at most one such X, up to unique isomorphism. To prove existence, as with colimits expressed as quotients of coproducts in [13.8], the pushout is a natural quotient of the coproduct: X is the quotient of $H^r \oplus \mathfrak{E}$ by the closure of the antidiagonal copy $\mathcal{D}^{-\Delta} = \{(\varphi, -\varphi) : \varphi \in \mathcal{D}\}$ of $\mathcal{D}$.

12.6 Example: Automorphic Green's Functions

Continue the situation of the previous section. From [12.4.8], the automorphic Dirac δ_z at $z \in \Gamma\backslash G/K$ is in $H^{-1-\varepsilon}$ for every $\varepsilon > 0$. By the previous section,

---

[3] In fact, such a diagram is a nondirected type of *colimit*, that is, with index set that is not a directed set.

the equation $(\Delta - \lambda_w)u_w = \delta_z$ has a solution $u_w = u_{w,z}$ in $H^{1-\varepsilon}$ for $\mathrm{Re}\,(w) > \frac{1}{2}$ and $w \neq 1$, with a meromorphic continuation in a topological vector space X large enough to include both $H^{1-\varepsilon}$ and Eisenstein series.

A traditional notation (slightly incompatibly with ours) is

$$G_w(z, z') = u_{w,z}(z') \qquad (\text{for } z, z' \in \Gamma\backslash G/K)$$

This is often called a *Green's function*. In traditional (partly heuristic) notation, the differential equation $(\Delta - \lambda_w)u_{w,z} = \delta_z$ would be written as

$$\Delta_{z'} G_w(z, z') = \lambda_w \cdot \delta_z(z') \qquad \text{or} \qquad \Delta_z G_w(z, z') = \lambda_w \cdot \delta_{z'}(z)$$

In contexts where distributions are not acknowledged, the description of $G_w(z, z')$ is considerably more awkward.

[12.6.1] Theorem: Use coordinates $z = x + iy$ on $\mathfrak{H} \approx G/K$. For $\mathrm{Re}\,(w) > \frac{1}{2}$ and $w \notin (\frac{1}{2}, 1]$, and for $a \geq \mathrm{Im}\,(z)$, the solution $u_w = u_{w,z}$ in $H^{1-\varepsilon}$ of the equation $(\Delta - \lambda_w)u_w = \delta_z$ has constant term

$$c_P u_w(ia) = \int_0^1 u_w(x + ia)\, dx = a^{1-w} \cdot \frac{E_w(z)}{1 - 2w}$$

Proof: From [12.2], since the orbits of $(N \cap \Gamma)\backslash N$ are compact and codimension 1, the distribution

$$\eta_a f = c_P f(ia) = \int_0^1 f(x + ia)\, dx$$

is in $H^{-\frac{1}{2}-\varepsilon}$ for every $\varepsilon > 0$, and, as was argued for δ_z in [12.4.8], has a corresponding spectral expansion

$$\eta_a = \frac{\eta_a(1) \cdot 1}{\langle 1, 1 \rangle} + \frac{1}{4\pi i} \int_{(\frac{1}{2})} \eta_a E_{1-s} \cdot E_s\, ds$$

Since $u_w \in H^{1-\varepsilon}$ for every $\varepsilon > 0$, η_a gives a continuous linear functional on a Sobolev space containing it, and by the extended asymmetrical form of Plancherel,

$$\eta_a(u_w) = \int_\Xi \mathcal{F}\eta_a \cdot \mathcal{F}u_w$$

$$= \frac{\eta_a(1) \cdot \delta_z(1)}{(\lambda_1 - \lambda_w) \cdot \langle 1, 1 \rangle} + \frac{1}{4\pi i} \int_{(\frac{1}{2})} \eta_a(E_{1-s}) \cdot \frac{\delta_z(E_s)}{\lambda_s - \lambda_w}\, ds$$

$$= \frac{1}{(\lambda_1 - \lambda_w) \cdot \langle 1, 1 \rangle} + \frac{1}{4\pi i} \int_{(\frac{1}{2})} (a^{1-s} + c_{1-s} a^s) \cdot \frac{E_s(z)}{\lambda_s - \lambda_w}\, ds$$

from the computation of the constant term [1.9.4]. Using the functional equation $c_{1-s}E_s = E_{1-s}$ and then replacing s by $1 - s$, the integral of $c_{1-s}a^s \cdot E_s(z)/(\lambda_s - \lambda_w)$ just produces another copy of the integral of $a^{1-s}E_s(z)/(\lambda_s - \lambda_w)$, and

$$\eta_a(u_w) = \frac{1}{(\lambda_1 - \lambda_w) \cdot \langle 1, 1 \rangle} + \frac{1}{2\pi i} \int_{(\frac{1}{2})} a^{1-s} \cdot \frac{E_s(z)}{\lambda_s - \lambda_w} \, ds$$

With $z = x + iy$, from the theory of the constant term [8.1], the Eisenstein series $E_s(z)$ is asymptotically dominated by its constant term $y^s + c_s y^{1-s}$. Thus, for $a \geq y$, by elementary estimates, the contour $\mathrm{Re}\,(s) = \frac{1}{2}$ can be pushed indefinitely to the right, picking up (negatives of) residues at $s = 1$ dues to the pole of E_s, and at $s = w$, due to the denominator. The constant-function term exactly cancels the residue at $s = 1$. Since $\lambda_s - \lambda_w = (s - w)(s - 1 + w)$,

$$\eta_a(u_w) = -\mathrm{Res}_{s=w} a^{1-s} \cdot \frac{E_s(z)}{\lambda_s - \lambda_w} = -a^{1-w} \cdot \frac{E_w(z)}{w - 1 + w}$$

as asserted. ///

[12.6.2] Corollary: For $E_w(z) = 0$, the function $u_w = u_{w,z}$ is of rapid decay.

Proof: In the region $\mathrm{Im}\,(z') > \mathrm{Im}\,(z)$, the function $z' \to u_{w,z}(z')$ is an eigenfunction for Δ. Thus, it is dominated by its constant term, just determined to be $\mathrm{Im}\,(z')^{1-w} \cdot E_w(z)/(1 - 2w)$. ///

[12.6.3] Remark: Similarly, the constant term of the meromorphic continuation of u_w in a topological vector space large enough to include both $H^{1-\varepsilon}$ and Eisenstein series is the meromorphic continuation of $a^{1-w} \cdot E_w(z)/(1 - 2w)$, assuming that the topology is fine enough so that η_a is a continuous linear functional on the larger space.

[12.6.4] Remark: As computed more generally in [2.C], finite linear combinations θ of automorphic Dirac δ's applied to Eisenstein series can be arranged to give values of L-functions. Thus, solutions $u_{w,\theta}$ to $(\Delta - \lambda_w)u = \theta$ are of rapid decay if and only if $\theta E_w = 0$.

12.7 Whittaker Models and a Subquotient Theorem

Asymptotics of solutions of second-order ordinary differential equations will imply that, for f either an Eisenstein series or a strong-sense cuspform, the representation generated by f has a common G-homomorphism image with an unramified principal series I_s. Specifically V_f be the closed subspace of

$C^\infty(\Gamma \backslash G)$ generated by f under right translations, where $C^\infty(\Gamma \backslash G)$ has the Fréchet-space structure given by sups of (Lie algebra) derivatives on compacts.

Let $\psi : N \to \mathbb{C}^\times$ be a nontrivial character on N, trivial on $\Gamma \cap N$, thus factoring through the (abelian) quotient $(\Gamma \cap N) \backslash N$. For example, one might take, with $x = (x_1, \ldots, x_{r-1}) \in \mathbb{R}^{r-1}$ and $n = \begin{pmatrix} 1 & x \\ 0 & 1 \end{pmatrix}$,

$$\psi(n) = e^{2\pi i(x_1 + \ldots + x_{r-1})}$$

The corresponding *Whittaker space* is

$$W_\psi = \{f \in C^\infty(G) : f(ng) = \psi_s(n) \cdot f(g) \text{ for } n \in N \text{ and } g \in G\}$$

with Fréchet space structure given by sups of (Lie algebra) derivatives on compacts. The natural G-homomorphism $\rho_\psi : V_f \longrightarrow W_\psi$ is given by a Gelfand-Pettis integral:

$$\rho_\psi(F) = \int_{(N \cap \Gamma) \backslash N} \overline{\psi}(n) \, F(ng) \, dn \qquad (\text{for } F \in V_f)$$

We will need to have the flexibility to choose ψ for given f so that $\rho_\psi(f) \neq 0$.

[12.7.1] Lemma: $\rho_\psi : C^\infty(\Gamma \backslash G) \to W_\psi$ is *continuous*.

Proof: Since functions in W_ψ are left ψ-equivariant, it suffices to show that a compact subset C of $N \backslash G / K$ is covered by a compact subset of G. Indeed, the *height* function η assumes a positive inf μ and finite sup σ on C. For a sufficiently large compact subset C_N of N, the compact set

$$\{g = n \cdot \begin{pmatrix} \sqrt{y} & 0 \\ 0 & 1/\sqrt{y} \end{pmatrix} \cdot K : n \in C_N, \ \mu \leq \eta(g) \leq \sigma\} \ \subset \ G$$

surjects to C. ///

[12.7.2] Claim: Let $\Delta f = \lambda \cdot f$ with $\lambda = (r-1)^2 \cdot s(s-1)$. The image $\rho_\psi(f)$ is a constant multiple of

$$\begin{pmatrix} 1 & x \\ 0 & 1 \end{pmatrix} \begin{pmatrix} \sqrt{y} & 0 \\ 0 & 1/\sqrt{t} \end{pmatrix} \cdot k \ \longrightarrow \ e^{2\pi i(x_1 + \ldots + x_{r-1})} \cdot u(y)$$

for $k \in K$, where u is the unique (up to scalars) not-rapidly-increasing solution (as $y \to +\infty$) of the differential equation

$$u'' - \frac{r-2}{y} \cdot u' - (4\pi^2 + \frac{\lambda}{y^2}) \cdot u = 0$$

That is, up to scalars, the image $\rho_\psi(f)$ in W_ψ is uniquely characterized (up to scalar multiples) by satisfaction of that differential equation, and (in fact) rapid decay as $y \to +\infty$.

Proof: The Casimir operator commutes with ρ_ψ, and on right K-invariant functions is Δ. On W_ψ, we can separate variables:

$$\lambda \cdot e^{2\pi i(x_1+\ldots+x_{r-1})} \cdot u(y)$$

$$= \Delta(e^{2\pi i(x_1+\ldots+x_{r-1})} \cdot u(y)$$

$$= \left(y^2\left(\frac{\partial^2}{\partial x_1^2} + \ldots + \frac{\partial^2}{\partial x_r^2}\right) - (r-2)y\frac{\partial}{\partial y}\right)(e^{2\pi i(x_1+\ldots+x_{r-1})} \cdot u(y)$$

$$= y^2\left((2\pi i)^2 \cdot (e^{2\pi i(x_1+\ldots+x_{r-1})} \cdot u(y) + (e^{2\pi i(x_1+\ldots+x_{r-1})} \cdot u''(y)\right)$$

$$- (r-2)y(e^{2\pi i(x_1+\ldots+x_{r-1})} \cdot u'(y)$$

$$= e^{2\pi i(x_1+\ldots+x_{r-1})} \cdot \left(-4\pi^2 y^2 \cdot u(y) + y^2 u''(y) - (r-2)yu'(y)\right)$$

Thus, the condition is

$$\lambda \cdot u = -4\pi^2 y^2 \cdot u + y^2 \cdot u'' - (r-2)y \cdot u'$$

or

$$u'' - \frac{r-2}{y} \cdot u' - \left(4\pi^2 + \frac{\lambda}{y^2}\right) \cdot u = 0$$

The point at $+\infty$ is an *irregular* singular point of a tractable sort, as in [16.10] and [16.B.1]. To see this most clearly, an equation of the form $u'' + pu' + qu = 0$ should be rearranged to have no first-derivative term by the standard procedure. Namely, let $u = v\varphi$ and determine φ so that no v' term appears in the corresponding differential equation for v:

$$0 = u'' + pu' + q = (v''\varphi + 2v'\varphi' + v\varphi'') + p(v'\varphi + v\varphi') + q(v\varphi)$$

$$= \varphi \cdot v'' + (2\varphi' + p\varphi) \cdot v' + (\varphi'' + p\varphi' + q\varphi) \cdot v$$

Thus, we require $2\varphi' + p\varphi = 0$, so $\varphi = e^{-\int p/2}$, and after dividing through by φ the equation is

$$v'' + \left(\frac{\varphi''}{\varphi} + p\frac{\varphi'}{\varphi} + q\right) \cdot v = 0$$

In the case at hand, $p(y) = -(r-2)/y$, so $\varphi(y) = e^{-\int p/2} = y^{(r-2)/2}$, and the equation is

$$v'' - \left(4\pi^2 + \frac{r-2}{2y^2}\right) \cdot v = 0$$

By *freezing* the coefficients at $y = \infty$, the solutions of the corresponding constant-coefficient differential equation give the correct leading-term asymptotics as $y \to +\infty$, up to powers of y. The frozen equation at $y = \infty$ is

$v'' - 4\pi^2 \cdot v = 0$. The solutions of the frozen equation are linear combinations of $e^{\pm 2\pi y}$. From [16.11], these are the leading terms in asymptotics for two linearly independent solutions v of the differential equation. Thus, two linearly independent solutions of the original have asymptotics

$$u = y^{\frac{r-2}{2}} \cdot e^{\pm 2\pi y} \qquad (\text{as } y \to +\infty)$$

Only the scalar multiples of $y^{\frac{r-2}{2}} \cdot e^{-2\pi y}$ alone, not involving $y^{\frac{r-2}{2}} \cdot e^{+2\pi y}$, are linear combinations decreasing as $y \to +\infty$.

By the theory of the constant term [8.1], since by assumption f is a moderate-growth eigenfunction for Δ, the asymptotic behavior of f as $y \to +\infty$ is dominated (in standard Siegel sets, as height goes to infinity) by its constant term $c_P f$, with a rapidly decreasing error term. In particular, applying the Whittaker map ρ to the constant term gives 0, so $\rho(f)$ is rapidly decreasing. This gives the assertion. ///

Thus, the image $\rho_\psi(f) \in W_\psi$ is uniquely determined up to constants, as is $\rho_\psi(V_f)$. It is important to note:

[12.7.3] Claim: For given nonconstant f, there is nontrivial ψ such that $\rho_\psi(f) \neq 0$.

Proof: If not, then in Iwasawa coordinates $N \cdot A^+$ the function f is constant along N, and is a function of the A^+ coordinate alone. But apart from constants, there is no such function on $\Gamma \backslash G / K$. ///

On the other hand, now we will identify the image in W_ψ of the corresponding principal series I_s. We will see that a G-homomorphism $I_s \to W_\psi$ from a principal series I_s also sends the spherical vector $\varphi_s^o \in I_s$ to a function satisfying the same differential equation and of rapid decay as $y \to +\infty$. First,

[12.7.4] Claim: On I_s, the Casimir operator Ω has eigenvalue $(r-1)^2 \cdot s(s-1)$.

Proof: This is a computation similar to those in [4.5–4.8]. The computation for $G = SL_2(\mathbb{R})$ suffices to illustrate the point. In the Lie algebra $\mathfrak{g}$, let

$$h = \begin{pmatrix} 1 & 0 \\ 0 & -1 \end{pmatrix} \qquad X = \begin{pmatrix} 0 & 1 \\ 0 & 0 \end{pmatrix} \qquad Y = \begin{pmatrix} 0 & 0 \\ 1 & 0 \end{pmatrix}$$

For comparison purposes, specify a normalization of the Casimir operator [4.2] $\Omega = \frac{1}{2}(\frac{1}{2}h^2 + XY + YX) \in U\mathfrak{g}$, so that by the computation in [4.5] in the Iwasawa coordinates x, y on G/K,

$$\Omega = y^2 \Big(\frac{\partial^2}{\partial x^2} + \frac{\partial^2}{\partial y^2} \Big) \qquad (\text{on } G/K)$$

Since I_s is defined by a left equivariance condition, it is reasonable to let Ω act on the *left*, as the derivative of the left translation action $(g \cdot f)(x) = f(g^{-1}x)$. In particular, X acts by 0 on $f \in I_s$. Thus, YX acts by 0. Using the commutation relation,

$$XY = YX + (XY - YX) = YX + [X, Y] = YX + h$$

Thus, XY acts by h. Thus, on I_s, Ω acts on the left by $\frac{1}{2}(\frac{1}{2}h^2 + h)$. On I_s, h acts by

$$(h \cdot f)(x) = \frac{\partial}{\partial t}\Big|_{t=0} f(e^{-t} \cdot x) = \frac{\partial}{\partial t}\Big|_{t=0} \left|\frac{e^{-t}}{e^t}\right|^s \cdot f(x)$$

$$= \frac{\partial}{\partial t}\Big|_{t=0} e^{-2ts} \cdot f(x) = -2s \cdot f(x)$$

Thus, $\frac{1}{2}(\frac{1}{2}h^2 + h)$ acts on I_s by

$$\tfrac{1}{2}(\tfrac{1}{2}(2s)^2 - (2s)) = s(s - 1)$$

as claimed. ///

At least when the integral converges suitably, the map

$$(\tau\varphi)(g) = \int_N \overline{\psi}(n) \cdot \varphi(wng)\, dn \qquad \left(\text{with } w = \begin{pmatrix} 0 & -1 \\ 1 & 0 \end{pmatrix}\right)$$

gives a natural G-homomorphism $\tau_{s,\psi} : I_s \to W_\psi$. On the spherical vector φ_s^o, it is completely determined by its values for g among a set of representatives for $N\backslash G/K$, namely, the Levi component, and by an explicit Iwasawa decomposition [1.3]

$$(\tau_{s,\psi}\varphi_s^o)\begin{pmatrix} \sqrt{y} & 0 \\ 0 & 1/\sqrt{y} \end{pmatrix} = \int_{\mathbb{R}^{r-1}} \overline{\psi}(x)\frac{y^{(r-1)s}}{(|x|^2 + y^2)^{(r-1)s}}\, dx$$

[12.7.5] Claim: The integral for $\tau_{s,\psi}$ on I_s converges absolutely for $\mathrm{Re}\,(s) > \frac{1}{2}$ and produces functions not of rapid growth.

Proof: It suffices to prove convergence for the spherical vector φ_s^o, since every other function in I_s is dominated by it. Since $|\psi| = 1$, letting $\sigma = \mathrm{Re}\,(s)$,

$$\left|\int_{\mathbb{R}^{r-1}} \overline{\psi}(x)\frac{y^{(r-1)s}}{(|x|^2 + y^2)^{(r-1)s}}\, dx\right| \leq y^{(r-2)\sigma} \int_{\mathbb{R}^{r-1}} \frac{dx}{(|x|^2 + y^2)^{(r-1)\sigma}}$$

$$= y^{(r-1)(1-\sigma)} \int_{\mathbb{R}^{r-1}} \frac{1}{(|x|^2 + 1)^{(r-1)\sigma}}\, dx$$

by replacing $x \in \mathbb{R}^{r-1}$ by $y \cdot x$. Converting to polar coordinates gives the desired convergence. Further, in that range, the bound is at worst of polynomial growth in y, so is not of rapid growth. ///

The following is necessary for the continuation.

[12.7.6] Theorem: The G-map $\tau_{s,\psi} : I_s \to W_\psi$ has a meromorphic continuation in $s \in C$, and $\tau_{s,\psi}(\varphi_s^o) \neq 0$ except for $s = 0, 1$. *(Proof for $G = SL_2(\mathbb{R})$ in the following section.)*

Granting the previous theorem, a sufficient subquotient theorem for our purposes follows:

[12.7.7] Theorem: let f be an Eisenstein series or a strong-sense cuspform, in particular generating an irreducible representation V_f of G under right translation on $\Gamma \backslash G$. This entails that f is a Δ-eigenfunction: let $\Delta f = \lambda_s \cdot f$ with $\lambda_s = (r-1)^2 \cdot s(s-1)$. Choose additive character ψ on $(N \cap \Gamma) \backslash N$ such that $\rho_\psi(f) \neq 0$. Then the image $\rho_\psi(V_f) \subset W_\psi$ is a subrepresentation of $\tau_{s,\psi}(I_s)$.

Proof: From [12.7.2], the image W_ψ contains a unique (up to scalars) right K-invariant function u of less than rapid growth with given Δ-eigenvalue λ_s. Since f is at worst of moderate growth, $\rho_\psi(f)$ must be a scalar multiple of of that function. Likewise, the image $\tau_{s,\psi}(\varphi_s^o)$ of the spherical vector in I_s is not of rapid growth and is nonzero. Thus, the irreducible $\rho_\psi(V_f)$ meets $\tau_{s,\psi}(I_s)$ at least in $\mathbb{C} \cdot u$. Thus, $\rho_\psi(V_f) \subset \tau_{s,\psi}(I_s)$. ///

12.8 Meromorphic Continuation of Intertwining Operators

The analytic continuation of $\tau_{s,\psi}$ to $\operatorname{Re}(s) = \frac{1}{2}$, for real and non-positive $\lambda_s = s(s-1)$ possible eigenvalue of Δ for eigenfunctions in the spectral expansion and Plancherel, is just beyond the range of convergence $\operatorname{Re}(s) > \frac{1}{2}$ of the integral giving $\tau_{s,\psi}$. To prove meromorphic continuation, take $G = SL_2(\mathbb{R})$.

[12.8.1] Theorem: The G-map $\tau_{s,\psi} : I_s \to W_\psi$ has a meromorphic continuation in $s \in C$, and $\tau_{s,\psi}(\varphi_s^o) \neq 0$ except for $s = 0, 1$.

Proof: First, we demonstrate meromorphic continuation of the value of τ on each of the vectors

$$\varphi_s^\ell \begin{pmatrix} 1 & x \\ 0 & 1 \end{pmatrix} \begin{pmatrix} \sqrt{y} & 0 \\ 0 & 1/\sqrt{y} \end{pmatrix} \begin{pmatrix} \cos\theta & \sin\theta \\ -\sin\theta & \cos\theta \end{pmatrix} = y^s \cdot e^{2i\ell\theta}$$

via Bochner's lemma [3.A]. Using the explicit Iwasawa decomposition [1.3] for $SL_2(\mathbb{R})$, after some typical minor rearrangements, up to irrelevant constants the integral for τ is

$$\int_{\mathbb{R}} e^{-ix} \frac{1}{(y+ix)^{s+\ell}(y-ix)^{s-\ell}} \, dx$$

absolutely convergent for $\mathrm{Re}(s) > 1$. Thus, for $\alpha, \beta \in \mathbb{C}$, on one hand consider

$$\int_{\mathbb{R}} e^{-ix} \frac{1}{(y+ix)^{\alpha}(y-ix)^{\beta}} \, dx$$

absolutely convergent for $\mathrm{Re}(\alpha + \beta) > \frac{1}{2}$. On the other hand, the identity $\int_0^{\infty} e^{-t(y+ix)} t^v \, dt/t = (iz)^{-v} \cdot \Gamma(v)$ for $y > 0$ can be viewed as a computation of a Fourier transform:

$$\int_{\mathbb{R}} e^{-itx} \cdot \left\{ \begin{array}{ll} t^{v-1}e^{-ty} & (\text{for } t > 0) \\ 0 & (\text{for } t < 0) \end{array} \right\} \, dt = (y+ix)^{-v} \cdot \Gamma(v)$$

By Fourier inversion, up to irrelevant constants,

$$\int_{\mathbb{R}} e^{ixt} \, (y+ix)^{-v} \, dx = \left\{ \begin{array}{ll} \frac{1}{\Gamma(v)} e^{-ty} t^{v-1} & (\text{for } t > 0) \\[2mm] 0 & (\text{for } t < 0) \end{array} \right.$$

Replacing x by $-x$ and t by $-t$ gives the corresponding identity for $(y-ix)^{-v}$:

$$\int_{\mathbb{R}} e^{ixt} \, (y-ix)^{-v} \, dx = \left\{ \begin{array}{ll} 0 & (\text{for } t > 0) \\[2mm] \frac{1}{\Gamma(v)} e^{-|t|y} |t|^{v-1} & (\text{for } t < 0) \end{array} \right.$$

The Fourier transform of a product is the convolution, so, up to irrelevant constants, for $\xi > 0$,

$$\int_{\mathbb{R}} e^{-ix\xi} \frac{1}{(1+ix)^{\alpha}(1-ix)^{\beta}} \, dx$$
$$= \frac{1}{\Gamma(\alpha) \cdot \Gamma(\beta)} \int_{\xi-t>0, \, t<0} e^{-|\xi-t|y} |\xi-t|^{\alpha-1} \cdot e^{-|t|y} |t|^{\beta-1} \, dt$$
$$= \frac{1}{\Gamma(\alpha) \cdot \Gamma(\beta)} \int_0^{\infty} e^{-|\xi+t|y} |\xi+t|^{\alpha-1} \cdot e^{-ty} t^{\beta-1} \, dt$$
$$= \frac{1}{\Gamma(\alpha) \cdot \Gamma(\beta)} \int_0^{\infty} e^{-(\xi+2t)y} (\xi+t)^{\alpha-1} t^{\beta} \, \frac{dt}{t}$$

Since $\xi > 0$, this is convergent for all $\alpha \in \mathbb{C}$ and for $\mathrm{Re}(\beta) > 0$. The convex hull of the union of the regions $\{(\alpha, \beta) : \mathrm{Re}(\alpha + \beta) > 1\}$ and $\{(\alpha, \beta) : \mathrm{Re}(\beta) > 0\}$ is all of $\mathbb{C}^2$ so Bochner's lemma [3.A] gives the meromorphic continuations of the functions $\tau_{s,\psi}(\varphi_s^{\ell})$. (The integral expressions show that the vertical growth in α and β is mild enough to allow application of Bochner's lemma.)

Via the Iwasawa decomposition $G = PK$, functions φ in the smooth principal series I_s can be identified with Fourier series on $K \approx SO_2(\mathbb{R})$ with rapidly decreasing coefficients. That is,

$$\varphi \begin{pmatrix} \cos\theta & \sin\theta \\ -\sin\theta & \cos\theta \end{pmatrix} = \sum_{\ell \in \mathbb{Z}} c_\ell \, \varphi_s^\ell \begin{pmatrix} \cos\theta & \sin\theta \\ -\sin\theta & \cos\theta \end{pmatrix} = \sum_{\ell \in \mathbb{Z}} c_\ell \, e^{2i\ell\theta}$$

with rapidly decreasing c_ℓ. We want to show that the image $\sum_{\ell \in \mathbb{Z}} c_\ell \, \tau_{s,\psi}(\varphi_s^\ell)$ is still convergent to a smooth function in W_ψ. The intertwining operator $\tau_{s,\psi}$ preserves the right K-equivariance. Thus, for some constants $C_{s,\ell,\psi}$,

$$\tau_{s,\psi}(\varphi_s^\ell) \begin{pmatrix} \cos\theta & \sin\theta \\ -\sin\theta & \cos\theta \end{pmatrix} = C_{s,\ell,\psi} \cdot e^{2i\ell\theta}$$

Thus, it suffices to show that the $C_{s,\ell,\psi}$ grow (in ℓ) at most polynomially.

In $\mathrm{Re}(s) \geq \frac{1}{2} + \delta$ for fixed small $\delta > 0$, the integral for $\tau_{s,\psi}(\varphi_s^\ell)$ converges absolutely, and is uniformly bounded in $\ell \in \mathbb{Z}$ and $\mathrm{Re}(s) \geq \frac{1}{2} + \varepsilon$. In the next section, we exhibit an intertwining operator $I_s \to I_{1-s}$ that is an *isomorphism* for all s with $|\mathrm{Re}(s) - \frac{1}{2}| < \delta$ with $0 < \delta < 1$, *and* that sends $\varphi_s^\ell \longrightarrow A_{s,\ell} \cdot \varphi_{1-s}^\ell$ with polynomial-growth (in ℓ) constants $A_{s,\ell}$. Thus, the analytic continuation demonstrated earlier extends to smooth vectors in $\frac{1}{4} \leq \mathrm{Re}(s) \leq \frac{1}{2} - \delta$ (for example). By Phragmén-Lindelöf, each individual polynomial growth bound extends to $\frac{1}{4} \leq \mathrm{Re}(s) \leq \frac{3}{4}$, giving the analytic continuation of $\tau_{s,\psi}$ to that region.   ///

12.9 Intertwining Operators among Principal Series

To have essentially elementary computations, we consider only $G = SL_2(\mathbb{R})$. The *standard intertwining operator* $T = T_s : I_s \to I_{1-s}$ is defined, for $\mathrm{Re}(s)$ sufficiently large, by the integral[4]

$$T_s f(g) = \int_N f(wn \cdot g) \, dn$$

with the longest Weyl element

$$w = \begin{pmatrix} 0 & -1 \\ 1 & 0 \end{pmatrix}$$

---

[4] *Why this integral?* This is an analogue of a finite-group method for writing formulas for intertwining operators from a representation induced from a subgroup A to a representation induced from a subgroup B, with intertwining operators roughly corresponding to double cosets $A\backslash G/B$. For finite groups, this goes by the name of *Mackey theory*, and Bruhat extended the idea to Lie groups and p-adic groups. For nonfinite groups, there are issues of convergence and analytic continuation.

Convergence will be clarified shortly. Since the map is an integration on the left, it does not disturb the right action of G. To verify that (assuming convergence) the image really does lie inside I_{1-s}, observe that $T_s f$ is left N-invariant by construction, and that for $m \in M$

$$(T_s f)(mg) = \int_N f(wn \cdot mg) \, dn = \int_N f(wm \, m^{-1} nm \cdot g) \, dn$$

$$= \chi_1(m) \cdot \int_N f(wm n \cdot g) \, dn$$

by replacing n by mnm^{-1}, taking into account the change of measure $d(mnm^{-1}) = \chi_1(m) \cdot dn$ coming from

$$\begin{pmatrix} a & 0 \\ 0 & a^{-1} \end{pmatrix} \begin{pmatrix} 1 & x \\ 0 & 1 \end{pmatrix} \begin{pmatrix} a & 0 \\ 0 & a^{-1} \end{pmatrix}^{-1} = \begin{pmatrix} 1 & a^2 x \\ 0 & 1 \end{pmatrix}$$

This is

$$\chi_1(m) \cdot \int_N f(wmw^{-1} \cdot w \, n \cdot g) \, dn = \chi_1(m) \cdot \int_N f(m^{-1} \cdot w \, n \cdot g) \, dn$$

$$= \chi_1(m)\chi_s(m^{-1}) \cdot \int_N f(w \, n \cdot g) \, dn = \chi_{1-s}(m) \cdot (T_s f)(g)$$

This verifies that $T_s : I_s \to I_{1-s}$.

Parametrize the maximal compact by

$$K = \{ \begin{pmatrix} \cos\theta & \sin\theta \\ -\sin\theta & \cos\theta \end{pmatrix} : \theta \in \mathbb{R}/2\pi i\mathbb{Z} \}$$

and note that the overlap is just $P \cap K = \pm 1$. Thus, a function f in I_s is completely determined by its values on K, in fact, on $\{\pm 1\} \backslash K$. Conversely, for fixed $s \in \mathbb{C}$, any smooth function f_o on $\{\pm 1\} \backslash K$ has a unique extension (depending upon s) to a function $f \in I_s$, by $f(pk) = \chi_s(p) \cdot f_o(k)$. Taking advantage of the simplicity of this situation, we may expand smooth functions on K in Fourier series

$$f \begin{pmatrix} \cos\theta & \sin\theta \\ -\sin\theta & \cos\theta \end{pmatrix} = \sum_{\ell \in \mathbb{Z}} c_\ell \, e^{2\pi i\ell\theta}$$

where the Fourier coefficients c_n are *rapidly decreasing* due to the smoothness of f.

Initially, we restrict our attention to functions $f \in I_s$ which are not merely smooth, but in fact *right K-finite* in the sense that the Fourier expansion of f restricted to K is *finite*. That is, these are finite sums of functions $\varphi_s^\ell(pk) = \chi_s(p) \cdot \rho_\ell(k)$ where

$$\rho_\ell \begin{pmatrix} \cos\theta & \sin\theta \\ -\sin\theta & \cos\theta \end{pmatrix} = e^{2\pi i\ell\theta}$$

For any function f on G with

$$f(pk) = f(g) \cdot \rho(k) \quad (\text{for } k \in K)$$

with ρ among the ρ_ℓ, say that f has (right) K-*type* ρ. From the Iwasawa decomposition $G = PK$

$$\dim_{\mathbb{C}}\{f \in I_s : f \text{ has right } K\text{-type } \rho_\ell\} = 1$$

The main computation: We will directly compute the effect of the intertwining operator T_s on φ_s^ℓ. Since the left integration over N cannot affect the right K-type, T_s preserves K-types. Since the dimensions of the subspaces of I_s and I_{1-s} with given K-type ρ are 1, necessarily T_s maps φ_s^ℓ to some multiple of φ_{1-s}^ℓ. To determine this constant, it suffices to evaluate $(T_s f)(1)$, that is, to evaluate the integral

$$(T_s f)(1) = \int_N f(wn)\, dn = \int_{\mathbb{R}} f(wn_x)\, dx$$

with $n_x = \begin{pmatrix} 1 & x \\ 0 & 1 \end{pmatrix}$. To evaluate $f(wn_x)$, we give the Iwasawa decomposition $wn_x = pk$. One convenient approach is to compute

$$(wn_x)(wn_x)^\top = (pk)(pk)^\top = pkk^{-1}p^\top = pp^\top$$

since k is orthogonal. Letting

$$p = \begin{pmatrix} a & b \\ 0 & a^{-1} \end{pmatrix}$$

and expanding $(wn_x)(wn_x)^\top$ gives

$$\begin{pmatrix} 1 & -x \\ -x & 1+x^2 \end{pmatrix} = \begin{pmatrix} a^2 + b^2 & b/a \\ b/a & 1/a^2 \end{pmatrix}$$

from which $a^{-2} = 1 + x^2$ and $b/a = -x$, so $a = 1/\sqrt{1+x^2}$ and $b = -x/\sqrt{1+x^2}$. Then $k = p^{-1}g$, so we find the Iwasawa decomposition

$$wn_x = \begin{pmatrix} \frac{1}{\sqrt{1+x^2}} & \frac{-x}{\sqrt{1+x^2}} \\ 0 & \sqrt{1+x^2} \end{pmatrix} \cdot \begin{pmatrix} \frac{1}{\sqrt{1+x^2}} & \frac{x}{\sqrt{1+x^2}} \\ \frac{-x}{\sqrt{1+x^2}} & \frac{1}{\sqrt{1+x^2}} \end{pmatrix}$$

Thus, with K-type ρ_ℓ,

$$\varphi_s^\ell(wn_x) = \varphi_s^\ell(pk) = \chi_s(p) \cdot \rho_\ell(k) = \left(\frac{1}{\sqrt{1+x^2}}\right)^{2s} \cdot \left(\frac{1+ix}{\sqrt{1+x^2}}\right)^{2\ell}$$

$$= (1+x^2)^{-s} \cdot \left(\frac{1+ix}{\sqrt{1+ix}\cdot\sqrt{1-ix}}\right)^{2\ell} = (1+x^2)^{-s} \cdot \left(\frac{1+ix}{1-ix}\right)^\ell$$

$$= (1+ix)^{-s+\ell}\,(1-ix)^{-s-\ell}$$

Thus, our intertwining operator when applied to a $f \in I_s$ with specified K-type ρ_ℓ, evaluated at $1 \in G$ is

$$(T_s f)(1) = \int_N f(wn)\, dn = \int_\mathbb{R} (1 + ix)^{-s+\ell} (1 - ix)^{-s-\ell}\, dx$$

To compute the latter, we use a standard trick employing the gamma function. That is, for complex z in the right half-plane, and for $\mathrm{Re}\,(s) > 0$,

$$\Gamma(s) \cdot z^{-s} = \int_0^\infty e^{-tz} t^{-s} \frac{dt}{t}$$

Thus,

$$(T_s \varphi_s^\ell)(1) = \int_\mathbb{R} (1 + ix)^{-s+\ell} (1 - ix)^{-s-\ell}\, dx$$

$$= \Gamma(s - \ell)^{-1} \Gamma(s + \ell)^{-1}$$

$$\times \int_\mathbb{R} \int_0^\infty \int_0^\infty e^{-u(1+ix)} u^{-(s-\ell)} e^{-v(1+ix)} v^{-(s+\ell)} \frac{du}{u} \frac{dv}{v}\, dx$$

Changing the order of integration and integrating in x first[5] gives an inner integral

$$\int_\mathbb{R} e^{ix(u-v)}\, dx = 2\pi \cdot \delta_{u-v}$$

where δ is the Dirac delta distribution. Thus, the whole integral becomes

$$(T_s \varphi_s^\ell)(1) = \frac{2\pi}{\Gamma(s - \ell)\Gamma(s + \ell)} \int_0^\infty e^{-u} u^{-(s-\ell)} e^{-u} u^{-(s+\ell)} u^{-1} u^{-1}\, du$$

$$= \frac{2\pi}{\Gamma(s - \ell)\Gamma(s + \ell)} \int_0^\infty e^{-2u} u^{-2s-1} \frac{du}{u} = \frac{2\pi\, 2^{1-2s}\, \Gamma(2s - 1)}{\Gamma(s - \ell)\Gamma(s + \ell)}$$

That is, under the intertwining $T_s : I_s \to I_{1-s}$, the function φ_s^ℓ is mapped to φ_{1-s}^ℓ multiplied by that last constant.

Subrepresentations: For brevity, let

$$\lambda(s, n) = \frac{2\pi\, 2^{1-2s}\, \Gamma(2s - 1)}{\Gamma(s - \ell)\Gamma(s + \ell)}$$

denote the constant computed above. The intertwining operator T_s is holomorphic at $s_o \in \mathbb{C}$ if for all integers ℓ the function $\lambda(s, \ell)$ is holomorphic at s_o.

---

[5] This is not legitimate from an elementary viewpoint. However, it is a compelling heuristic, correctly suggests the true conclusion, and can immediately be justified by *Fourier inversion*, as is done in the appendix.

The numerator $\Gamma(2s - 1)$ has poles at

$$\frac{1}{2}, \; 0, \; -\frac{1}{2}, \; -1, \; -\frac{3}{2}, \; -2, \; \ldots$$

The half-integer poles are not canceled by the poles of the denominator, so T_s has poles at these half-integers. At the nonpositive integers, regardless of the value of ℓ the poles of the denominator cancel the pole of the numerator. That is,

[12.9.1] Claim: $T_s : I_s \to I_{1-s}$ is holomorphic away from

$$s = \frac{1}{2}, \; -\frac{1}{2}, \; -\frac{3}{2}, \; -\frac{5}{2}, \; -\frac{7}{2}, \; \ldots$$

at which it has simple poles.   ///

For s *not* an integer, the denominator has no poles, so (away from the half-integers at which the numerator has a pole) $\lambda(s, \ell) \neq 0$ for *all* K-types ρ_{2n}. Thus,

[12.9.2] Claim: The intertwining operator $T_s : I_s \to I_{1-s}$ has trivial kernel for s not an integer (and away from its poles).   ///

Consider $s = m$ with $0 < m \in \mathbb{Z}$. The numerator has no pole at m, while the denominator has a pole, yielding a $\lambda(m, \ell) = 0$ for all integers

$$n = \pm m, \; \pm(m+1), \; \pm(m+2), \; \pm(m+3), \; \ldots$$

Thus, for $0 < m \in \mathbb{Z}$, I_m has a nontrivial infinite-dimensional subrepresentation[6] consisting of these K-types in the kernel of $T_m : I_m \to I_{1-m}$.

Consider $s = -m$ with $0 \geq -m \in \mathbb{Z}$. The numerator has a pole at $-m$, and the denominator has a *double* pole for integers

$$\ell = 0, \; \pm 1, \; \pm 2, \; \ldots, \; \pm m$$

and a *single* pole for integers

$$\ell = \pm(m+1), \; \pm(m+2), \; \pm(m+3), \; \ldots$$

Thus, $\lambda(-m, \ell) = 0$ for the double poles, and the single poles cancel. Thus, for $0 \geq m \in \mathbb{Z}$, I_m has a nontrivial subrepresentation consisting of the finitely many K-types at which the denominator has a double pole. These are (therefore) *finite-dimensional* representations, the kernels of $T_{-m} : I_{-m} \to I_{1+m}$.

---

[6] These subrepresentations have names, based on how they arose in other circumstances: they are the sum of the *holomorphic discrete series* and *anti-holomorphic discrete series* representations.

Smooth vectors: The explicit computation of the scalar $\lambda(s, \ell) = (T_s \varphi_s^\ell)(1)$ for φ_s^ℓ also shows that T_s has an analytic continuation on *smooth* vectors in I_s, not merely K-finite vectors, as follows. From $\Gamma(s) \cdot s = \Gamma(s+1)$,

$$(T_s \varphi_s^\ell)(1) = \frac{2\pi \, 2^{1-2s} \, \Gamma(2s-1)}{\Gamma(s-\ell)\Gamma(s+\ell)} = \text{polynomial growth in } \ell$$

By asymptotics of $\Gamma(s)$, for example, from Stirling's formula in [16.1]. Let

$$f = \sum_{\ell \in \mathbb{Z}} c_\ell \cdot \varphi_s^\ell$$

be smooth in I_s. Smoothness is equivalent to the *rapid decrease* of the Fourier coefficients. Then

$$T_s f = \sum_{\ell \in \mathbb{Z}} \lambda(s, \ell) \cdot c_\ell \cdot \varphi_s^\ell$$

still has rapidly decreasing coefficients, so is a smooth vector in I_{1-s}. That is, away from the poles, the intertwining operator T_s when analytically continued *is* defined on all smooth vectors in I_{1-s}, not merely K-finite ones.   ////

12.A Appendix: A Usual Trick with $\Gamma(s)$

The preceding property of $\Gamma(s)$ is sufficiently important that we review it. The gamma function is given for $\text{Re}(s) > 0$ by Euler's integral

$$\Gamma(s) = \int_0^\infty e^{-t} t^s \frac{dt}{t}$$

Replacing t by ty with $y > 0$

$$\Gamma(s) \cdot y^{-s} = \int_0^\infty e^{-ty} t^s \frac{dt}{t}$$

By analytic continuation to the right complex half-plane, for $y > 0$ and $x \in \mathbb{R}$

$$\Gamma(s) \cdot (y + 2\pi i x)^{-s} = \int_0^\infty e^{-t(y + 2\pi i x)} t^s \frac{dt}{t}$$

Having analytically continued, we may let $y = 1$ again, obtaining

$$\Gamma(s) \cdot (1 + 2\pi i x)^{-s} = \int_0^\infty e^{-t(1 + 2\pi i x)} t^s \frac{dt}{t} = \int_0^\infty e^{-2\pi i x t} e^{-t} t^s \frac{dt}{t}$$

which is the Fourier transform of

$$\varphi_s(t) = \begin{cases} e^{-t} t^{s-1} & (t > 0) \\ 0 & (t < 0) \end{cases}$$

To compute the concrete integral for $(T_s f)(1)$ invoke the Plancherel theorem, that

$$\int_{\mathbb{R}} f(x)\,\overline{\varphi(x)}\,dx = \int_{\mathbb{R}} \hat{f}(x)\,\overline{\hat{\varphi}(x)}\,dx$$

and Fourier inversion. With real $s \gg 0$, replacing x by $2\pi x$ at the first step, and with real s,

$$\int_{\mathbb{R}} (1 + ix)^{-s+n}\,(1 - ix)^{-s-n}\,dx$$

$$= 2\pi \int_{\mathbb{R}} (1 + 2\pi ix)^{-s+n}\,(1 - 2\pi ix)^{-s-n}\,dx$$

$$= 2\pi \int_{\mathbb{R}} \hat{\varphi}_{s-n}(x)\,\overline{\hat{\varphi}_{s+n}(x)}\,dx = 2\pi \int_{\mathbb{R}} \varphi_{s-n}(x)\,\overline{\varphi_{s+n}(x)}\,dx$$

$$= \frac{2\pi}{\Gamma(s-n)\,\Gamma(s+n)} \int_0^\infty e^{-u} u^{-(s-n)-1} \cdot e^{-u} u^{-(s+n)-1}\,du$$

$$= \frac{2\pi\,\Gamma(2s-1)}{\Gamma(s-n)\,\Gamma(s+n)}$$

as computed heuristically earlier. This also exhibits the constant 2π.

13

Examples: Topologies on Natural Function Spaces

We review natural topological vectorspaces of functions on relatively simple geometric objects, such as $\mathbb{R}$, as opposed to the automorphic examples $\Gamma \backslash G$ and $\Gamma \backslash X$, to separate the geometric and group-theoretic complications from the topological-analytical.

In all cases, we specify a natural topology, in which differentiation and other natural operators are *continuous*, so that the space is *complete*.

Many familiar and useful spaces of continuous or differentiable functions, such as $C^k[a, b]$, have natural metric structures and are *complete*. Often, the metric $d(,)$ comes from a *norm* $|\cdot|$, on the functions, giving Banach spaces.

Other natural function spaces, such as $C^\infty[a, b]$, $C^o(\mathbb{R})$, are *not* Banach but still do have a metric topology and are complete: these are *Fréchet spaces*, appearing as (projective) *limits* of Banach spaces, as we will see here. These lack some of the conveniences of Banach spaces, but their expressions as *limits* of Banach spaces is often sufficient.

Other important spaces, such as compactly supported continuous functions $C_c^o(\mathbb{R})$ on $\mathbb{R}$, or compactly supported smooth functions (test functions) $\mathcal{D}(\mathbb{R}) = C_c^\infty(\mathbb{R})$ on $\mathbb{R}$, are not metrizable so as to be *complete*. Nevertheless, some are expressible as *colimits* (sometimes called *inductive limits*) of Banach or Fréchet spaces, and such descriptions suffice for many applications. An *LF-space* is a countable ascending union of Fréchet spaces with each Fréchet subspace *closed* in the next. These are *strict colimits* or *strict inductive limits* of Fréchet spaces. These are generally *not* complete in the strongest sense but nevertheless, as demonstrated in [13.12], are *quasi-complete*, and this suffices for applications.

13.1 Banach Spaces $C^k[a, b]$

We give the vector space $C^k[a, b]$ of k-times continuously differentiable functions on an interval $[a, b]$ a metric that makes it *complete*. Mere *pointwise* limits of continuous functions easily fail to be continuous. First recall the standard

[13.1.1] Claim: The set $C^o(K)$ of complex-valued continuous functions on a compact set K is *complete* with the metric $|f - g|_{C^o}$, with the C^o-*norm* $|f|_{C^o} = \sup_{x \in K} |f(x)|$.

Proof: This is a typical three-epsilon argument. To show that a Cauchy sequence $\{f_i\}$ of continuous functions has a *pointwise* limit that is a continuous function, first argue that f_i has a pointwise limit at every $x \in K$. Given $\varepsilon > 0$, choose N large enough such that $|f_i - f_j| < \varepsilon$ for all $i, j \geq N$. Then $|f_i(x) - f_j(x)| < \varepsilon$ for any x in K. Thus, the sequence of values $f_i(x)$ is a Cauchy sequence of complex numbers, so has a limit $f(x)$. Further, given $\varepsilon' > 0$ choose $j \geq N$ sufficiently large such that $|f_j(x) - f(x)| < \varepsilon'$. For $i \geq N$

$$|f_i(x) - f(x)| \leq |f_i(x) - f_j(x)| + |f_j(x) - f(x)| < \varepsilon + \varepsilon'$$

This is true for every positive ε', so $|f_i(x) - f(x)| \leq \varepsilon$ for every x in K. That is, the pointwise limit is approached uniformly in $x \in [a, b]$.

To prove that $f(x)$ is continuous, for $\varepsilon > 0$, take N be large enough so that $|f_i - f_j| < \varepsilon$ for all $i, j \geq N$. From the previous paragraph $|f_i(x) - f(x)| \leq \varepsilon$ for every x and for $i \geq N$. Fix $i \geq N$ and $x \in K$, and choose a small enough neighborhood U of x such that $|f_i(x) - f_i(y)| < \varepsilon$ for any y in U. Then

$$|f(x) - f(y)| \leq |f(x) - f_i(x)| + |f_i(x) - f_i(y)| + |f(y) - f_i(y)|$$
$$\leq \varepsilon + |f_i(x) - f_i(y)| + \varepsilon < \varepsilon + \varepsilon + \varepsilon$$

Thus, the pointwise limit f is continuous at every x in U. ///

Unsurprisingly, but significantly,

[13.1.2] Claim: For $x \in [a, b]$, the *evaluation* map $f \to f(x)$ is a continuous linear functional on $C^o[a, b]$.

Proof: For $|f - g|_{C^o} < \varepsilon$, we have

$$|f(x) - g(x)| \leq |f - g|_{C^o} < \varepsilon$$

proving the continuity. ///

As usual, a real-valued or complex-valued function f on a closed interval $[a, b] \subset \mathbb{R}$ is *continuously differentiable* when it has a derivative which is itself a continuous function. That is, the limit

$$f'(x) = \lim_{h \to 0} \frac{f(x + h) - f(x)}{h}$$

exists for all $x \in [a, b]$, and the function $f'(x)$ is in $C^o[a, b]$. Let $C^k[a, b]$ be the collection of k-times continuously differentiable functions on $[a, b]$, with the C^k-*norm*

$$|f|_{C^k} = \sum_{0 \leq i \leq k} \sup_{x \in [a,b]} |f^{(i)}(x)| = \sum_{0 \leq i \leq k} |f^{(i)}|_\infty$$

where $f^{(i)}$ is the ith derivative of f. The *associated metric* on $C^k[a, b]$ is $|f - g|_{C^k}$.

Similar to the assertion about evaluation on $C^o[a, b]$,

[13.1.3] Claim: For $x \in [a, b]$ and $0 \leq j \leq k$, the *evaluation* map $f \to f^{(j)}(x)$ is a continuous linear functional on $C^k[a, b]$.

Proof: For $|f - g|_{C^k} < \varepsilon$,

$$|f^{(j)}(x) - g^{(j)}(x)| \leq |f - g|_{C^k} < \varepsilon$$

proving the continuity. ///

We see that $C^k[a, b]$ is a Banach space:

[13.1.4] Theorem: The normed metric space $C^k[a, b]$ is complete.

Proof: For a Cauchy sequence $\{f_i\}$ in $C^k[a, b]$, all the pointwise limits $\lim_i f_i^{(j)}(x)$ of j-fold derivatives exist for $0 \le j \le k$, and are uniformly continuous. The issue is to show that $\lim_i f^{(j)}$ is differentiable, with derivative $\lim_i f^{(j+1)}$. It suffices to show that, for a Cauchy sequence f_n in $C^1[a, b]$, with pointwise limits $f(x) = \lim_n f_n(x)$ and $g(x) = \lim_n f_n'(x)$ we have $g = f'$. By the fundamental theorem of calculus, for any index i,

$$f_i(x) - f_i(a) = \int_a^x f_i'(t)\,dt$$

Since the f_i' *uniformly* approach g, given $\varepsilon > 0$ there is i_o such that $|f_i'(t) - g(t)| < \varepsilon$ for $i \ge i_o$ and for *all* t in the interval, so for such i

$$\left| \int_a^x f_i'(t)\,dt - \int_a^x g(t)\,dt \right| \le \int_a^x |f_i'(t) - g(t)|\,dt \le \varepsilon \cdot |x - a| \longrightarrow 0$$

Thus,

$$\lim_i f_i(x) - f_i(a) = \lim_i \int_a^x f_i'(t)\,dt = \int_a^x g(t)\,dt$$

from which $f' = g$.   ///

By design, we have

[13.1.5] Theorem: The map $\frac{d}{dx} : C^k[a, b] \to C^{k-1}[a, b]$ is continuous.

Proof: As usual, for a linear map $T : V \to W$, by linearity $Tv - Tv' = T(v - v')$ it suffices to check continuity at 0. For Banach spaces the homogeneity $|\sigma \cdot v|_V = |\alpha| \cdot |v|_V$ shows that continuity is equivalent to existence of a constant B such that $|Tv|_W \le B \cdot |v|_V$ for $v \in V$. Then

$$|\frac{d}{dx} f|_{C^{k-1}} = \sum_{0 \le i \le k-1} \sup_{x \in [a,b]} |(\frac{df}{dx})^{(i)}(x)|$$

$$= \sum_{1 \le i \le k} \sup_{x \in [a,b]} |f^{(i)}(x)| \le 1 \cdot |f|_{C^k}$$

as desired.   ///

13.2 Non-Banach Limit $C^\infty[a, b]$ of Banach Spaces $C^k[a, b]$

The space $C^\infty[a, b]$ of infinitely differentiable complex-valued functions on a (finite) interval $[a, b]$ in $\mathbb{R}$ is not a Banach space.[1] Nevertheless, the topology

---

[1] It is not essential to prove that there is no reasonable Banach space structure on $C^\infty[a, b]$, but this can be readily proven in a suitable context.

is *completely determined* by its relation to the Banach spaces $C^k[a, b]$. That is, there is a *unique* reasonable topology on $C^\infty[a, b]$. After explaining and proving this uniqueness, we also show that this topology is *complete metric*.

This function space can be presented as

$$C^\infty[a, b] = \bigcap_{k \geq 0} C^k[a, b]$$

and we reasonably require that whatever topology $C^\infty[a, b]$ should have, each inclusion $C^\infty[a, b] \longrightarrow C^k[a, b]$ is continuous.

At the same time, given a family of *continuous linear* maps $Z \to C^k[a, b]$ from a vector space Z in some reasonable class (specified in the next section), with the *compatibility* condition of giving commutative diagrams

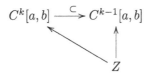

the image of Z actually lies in the intersection $C^\infty[a, b]$. Thus, diagrammatically, for every family of compatible maps $Z \to C^k[a, b]$, there is a *unique* $Z \to C^\infty[a, b]$ fitting into a commutative diagram

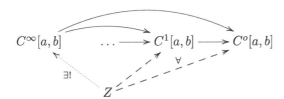

We require that this induced map $Z \to C^\infty[a, b]$ is *continuous*.

When we know that these conditions are met, we would say that $C^\infty[a, b]$ is the (projective) *limit* of the spaces $C^k[a, b]$, written

$$C^\infty[a, b] = \lim_k C^k[a, b]$$

with implicit reference to the inclusions $C^{k+1}[a, b] \to C^k[a, b]$ and $C^\infty[a, b] \to C^k[a, b]$.

[13.2.1] Claim: Up to unique isomorphism, there exists at most one topology on $C^\infty[a, b]$ such that to every compatible family of continuous linear maps $Z \to C^k[a, b]$ from a topological vector space Z, there is a unique continuous linear $Z \to C^\infty[a, b]$ fitting into a commutative diagram as just above.

Proof: Let X, Y be $C^\infty[a, b]$ with two topologies fitting into such diagrams, and show $X \approx Y$, and for a unique isomorphism. First, claim that the identity map $\mathrm{id}_X : X \to X$ is the only map $\varphi : X \to X$ fitting into a commutative diagram

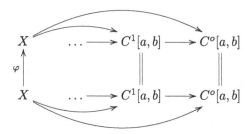

Indeed, given a compatible family of maps $X \to C^k[a, b]$, there is *unique* φ fitting into

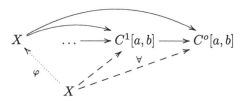

Since the identity map id_X fits, necessarily $\varphi = \mathrm{id}_X$. Similarly, given the compatible family of inclusions $Y \to C^k[a, b]$, there is unique $f : Y \to X$ fitting into

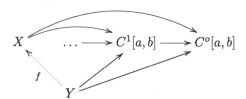

Similarly, given the compatible family of inclusions $X \to C^k[a, b]$, there is unique $g : X \to Y$ fitting into

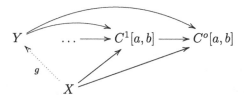

Then $f \circ g : X \to X$ fits into a diagram

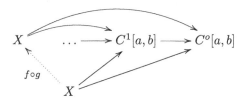

Therefore, $f \circ g = \mathrm{id}_X$. Similarly, $g \circ f = \mathrm{id}_Y$. That is, f, g are mutual inverses, so are isomorphisms of topological vector spaces.

Existence of a topology on $C^\infty[a, b]$ satisfying the foregoing condition will be proven by identifying $C^\infty[a, b]$ as the obvious diagonal *closed subspace* of the *topological product* of the *limitands* $C^k[a, b]$:

$$C^\infty[a, b] = \{\{f_k : f_k \in C^k[a, b]\} : f_k = f_{k+1} \text{ for all } k\}$$

An arbitrary *product* of topological spaces X_α for α in an index set A is a topological space X with (*projections*) $p_\alpha : X \to X_\alpha$, such that every family $f_\alpha : Z \to X_\alpha$ of maps from any other topological space Z *factors through* the p_α *uniquely*, in the sense that there is a unique $f : Z \to X$ such that $f_\alpha = p_\alpha \circ f$ for all α. Pictorially, *all triangles commute* in the diagram

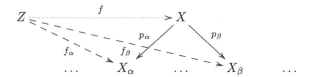

A similar argument to that for uniqueness of limits proves *uniqueness* of products up to unique isomorphism. *Construction* of products is by putting the usual product topology with basis consisting of products $\prod_\alpha Y_\alpha$ with $Y_\alpha = X_\alpha$ for all but finitely many indices, on the Cartesian product of the *sets* X_α, whose existence we grant ourselves. Proof that this usual is *a* product amounts to unwinding the definitions. By uniqueness, in particular, despite the plausibility of the *box topology* on the product, it cannot function as a product topology since it differs from the standard product topology in general.

[13.2.2] Claim: Giving the diagonal copy of $C^\infty[a, b]$ inside $\prod_k C^k[a, b]$ the subspace topology yields a (projective) limit topology.

Proof: The projection maps $p_k : \prod_j C^j[a, b] \to C^k[a, b]$ from the whole product to the factors $C^k[a, b]$ are continuous, so their restrictions to the diagonally imbedded $C^\infty[a, b]$ are continuous. Further, letting $i_k : C^k[a, b] \to C^{k-1}[a, b]$ be the inclusion, on that diagonal copy of $C^\infty[a, b]$ we have $i_k \circ p_k = p_{k-1}$ as required.

On the other hand, *any* family of maps $\varphi_k : Z \to C^k[a, b]$ induces a map $\widetilde{\varphi} : Z \to \prod C^k[a, b]$ such that $p_k \circ \widetilde{\varphi} = \varphi_k$, by the property of the product. *Compatibility* $i_k \circ \varphi_k = \varphi_{k-1}$ implies that the image of $\widetilde{\varphi}$ is inside the diagonal, that is, inside the copy of $C^\infty[a, b]$.   ///

A *countable* product of *metric* spaces X_k with metrics d_k has no canonical single metric but is *metrizable*. One of many topologically equivalent metrics is the usual

$$d(\{x_k\}, \{y_k\}) = \sum_{k=0}^{\infty} 2^{-k} \frac{d_k(x_k, y_k)}{d_k(x_k, y_k) + 1}$$

When the metric spaces X_k are *complete*, the product is complete. A closed subspace of a complete metrizable space is complete metrizable, so we have

[13.2.3] Corollary: $C^\infty[a, b]$ is complete metrizable.   ///

Abstracting the foregoing, for a (not necessarily countable) family

$$\cdots \xrightarrow{\varphi_2} B_1 \xrightarrow{\varphi_1} B_o$$

of Banach spaces with continuous linear transition maps as indicated, *not* necessarily requiring the continuous linear maps to be injective (or surjective), a *(projective) limit* $\lim_i B_i$ is a topological vector space with continuous linear maps $\lim_i B_i \to B_j$ such that, for every compatible family of continuous linear maps $Z \to B_i$ there is unique continuous linear $Z \to \lim_i B_i$ fitting into

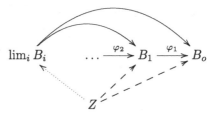

The same *uniqueness* proof as already described shows that there is at most one topological vector space $\lim_i B_i$. For *existence* by *construction*, the earlier argument needs only minor adjustment. The conclusion of complete metrizability would hold when the family is countable.

Before declaring $C^\infty[a, b]$ to be a *Fréchet* space, we must certify that it is *locally convex*, in the sense that every point has a local basis of *convex* opens.

Normed spaces are immediately locally convex because open balls are convex: for $0 \le t \le 1$ and x, y in the ε-ball at 0 in a normed space,

$$|tx + (1 - t)y| \le |tx| + |(1 - t)y| \le t|x| + (1 - t)|y| < t \cdot \varepsilon + (1 - t) \cdot \varepsilon = \varepsilon$$

Product topologies of locally convex vectorspaces are locally convex, from the *construction* of the product. The construction of the limit as the diagonal in the product, with the subspace topology, shows that it is locally convex. In particular, *countable limits of Banach spaces are locally convex, hence, are Fréchet*. All spaces of practical interest are locally convex for simple reasons, so demonstrating local convexity is rarely interesting.

[13.2.4] Theorem: $\frac{d}{dx} : C^\infty[a, b] \to C^\infty[a, b]$ is continuous.

Proof: In fact, the differentiation operator is characterized via the expression of $C^\infty[a, b]$ as a limit. We already know that differentiation d/dx gives a continuous map $C^k[a, b] \to C^{k-1}[a, b]$. Differentiation is compatible with the inclusions among the $C^k[a, b]$. Thus, we have a commutative diagram

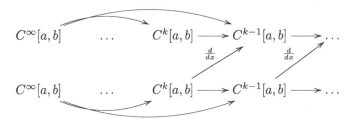

Composing the projections with d/dx gives (dashed) induced maps from $C^\infty[a, b]$ to the limitands, inducing a unique (dotted) continuous linear map to the limit, as in

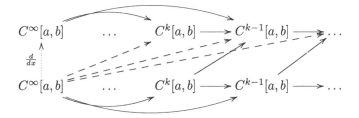

This proves the continuity of differentiation in the limit topology.   ///

In a slightly different vein, we have

[13.2.5] Claim: For all $x \in [a, b]$ and for all nonnegative integers k, the evaluation map $f \to f^{(k)}(x)$ is a continuous linear map $C^\infty[a, b] \to \mathbb{C}$.

Proof: The inclusion $C^\infty[a, b] \to C^k[a, b]$ is continuous, and the evaluation of the kth derivative is continuous.

13.3 Sufficient Notion of Topological Vectorspace

To describe a (projective) limit by characterizing its behavior in relation to *all* topological vectorspaces requires specification of what a topological vectorspace *should be*.

A *topological vector space* V (over $\mathbb{C}$) is a $\mathbb{C}$-vectorspace V with a topology on V in which *points are closed*, and so that *scalar multiplication*

$$x \times v \longrightarrow xv \qquad (\text{for } x \in k \text{ and } v \in V)$$

and *vector addition*

$$v \times w \longrightarrow v + w \qquad (\text{for } v, w \in V)$$

are *continuous*. For subsets X, Y of V, let

$$X + Y = \{x + y : x \in X, y \in Y\}$$

and

$$-X = \{-x : x \in X\}$$

The following trick is elementary, but indispensable. Given an open neighborhood U of 0 in a topological vectorspace V, continuity of vector addition yields an open neighborhood U' of 0 such that

$$U' + U' \subset U$$

Since $0 \in U'$, necessarily $U' \subset U$. This can be repeated to give, for any positive integer n, an open neighborhood U_n of 0 such that

$$\underbrace{U_n + \cdots + U_n}_{n} \subset U$$

In a similar vein, for fixed $v \in V$ the map $V \to V$ by $x \to x + v$ is a *homeomorphism*, being invertible by the obvious $x \to x - v$. Thus, *the open neighborhoods of v are of the form $v + U$ for open neighborhoods U of* 0. In particular, *a local basis at 0 gives the topology on a topological vectorspace.*

[13.3.1] Lemma: Given a compact subset K of a topological vectorspace V and a closed subset C of V not meeting K, there is an open neighborhood U of 0 in V such that

$$\text{closure}(K + U) \cap (C + U) = \phi$$

Proof: Since C is closed, for $x \in K$ there is a neighborhood U_x of 0 such that the neighborhood $x + U_x$ of x does not meet C. By continuity of vector addition

$$V \times V \times V \to V \quad \text{by} \quad v_1 \times v_2 \times v_3 \to v_1 + v_2 + v_3$$

there is a smaller open neighborhood N_x of 0 so that

$$N_x + N_x + N_x \subset U_x$$

By replacing N_x by $N_x \cap -N_x$, which is still an open neighborhood of 0, suppose that N_x is *symmetric* in the sense that $N_x = -N_x$.

Using this symmetry,

$$(x + N_x + N_x) \cap (C + N_x) = \phi$$

Since K is compact, there are finitely many $x_1, \ldots, x_n$ such that

$$K \subset (x_1 + N_{x_1}) \cup \ldots \cup (x_n + N_{x_n})$$

Let $U = \bigcap_i N_{x_i}$. Since the intersection is finite, U is open. Then

$$K + U \subset \bigcup_{i=1,\ldots,n} (x_i + N_{x_i} + U) \subset \bigcup_{i=1,\ldots,n} (x_i + N_{x_i} + N_{x_i})$$

These sets do not meet $C + U$, by construction, since $U \subset N_{x_i}$ for all i. Finally, since $C + U$ is a union of opens $y + U$ for $y \in C$, it is open, so even the *closure* of $K + U$ does not meet $C + U$.   ///

Conveniently, Hausdorff-ness of topological vectorspaces follows from the weaker assumption that points are closed:

[13.3.2] Corollary: A topological vectorspace is *Hausdorff*.

Proof: Take $K = \{x\}$ and $C = \{y\}$ in the lemma.

[13.3.3] Corollary: The topological closure $\bar{E}$ of a subset E of a topological vectorspace V can be expressed as

$$\bar{E} = \bigcap_U (E + U) \qquad \text{(where U ranges over a local basis at 0)}$$

Proof: In the lemma, take $K = \{x\}$ and $C = \bar{E}$ for a point x of V not in C. Then we obtain an open neighborhood U of 0 so that $x + U$ does not meet $\bar{E} + U$. The latter contains $E + U$, so certainly $x \notin E + U$. That is, for x not in the closure, there is an open U containing 0 so that $x \notin E + U$.   ///

As usual, for two topological vectorspaces V, W over $\mathbb{C}$, a function $f : V \longrightarrow W$ is *(k-)linear* when $f(\alpha x + \beta y) = \alpha f(x) + \beta f(y)$ for all $\alpha, \beta \in k$ and $x, y \in V$. Almost without exception we care about *continuous* linear maps, meaning linear maps continuous for the topologies on V, W. As expected, the *kernel* $\ker f$ of a linear map is

$$\ker f = \{v \in V : f(v) = 0\}$$

Being the inverse image of a closed set by a continuous map, the kernel is a *closed* subspace of V.

For a *closed* subspace H of a topological vectorspace V, the *quotient* V/H is *characterized* as topological vectorspace with linear quotient map $q : V \to V/H$ through which any continuous $f : V \to W$ with $\ker f \supset H$ *factors*, in the sense that there is a unique continuous linear $\overline{f} : V/H \to W$ giving a commutative diagram

$$
\begin{array}{ccc}
V/H & & \\
\Big\uparrow{\scriptstyle q} & \searrow^{\overline{f}} & \\
V & \xrightarrow{\;\;f\;\;} & W
\end{array}
$$

Uniqueness of the quotient $q : V \to V/H$, up to unique isomorphism, follows by the usual categorical arguments, as with the foregoing limits and products. The *existence* of the quotient is proven by the usual construction of V/H as the collection of cosets $v + H$, with q given as usual by $q : v \longrightarrow v + H$. We verify that this construction succeeds in the proposition below.

The *quotient topology* on V/H is the *finest* topology such that the quotient map $q : V \to V/H$ is continuous, namely, a subset E of V/H is open if and only if $q^{-1}(E)$ is open.

For *nonclosed* subspaces H, the quotient topology on the collection of cosets $\{v + H\}$ would *not* be Hausdorff. Thus, the proper categorical notion of topological vectorspace quotient, by nonclosed subspace, would produce the collection of cosets $v + \overline{H}$ for the *closure* $\overline{H}$ of H.

[13.3.4] Claim: For a closed subspace W of a topological vectorspace V, the collection $Q = \{v + W : v \in V\}$ of cosets by W with map $q(v) = v + W$ is a topological vectorspace and q is a quotient map.

Proof: The *algebraic* quotient $Q = V/W$ of cosets $v + W$ and $q(v) = v + W$ constructs a vectorspace quotient without any topological hypotheses on W. Since W is closed, and since vector addition is a homeomorphism, $v + W$ is closed as well. Thus, its complement $V - (v + W)$ is open, so $q(V - (v + W))$ is open, by definition of the quotient topology. Thus, the complement

$$q(v) = v + W = q(v + W) = V/W - q(V - (v + W))$$

of the open set $q(V - (v + W))$ is closed.   ///

Unlike general topological quotient maps,

[13.3.5] Claim: For a closed subspace H of a topological vector space V, the quotient map $q : V \to V/H$ is *open*, that is, carries open sets to open sets.

Proof: For U open in V,

$$q^{-1}(q(U)) = q^{-1}(U + H) = U + H = \bigcup_{h \in H} h + U$$

This is a union of opens.   ///

[13.3.6] Corollary: For $f : V \to X$ a linear map with a closed subspace W of V contained in $\ker f$, and $\bar{f}$ the induced map $\bar{f} : V/W \to X$ defined by $\bar{f}(v + W) = f(v)$, f is continuous if and only if $\bar{f}$ is continuous.

Proof: Certainly if $\bar{f}$ is continuous then $f = \bar{f} \circ q$ is continuous. The converse follows from the fact that q is *open*.   ///

This proves that the *construction* of the quotient by cosets succeeds in producing a quotient: a continuous linear map $f : V \to X$ *factors through* any quotient V/W for W a closed subspace contained in the kernel of f.

The notions of *balanced subset, absorbing subset, directed set, Cauchy net,* and *completeness* are necessary:

A subset E of V is *balanced* when $xE \subset E$ for every $x \in \mathbb{C}$ with $|x| \leq 1$.

[13.3.7] Lemma: Every neighborhood u of 0 in a topological vectorspace V over k contains a *balanced* neighborhood N of 0.

Proof: By continuity of scalar multiplication, there is $\varepsilon > 0$ and a neighborhood U' of $0 \in V$ so that if $|x| < \varepsilon$ and $v \in U'$ then $xv \in U$. Since $\mathbb{C}$ is not discrete, there is $x_o \in \mathbb{C}$ with $0 < |x_o| < \varepsilon$. Since scalar multiplication by a nonzero element is a homeomorphism, $x_o U'$ is a neighborhood of 0 and $x_o U' \subset U$. Put

$$N = \bigcup_{|y| \leq 1} y x_o U'$$

For $|x| \leq 1$, $|xy| \leq |y| \leq 1$, so

$$xN = \bigcup_{|y| \leq 1} x(y x_o U') \subset \bigcup_{|y| \leq 1} y x_o U' = N$$

producing the desired N.   ///

A subset E of vectorspace V over k is *absorbing* when for every $v \in V$ there is $t_o \in R$ so that $v \in \alpha E$ for every $\alpha \in k$ so that $|\alpha| \geq t_o$.

[13.3.8] Lemma: Every neighborhood U of 0 in a topological vectorspace is *absorbing*.

Proof: We may *shrink U* to assume U is *balanced*. By continuity of the map $k \to V$ given by $\alpha \to \alpha v$, there is $\varepsilon > 0$ so that $|\alpha| < \varepsilon$ implies $\alpha v \in U$. By the *nondiscreteness* of $\mathbb{C}$, there is nonzero $\alpha \in \mathbb{C}$ satisfying any such inequality. Then $v \in \alpha^{-1}U$, as desired.   ///

A *poset S*, $\leq$ is a partially ordered set. A *directed set* is a poset S such that, for any two elements $s, t \in S$, there is $z \in S$ so that $z \geq s$ and $z \geq t$.

A *net* in V is a subset $\{x_s : s \in S\}$ of V indexed by a directed set S. A net $\{x_s : s \in S\}$ in a topological vectorspace V is a *Cauchy net* if, for every neighborhood U of 0 in V, there is an index s_o so that for $s, t \geq s_o$ we have $x_s - x_t \in U$. A net $\{x_s : s \in S\}$ is *convergent* if there is $x \in V$ so that, for every neighborhood U of 0 in V there is an index s_o so that for $s \geq s_o$ we have $x - x_s \in U$. Since points are closed, there can be *at most* one point to which a net converges. Thus, *a convergent net is Cauchy*. Oppositely, a topological vectorspace is *complete* if every Cauchy net is convergent.

[13.3.9] Lemma: Let Y be a vector subspace of a topological vector space X, *complete* when given the subspace topology from X. Then Y is a *closed* subset of X.

Proof: Let $x \in X$ be in the closure of Y. Let S be a local basis of opens at 0, where we take the partial ordering so that $U \geq U'$ if and only if $U \subset U'$. For each $U \in S$ choose $y_U \in (x + U) \cap Y$. The net $\{y_U : U \in S\}$ converges to x, so is Cauchy. It must converge to a point in Y, so by uniqueness of limits of nets it must be that $x \in Y$. Thus, Y is closed.

Unfortunately, *completeness* as above is too strong a condition for general topological vectorspaces, beyond Fréchet spaces. A slightly weaker version of completeness, *quasi-completeness* or *local* completeness, *does* hold for most important natural spaces, as discussed in [13.12].

13.4 Unique Vectorspace Topology on $\mathbb{C}^n$

Finite-dimensional topological vectorspaces, and their interactions with other topological vectorspaces, are especially simple:

[13.4.1] Theorem: A *finite-dimensional* complex vectorspace V has just one topological vectorspace topology, that of the product topology on $\mathbb{C}^n$ for $n = \dim V$. A finite-dimensional subspace V of a topological vectorspace W is *closed*. A $\mathbb{C}$-linear map $X \to V$ to a finite-dimensional space V is continuous if and only if the kernel is closed.

Proof: The argument is by induction. First treat the one-dimensional situation:

[13.4.2] Claim: For a one-dimensional topological vectorspace V with basis e the map $\mathbb{C} \to V$ by $x \to xe$ is a *homeomorphism*.

Proof: Since scalar multiplication is continuous, we need only show that the map is *open*. We need only do this at 0, since translation addresses other points. Given $\varepsilon > 0$, by the nondiscreteness of $\mathbb{C}$ there is x_o in $\mathbb{C}$ so that $0 < |x_o| < \varepsilon$. Since V is Hausdorff, there is a neighborhood U of 0 so that $x_o e \notin U$. Shrink U so it is *balanced*. Take $x \in k$ so that $xe \in U$. For $|x| \geq |x_o|$, $|x_o x^{-1}| \leq 1$, so

$$x_o e = (x_o x^{-1})(xe) \in U$$

by balanced-ness of U, contradiction. Thus, $xe \in U$ implies that $|x| < |x_o| < \varepsilon$. ///

[13.4.3] Corollary: For fixed $x_o \in \mathbb{C}$, a not-identically-zero $\mathbb{C}$-linear $\mathbb{C}$-valued function f on V is *continuous* if and only if the *affine hyperplane* $H = \{v \in V : f(v) = x_o\}$ is *closed* in V.

Proof: Certainly if f is continuous then H is closed. For the converse, consider only the case $x_o = 0$, since translations (vector additions) are homeomorphisms of V to itself.

For v_o with $f(v_o) \neq 0$ and for any other $v \in V$

$$f(v - f(v)f(v_o)^{-1}v_o) = f(v) - f(v)f(v_o)^{-1}f(v_o) = 0$$

Thus, V/H is one-dimensional. The induced $\mathbb{C}$-linear map $\bar{f} : V/H \to k$ so that $f = \bar{f} \circ q$, that is, $\bar{f}(v + H) = f(v)$, is a homeomorphism to $\mathbb{C}$, by the previous result, so f is continuous. ///

For the theorem, for uniqueness of the topology it suffices to prove that for any $\mathbb{C}$-basis $e_1, \ldots, e_n$ for V, the map $\mathbb{C} \times \ldots \times \mathbb{C} \longrightarrow V$ by

$$(x_1, \ldots, x_n) \longrightarrow x_1 e_1 + \ldots + x_n e_n$$

is a homeomorphism. Prove this by induction on the dimension n, that is, on the number of generators for V as a free $\mathbb{C}$-module. The case $n = 1$ was treated. Since $\mathbb{C}$ is complete, the lemma asserting the closed-ness of complete subspaces shows that any one-dimensional subspace is closed. For $n > 1$, let $H = \mathbb{C}e_1 + \ldots + \mathbb{C}e_{n-1}$. By induction, H is closed in V, so the quotient $q : V \to V/H$ is constructed as expected, as the set of cosets $v + H$. The space V/H is a one-dimensional topological vectorspace over $\mathbb{C}$, with basis $q(e_n)$. By induction, $\phi : xq(e_n) = q(xe_n) \longrightarrow x$ is a homeomorphism $V/H \to \mathbb{C}$.

Likewise, $\mathbb{C}e_n$ is a closed subspace and we have the quotient map

$$q' : V \longrightarrow V/\mathbb{C}e_n$$

The image has basis $q'(e_1), \ldots, q'(e_{n-1})$, and by induction

$$\phi' : x_1 q'(e_1) + \cdots + x_{n-1} q'(e_{n-1}) \to (x_1, \ldots, x_{n-1})$$

is a homeomorphism. By the induction hypothesis,

$$v \longrightarrow (\phi \circ q)(v) \times (\phi' \circ q')(v)$$

is continuous to $\mathbb{C}^{n-1} \times \mathbb{C} \approx \mathbb{C}^n$. On the other hand, by the continuity of scalar multiplication and vector addition,

$$\mathbb{C}^n \longrightarrow V \quad \text{by} \quad x_1 \times \cdots \times x_n \longrightarrow x_1 e_1 + \cdots + x_n e_n$$

is continuous. These two maps are mutual inverses, certifying the homeomorphism.

Thus, an n-dimensional subspace is homeomorphic to $\mathbb{C}^n$ with its *product* topology, so is complete, since a finite product of complete spaces is complete. By the closedness of complete subspaces, it is closed.

Continuity of a linear map $f : X \to \mathbb{C}^n$ implies that the kernel $N = \ker f$ is closed. On the other hand, for N closed, the set of cosets $x + N$ constructs a quotient and is a topological vectorspace of dimension at most n. Therefore, the induced map $\bar{f} : X/N \to V$ is unavoidably continuous. Then $f = \bar{f} \circ q$ is continuous, where q is the quotient map. This completes the induction step. ///

13.5 Non-Banach Limits $C^k(\mathbb{R})$, $C^\infty(\mathbb{R})$ of Banach Spaces $C^k[a, b]$

For a *noncompact* topological space such as $\mathbb{R}$, the space $C^o(\mathbb{R})$ of continuous functions is *not* a Banach space with sup norm because the sup of the absolute value of a continuous function may be $+\infty$.

But, $C^o(\mathbb{R})$ has a Fréchet-space structure: express $\mathbb{R}$ as a *countable union of compact subsets* $K_n = [-n, n]$. Despite the likely non-injectivity of the map $C^o(\mathbb{R}) \to C^o(K_i)$, giving $C^o(\mathbb{R})$ the (projective) limit topology $\lim_i C^o(K_i)$ is reasonable: certainly the restriction map $C^o(\mathbb{R}) \to C^o(K_i)$ should be continuous, as should all the restrictions $C^o(K_i) \to C^o(K_{i-1})$, whether or not these are *surjective*.

The argument in favor of giving $C^o(\mathbb{R})$ the limit topology is that a *compatible* family of maps $f_i : Z \to C^o(K_i)$ gives *compatible fragments* of functions F on $\mathbb{R}$. That is, for $z \in Z$, given $x \in \mathbb{R}$ take K_i such that x is in the interior of K_i. Then for all $j \geq i$ the function $x \to f_j(z)(x)$ is continuous near x, and the compatibility ensures that all these functions are the same.

That is, the compatibility of these fragments is exactly the assertion that they fit together to make a function $x \to F_z(x)$ on the whole space X. Since continuity is a *local* property, $x \to F_z(x)$ is in $C^o(X)$. Further, there is *just one* way to piece the fragments together. Thus, diagrammatically,

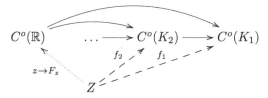

Thus, $C^o(X) = \lim_n C^o(K_n)$ is a Fréchet space. Similarly, $C^k(\mathbb{R}) = \lim_n C^k(K_n)$ is a Fréchet space.

[13.5.1] Remark: The question of whether the restriction maps $C^o(K_n) \to C^o(K_{n-1})$ or $C^o(\mathbb{R}) \to C^o(K_n)$ are *surjective* need not be addressed.

Unsurprisingly, we have

[13.5.2] Theorem: $\frac{d}{dx} : C^k(\mathbb{R}) \to C^{k-1}(\mathbb{R})$ is continuous.

Proof: The argument is structurally similar to the argument for $\frac{d}{dx} : C^\infty[a, b] \to C^\infty[a, b]$. The differentiations $\frac{d}{dx} : C^k(K_n) \to C^{k-1}(K_n)$ are a compatible family, fitting into a commutative diagram

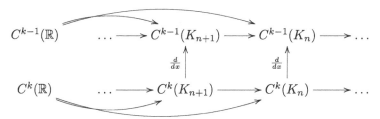

Composing the projections with d/dx gives (dashed) induced maps from $C^k(\mathbb{R})$ to the limitands, inducing a unique (dotted) continuous linear map to the limit, as in

$C^{k-1}(\mathbb{R}) \quad \cdots \longrightarrow C^{k-1}(K_{n+1}) \longrightarrow C^{k-1}(K_n) \longrightarrow \cdots$

$\frac{d}{dx}$

$C^k(\mathbb{R}) \quad \cdots \longrightarrow C^k(K_{n+1}) \longrightarrow C^{k-1}(K_n) \longrightarrow \cdots$

That is, there is a unique continuous linear map $\frac{d}{dx} : C^k(\mathbb{R}) \to C^{k-1}(\mathbb{R})$ compatible with the differentiations on finite intervals.   ///

Similarly,

[13.5.3] Theorem: $C^\infty(\mathbb{R}) = \lim_k C^k(\mathbb{R})$, $C^\infty(\mathbb{R}) = \lim_n C^\infty(K_n)$, and $\frac{d}{dx} : C^\infty(\mathbb{R}) \to C^\infty(\mathbb{R})$ is continuous.

Proof: From $C^\infty(\mathbb{R}) = \lim_k C^k(\mathbb{R})$ we can obtain the induced map d/dx, as follows. Starting with the commutative diagram

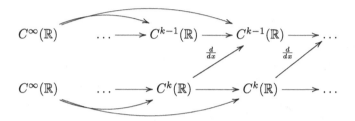

Composing the projections with d/dx gives (dashed) induced maps from $C^k(\mathbb{R})$ to the limitands, inducing a unique (dotted) continuous linear map to the limit, as in

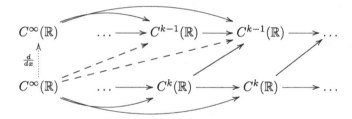

A novelty is the assertion that (projective) limits *commute* with each other, so that the limits of $C^k(K_n)$ in k and in n can be taken in either order. Generally, in a situation with maps $V_{ij} \to V_{i-1,j}$ and $V_{ij} \to V_{i,j-1}$, the maps $\lim_j(\lim_i V_{ij}) \to V_{k\ell}$ induce a map $\lim_j(\lim_i V_{ij}) \to \lim_\ell V_{k\ell}$, which induce a unique $\lim_j(\lim_i V_{ij}) \to \lim_k(\lim_\ell V_{k\ell})$. Similarly, a unique map is induced in the opposite direction, and, for the usual reason, these are mutual inverses. ///

[13.5.4] Claim: For fixed $x \in \mathbb{R}$ and fixed nonnegative integer k, the *evaluation map* $f \to f^{(k)}(x)$ is *continuous*.

Proof: Take n large enough so that $x \in [-n, n]$. Evaluation $f \to f^{(k)}(x)$ was shown in [13.1] to be continuous on $C^k[-n, n]$. Composing with the continuous $C^\infty(\mathbb{R}) \to C^k(\mathbb{R}) \to C^k[-n, n]$ gives the continuity.   ///

13.6 Banach Completion $C_o^k(\mathbb{R})$ of $C_c^k(\mathbb{R})$

It is reasonable to ask about the completion of the space $C_c^o(\mathbb{R})$ of compactly supported continuous functions in the metric given by the sup norm, and, more generally, about the completion of the space $C_c^k(\mathbb{R})$ of compactly supported k-times continuously differentiable functions in the metric given by the sum of the sups of the k derivatives.

The spaces $C_c^k(\mathbb{R})$ are *not complete* with those norms because supports can *leak* out to infinity: for example, in fix any u such that $u(x) = 1$ for $|x| \leq 1$, $0 \leq u(x) \leq 1$ for $1 \leq |x| \leq 2$, and $u(x) = 0$ for $|x| \geq 2$. Then

$$f(x) = \sum_{n=0}^{\infty} \frac{u(x-n)}{n^2}$$

converges in sup norm, the partial sums have compact support, but the whole does not have compact support.

[13.6.1] Claim: The completion of the space $C_c^o(\mathbb{R})$ of compactly supported continuous functions in the metric given by the sup norm $|f|_{C^o} = \sup_{x \in \mathbb{R}} |f(x)|$ is the space $C_o^o(\mathbb{R})$ of continuous functions f *vanishing at infinity*, in the sense that, given $\varepsilon > 0$, there is a compact interval $K = [-N, N] \subset X$ such that $|f(x)| < \varepsilon$ for $x \notin K$.

[13.6.2] Remark: Since we need to distinguish compactly supported functions $C_c^o(\mathbb{R})$ from functions $C_o^o(\mathbb{R})$ going to 0 at infinity, we cannot use the latter notation for the former, unlike some sources.

Proof: This is almost a tautology. Given $f \in C_o^o(\mathbb{R})$, given $\varepsilon > 0$, let $K = [-N, N] \subset X$ be compact such that $|f(x)| < \varepsilon$ for $x \notin K$. It is easy to make an auxiliary function φ that is continuous, *compactly supported*, real-valued function such that $\varphi = 1$ on K and $0 \leq \varphi \leq 1$ on X. Then $f - \varphi \cdot f$ is 0 on K, and of absolute value $|\varphi(x) \cdot f(x)| \leq |f(x)| < \varepsilon$ off K. That is, $\sup_{\mathbb{R}} |f - \varphi \cdot f| < \varepsilon$, so $C_c^o(\mathbb{R})$ is dense in $C_o^o(\mathbb{R})$.

On the other hand, a sequence f_i in $C_c^o(\mathbb{R})$ that is a Cauchy sequence with respect to sup norm gives a Cauchy sequence in each $C^o[a, b]$, and converges uniformly pointwise to a continuous function on $[a, b]$ for every $[a, b]$. Let f be the pointwise limit. Given $\varepsilon > 0$ take i_o such that $\sup_x |f_i(x) - f_j(x)| < \varepsilon$ for all $i, j \geq i_o$. With K the support of f_{i_o},

$$\sup_{x \notin K} |f(x)| \leq \sup_{x \notin K} |f(x) - f_{i_o}(x)| + \sup_{x \notin K} |f_{i_o}(x)|$$

$$= \sup_{x \notin K} |f(x) - f_{i_o}(x)| + 0 \leq \varepsilon < 2\varepsilon$$

showing that f goes to 0 at infinity.   ///

[13.6.3] Corollary: Continuous functions vanishing at infinity are *uniformly* continuous.

Proof: For $f \in C_o^o(\mathbb{R})$, given $\varepsilon > 0$, let $g \in C_c^o(\mathbb{R})$ be such that $\sup |f - g| < \varepsilon$. By the uniform continuity of g, there is $\delta > 0$ such that $|x - y| < \delta$ implies $|g(x) - g(y)| < \varepsilon$, and

$$|f(x) - f(y)| \leq |f(x) - g(x)| + |f(y) - g(y)| + |g(x) - g(y)| < 3\varepsilon$$

as desired.   ///

The arguments for $C^k(\mathbb{R})$ are completely parallel: the completion of the space $C_c^k(\mathbb{R})$ of compactly supported k-times continuously differentiable functions is the space $C_c^k(\mathbb{R})$ of k-times continuously differentiable functions whose k derivatives go to zero at infinity. Similarly,

[13.6.4] Corollary: The space of C^k functions whose k derivatives all vanish at infinity have *uniformly* continuous derivatives.   ///

[13.6.5] Claim: The limit $\lim_k C_o^k(\mathbb{R})$ is the space $C_o^\infty(\mathbb{R})$ of smooth functions all whose derivatives go to 0 at infinity. All those derivatives are *uniformly* continuous.

Proof: As with $C^\infty[a, b] = \bigcap_k C^k[a, b] = \lim_k C^k[a, b]$, by its very definition $C_o^\infty(\mathbb{R})$ is the intersection of the Banach spaces $C_o^k(\mathbb{R})$. For any compatible family $Z \to C_o^k(\mathbb{R})$, the compatibility implies that the image of Z is in that intersection.   ///

[13.6.6] Corollary: The space $C_o^\infty(\mathbb{R})$ is a Fréchet space, so is *complete*.

Proof: As earlier, countable limits of Banach spaces are Fréchet.

[13.6.7] Remark: In contrast, the space of merely *bounded* continuous functions does not behave so well. Functions such as $f(x) = \sin(x^2)$ are not *uniformly* continuous. This has the bad side effect that $\sup_x |f(x + h) - f(x)| = 1$ for all $h \neq 0$, which means that the *translation action* of $\mathbb{R}$ on that space of functions is *not continuous*.

13.7 Rapid-Decay Functions, Schwartz Functions

A continuous function f on $\mathbb{R}$ is *of rapid decay* when

$$\sup_{x \in \mathbb{R}} (1 + x^2)^n \cdot |f(x)| < +\infty \qquad \text{(for every } n = 1, 2, \ldots)$$

With norm $\nu_n(f) = \sup_{x \in \mathbb{R}} (1 + x^2)^n \cdot |f(x)|$, let the Banach space B_n be the completion of $C_c^o(\mathbb{R})$ with respect to the metric $\nu_n(f - g)$ associated to ν_n.

[13.7.1] Lemma: The Banach space B_n is isomorphic to $C_o^o(\mathbb{R})$ by the map $T : f \to (1 + x^2)^n \cdot f$. Thus, B_n is the space of continuous functions f such that $(1 + x^2)^n \cdot f(x)$ goes to 0 at infinity.

Proof: By design, $\nu_n(f)$ is the sup norm of Tf. Thus, the result [13.6] for $C_o^o(\mathbb{R})$ under sup norm gives this lemma.

[13.7.2] Remark: Just as we want the completion $C_o^o(\mathbb{R})$ of $C_c^o(\mathbb{R})$, rather than the space of all *bounded* continuous functions, we want B_n rather than the space of all continuous functions f with $\sup_x (1 + x^2) \cdot |f(x)| < \infty$. This distinction disappears in the limit, but it is only via the density of $C_c^o(\mathbb{R})$ in every B_n that it follows that $C_c^o(\mathbb{R})$ is dense in the space of continuous functions of rapid decay, in the corollary below.

[13.7.3] Claim: The space of continuous functions of rapid decay on $\mathbb{R}$ is the nested *intersection*, thereby the *limit*, of the Banach spaces B_n, so is Fréchet.

Proof: The key issue is to show that rapid-decay f is a ν_n-limit of *compactly supported* continuous functions for every n. For each fixed n the function $f_n = (1 + x^2)^n f$ is continuous and goes to 0 at infinity. From [13.6], f_n is the sup-norm limit of *compactly supported* continuous functions F_{nj}. Then $(1 + x^2)^{-n} F_{nj} \to f$ in the topology on B_n, and $f \in B_n$. Thus, the space of rapid-decay functions lies *inside* the intersection.

On the other hand, a function $f \in \bigcap_k B_k$ is continuous. For each n, since $(1 + x^2)^{n+1} |f(x)|$ is continuous and goes to 0 at infinity, it has a finite sup σ, and

$$\sup_x (1 + x^2)^n \cdot |f(x)| = \sup_x (1 + x^2)^{-1} \cdot (1 + x^2)^{n+1} |f(x)|$$

$$\leq \sup_x (1 + x^2)^{-1} \cdot \sigma < +\infty$$

This holds for all n, so f is of rapid decay. ///

[13.7.4] Corollary: The space $C_c^o(\mathbb{R})$ is *dense* in the space of continuous functions of rapid decay.

Proof: That every B_n is a completion of $C_c^o(\mathbb{R})$ is essential for this argument.

Use the model of the limit $X = \lim_n B_n$ as the diagonal in $\prod_n B_n$, with the product topology restricted to X. Let $p_n : \prod_k B_k \to B_n$ be the projection. Thus, given $x \in X$, there is a basis of neighborhood N of x in X of the form $N = X \cap U$ for an open U in the product of the form $U = \prod_n U_n$ with all but finitely many $U_n = B_n$. Thus, for $y \in C_c^o(\mathbb{R})$ such that $p_n(y) \in p_n(N) = p_n(U)$ for the finitely many indices such that $U_n \neq B_n$, we have $y \in N$. That is, approximating x in only *finitely many* of the limitands B_n suffices to approximate x in the limit. Thus, density in the limitands B_n implies density in the limit.   ///

[13.7.5] Remark: The previous argument applies generally, showing that a common subspace dense in all limitands is dense in the limit.

Certainly the operator of multiplication by $1 + x^2$ preserves $C_c^o(\mathbb{R})$ and is a continuous map $B_n \to B_{n-1}$. Much as d/dx was treated earlier,

[13.7.6] Claim: Multiplication by $1 + x^2$ is a continuous map of the space of continuous rapidly decreasing functions to itself.

Proof: Let T denote the multiplication by $1 + x^2$, and let $B = \lim_n B_n$ be the space of rapid-decay continuous functions. From the commutative diagram

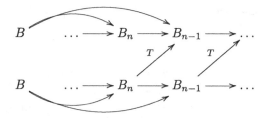

composing the projections with T giving (dashed) induced maps from B to the limitands, inducing a unique (dotted) continuous linear map to the limit, as in

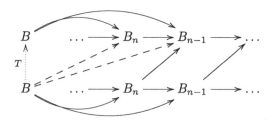

giving the *continuous* multiplication map on the space of rapid-decay continuous functions.

Similarly, adding differentiability conditions, the space of *rapidly decreasing* C^k functions is the space of k-times continuously differentiable functions f such that, for every $\ell = 0, 1, 2, \ldots, k$ and for every $n = 1, 2, \ldots$,

$$\sup_{x \in \mathbb{R}} (1 + x^2)^n \cdot |f^{(\ell)}(x)| < +\infty$$

Let B_n^k be the completion of $C_c^k(\mathbb{R})$ with respect to the metric from the norm

$$\nu_n^k(f) = \sum_{0 \le \ell \le k} \sup_{x \in \mathbb{R}} (1 + x^2)^n |f^{(\ell)}(x)|$$

Essentially identical arguments give

[13.7.7] Claim: The space of C^k functions of rapid decay on $\mathbb{R}$ is the *nested intersection*, thereby the *limit*, of the Banach spaces B_n^k, so is Fréchet.   ///

[13.7.8] Corollary: The space $C_c^k(\mathbb{R})$ is *dense* in the space of C^k functions of rapid decay.

Identifying B_n^k as a space of C^k functions with additional decay properties at infinity gives the obvious map $\frac{d}{dx} : B_n^k \to B_n^{k-1}$.

[13.7.9] Claim: $\frac{d}{dx} : B_n^k \to B_n^{k-1}$ is *continuous*.

Proof: Since B_n^k is the closure of $C_c^k(\mathbb{R})$, it suffices to check the continuity of $\frac{d}{dx} : C_c^k(\mathbb{R}) \to C_c^{k-1}(\mathbb{R})$ for the B_n^k and B_n^{k-1} topologies. As usual, that continuity was designed into the situation.   ///

The space $\mathscr{S}(\mathbb{R})$ of *Schwartz functions* consists of smooth functions f all whose derivatives are of rapid decay. One reasonable topology on $\mathscr{S}(\mathbb{R})$ is as a limit

$$\mathscr{S}(\mathbb{R}) = \bigcap_k \{C^k \text{ functions of rapid decay}\}$$

$$= \lim_k \{C^k \text{ functions of rapid decay}\}$$

As a countable limit of Fréchet spaces, this makes $\mathscr{S}(\mathbb{R})$ Fréchet.

[13.7.10] Corollary: $\frac{d}{dx} : \mathscr{S}(\mathbb{R}) \to \mathscr{S}(\mathbb{R})$ is *continuous*.

Proof: This is structurally the same as before: from the commutative diagram

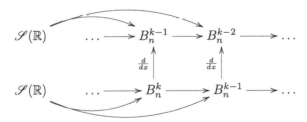

composing the projections with d/dx to give (dashed) induced maps from $\mathscr{S}(\mathbb{R})$ to the limitands, inducing a unique (dotted) continuous linear map to the limit:

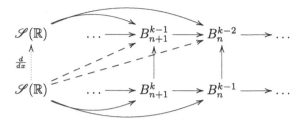

as desired.

Finally, to induce a canonical continuous map $T : \mathscr{S}(\mathbb{R}) \to \mathscr{S}(\mathbb{R})$ by multiplication by $1 + x^2$, examine the behavior of this multiplication map on the auxiliary spaces B_n^k and its interaction with $\frac{d}{dx}$:

[13.7.11] Claim: $T : B_n^k \to B_{n-1}^{k-1}$ is continuous.

Proof: Of course,

$$\left| \frac{d}{dx}\left((1+x^2) \cdot f(x) \right) \right| = \left| 2x \cdot f(x) + (1+x^2) \cdot f'(x) \right|$$

$$\leq 2 \cdot (1+x^2) \cdot |f(x)| + (1+x^2) \cdot |f'(x)|$$

Thus, $T : C_c^k(\mathbb{R}) \to C_c^{k-1}(\mathbb{R})$ is continuous with the B_n^k and B_{n-1}^{k-1} topologies. For general reasons, cofinal limits are isomorphic, so the same argument gives a unique continuous linear map $\mathscr{S}(\mathbb{R})$. ///

It is worth noting

[13.7.12] Claim: Compactly supported smooth functions are *dense* in $\mathscr{S}$.

Proof: At least up to rearranging the order of limit-taking, the description of $\mathscr{S}$ above is as a limit of spaces in each of which compactly supported smooth functions are dense. Thus, we claim a general result: for a limit $X = \lim_i X_i$ and compatible maps $f_i : V \to X_i$ with dense image, the induced map $f : V \to X$ has dense image. As in [13.2], the limit is the diagonal

$$D = \{\{x_i\} \in \prod_i X_i : x_i \mapsto x_{i-1}, \text{ for all } i\} \subset \prod_i X_i$$

with the subspace topology from the product. Suppose we are given a finite collection of neighborhoods $x_{i_1} \in U_{i_1} \subset X_{i_1}, \ldots, x_{i_n} \in U_{i_n} \subset X_{i_n}$, with $x_{i_j} \mapsto x_{i_k}$ if $i_j \geq i_k$. Take $i = \max_j i_j$, and U a neighborhood of x_i such that the image of U is inside every U_{i_j}, by continuity. Since the image of V is dense in X_i, there is $v \in V$ such that $f_i(v) \in U$. By compatibility, $f_{i_j}(v) \in U_{i_j}$ for all j. Thus, the image of V is dense in the limit. ////

13.8 Non-Fréchet Colimit $\mathbb{C}^{\infty}$ of $\mathbb{C}^{n}$, Quasi-Completeness

Toward topologies in which $C_c^o(\mathbb{R})$ and $C_c^{\infty}(\mathbb{R})$ could be *complete*, we consider first

$$\mathbb{C}^{\infty} = \bigcup_n \mathbb{C}^n$$

where $i_n : \mathbb{C}^n \subset \mathbb{C}^{n+1}$ by $i_n : (x_1, \ldots, x_n) \to (x_1, \ldots, x_n, 0)$. We want to topologize $\mathbb{C}^{\infty}$ so that it is *complete*, in a suitable sense. We have already seen that finite-dimensional complex vectorspaces have *unique* vectorspace topologies, so the only question is how to fit them together.

A countable ascending union of complete metric topological vector spaces, each a proper *closed* subspace of the next, such as $C^{\infty} = \bigcup \mathbb{C}^n$, *cannot* be a complete *metric* space, because it is exactly *presented* as a countable union of nowhere-dense closed subsets, contradicting the conclusion of the Baire category theorem. The function spaces $C_c^o(\mathbb{R})$ and $C_c^{\infty}(\mathbb{R})$ are also of this type, being the ascending unions of spaces C_K^o or C_K^{∞}, continuous or smooth functions with supports inside compact $K \subset \mathbb{R}$.

Thus, we cannot hope to give such space *metric* topologies for which they are *complete*.

Nevertheless, ascending unions are a type of *colimit*, just as descending intersections are a type of *limit*. That is, the topology on $\mathbb{C}^{\infty}$ is characterized by a universal property: for every collection of maps $f_n : \mathbb{C}^n \to Z$ with the compatibility $i_n \circ f_n = f_{n+1}$, there is a unique $f : \mathbb{C}^{\infty} \to Z$ through which all f_n's *factor*. That is, given a commutative diagram

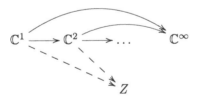

there is a *unique* (dotted) map $\mathbb{C}^{\infty} \to Z$ giving a commutative diagram

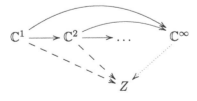

To argue that an ascending union $X = \bigcup_n X_n$ with $X_1 \subset X_2 \subset \ldots$ is an example of a colimit, observe that every $x \in X$ lies in some X_n, so all values $f(x)$ for

a map $f : X \to Z$ are completely determined by the restrictions of f to the limitands X_n. Thus, on one hand, given a compatible family $f_n : X_n \to Z$, there is *at most one* compatible $f : X \to Z$. On the other hand, a compatible family $f_n : X_n \to Z$ *defines* a map $X \to Z$: given $x \in X$, take n sufficiently large so that $x \in X_n$, and define $f(x) = f_n(x)$. The compatibility ensures that it does not matter which sufficiently large n we use.

For the topology of $\mathbb{C}^\infty$, the colimit characterization has a possibly counterintuitive consequence:

[13.8.1] Claim: *Every* linear map from the space $\mathbb{C}^\infty = \mathrm{colim}_n \mathbb{C}^n$ with the colimit topology to *any* topological vectorspace is *continuous*.

Proof: Given arbitrary linear $f : \mathbb{C}^\infty \to Z$, composition with inclusion gives a compatible family of linear maps $f_n : \mathbb{C}^n \to Z$. From [13.4], *every* linear map from a finite-dimensional space is *continuous*. The collection $\{f_n\}$ induces a unique *continuous* map $F : \mathbb{C}^\infty \to Z$ such that $F \circ i_n : \mathbb{C}^n \to Z$ is the same as $f \circ i_n$. In general, this might not be force $f = F$. However, because X is an ascending union, the values of both F and f are completely determined by their values on the limitands, and these are the same. Thus, $f = F$.   ///

The *uniqueness* argument for locally convex colimits of locally convex topological vectorspaces, that there is *at most one* such topology, is identical to the uniqueness argument for *limits* in [13.2] with arrows reversed.

[13.8.2] Remark: The fact that a colimit of finite-dimensional spaces has a unique canonical topology, from which *every* linear map from such a colimit is *continuous*, is often misunderstood and misrepresented as suggesting that there is *no* topology on that colimit. Again, there is a *unique canonical* topology, from which every *linear* map is *continuous*.

To prove *existence* of colimits, just as limits are *subobjects* of products, colimits are *quotients* of *coproducts*, as follows. A *locally convex colimit* of topological vector spaces X_α with *transition maps* $j_\beta^\alpha : X_\alpha \to X_\beta$ is the *quotient* of the *locally convex coproduct* X of the X_α by the *closure* of the subspace Z spanned by vectors

$$j_\alpha(x_\alpha) - (j_\beta \circ j_\beta^\alpha)(x_\alpha) \qquad \text{(for all } \alpha < \beta \text{ and } x_\alpha \in X_\alpha)$$

Annihilation of these differences in the quotient forces the desired compatibility relations. Obviously, quotients of locally convex spaces are locally convex.

Locally convex coproducts X of topological vector spaces X_α are coproducts (also called *direct sums*) of the *vector spaces* X_α topologized by the *diamond topology*, described as follows.[2] For a collection U_α of convex neighborhoods of 0 in the X_α, with $j_\alpha : X_\alpha \to X$ the αth canonical map, let

$$U = \text{convex hull in } X \text{ of the union of } j_\alpha(U_\alpha)$$

The diamond topology has local basis at 0 consisting of such U. Thus, it is locally convex by construction. *Closedness of points* follows from the corresponding property of the X_α. Thus, *existence* of a *locally convex* coproduct of locally convex spaces is ensured by the *construction*.

A *countable* colimit of a family $V_1 \to V_2 \to \cdots$ of topological vectorspaces is a *strict* colimit, or *strict inductive limit*, when each $V_i \to V_{i+1}$ is an isomorphism to its image, and each image is closed. A strict colimit of *Fréchet* spaces is called an *LF-space*.

Just to be sure:

[13.8.3] Claim: In a colimit indexed by positive integers $V = \text{colim}V_i$, if every transition $V_i \to V_{i+1}$ is *injective*, then every limitand V_i *injects* to the colimit V. Further, the colimit is the ascending union of the limitands V_i, suitably topologized. ///

Proof: In effect, the argument presents the colimit corresponding to an ascending union more directly, not as a quotient of the coproduct, although it is convenient to already have *existence* of the colimit. Certainly each V_i injects to $W = \bigcup_n V_n$. We will give W a locally convex topology so that every inclusion $V_i \to W$ is continuous. The universal property of the colimit produces a unique compatible map $V \to W$, so every V_i must inject to V itself.

Since the maps j_i of V_i to the colimit V are injections, the ascending union W injects to V by $j(w) = j_i(w)$ for any index i large enough so that $w \in V_i$. The compatibility of the maps among the V_i ensures that j is well-defined. We claim that $j(W)$ with the subspace topology from V, and the inclusions $V_i \to j_i(V_i) \subset j(W)$, give a colimit of the V_i. Indeed for any compatible, family $f_i : V_i \to Z$ and induced $f : V \to Z$, the restriction of f to $j(W)$ gives a map $j(W) \to Z$ through which the f_i factor. Thus, in fact, such a colimit is the ascending union with a suitable topology.

---

[2] The *product* topology of locally convex topological vector spaces is locally convex, whether in the category of locally convex topological vector spaces or in the larger category of not-necessarily-locally-convex topological vector spaces. However, *coproducts* behave differently: the locally convex coproduct of *uncountably many* locally convex spaces is *not* a coproduct in the larger category of not-necessarily-locally-convex spaces. This already occurs with an uncountable coproduct of *lines*.

Now we describe a topology on the ascending union W so that all inclusions $V_i \to W$ are continuous. Give W a local basis $\{U\}$ at 0, by taking arbitrary convex opens $U_i \subset V_i$ containing 0, and letting U be the convex hull of $\bigcup_i U_i$. Every injection $V_i \to W$ is continuous because the inverse image of such $U \cap V_i$ contains U_i, giving continuity at 0.

To be sure that *points are closed* in W, given $0 \neq x \in W$, we find a neighborhood of 0 in W not containing x. Let i_o be the first index such that $x \in V_{i_o}$. By Hahn-Banach, there is a continuous linear functional λ_{i_o} on V_{i_o} such that $\lambda_{i_o}(x) \neq 0$. Without loss of generality, $\lambda_{i_o}(x) = 1$ and $|\lambda_{i_o}| = 1$. Use Hahn-Banach to extend λ_{i_o} to a continuous linear functional λ_i on V_i for every $i \geq i_o$, with $|\lambda_i| \leq 1$. λ_{i_o} gives a continuous linear functional on V_i for $i < i_o$ by composition with the injection $V_i \to V_{i_o}$. Then $U_i = \{y \in V_i : |\lambda_i(y)| < 1\}$ is open in V_i and does not contain x, for all i. The convex hull of the ascending union $\bigcup_i U_i$ is just $\bigcup_i U_i$ itself, so does not contain x.

We did not quite prove that this topology is exactly the colimit topology, but we will never need that fact.   ///

Typical *colimit* topologies are *not* complete in the strongest possible sense (discussed subsequently), but are *quasi-complete*, a property sufficient for all applications. To describe quasi-completeness, we need a notion of *boundedness* in general topological vectorspaces, not merely metrizable ones. A subset B of a topological vector space V is *bounded* when, for every open neighborhood N of 0 there is $t_o > 0$ such that $B \subset tN$ for every $t \geq t_o$. A space is *quasi-complete* when every *bounded* Cauchy *net* is *convergent*.

Nothing new for metric spaces:

[13.8.4] Lemma: Complete metric spaces are quasi-complete. In particular, Cauchy nets converge and contain cofinal *sequences* converging to the same limit.

Proof: Let $\{s_i : i \in I\}$ be a Cauchy net in X. Given a natural number n, let $i_n \in I$ be an index such that $d(x_i, x_j) < \frac{1}{n}$ for $i, j \geq i_n$. Then $\{x_{i_n} : n = 1, 2, \ldots\}$ is a Cauchy sequence, with limit x. Given $\varepsilon > 0$, let $j \geq i_n$ be also large enough such that $d(x, x_j) < \varepsilon$. Then

$$d(x, x_{i_n}) \leq d(x, x_j) + d(x_j, x_{i_n}) < \varepsilon + \frac{1}{n} \qquad \text{(for every } \varepsilon > 0)$$

Thus, $d(x, x_{i_n}) \leq \frac{1}{n}$. The original Cauchy net also converges to x: given $\varepsilon > 0$, for n large enough so that $\varepsilon > \frac{1}{n}$,

$$d(x_i, x) \leq d(x_i, x_{i_n}) + d(x_{i_n}, x) < \varepsilon + \varepsilon \qquad \text{(for } i \geq i_n)$$

with the strict inequality coming from $d(x_{i_n}, x) < \varepsilon$.   ///

[13.8.5] Theorem: A bounded subset of an LF-space $X = \mathrm{colim}_n X_n$ lies in some limitand X_n. An LF-space is *quasi-complete*.

Proof: Let B be a bounded subset of X. Suppose B does *not* lie in any X_i. Then there is a sequence $i_1, i_2, \ldots$ of positive integers and x_{i_ℓ} in $X_{i_\ell} \cap B$ with x_{i_ℓ} *not* lying in $X_{i_\ell - 1}$. Using $X = \bigcup_j X_{i_\ell}$, without loss of generality, suppose that $i_\ell = \ell$.

By the Hahn-Banach theorem and induction, using the closedness of X_{i-1} in X_i, there are continuous linear functionals λ_i on X_i's such that $\lambda_i(x_i) = i$ and the restriction of λ_i to X_{i-1} is λ_{i-1}, for example. Since X is the colimit of the X_i, this collection of functionals exactly describes a unique compatible continuous linear functional λ on X.

But $\lambda(B)$ is *bounded* since B is bounded and λ is continuous, precluding the possibility that λ takes on all positive integer values at the points x_i of B. Thus, it could *not* have been that B failed to lie inside some single X_i. The strictness of the colimit implies that B is bounded as a subset of X_i, proving one direction of the equivalence. The other direction of the equivalence is less interesting.

Thus a *bounded* Cauchy net lies in some limitand Fréchet space X_n, so is convergent there, since Fréchet spaces are complete.   ///

[13.8.6] Remark: Strict inductive limits of finite-dimensional spaces do appear as natural function spaces, for example, the Schwartz space on $\mathbb{Q}_p$, as in [13.17].

13.9 Non-Fréchet Colimit $C_c^\infty(\mathbb{R})$ of Fréchet Spaces

The space of compactly supported continuous functions

$$C_c^o(\mathbb{R}) = \text{compactly supported continuous functions on } \mathbb{R}$$

is an *ascending union* of the subspaces

$$C_{[-n,n]}^o = \{f \in C^o(\mathbb{R}) \ : \ \mathrm{spt} f \subset [-n, n]\}$$

Each space $C_{[-n,n]}^o$ is a Banach space, being a closed subspace of the Banach space $C^o[-n, n]$, further requiring vanishing of the functions on the boundary of $[-n, n]$. A closed subspace of a Banach space is a Banach space. Thus, $C_c^o(\mathbb{R})$ is an LF-space and is *quasi-complete*.

Similarly,

$$C_c^k(\mathbb{R}) = \text{compactly supported } C^k \text{ functions on } \mathbb{R}$$

is an *ascending union* of the subspaces

$$C_{[-n,n]}^k = \{f \in C^k(\mathbb{R}) \ : \ \mathrm{spt} f \subset [-n, n]\}$$

Each space $C^k_{[-n,n]}$ is a Banach space, being a closed subspace of the Banach space $C^k[-n, n]$, further requiring vanishing of the functions and derivatives on the boundary of $[-n, n]$. A closed subspace of a Banach space is a Banach space. Thus, $C^k_c(\mathbb{R})$ is an LF-space and is *quasi-complete*.

The space of *test functions* is

$$\mathcal{D}(\mathbb{R}) = C^\infty_c(\mathbb{R}) = \text{compactly supported } C^\infty \text{ functions on } \mathbb{R}$$

is an *ascending union* of the subspaces

$$\mathcal{D}_{[-n,n]} = C^\infty_{[-n,n]} = \{f \in C^\infty(\mathbb{R}) : \text{spt} f \subset [-n, n]\}$$

Each space $\mathcal{D}_{[-n,n]}$ is a Fréchet space, being a closed subspace of the Fréchet space $C^\infty[-n, n]$, by further requiring vanishing of the functions and derivatives on the boundary of $[-n, n]$. A closed subspace of a Fréchet space is a Fréchet space. Thus, $\mathcal{D}(\mathbb{R}) = C^\infty_c(\mathbb{R})$ is an LF-space and is *quasi-complete*.

The operator $\frac{d}{dx} : C^k[-n, n] \to C^{k-1}[-n, n]$ is continuous and preserves the vanishing conditions at the endpoints, so restricts to a continuous map $\frac{d}{dx} : C^k_{[-n,n]} \to C^{k-1}_{[-n,n]}$ on the Banach sub-spaces of functions vanishing suitably at the endpoints. Composing with the inclusions $C^{k-1}_{[-n,n]} \to C^{k-1}_c(\mathbb{R})$ gives a compatible family of continuous maps $\frac{d}{dx} : C^k_{[-n,n]} \to C^{k-1}_c(\mathbb{R})$. This induces a unique continuous map on the colimit: $\frac{d}{dx} : C^k_c(\mathbb{R}) \to C^{k-1}_c(\mathbb{R})$.

Similarly, $\frac{d}{dx} : C^\infty[-n, n] \to C^\infty[-n, n]$ is continuous and preserves the vanishing conditions at the endpoints, so restricts to a continuous map $\frac{d}{dx} : \mathcal{D}_{[-n,n]} \to \mathcal{D}_{[-n,n]}$ on the Frechet subspaces of functions vanishing to all orders at the endpoints. Composing with the inclusions $\mathcal{D}_{[-n,n]} \to \mathcal{D}(\mathbb{R})$ gives a compatible family of continuous maps $\frac{d}{dx} : \mathcal{D}_{[-n,n]} \to \mathcal{D}(\mathbb{R})$. This induces a unique continuous map on the colimit: $\frac{d}{dx} : \mathcal{D}(R) \to \mathcal{D}(\mathbb{R})$. Diagrammatically,

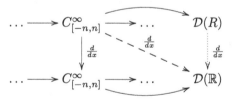

That is, $\frac{d}{dx}$ is continuous in the LF-space topology on test functions $\mathcal{D}(\mathbb{R}) = C^\infty_c(\mathbb{R})$.

[13.9.1] Claim: For fixed $x \in \mathbb{R}$ and nonnegative integer k, the evaluation map $f \to f^{(k)}(x)$ on $\mathcal{D}(\mathbb{R}) = C^\infty_c(\mathbb{R})$ is *continuous*.

Proof: This evaluation map is continuous on $C^\infty[-n, n]$ for every large-enough n so that $x \in [-n, n]$, so is continuous on the closed subspace $\mathcal{D}_{[-n,n]}$ of $C^\infty[-n, n]$. The inclusions among these spaces are extend-by-0, so the evaluation map is the 0 map on $\mathcal{D}_{[-n,n]}$ if $|x| \geq n$. These maps to $\mathbb{C}$ fit together into a compatible family, so extend uniquely to a continuous linear map of the colimit $\mathcal{D}(\mathbb{R})$ to $\mathbb{C}$.   ///

[13.9.2] Claim: For $F \in C^\infty(\mathbb{R})$, the map $f \to F \cdot f$ is a continuous map of $\mathcal{D}(\mathbb{R})$ to itself.

Proof: By the colimit characterization, it suffices to show that such a map is continuous on $C^\infty_{[-n,n]}$ or on the larger Fréchet space $C^\infty[-n, n]$ without vanishing conditions on the boundary. This is the limit of $C^k[-n, n]$, so it suffices to show that $f \to F \cdot f$ is a continuous map $C^k[-n, n] \to C^k[-n, n]$ for every k. The sum of sups of derivatives is

$$\sum_{0 \leq i \leq k} \sup_{|x| \leq n} \left| \left(\frac{d}{dx}\right)^i (Ff)(x) \right|$$

$$\leq 2^k \left(\sum_{0 \leq i \leq k} \sup_{|x| \leq n} |F^{(i)}(x)| \right) \cdot \left(\sum_{0 \leq i \leq k} \sup_{|x| \leq n} |f^{(i)}(x)| \right)$$

Although F and its derivatives need not be bounded, this estimate only uses their boundedness on $[-n, n]$. This is a bad estimate but sufficient for continuity.

[13.9.3] Claim: The inclusion $\mathcal{D}(\mathbb{R}) \to \mathscr{S}(\mathbb{R})$ is *continuous*, and the image is *dense*.

Proof: At least after changing order of limits, $\mathscr{S}(\mathbb{R})$ is described as a limit of spaces in which $\mathcal{D}(\mathbb{R})$ is dense, so $\mathcal{D}(\mathbb{R})$ is dense in that limit.

The slightly more serious issue is that $\mathcal{D}(\mathbb{R})$ with its LF-space topology maps continuously to $\mathscr{S}(\mathbb{R})$. Since $\mathcal{D}(\mathbb{R})$ is a colimit, we need only check that the limitands (compatibly) map continuously. On a limitand $C^\infty_{[-n,n]}$, the norms

$$\nu_{N,k}(f) = \sup_x (1 + x^2)^N \cdot |f^{(k)}(x)|$$

differ from the norms $\sup_x |f^{(k)}(x)|$ defining the topology on $C^\infty_{[-n,n]}$ merely by *constants*, namely, the sups of $(1 + x^2)^N$ on $[-n, n]$. Thus, we have the desired continuity on the limitands.   ///

13.10 LF-Spaces of Moderate-Growth Functions

The space $C^o_{\text{mod}}(\mathbb{R})$ of continuous functions of *moderate growth* on $\mathbb{R}$ is

$$C^o_{\text{mod}}(\mathbb{R})$$

$$= \{ f \in C^o(\mathbb{R}) \; : \; \sup_{x \in \mathbb{R}} (1 + x^2)^{-N} \cdot |f(x)| < +\infty \text{ for some } N \}$$

Literally, it is an ascending union

$$C^o_{\mathrm{mod}}(\mathbb{R}) = \bigcup_N \left\{ f \in C^o(\mathbb{R}) \; : \; \sup_{x \in \mathbb{R}} (1 + x^2)^{-N} \cdot |f(x)| < +\infty \right\}$$

However, it is ill-advised to use the individual spaces

$$B_N = \left\{ f \in C^o(\mathbb{R}) \; : \; \sup_{x \in \mathbb{R}} (1 + x^2)^{-N} \cdot |f(x)| < +\infty \right\}$$

with norms $\nu_N(f) = \sup_{x \in \mathbb{R}} (1 + x^2)^{-N} \cdot |f(x)|$ because $C^o_c(\mathbb{R})$ is not *dense* in these spaces B_N. Indeed, in the simple case $N = 0$, the norm ν_0 is the sup norm, and the sup-norm closure of $C^o_c(\mathbb{R})$ is continuous functions *going to* 0 *at infinity*, which excludes many *bounded* continuous functions.

In particular, there are many bounded continuous functions f which are not *uniformly* continuous, and the translation action of $\mathbb{R}$ on such functions cannot be continuous: no matter how small $\delta > 0$ is, $\sup_{x \in \mathbb{R}} |f(x + \delta) - f(x)|$ may be large. For example, $f(x) = \sin(x^2)$ has this feature.

This difficulty does not mean that the characterization of the whole set of moderate-growth functions is incorrect, nor that the norms ν_N are inappropriate, but only that the Banach spaces B_N are too large and that the topology of the whole should *not* be the strict colimit of the Banach spaces B_N. Instead, take the smaller

$$V_N = \text{completion of } C^o_c(\mathbb{R}) \text{ with respect to } \nu_N$$

As in the case of completion of $C^o_c(\mathbb{R})$ with respect to sup norm ν_0,

[13.10.1] Claim: V_N is the set of continuous f such that $(1 + x^2)^{-N} f(x)$ goes to 0 at infinity.   ///

Of course, if $(1 + x^2)^{-N} f(x)$ is merely *bounded*, then $(1 + x^2)^{-(N+1)} f(x)$ then goes to 0 at infinity. Thus, as *sets*, $B_N \subset V_{N+1}$, but this inclusion *cannot be continuous*, since $C^o_c(\mathbb{R})$ is dense in V_{N+1}, but not in B_N. That is, there is a nontrivial effect on the topology in setting

$$C^o_{\mathrm{mod}} = \mathrm{colim}_N V_N$$

instead of the colimit of the too large spaces B_N.

13.11 Seminorms and Locally Convex Topologies

So far, the vectorspace topologies have been described as Banach spaces, limits of Banach spaces, and colimits of limits of Banach spaces. By design, these descriptions facilitate proof of (quasi-)*completeness*. *Weaker* topologies are not usually described in this fashion. For example, for a topological vectorspace V, with (continuous) *dual*

$$V^* = \{\text{continuous linear maps } V \to \mathbb{C}\}$$

the *weak dual topology*[3] on V^* has a local subbasis at 0 consisting of sets

$$U = U_{v,\varepsilon} = \{\lambda \in V^* : |\lambda(v)| < \varepsilon\} \quad \text{(for fixed } v \in V \text{ and } \varepsilon > 0)$$

Unless V is finite-dimensional, this topology on V^* is much coarser than a Banach, Fréchet, or LF topology. The map $\lambda \to |\lambda(v)|$ is a natural example of a *seminorm*. It is not a norm because $\lambda(v) = 0$ can easily happen.

Seminorms are a general device to describe topologies on vectorspaces. These topologies are invariably *locally convex*, in the sense of having a local basis at 0 consisting of *convex* sets.

Description of a vectorspace topology by seminorms does *not* generally give direct information about *completeness*. Nevertheless, we can prove *quasi-completeness* for an important class of examples, just below.

A *seminorm* v on a complex vectorspace V is a real-valued function on V so that $v(x) \geq 0$ for all $x \in V$ *(nonnegativity)*, $v(\alpha x) = |\alpha| \cdot v(x)$ for all $\alpha \in \mathbb{C}$ and $x \in V$ *(homogeneity)*, and $v(x + y) \leq v(x) + v(y)$ for all $x, y \in V$ *(triangle inequality)*. This differs from the notion of *norm* only in the significant point that we allow $v(x) = 0$ for $x \neq 0$.

To compensate for the possibility that an individual seminorm can be 0 on a particular nonzero vector, since we want Hausdorff topologies, we mostly care about *separating* families $\{v_i : i \in I\}$ of seminorms: for every $0 \neq x \in V$ there is v_i so that $v_i(x) \neq 0$.

[13.11.1] Claim: The collection Φ of all *finite intersections* of sets

$$U_{i,\varepsilon} = \{x \in V : v_i(x) < \varepsilon\} \quad \text{(for } \varepsilon > 0 \text{ and } i \in I)$$

is a *local basis* at 0 for a locally convex topology on V.

Proof: As expected, we intend to define a topological vector space topology on V by saying a set U is *open* if and only if for every $x \in U$ there is some $N \in \Phi$ so that $x + N \subset U$. This would be the *induced topology* associated to the family of seminorms.

That we have a *topology* does not use the hypothesis that the family of seminorms is *separating*, although points will not be closed without the separating

---

[3] The weak dual topology is traditionally called the *weak-*-topology* but replacing * by *dual* is more explanatory.

property. Arbitrary unions of sets containing sets of the form $x + N$ containing each point x have the same property. The empty set and the whole space V are visibly in the collection. The least trivial issue is to check that finite intersections of such sets are again of the same form. Looking at each point x in a given finite intersection, this amounts to checking that finite intersections of sets in Φ are again in Φ. But Φ is *defined* to be the collection of all finite intersections of sets $U_{i,\varepsilon}$, so this succeeds: we have closure under finite intersections and a topology on V.

To verify that this topology makes V a topological vectorspace is to verify the continuity of vector addition and continuity of scalar multiplication and closed-ness of points. None of these verifications is difficult:

The *separating* property implies that for each $x \in V$ the intersection of *all* the sets $x + N$ with $N \in \Phi$ is just x. Given $y \in V$, for each $x \neq y$ let U_x be an open set containing x but not y. Then

$$U = \bigcup_{x \neq y} U_x$$

is *open* and has complement $\{y\}$, so the singleton $\{y\}$ is *closed*.

For continuity of vector addition, it suffices to prove that, given $N \in \Phi$ and given $x, y \in V$ there are $U, U' \in \Phi$ so that

$$(x + U) + (y + U') \subset x + y + N$$

The triangle inequality implies that for a fixed index i and for $\varepsilon_1, \varepsilon_2 > 0$

$$U_{i,\varepsilon_1} + U_{i,\varepsilon_2} \subset U_{i,\varepsilon_1+\varepsilon_2}$$

Then

$$(x + U_{i,\varepsilon_1}) + (y + U_{i,\varepsilon_2}) \subset (x + y) + U_{i,\varepsilon_1+\varepsilon_2}$$

Thus, given

$$N = U_{i_1,\varepsilon_1} \cap \ldots \cap U_{i_n,\varepsilon_n}$$

take

$$U = U' = U_{i_1,\varepsilon_1/2} \cap \ldots \cap U_{i_n,\varepsilon_n/2}$$

proving continuity of vector addition.

For continuity of scalar multiplication, prove that for given $\alpha \in k, x \in V$, and $N \in \Phi$ there are $\delta > 0$ and $U \in \Phi$ so that

$$(\alpha + B_\delta) \cdot (x + U) \subset \alpha x + N \qquad \text{(with } B_\delta = \{\beta \in k : |\alpha - \beta| < \delta\})$$

Since N is an intersection of the sub-basis sets $U_{i,\varepsilon}$, it suffices to consider the case that N is such a set. Given α and x, for $|\alpha' - \alpha| < \delta$ and for $x - x' \in U_{i,\delta}$,

$$\nu_i(\alpha x - \alpha' x') = \nu_i((\alpha - \alpha')x + (\alpha'(x - x')))$$

$$\leq \nu_i((\alpha - \alpha')x) + \nu_i(\alpha'(x - x'))$$

$$= |\alpha - \alpha'| \cdot \nu_i(x) + |\alpha'| \cdot \nu_i(x - x')$$

$$\leq |\alpha - \alpha'| \cdot \nu_i(x) + (|\alpha| + \delta) \cdot \nu_i(x - x')$$

$$\leq \delta \cdot (\nu_i(x) + |\alpha| + \delta)$$

Thus, for the joint continuity, take $\delta > 0$ small enough so that

$$\delta \cdot (\delta + \nu_i(x) + |\alpha|) < \varepsilon$$

Taking finite intersections presents no further difficulty, taking the corresponding finite intersections of the sets B_δ and $U_{i,\delta}$, finishing the demonstration that separating families of seminorms give a structure of topological vectorspace.

Last, check that finite intersections of the sets $U_{i,\varepsilon}$ are *convex*. Since intersections of convex sets are convex, it suffices to check that the sets $U_{i,\varepsilon}$ themselves are convex, which follows from the homogeneity and the triangle inequality: with $0 \leq t \leq 1$ and $x, y \in U_{i,\varepsilon}$,

$$\nu_i(tx + (1 - t)y) \leq \nu_i(tx) + \nu_i((1 - t)y) = t\nu_i(x) + (1 - t)\nu_i(y)$$

$$\leq t\varepsilon + (1 - t)\varepsilon = \varepsilon$$

Thus, the set $U_{i,\varepsilon}$ is convex. ///

The converse, that *every* locally convex topology is given by a family of seminorms, is more difficult:

Let U be a *convex* open set containing 0 in a topological vectorspace V. Every open neighborhood of 0 contains a *balanced* neighborhood of 0, so shrink U if necessary so it is balanced, that is, $\alpha v \in U$ for $v \in U$ and $|\alpha| \leq 1$. The *Minkowski functional* ν_U associated to U is

$$\nu_U(v) = \inf\{t \geq 0 : v \in tU\}$$

[13.11.2] Claim: The Minkowski functional ν_U associated to a balanced convex open neighborhood U of 0 in a topological vectorspace V is a *seminorm* on V and is *continuous* in the topology on V.

Proof: The argument is as expected:

By continuity of scalar multiplication, *every* neighborhood U of 0 is *absorbing*, in the sense that every $v \in V$ lies inside tU for large enough $|t|$. Thus, the set over which we take the infimum to define the Minkowski functional is *nonempty*, so the infimum exists.

Let α be a scalar, and let $\alpha = s\mu$ with $s = |\alpha|$ and $|\mu| = 1$. The balanced-ness of U implies the balanced-ness of tU for any $t \geq 0$, so for $v \in tU$ also

$$\alpha v \in \alpha t U = s\mu t U = st U$$

From this,

$$\{t \geq 0 : \alpha v \in \alpha U\} = |\alpha| \cdot \{t \geq 0 : \alpha v \in tU\}$$

from which follows the *homogeneity* property required of a seminorm:

$$\nu_U(\alpha v) = |\alpha| \cdot \nu_U(v) \qquad \text{(for scalar } \alpha)$$

For the triangle inequality use the convexity. For $v, w \in V$ and $s, t > 0$ such that $v \in sU$ and $w \in tU$,

$$v + w \in sU + tU = \{su + tu' \; : \; u, u' \in U\}$$

By convexity,

$$su + tu' = (s+t) \cdot \left(\frac{s}{s+t} \cdot u + \frac{t}{s+t} \cdot u' \right) \in (s+t) \cdot U$$

Thus,

$$\nu_U(v + w) = \inf\{r \geq 0 : v + w \in rU\}$$

$$\leq \inf\{r \geq 0 : v \in rU\} + \inf\{r \geq 0 : w \in rU\} = \nu_U(v) + \nu_U(w)$$

Thus, the Minkowski functional ν_U attached to balanced, convex U is a continuous seminorm.   ///

[13.11.3] Theorem: The topology of a *locally convex* topological vectorspace V is given by the collection of seminorms obtained as Minkowski functionals ν_U associated to a local basis at 0 consisting of convex, balanced opens.

Proof: The proof is straightforward, once we decide to tolerate an extravagantly large collection of seminorms. With or without local convexity, every neighborhood of 0 contains a *balanced* neighborhood of 0. Thus, a locally convex topological vectorspace has a local basis X at 0 of *balanced convex* open sets.

We claim that every open $U \in X$ can be recovered from the corresponding seminorm ν_U by

$$U = \{v \in V \; : \; \nu_U(v) < 1\}$$

Indeed, for $v \in U$, the continuity of scalar multiplication gives $\delta > 0$ and a neighborhood N of v such that $z \cdot v - 1 \cdot v \in U$ for $|1 - z| < \delta$. Thus, $v \in (1 + \delta)^{-1} \cdot U$, so

$$\nu_U(v) = \inf\{t \geq 0 \ : \ v \in t \cdot U\} \leq \frac{1}{1+\delta} < 1$$

On the other hand, for $\nu_U(v) < 1$, there is $t < 1$ such that $v \in tU \subset U$, since U is convex and contains 0. Thus, the seminorm topology is at least as fine as the original.

Oppositely, the same argument shows that every seminorm local basis open

$$\{v \in V \ : \ \nu_U(v) < t\}$$

is simply tU. Thus, the original topology is at least as fine as the seminorm topology.   ///

The comparison of descriptions of topologies is straightforward, as follows. For a seminorm ν on a topological vectorspace V, we can form a Banach space *completing* with respect to the *pseudo-metric* $\nu(x - y)$. In particular, unlike completions with respect to genuine metrics, there can be *collapsing*, so that the natural map of V to this completion need not be an injection.

[13.11.4] Claim: Let V be a topological vectorspace with topology given by a (separating) family of seminorms $S = \{\nu\}$. Order the set of finite subsets of S by inclusion, and

$$\nu_F = \sum_{\nu \in F} \nu \qquad \text{(for finite subset } F \text{ of } S)$$

Then V with its seminorm topology is a dense subspace of the limit $\lim_{F \in \Phi} V_F$ of the Banach-space completions V_F with respect to ν_F.

Proof: As earlier, the seminorm topology is literally the subspace topology on the diagonal copy of V in the product of the V_F.

Of course, the poset of finite subsets of S is more complicated than the poset of positive integers, so such a limit can be large. Certainly V has a natural map to every V_F. Indeed, by definition of the seminorm topology, the open sets in V are exactly the inverse images in V of open sets in the various V_F.

For $F \subset F'$, since $\nu_{F'} \geq \nu_F$, there is a natural continuous linear map $V_{F'} \to V_F$. The maps $V \to V_F$ are compatible, in the sense that the composite $V \to V_{F'} \to V_F$ is the same as $V \to V_F$, for $F \subset F'$. This induces a unique continuous linear map of V to the limit of the V_F.

As in [13.2], the limit is the diagonal

$$D = \{\{v_F\} \in \prod_F V_F : v_{F'} \to v_F, \text{ for all } F' \supset F\} \subset \prod_F V_F$$

with the subspace topology. Repeating part of an earlier argument, given a finite collection of finite subsets $F_1, \ldots, F_n$ of S, for $\{v_F\} \in D$, take neighborhoods $U_i \subset V_{F_i}$ containing v_{F_i}. Let $\Phi = \bigcup_i F_i$. The compatibility implies that there is $v_\Phi \in V_\Phi$ such that $v_\Phi \to v_{F_i}$ for all i. Also, there is a sufficiently small neighborhood U of v_Φ such that its image in every V_{F_i} is inside the neighborhood U_i of v_{F_i}. Since the image of V is dense in V_Φ, take $v \in V$ with image inside U. Then the image of v is inside U_i for all i. Thus, the image of V is dense in the limit.   ///

Although it turns out that we only care about locally convex topological vectorspaces, there *do exist* complete-metric topological vectorspaces which *fail* to be locally convex. This underscores the need to explicitly specify that a Fréchet space should be locally convex. The usual example of a not-locally-convex complete-metric space is the sequence space

$$\ell^p = \{x = (x_1, x_2, \ldots) : \sum_i |x_i|^p < \infty\}$$

for $0 < p < 1$ with metric

$$d(x, y) = \sum_i |x_i - y_i|^p \qquad (\text{no } p^{\text{th}} \text{ root, unlike the } p \geq 1 \text{ case})$$

This example's interest is mostly as a counterexample to a naive presumption that local convexity is automatic.

13.12 Quasi-Completeness Theorem

We have already seen that LF-spaces such as the space of test functions $\mathcal{D}(\mathbb{R}) = C_c^\infty(\mathbb{R})$, although *not* complete metrizable, are *quasi-complete*. It is fortunate that most important topological vector spaces are quasi-complete.

At the end of this section, we show that the fullest notion of completeness easily fails to hold, even for quasi-complete spaces.

It is clear that *closed subspaces* of quasi-complete spaces are quasi-complete. Products and finite sums of quasi-complete spaces are quasi-complete.

Let $\text{Hom}(X, Y)$ be the space of continuous linear functions from a topological vectorspace X to another topological vectorspace Y. Give $\text{Hom}(X, Y)$ the topology by seminorms $p_{x,U}$ where $x \in X$ and U is a convex, balanced neighborhood of 0 in Y, defined by

$$p_{x,U}(T) = \inf\{t > 0 : Tx \in tU\} \qquad (\text{for } T \in \text{Hom}(X, Y))$$

For $Y = \mathbb{C}$, this gives the weak dual topology on X^*.

[13.12.1] Theorem: For X a Fréchet space or LF-space, and Y quasi-complete, the space $\text{Hom}(X, Y)$, with the topology induced by the seminorms $p_{x,U}$, is *quasi-complete*.

Proof: Some preparation is required. A set E of continuous linear maps from one topological vectorspace X to another topological vectorspace Y is *equicontinuous* when, for every neighborhood U of 0 in Y, there is a neighborhood N of 0 in X so that $T(N) \subset U$ for every $T \in E$.

[13.12.2] Claim: Let V be a strict colimit of a well-ordered countable collection of locally convex closed subspaces V_i. Let Y be a locally convex topological vectorspace. Let E be a set of continuous linear maps from V to Y. Then E is *equicontinuous* if and only if for each index i the collection of continuous linear maps $\{T|_{V_i} : T \in E\}$ is equicontinuous.

Proof: Given a neighborhood U of 0 in Y, shrink U if necessary so that U is convex and balanced. For each index i, let N_i be a convex, balanced neighborhood of 0 in V_i so that $TN_i \subset U$ for all $T \in E$. Let N be the convex hull of the union of the N_i. By the convexity of N, still $TN \subset U$ for all $T \in E$. By the construction of the diamond topology, N is an open neighborhood of 0 in the coproduct, hence in the colimit, which is a quotient of the coproduct. This gives the equicontinuity of E. The other direction of the implication is easy. ///

[13.12.3] Claim: (*Banach-Steinhaus/uniform boundedness theorem*) Let X be a Fréchet space or LF-space and Y a locally convex topological vector space. A set E of linear maps $X \to Y$, such that every set $Ex = \{Tx : T \in E\}$ is *bounded* in Y, is *equicontinuous*.

Proof: First consider X Fréchet. Given a neighborhood U of 0 in Y, let $A = \bigcap_{T \in E} T^{-1}\overline{U}$. By assumption, $\bigcup_n nA = X$. By the Baire category theorem, the complete metric space X is not a countable union of nowhere dense subsets, so at least one of the closed sets nA has nonempty interior. Since (nonzero) scalar multiplication is a homeomorphism, A itself has nonempty interior, containing some $x + N$ for a neighborhood N of 0 and $x \in A$. For every $T \in E$,

$$TN \subset T\{a - x : a \in A\} \subset \{u_1 - u_2 \ : \ u_1, u_2 \in \overline{U}\} = \overline{U} - \overline{U}$$

By continuity of addition and scalar multiplication in Y, given an open neighborhood U_o of 0, there is U such that $\overline{U} - \overline{U} \subset U_o$. Thus, $TN \subset U_o$ for every $T \in E$, and E is equicontinuous.

For $X = \bigcup_i X_i$ an LF-space, this argument already shows that E restricted to each X_i is equicontinuous. As in the previous claim, this gives equicontinuity on the strict colimit.	///

For the proof of the theorem on quasi-completeness, let $E = \{T_i : i \in I\}$ be a bounded Cauchy net in $\operatorname{Hom}(X, Y)$, where I is a directed set. Of course, attempt to define the limit of the net by $Tx = \lim_i T_i x$. For $x \in X$ the evaluation map $S \to Sx$ from $\operatorname{Hom}(X, Y)$ to Y is continuous. In fact, the topology on $\operatorname{Hom}(X, Y)$ is the coarsest with this property. Therefore, by the quasi-completeness of Y, for each fixed $x \in X$ the net $T_i x$ in Y is bounded and Cauchy, so converges to an element of Y suggestively denoted Tx.

To prove *linearity* of T, fix x_1, x_2 in X, $a, b \in \mathbb{C}$ and fix a neighborhood U_o of 0 in Y. Since T is in the closure of E, for any open neighborhood N of 0 in $\operatorname{Hom}(X, Y)$, there exists

$$T_i \in E \cap (T + N)$$

In particular, for any neighborhood U of 0 in Y, take

$$N = \{S \in \operatorname{Hom}(X, Y) : S(ax_1 + bx_2) \in U, \ S(x_1) \in U, \ S(x_2) \in U\}$$

Then

$$T(ax_1 + bx_2) - aT(x_1) - bT(x_2)$$

$$= (T(ax_1 + bx_2) - aT(x_1) - bT(x_2))$$

$$- (T_i(ax_1 + bx_2) - aT_i(x_1) - bT_i(x_2))$$

since T_i is linear. The latter expression is

$$T(ax_1 + bx_2) - (ax_1 + bx_2) + a(T(x_1) - T_i(x_1)) + b(T(x_2) - T_i(x_2)$$

$$\in U + aU + bU$$

By choosing U small enough so that

$$U + aU + bU \subset U_o$$

we find that

$$T(ax_1 + bx_2) - aT(x_1) - bT(x_2) \ \in \ U_o$$

Since this is true for every neighborhood U_o of 0 in Y,

$$T(ax_1 + bx_2) - aT(x_1) - bT(x_2) = 0$$

which proves linearity.

Continuity of the limit operator T exactly requires *equicontinuity* of $E = \{T_i x : i \in I\}$. Indeed, for each $x \in X$, $\{T_i x : i \in I\}$ is *bounded* in Y, so by Banach-Steinhaus, $\{T_i : i \in I\}$ is equicontinuous.

Fix a neighborhood U of 0 in Y. Invoking the equicontinuity of E, let N be a small enough neighborhood of 0 in X so that $T(N) \subset U$ for all $T \in E$. Let $x \in N$. Choose an index i sufficiently large so that $Tx - T_i x \in U$, via the definition of the topology on $\text{Hom}(X, Y)$. Then

$$Tx \in U + T_i x \subset U + U$$

The usual rewriting, replacing U by U' such that $U' + U' \subset U$, shows that T is continuous. ///

Finally, we demonstrate that weak duals of reasonable topological vector spaces, such as infinite-dimensional Hilbert, Banach, or Fréchet spaces, are definitely *not* complete in the strongest sense. That is, in these weak duals there are Cauchy *nets* that *do not converge*.

[13.12.4] Theorem: The weak dual of a locally convex topological vector space V is *complete* if and only if every linear functional on V is *continuous*.

Proof: A vectorspace V can be (re-)topologized as the colimit V_{init} of all its finite-dimensional subspaces. Although the poset of finite-dimensional subspaces is much larger than the poset of positive integers, the earlier argument still applies: this colimit really is the ascending union with a suitable topology.

[13.12.5] Claim: For a locally convex topological vector space V the identity map $V_{\text{init}} \to V$ is *continuous*. That is, V_{init} is the finest locally convex topological vector space topology on V.

Proof: Finite-dimensional topological vector spaces have unique topologies [13.4]. Thus, for any finite-dimensional vector subspace X of V the inclusion $X \to V$ is continuous with that unique topology on X. These inclusions form a compatible family of maps to V, so by the characterization of colimit there is a *unique* continuous map $V_{\text{init}} \to V$. This map is the identity on every finite-dimensional subspace, so is the identity on the underlying set V. ///

[13.12.6] Claim: Every linear functional $\lambda : V_{\text{init}} \to \mathbb{C}$ is *continuous*.

Proof: The restrictions of a given linear function λ on V to finite-dimensional subspaces are compatible with the inclusions among finite-dimensional subspaces. Every linear functional on a finite-dimensional space is continuous, so the characterizing property of the colimit implies that λ is continuous on V_{init}.

[13.12.7] Claim: The weak dual V^* of a locally convex topological vector space V injects continuously to the limit of the finite-dimensional Banach spaces

$$V_\Phi^* = \text{completion of } V^* \text{ under seminorm } p_\Phi(\lambda) = \sum_{v \in \Phi} |\lambda(v)|$$

where Φ ranges over finite subsets of V, and the weak dual topology is the subspace topology.

Proof: The weak dual topology on the continuous dual V^* of a topological vector space V is given by the seminorms

$$p_v(\lambda) = |\lambda(v)| \qquad (\text{for } \lambda \in V^* \text{ and } v \in V)$$

The corresponding local basis is finite intersections, for arbitrary finite sets $\Phi \subset V$:

$$\{\lambda \in V^* \ : \ |\lambda(v)| < \varepsilon, \text{ for all } v \in \Phi\}$$

These sets contain, and are contained in, sets of the form

$$\{\lambda \in V^* \ : \ \sum_{v \in \Phi} |\lambda(v)| < \varepsilon\} \qquad (\text{for arbitrary finite sets } \Phi \subset V)$$

Therefore, the weak dual topology on V^* is also given by seminorms

$$p_\Phi(\lambda) = \sum_{v \in \Phi} |\lambda(v)| \qquad (\text{finite } \Phi \subset V)$$

These have the convenient feature that they form a projective family, indexed by (reversed) inclusion. Let $V^*(\Phi)$ be V^* with the p_Φ-topology: this is not Hausdorff, so continuous linear maps $V^* \to V^*(\Phi)$ descend to maps $V^* \to V_\Phi^*$ to the *completion* V_Φ^* of V^* with respect to the pseudo-metric attached to p_Φ. The quotient map $V^*(\Phi) \to V_\Phi^*$ typically has a large kernel, since

$$\dim_{\mathbb{C}} V_\Phi^* = \text{card } \Phi \qquad (\text{for finite } \Phi \subset V)$$

The maps $V^* \to V_\Phi^*$ are compatible with respect to (reverse) inclusion $\Phi \supset Y$, so V^* has a natural induced map to the $\lim_\Phi V_\Phi^*$. Since V separates points in V^*, V^* *injects* to the limit. The weak topology on V^* is exactly the subspace topology from that limit.   ///

[13.12.8] Claim: The weak dual V_{init}^* of V_{init} is the limit of the finite-dimensional Banach spaces

$$V_\Phi^* = \text{completion of } V_{\text{init}}^* \text{ under seminorm } p_\Phi(\lambda) = \sum_{v \in \Phi} |\lambda(v)|$$

for finite $\Phi \subset V$.

Proof: The previous proposition shows that V^*_{init} *injects* to the limit, and that the subspace topology from the limit is the weak dual topology. On the other hand, the limit consists of linear functionals on V, without regard to topology or continuity. Since *all* linear functionals are continuous on V_{init}, the limit is naturally a subspace of V^*_{init}. ///

Returning to the proof of the theorem, $\lim_\Phi V^*_\Phi$ is a closed subspace of the corresponding *product*, so is *complete* in the fullest sense. Any other locally convex topologization V_τ of V has weak dual $(V_\tau)^* \subset (V_{\text{init}})^*$ with the subspace topology, and the image is *dense* in $(V_{\text{init}})^*$. Thus, unless $(V_\tau)^* = (V_{\text{init}})^*$, the weak dual V^*_τ *is not complete.* ///

13.13 Strong Operator Topology

For X and Y Hilbert spaces, the topology on $\text{Hom}(X, Y)$ given by seminorms

$$p_{x,U}(T) = \inf\{t > 0 : Tx \in tU\} \qquad (\text{for } T \in \text{Hom}(X, Y))$$

where $x \in X$ and U is a convex, balanced neighborhood of 0 in Y, is the *strong operator topology*. Indeed, every neighborhood of 0 in Y contains an open *ball*, so this topology can also be given by seminorms

$$q_x(T) = |Tx|_Y \qquad (\text{for } T \in \text{Hom}(X, Y))$$

where $x \in X$. The strong operator topology is weaker than the *uniform* topology given by the operator norm $|T| = \sup_{|x| \leq 1} |Tx|_Y$.

The *uniform* operator-norm topology makes the space of operators a Banach space, certainly simpler than the strong operator topology, but the uniform topology is too strong for many purposes.

For example, *group actions* on Hilbert spaces are rarely continuous for the uniform topology: letting $\mathbb{R}$ act on $L^2(\mathbb{R})$ by $T_x f(y) = f(x + y)$, no matter how small $\delta > 0$ is, there is an L^2 function f with $|f|_{L^2} = 1$ such that $|T_\delta f - f|_{L^2} = \sqrt{2}$.

Despite the strong operator topology being less elementary than the uniform topology, the theorem of the previous section [13.12] shows that $\text{Hom}(X, Y)$ with the strong operator topology is *quasi-complete*.

13.14 Generalized Functions (Distributions) on $\mathbb{R}$

The most immediate definition of the space of *distributions* or *generalized functions* on $\mathbb{R}$ is as the dual $\mathcal{D}^* = \mathcal{D}(\mathbb{R})^* = C^\infty_c(\mathbb{R})^*$ to the space $\mathcal{D} = \mathcal{D}(\mathbb{R})$ of test functions, with the *weak dual topology* by seminorms $\nu_f(u) = |u(f)|$ for test functions f and distributions u.

Similarly, the *tempered* distributions are the weak dual $\mathscr{S}^* = \mathscr{S}(\mathbb{R})^*$, and the *compactly supported* distributions are the weak dual $\mathcal{E}^* = \mathcal{E}(\mathbb{R})^*$, in this context writing $\mathcal{E}(\mathbb{R}) = C^\infty(\mathbb{R})$. Naming $\mathcal{E}^*$ *compactly supported* will be justified below.

By dualizing, the continuous containments $\mathcal{D} \subset \mathscr{S} \subset \mathcal{E}$ give continuous maps $\mathcal{E}^* \to \mathscr{S}^* \to \mathcal{D}^*$. When we know that $\mathcal{D}$ is *dense* in $\mathscr{S}$ and in $\mathcal{E}$, it will follow that these are *injections*. The most straightforward argument for density uses Gelfand-Pettis integrals, as in [14.5]. Thus, for the moment, we cannot claim that $\mathcal{E}^*$ and $\mathscr{S}^*$ *are* distributions, but only that they naturally *map to* distributions.

[13.12] shows that $\mathcal{D}^*$, $\mathscr{S}^*$, and $\mathcal{E}^*$ are *quasi-complete*, despite not being complete in the strongest possible sense.

The description of the space of distributions as the weak dual to the space of test functions falls far short of explaining its utility. There is a natural imbedding $\mathcal{D}(\mathbb{R}) \to \mathcal{D}(\mathbb{R})^*$ of test functions into distributions, for $f, g \in \mathcal{D}(\mathbb{R})$, by

$$f \to u_f \qquad \text{by} \qquad u_f(g) = \int_{\mathbb{R}} f(x)\, g(x)\, dx$$

That is, via this imbedding we consider distributions to be *generalized functions*. Indeed, [14.5] shows that test functions $\mathcal{D}(\mathbb{R})$ are *dense* in $\mathcal{D}(\mathbb{R})^*$.

The simplest example of a distribution not obtained by integration against a test function on $\mathbb{R}$ is the *Dirac delta*, the evaluation map $\delta(f) = f(0)$. From [13.9] and other earlier results, this is *continuous* for the LF-space topology on test functions.

This imbedding, and integration by parts, explains how to define $\frac{d}{dx}$ on distributions in a form compatible with the imbedding $\mathcal{D} \subset \mathcal{D}^*$: noting the sign, due to integration by parts,

$$\left(\frac{d}{dx}u\right)(f) = -u\left(\frac{d}{dx}f\right) \qquad (\text{for } u \in \mathcal{D}^* \text{ and } f \in \mathcal{D})$$

[13.14.1] Claim: $\frac{d}{dx} : \mathcal{D}^* \to \mathcal{D}^*$ is continuous.

Proof: By the nature of the weak dual topology, it suffices to show that for each $f \in \mathcal{D}$ and $\varepsilon > 0$ there are $g \in \mathcal{D}$ and $\delta > 0$ such that $|u(g)| < \delta$ implies $|(\frac{d}{dx}u)(f)| < \varepsilon$. Taking $g = \frac{d}{dx}f$ and $\delta = \varepsilon$ succeeds.   ///

Multiplications by $F \in C^\infty(\mathbb{R})$ give continuous maps $\mathcal{D}$ to itself, compatible with the imbedding $\mathcal{D} \to \mathcal{D}^*$ in the sense that

$$\int_{\mathbb{R}} (F \cdot u)(x)\, f(x)\, dx = \int_{\mathbb{R}} u(x)\, (F \cdot f)(x)\, dx$$

for $F \in C^\infty(\mathbb{R})$ and $u, f \in \mathcal{D}(\mathbb{R})$. Extend this to a map $\mathcal{D}^* \to \mathcal{D}^*$ by

$$(F \cdot u)(f) = u(F \cdot f) \qquad (\text{for } F \in C^\infty, u \in \mathcal{D}^*, \text{ and } f \in \mathcal{D})$$

[13.14.2] Claim: Multiplication operators $\mathcal{D}^* \to \mathcal{D}^*$ by $F \in C^\infty$ are continuous.

Proof: By the nature of the weak dual topology, it suffices to show that for each $f \in \mathcal{D}$ and $\varepsilon > 0$ there are $g \in \mathcal{D}$ and $\delta > 0$ such that $|u(g)| < \delta$ implies $|F \cdot u)(f)| < \varepsilon$. Taking $g = F \cdot f$ and $\delta = \varepsilon$ succeeds.   ///

Since $\mathscr{S}$ is mapped to itself by Fourier transform [13.13], this gives a way to define Fourier transform on $\mathscr{S}^*$, by a duality extending the Plancherel theorem:

$$\widehat{u}(f) = u(\widehat{f}) \qquad (\text{for } f \in \mathscr{S} \text{ and } u \in Schw^*)$$

Recall that the support of a *function* is the *closure* of the set on which it is nonzero, slightly complicating the notion of support for a *distribution* u: support of u is the *complement* of the *union* of all open sets U such that $u(f) = 0$ for all test functions f with support inside U.

[13.14.3] Theorem: A distribution with *support* $\{0\}$ is a finite linear combination of Dirac's δ and its derivatives.

Proof: Since $\mathcal{D}$ is a colimit of $\mathcal{D}_K$ over $K = [-n, n]$, it suffices to classify u in $\mathcal{D}_K^*$ with support $\{0\}$. We claim that a continuous linear functional on $\mathcal{D}_K = \lim_k C_K^k$ factors through some limitand

$$C_K^k = \{f \in C^k(K) : f^{(i)} \text{ vanishes on } \partial K \text{ for } 0 \leq i \leq k\}$$

This is a special case of

[13.14.4] Claim: Let $X = \lim_n B_n$ be a limit of Banach spaces, with the image of X *dense* in each B_n. A continuous linear map $T : \lim_n B_n \to Z$ from a, to a *normed* space Z factors through some limitand B_n. For $Z = \mathbb{C}$, the same conclusion holds without the density assumption.

Proof: Let $X = \lim_i B_i$ with projections $p_i : X \to B_i$. Each B_i is the closure of the image of X. By the continuity of T at 0, there is an open neighborhood U of 0 in X such that TU is inside the open unit ball at 0 in Z. By the description of the limit topology as the product topology restricted to the diagonal, there are finitely many indices $i_1, \ldots, i_n$ and open neighborhoods V_{i_t} of 0 in B_{i_t} such that

$$\bigcap_{t=1}^{n} p_{i_t}^{-1}\left(p_{i_t} X \cap V_{i_t}\right) \subset U$$

We can make a *smaller* open in X by a condition involving a single limitand, as follows. Let j be *any* index with $j \geq i_t$ for all t, and

$$N = \bigcap_{t=1}^{n} p_{i_t,j}^{-1}\left(p_{i_t,j}B_j \cap V_{i_t}\right) \subset B_j$$

By the compatibility $p_{i_t}^{-1} = p_j^{-1} \circ p_{i_t,j}^{-1}$, we have $p_{i_t,j}N \subset V_{i_t}$ for $i_1, \ldots, i_n$, and $p_j^{-1}(p_jX \cap N) \subset U$. By the linearity of T, for any $\varepsilon > 0$,

$$T(\varepsilon \cdot p_j^{-1}\left(p_jX \cap N\right)) = \varepsilon \cdot T(p_j^{-1}\left(p_jX \cap N\right)) \subset \varepsilon\text{-ball in } Z$$

We claim that T factors through p_jX with the subspace topology from B_j. One potential issue in general is that $p_j : X \to B_j$ can have a nontrivial kernel, and we must check that $\ker p_j \subset \ker T$. By the linearity of T,

$$T(\frac{1}{n} \cdot p_j^{-1}(p_j \cap N)) \subset \frac{1}{n}\text{-ball in } Z$$

so

$$T\left(\bigcap_n \frac{1}{n} \cdot p_j^{-1}(p_jX \cap N)\right) \subset \frac{1}{m}\text{-ball in } Z \qquad \text{(for all } m\text{)}$$

and then

$$T\left(\bigcap_n \frac{1}{n} \cdot p_j^{-1}(p_j \cap N)\right) \subset \bigcap_m \frac{1}{m}\text{-ball in } Z = \{0\}$$

Thus,

$$\bigcap_n p_j^{-1}(p_jX \cap \frac{1}{n} \cdot N) = \bigcap_n \frac{1}{n} \cdot p^{-1}(p_jX \cap N) \subset \ker T$$

Thus, for $x \in X$ with $p_jx = 0$, certainly $p_jx \in \frac{1}{n}N$ for all $n = 1, 2, \ldots$, and

$$x \in \bigcap_n p_j^{-1}(p_jX \cap \frac{1}{n}N) \subset \ker T$$

This proves the subordinate claim that T factors through $p_j : X \to B_j$ via a (not necessarily continuous) linear map $T' : p_jX \to Z$. The continuity follows from continuity at 0, which is

$$T(\varepsilon \cdot p_j^{-1}\left(p_jX \cap N\right)) = \varepsilon \cdot T(p_j^{-1}\left(p_jX \cap N\right)) \subset \varepsilon\text{-ball in } Z$$

Then $T' : p_jX \to Z$ extends to a map $B_j \to Z$ by continuity: given $\varepsilon > 0$, take symmetric convex neighborhood U of 0 in B_j such that $|T'y|_Z < \varepsilon$ for $y \in p_jX \cap U$. Let y_i be a Cauchy net in p_jX approaching $b \in B_j$. For y_i and y_j inside $b + \frac{1}{2}U$, $|T'y_i - T'y_j| = |T'(y_i - y_j)| < \varepsilon$, since $y_i - y_j \in \frac{1}{2} \cdot 2U = U$.

Then unambiguously define $T'b$ to be the Z-limit of the $T'y_i$. The closure of $p_j X$ in B_j is B_j, giving the desired map.

When u is a *functional*, that is, a map to $\mathbb{C}$, we can extend it by Hahn-Banach.

Returning to the proof of the theorem: thus, there is $k \geq 0$ such that u factors through a limit and C_K^k. In particular, u is continuous for the C^k topology on $\mathcal{D}_K$.

We need an auxiliary gadget. Fix a test function ψ identically 1 on a neighborhood of 0, bounded between 0 and 1, and (necessarily) identically 0 outside some (larger) neighborhood of 0. For $\varepsilon > 0$ let

$$\psi_\varepsilon(x) = \psi(\varepsilon^{-1}x)$$

Since the support of u is just $\{0\}$, for all $\varepsilon > 0$ and for all $f \in \mathcal{D}(\mathbb{R}^n)$ the support of $f - \psi_\varepsilon \cdot f$ does not include 0, so

$$u(\psi_\varepsilon \cdot f) = u(f)$$

Thus, for implied constant depending on k and K, but not on f,

$$|\psi_\varepsilon f|_k = \sup_{x \in K} \sum_{0 \leq i \leq k} |(\psi_\varepsilon f)^{(i)}(x)|$$

$$\ll \sum_{i \leq k} \sum_{0 \leq j \leq i} \sup_x \varepsilon^{-j} \left| \psi^{(j)}(\varepsilon^{-1}x) f^{(i-j)}(x) \right|$$

For test function f vanishing to order k at 0, that is, $f^{(i)}(0) = 0$ for all $0 \leq i \leq k$, on a fixed neighborhood of 0, by a Taylor-Maclaurin expansion, $|f(x)| \ll |x|^{k+1}$, and, generally, for i^{th} derivatives with $0 \leq i \leq k$, $|f^{(i)}(x)| \ll |x|^{k+1-i}$. By design, all derivatives $\psi', \psi'', \ldots$ are identically 0 in a neighborhood of 0, so, for suitable implied constants independent of ε,

$$|\psi_\varepsilon f|_k \ll \sum_{0 \leq i \leq k} \sum_{0 \leq j \leq i} \varepsilon^{-j} \cdot \left| \psi^{(j)}(\varepsilon^{-1}x) f^{(i-j)}(x) \right|$$

$$\ll \sum_{0 \leq i \leq k} \sum_{j=0} \varepsilon^{-j} \cdot 1 \cdot \varepsilon^{k+1-i} = \sum_{0 \leq i \leq k} \varepsilon^{k+1-i} \ll \varepsilon^{k+1-k} = \varepsilon$$

Thus, for sufficiently small $\varepsilon > 0$, for smooth f vanishing to order k at 0, $|u(f)| = |u(\psi_\varepsilon f)| \ll \varepsilon$, and $u(f) = 0$. That is,

$$\ker u \supset \bigcap_{0 \leq i \leq k} \ker \delta^{(i)}$$

The conclusion, that u is a linear combination of the distributions $\delta, \delta', \delta^{(2)}, \ldots, \delta^{(k)}$, follows from

[13.14.5] Claim: A linear functional $\lambda \in V^*$ vanishing on the intersection $\bigcap_i \ker \lambda_i$ of kernels of a finite collection $\lambda_1, \ldots, \lambda_n \in V^*$ is a *linear combination* of the λ_i.

Proof: The linear map

$$q : V \longrightarrow \mathbb{C}^n \quad \text{by} \quad v \longrightarrow (\lambda_1 v, \ldots, \lambda_n v)$$

is *continuous* since each λ_i is continuous, and λ factors through q, as $\lambda = L \circ q$ for some linear functional L on $\mathbb{C}^n$. We know all the linear functionals on $\mathbb{C}^n$, namely, L is of the form

$$L(z_1, \ldots, z_n) = c_1 z_1 + \cdots + c_n z_n \qquad \text{(for some } c_i \in \mathbb{C})$$

Thus,

$$\lambda(v) = (L \circ q)(v) = L(\lambda_1 v, \ldots, \lambda_n v) = c_1 \lambda_1(v) + \cdots + c_n \lambda_n(v)$$

expressing λ as a linear combination of the λ_i. ///

The *order* of a distribution $u : \mathcal{D} \to \mathbb{C}$ is the integer k, if such exists, such that u is continuous when $\mathcal{D}$ is given the weaker topology from $\mathrm{colim}_K C_K^k$. Not every distribution has finite order, but there is a useful technical application of the previous discussion:

[13.14.6] Corollary: A distribution $u \in \mathcal{D}^*$ with *compact support* has *finite order*.

Proof: Let ψ be a test function that is identically 1 on an open containing the support of u. Then

$$u(f) = u((1 - \psi) \cdot f) + u(\psi \cdot f) = 0 + u(\psi \cdot f)$$

since $(1 - \psi) \cdot f$ is a test function with support not meeting the support of u. With $K = \mathrm{spt}\, \psi$, this suggests that u factors through a subspace of $\mathcal{D}_K$ via $f \to \psi \cdot f \to u(\psi \cdot f)$, but there is the issue of continuity. Distinguishing things a little more carefully, the compatibility embodied in the commutative diagram

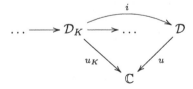

gives

$$u(f) = u(\psi \cdot f) = u\big(i(\psi f)\big) = u_K(\psi f)$$

The map u_K is continuous, as is the multiplication $f \to \psi f$. The map u_K is from the limit $\mathcal{D}_K$ of Banach spaces C_K^k to the normed space $\mathbb{C}$, so factors through some limitand C_K^k, by [13.4.4]. As in proof that multiplication is continuous in the C^∞ topology, by Leibniz's rule, the C^k norm of ψf is

$$|\psi f|_k = \sum_{0 \le i \le k} \sup_{x \in K} |(\psi f)^{(i)}(x)| \ll \sum_{i \le k} \sum_{0 \le j \le i} \sup_x |\psi^{(j)}(x) f^{(i-j)}(x)|$$

$$\ll \sum_{0 \le i \le k} \sup_{x \in K} |f^{(i)}(x)| \cdot \sum_{j \le k} \sup_x |\psi^{(j)}(x)| = |f|_{C^k} \cdot |\psi|_{C^k}$$

Since ψ is fixed, this gives continuity in f in the C^k topology.   ///

[13.14.7] Claim: In the inclusion $\mathcal{E}^* \subset \mathscr{S}^* \subset \mathcal{D}^*$, the image of $\mathcal{E}^*$ really is the collection of distributions with compact support.

Proof: On one hand the previous shows that $u \in \mathcal{D}^*$ with compact support can be composed as $u(f) = u_K(\psi f)$ for suitable $\psi \in \mathcal{D}$. The map $f \to \psi \cdot f$ is also continuous as a map $\mathcal{E} \to \mathcal{D}$, so the same expression $f \to \psi f \to u_K(\psi f)$ *extends* $u \in \mathcal{D}^*$ to a continuous linear functional on $\mathcal{E}$.

On the other hand, let $u \in \mathcal{E}^*$. Composition of u with $\mathcal{D} \to \mathcal{E}$ gives an element of $\mathcal{D}^*$, which we must check has compact support. From [13.5], $\mathcal{E}$ is a limit of the Banach spaces $C^k(K)$ with $K = [-n, n]$, *without* claiming that the image of $\mathcal{E}$ is necessarily dense in any of these. By [13.4.4], u factors through some limitand $C^k(K)$. The map $\mathcal{D} \to \mathcal{E}$ is compatible with the *restriction* maps $\mathrm{Res}_K : \mathcal{D} \to C^k(K)$: the diagram commutes. For $f \in \mathcal{D}$ with support

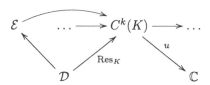

disjoint from K, $\mathrm{Res}_K(f) = 0$, and $u(f) = 0$. This proves that the support of the (induced) distribution is contained in K, so is compact.   ///

13.15 Tempered Distributions and Fourier Transforms on $\mathbb{R}$

One normalization of the *Fourier transform* integral is

$$\widehat{f}(\xi) = \mathcal{F}f(\xi) = \int_{\mathbb{R}} \overline{\psi}_\xi(x) f(x) \, dx \qquad (\text{with } \psi_\xi(x) = e^{2\pi i \xi x})$$

This converges nicely for f in the space $\mathscr{S}(\mathbb{R})$ of Schwartz functions.

[13.15.1] Theorem: Fourier transform is a topological isomorphism of $\mathscr{S}(\mathbb{R})$ to itself, with Fourier inversion map $\varphi \to \check{\varphi}$ given by

$$\check{\varphi}(x) = \int_{\mathbb{R}} \psi_{\xi}(x)\, \widehat{f}(\xi)\, d\xi$$

Proof: Using the idea [14.3] that Schwartz functions extend to smooth functions on a suitable one-point compactification of $\mathbb{R}$ vanishing to infinite order at the point at infinity, Gelfand-Pettis integrals justify moving a differentiation under the integral,

$$\frac{d}{d\xi}\widehat{f}(\xi) = \frac{d}{d\xi}\int_{\mathbb{R}} \overline{\psi}_{\xi}(x)\, f(x)\, dx = \int_{\mathbb{R}} \frac{\partial}{\partial \xi}\overline{\psi}_{\xi}(x)\, f(x)\, dx$$

$$= \int_{\mathbb{R}} (-2\pi i x)\,\overline{\psi}_{\xi}(x)\, f(x)\, dx = (-2\pi i)\int_{\mathbb{R}} \overline{\psi}_{\xi}(x)\, x f(x)\, dx$$

$$= (-2\pi i)\widehat{xf}(\xi)$$

Similarly, with an integration by parts,

$$-2\pi i \xi \cdot \widehat{f}(\xi) = \int_{\mathbb{R}} \frac{\partial}{\partial x}\overline{\psi}_{\xi}(x) \cdot f(x)\, dx = -\mathcal{F}\frac{df}{dx}(\xi)$$

Thus, $\mathcal{F}$ maps $\mathscr{S}(\mathbb{R})$ to itself.

The natural idea to prove Fourier inversion for $\mathscr{S}(\mathbb{R})$, that unfortunately begs the question, is the obvious:

$$\int_{\mathbb{R}} \psi_{\xi}(x)\, \widehat{f}(\xi)\, d\xi = \int_{\mathbb{R}} \psi_{\xi}(x)\left(\int_{\mathbb{R}} \overline{\psi}_{\xi}(t)\, f(t)\, dt \right) d\xi$$

$$= \int_{\mathbb{R}} f(t)\left(\int_{\mathbb{R}} \psi_{\xi}(x-t)\, dt \right) dt$$

If we could *justify* asserting that the inner integral is $\delta_x(t)$, which it *is*, then Fourier inversion follows. However, Fourier inversion for $\mathscr{S}(\mathbb{R})$ is used to make sense of that inner integral in the first place.

Despite that issue, a dummy *convergence factor* will legitimize the idea. For example, let $g(x) = e^{-\pi x^2}$ be the usual Gaussian. Various computations show that it is its own Fourier transform. For $\varepsilon > 0$, as $\varepsilon \to 0^+$, the dilated Gaussian $g_{\varepsilon}(x) = g(\varepsilon \cdot x)$ approaches 1 uniformly on compacts. Thus,

$$\int_{\mathbb{R}} \psi_{\xi}(x)\, \widehat{f}(\xi)\, d\xi = \int_{\mathbb{R}} \lim_{\varepsilon \to 0^+} g(\varepsilon\xi)\, \psi_{\xi}(x)\, \widehat{f}(\xi)\, d\xi$$

$$= \lim_{\varepsilon \to 0^+} \int_{\mathbb{R}} g(\varepsilon\xi)\, \psi_{\xi}(x)\, \widehat{f}(\xi)\, d\xi$$

by *monotone convergence* or more elementary reasons. Then the iterated integral is legitimately rearranged:

$$\int_{\mathbb{R}} g(\varepsilon\xi)\,\psi_\xi(x)\,\widehat{f}(\xi)\,d\xi = \int_{\mathbb{R}}\int_{\mathbb{R}} g(\varepsilon\xi)\,\psi_\xi(x)\,\overline{\psi}_\xi(t)\,f(t)\,dt\,d\xi$$

$$= \int_{\mathbb{R}}\int_{\mathbb{R}} g(\varepsilon\xi)\,\psi_\xi(x-t)\,f(t)\,d\xi\,dt$$

By changing variables in the definition of Fourier transform, $\widehat{g_\varepsilon} = \frac{1}{\varepsilon}g_{1/\varepsilon}$. Thus,

$$\int_{\mathbb{R}} \psi_\xi(x)\,\widehat{f}(\xi)\,d\xi = \int_{\mathbb{R}} \frac{1}{\varepsilon}\,g\Big(\frac{x-t}{\varepsilon}\Big)\,f(t)\,dt = \int_{\mathbb{R}} \frac{1}{\varepsilon}\,g\Big(\frac{t}{\varepsilon}\Big)\cdot f(x+t)\,dt$$

The sequence of function $g_{1/\varepsilon}/\varepsilon$ is not an *approximate identity* in the strictest sense, since the supports are the entire line. Nevertheless, the integral of each is 1, and as $\varepsilon \to 0^+$, the mass is concentrated on smaller and smaller neighborhoods of $0 \in \mathbb{R}$. Thus, for $f \in \mathscr{S}(\mathbb{R})$,

$$\lim_{\varepsilon\to 0^+} \int_{\mathbb{R}} \frac{1}{\varepsilon}\,g\Big(\frac{t}{\varepsilon}\Big)\cdot f(x+t)\,dt = f(x)$$

This proves Fourier inversion. In particular, this proves that Fourier transform *bijects* the Schwartz space to itself.   ///

With Fourier inversion in hand, we can prove the Plancherel identity for Schwartz functions:

[13.15.2] Corollary: For $f, g \in \mathscr{S}$, the Fourier transform is an isometry in the $L^2(\mathbb{R})$ topology, that is, $\langle \widehat{f}, \widehat{g} \rangle = \langle f, g \rangle$.

Proof: There is an immediate preliminary identity:

$$\int_{\mathbb{R}} \widehat{f}(\xi)\,h(\xi)\,d\xi = \int_{\mathbb{R}}\int_{\mathbb{R}} e^{-2\pi i\xi x}\,f(x)\,h(\xi)\,d\xi\,dx$$

$$= \int_{\mathbb{R}}\int_{\mathbb{R}} e^{-2\pi i\xi x}\,f(x)\,h(\xi)\,dx\,d\xi = \int_{\mathbb{R}} f(x)\,\widehat{h}(x)\,dx$$

To get from this identity to Plancherel requires, given $g \in \mathscr{S}$, existence of $h \in \mathscr{S}$ such that $\widehat{h} = \overline{g}$, with complex conjugation. By Fourier inversion on Schwartz functions, $h = (\overline{g})^\vee$ succeeds.   ///

[13.15.3] Corollary: Fourier transform extends by continuity to an isometry $L^2(\mathbb{R}) \to L^2(\mathbb{R})$.

Proof: Schwartz functions are dense in in $L^2(\mathbb{R})$.   ///

[13.15.4] Corollary: Fourier transform extends to give a bijection of the space tempered distributions $\mathscr{S}^*$ to itself, by

$$\widehat{u}(\varphi) = u(\widehat{\varphi}) \qquad (\text{for all } \varphi \in \mathscr{S})$$

Proof: Fourier transform is a topological isomorphism of $\mathscr{S}$ to itself.   ///

13.16 Test Functions and Paley-Wiener Spaces

Of course, the original [Paley-Wiener 1934] referred to L^2 functions, not distributions. The distributional aspect is from [Schwartz 1952]. An interesting point is that rate-of-growth of the Fourier transforms in the imaginary part determines the support of the inverse Fourier transforms.

The class *PW* of entire functions appearing in the following theorem is the *Paley-Wiener space* in one complex variable. The assertion is that, in contrast to the fact that Fourier transform maps the Schwartz space to itself, on test functions the Fourier transform has less symmetrical behavior, bijecting to the Paley-Wiener space.

[13.16.1] Theorem: A test function f supported on $[-r, r] \subset \mathbb{R}$ has Fourier transform $\widehat{f}$ extending to an entire function on $\mathbb{C}$, with

$$|\widehat{f}(z)| \ll_N (1 + |z|)^{-N} e^{r \cdot |y|} \qquad \text{(for } z = x + iy \in \mathbb{C}, \text{ for every } N\text{)}$$

Conversely, an entire function satisfying such an estimate has (inverse) Fourier transform which is a test function supported in $[-r, r]$.

Proof: First, the integral for $\widehat{f}(z)$ is the integral of the compactly supported, continuous, entire-function-valued function,

$$\xi \longrightarrow \left(z \to f(\xi) \cdot e^{-i\xi z} \right)$$

where the space of entire functions is given the sups-on-compacts semi-norms $\sup_{z \in K} |f(z)|$. Since $\mathbb{C}$ can be covered by countably many compacts, this topology is metrizable. Cauchy's integral formula proves *completeness*, so this space is Fréchet. Thus, the Gelfand-Pettis integral exists and is entire. Multiplication by z is converted to differentiation inside the integral,

$$(-iz)^N \cdot \widehat{f}(z) = \int_{|\xi| \leq r} \frac{\partial^N}{\partial \xi^N} e^{-iz \cdot \xi} \cdot f(\xi) \, d\xi$$

$$= (-1)^N \int_{|\xi| \leq r} e^{-iz \cdot \xi} \cdot \frac{\partial^N}{\partial \xi^N} f(\xi) \, d\xi$$

by integration by parts. Differentiation does not enlarge support, so

$$|\widehat{f}(z)| \ll_N (1 + |z|)^{-N} \cdot \left| \int_{|\xi| \leq r} e^{-iz \cdot \xi} f^{(N)}(\xi) \, d\xi \right|$$

$$\leq (1 + |z|)^{-N} \cdot e^{r \cdot |y|} \cdot \left| \int_{|\xi| \leq r} e^{-ix \cdot \xi} f^{(N)}(\xi) \, d\xi \right|$$

$$\leq (1 + |z|)^{-N} \cdot e^{r \cdot |y|} \cdot \int_{|\xi| \leq r} |f^{(N)}(\xi)| \, d\xi \ll_{f,N} (1 + |z|)^{-N} \cdot e^{r \cdot |y|}$$

Conversely, for an entire function F with the indicated growth and decay property, we show that

$$\varphi(\xi) = \int_{\mathbb{R}} e^{ix\xi} \, F(x) \, dx$$

is a test function with support inside $[-r, r]$. The assumptions on F do *not* directly include any assertion that F is Schwartz, so we cannot directly conclude that φ is smooth. Nevertheless, a similar obvious computation would give

$$\int_{\mathbb{R}} (ix)^N \cdot e^{ix\xi} \, F(x) \, dx = \int_{\mathbb{R}} \frac{\partial^N}{\partial \xi^N} e^{ix\xi} \, F(x) \, dx = \frac{\partial^N}{\partial \xi^N} \int_{\mathbb{R}} e^{ix\xi} \, F(x) \, dx$$

Moving the differentiation outside the integral is *necessary*, justified via Gelfand-Pettis integrals by a compactification device, as in [14.3], as follows. Since F strongly vanishes at ∞, the integrand extends continuously to the stereographic-projection one-point compactification of $\mathbb{R}$, giving a compactly supported smooth-function-valued function on this compactification. The measure on the compactification can be adjusted to be finite, taking advantage of the rapid decay of F:

$$\varphi(\xi) = \int_{\mathbb{R}} e^{ix\xi} \, F(x) \, dx = \int_{\mathbb{R}} e^{ix\xi} \, F(x) \, (1 + x^2)^N \, \frac{dx}{(1 + x^2)^N}$$

Thus, the Gelfand-Pettis integral exists, and φ is smooth. Thus, in fact, the justification proves that such an integral of smooth functions is smooth without necessarily producing a formula for derivatives.

To see that φ is supported inside $[-r, r]$, observe that, taking y of the same sign as ξ,

$$\left| F(x + iy) \cdot e^{i\xi(x+iy)} \right| \ll_N (1 + |z|)^{-N} \cdot e^{(r-|\xi|)\cdot|y|}$$

Thus,

$$|\varphi(\xi)| \ll_N \int_{\mathbb{R}} (1 + |z|)^{-N} \cdot e^{(r-|\xi|)\cdot|y|} \, dx \le e^{(r-|\xi|)\cdot|y|} \cdot \int_{\mathbb{R}} \frac{dx}{(1 + |x|)^{-N}}$$

For $|\xi| > r$, letting $|y| \to +\infty$ shows that $\varphi(\xi) = 0$.

[13.16.2] Corollary: We can topologize PW by requiring that the linear bijection $\mathcal{D} \to PW$ be a topological vector space isomorphism. ///

[13.16.3] Remark: The latter topology on PW is finer than the sups-on-compacts topology on all entire functions, since the latter cannot detect growth properties.

[13.16.4] Corollary: Fourier transform can be defined on *all* distributions $u \in \mathcal{D}^*$ by $\widehat{u}(\varphi) = u(\widehat{\varphi})$ for $\varphi \in PW$, giving an isomorphism $\mathcal{D}^* \to PW^*$ to the dual of the Paley-Wiener space.   ///

For example, the exponential $t \to e^{iz \cdot t}$ with $z \in \mathbb{C}$ but $z \notin \mathbb{R}$ is *not* a tempered distribution, but is a distribution, and its Fourier transform is the Dirac delta $\delta_z \in PW'$.

Compactly supported distributions have a similar characterization:

[13.16.5] Theorem: The Fourier transform $\widehat{u}$ of a distribution u supported in $[-r, r]$, of order N, is (integration against) the function $x \to u(\xi \to e^{-ix\xi})$, which is *smooth* and extends to an *entire* function satisfying

$$|\widehat{u}(z)| \ll (1 + |z|)^N \cdot e^{r \cdot |y|}$$

Conversely, an entire function meeting such a bound is the Fourier transform of a distribution of order N supported inside $[-r, r]$.

Proof: The Fourier transform $\widehat{u}$ is the tempered distribution defined for Schwartz functions φ by

$$\widehat{u}(\varphi) = u(\widehat{\varphi}) = u\left(\xi \to \int_{\mathbb{R}} e^{-ix\xi} \, \varphi(x) \, dx\right)$$

$$= \int_{\mathbb{R}} u(\xi \to e^{-ix\xi}) \, \varphi(x) \, dx$$

since $x \to (\xi \to e^{-ix\xi} \varphi(\xi))$ extends to a continuous smooth-function-valued function on the one-point compactification of $\mathbb{R}$, and Gelfand-Pettis applies. Thus, as expected, $\widehat{u}$ is integration against $x \to u(\xi \to e^{-ix\xi})$.

The smooth-function-valued function $z \to (\xi \to e^{-iz\xi})$ is holomorphic in z. Compactly supported distributions constitute the dual of $C^{\infty}(\mathbb{R})$. Application of u gives a holomorphic *scalar*-valued function $z \to u(\xi \to e^{-iz\xi})$.

Let ν_N be the Nth-derivative seminorm on $C^{\infty}[-r, r]$, so

$$|u(\varphi)| \ll_{\varepsilon} \nu_N(\varphi)$$

Then

$$|\widehat{u}(z)| = |u(\xi \to e^{-iz\xi})| \ll_{\varepsilon} \nu_N(\xi \to e^{-iz\xi}) \ll \sup_{[-r,r]} \left|(1 + |z|)^N e^{-iz\xi}\right|$$

$$\leq (1 + |z|)^N e^{r \cdot |y|}$$

Conversely, let F be an entire function with $|F(z)| \ll (1 + |z|)^N e^{r \cdot |y|}$. Certainly F is a tempered distribution, so $F = \widehat{u}$ for a tempered distribution. We show that u is of order at most N and has support in $[-r, r]$.

With η supported on $[-1, 1]$ with $\eta \geq 0$ and $\int \eta = 1$, make an *approximate identity* $\eta_\varepsilon(x) = \eta(x/\varepsilon)/\varepsilon$ for $\varepsilon \to 0^+$. By the easy half of Paley-Wiener for test functions, $\widehat{\eta}_\varepsilon$ is entire and satisfies

$$|\widehat{\eta}_\varepsilon(z)| \ll_{\varepsilon, N} (1 + |z|)^{-N} \cdot e^{\varepsilon \cdot |y|} \qquad \text{(for all } N\text{)}$$

Note that $\widehat{\eta}_\varepsilon(x) = \widehat{\eta}(\varepsilon \cdot x)$ goes to 1 as tempered distribution.

By the more difficult half of Paley-Wiener for test functions, $F \cdot \widehat{\eta}_\varepsilon$ is $\widehat{\varphi}_\varepsilon$ for some test function φ_ε supported in $[-(r + \varepsilon), r + \varepsilon]$. Note that $F \cdot \widehat{\eta}_\varepsilon \to F$.

For Schwartz function g with the support of $\widehat{g}$ not meeting $[-r, r]$, $\widehat{g} \cdot \varphi_\varepsilon$ for sufficiently small $\varepsilon > 0$. Since $F \cdot \widehat{\eta}_\varepsilon$ is a Cauchy net as tempered distributions,

$$u(\widehat{g}) = \widehat{u}(g) = \int F \cdot g = \int \lim_\varepsilon (F \cdot \widehat{\eta}_\varepsilon)\, g = \lim_\varepsilon \int (F \cdot \widehat{\eta}_\varepsilon)\, g$$

$$= \lim_\varepsilon \int \widehat{\varphi}_\varepsilon\, g = \lim_\varepsilon \int \varphi_\varepsilon\, \widehat{g} = 0$$

This shows that the support of u is inside $[-r, r]$.   ///

13.17 Schwartz Functions and Fourier Transforms on $\mathbb{Q}_p$

For simplicity, we only look at Fourier analysis on $\mathbb{Q}_p$, rather than on general p-adic fields. The same ideas apply to the general case, with minor modifications.

Fix a prime p, let $\mathbb{Q}_p$ be the p-adic field and $\mathbb{Z}_p$ the p-adic integers. Give $\mathbb{Q}_p$ the additive Haar measure that gives $\mathbb{Z}_p$ total measure 1. This determines the measure of every set $x + p^n \mathbb{Z}_p$ with $n \geq 0$, by translation-invariance, and the fact that $\mathbb{Z}_p$ is a disjoint union of such translates, as x ranges over $\mathbb{Z}_p / p^n \mathbb{Z}_p \approx \mathbb{Z}/p^n \mathbb{Z}$. The standard choice of additive character, trivial on $\mathbb{Z}_p$, is $\psi_1(x) = e^{-2\pi i x'}$, where $x' \in \mathbb{Z}[\frac{1}{p}]$ is such that $x - x' \in \mathbb{Z}_p$. Parametrize additive characters by $\psi_\xi(x) = \psi_1(\xi \cdot x)$.

Unsurprisingly, the Fourier transform on $\mathbb{C}$-valued L^1 functions on $\mathbb{Q}_p$ is

$$\mathscr{F} f(\xi) = \widehat{f}(\xi) = \int_{\mathbb{Q}_p} \overline{\psi}_\xi(x)\, f(x)\, dx$$

The space of *Schwartz functions* $\mathscr{S}(\mathbb{Q}_p)$ on $\mathbb{Q}_p$ should be mapped to itself homeomorphically under Fourier transform, should consist of very simple functions and should be dense in $L^2(\mathbb{Q}_p)$. We will show that the following choice succeeds: take

$$\mathscr{S}(\mathbb{Q}_p) = \{\text{compactly supported, locally constant functions}\}$$

($\mathbb{C}$-valued), where f being *locally constant* means that every $x \in \mathbb{Q}_p$ has a neighborhood U such that $f(x') = f(x)$ for $x' \in U$.

[13.17.1] Remark: The local constancy turns out to be the appropriate p-adic notion of *smoothness*. Unlike the archimedean case, p-adic Schwartz functions are *compactly supported*. That is, in the p-adic case, *test functions* and *Schwartz functions* are the same classes of functions.

[13.17.2] Claim: $f \in \mathscr{S}(\mathbb{Q}_p)$ is *uniformly* locally constant: there is a (compact, open) subgroup $U = p^n \mathbb{Z}_p$ such that $f(x + u) = f(x)$ for all $x \in \mathbb{Q}_p$, and for all $u \in U$.

Proof: Since $\bigcup_{m \geq 0} p^{-m} \mathbb{Z}_p = \mathbb{Q}_p$, the support of a given $f \in \mathscr{S}(\mathbb{Q}_p)$ is contained in some $p^{-m} \mathbb{Z}_p$. For each $x \in p^{-m} \mathbb{Z}_p$, there is a neighborhood $x + p^{n_x} \mathbb{Z}_p$ on which f is constant. By compactness of $p^{-m} \mathbb{Z}_p$, there are finitely many points $x_1, \dots, x_\ell$ so that the corresponding neighborhoods cover $p^{-m} \mathbb{Z}_p$. Let $n = \max_{1 \leq i \leq \ell} n_{x_i}$ and $U = p^n \mathbb{Z}_p$. A given $x \in p^{-m} \mathbb{Z}_p$ lies in $x_j + p^{n_{x_j}} \mathbb{Z}_p$ for some j, and

$$x + U \subset x_j + p^{n_{x_j}} \mathbb{Z}_p + U \subset x_j + p^{n_{x_j}} \mathbb{Z}_p + p^{n_{x_j}} \mathbb{Z}_p + U = x_j + p^{n_{x_j}} \mathbb{Z}_p$$

since every $p^n \mathbb{Z}_p$ is closed under addition. Thus, f is locally constant on $x + U$. ///

[13.17.3] Corollary: $\mathscr{S}(\mathbb{Q}_p)$ is a strict colimit of the finite-dimensional subspaces

$$V_{m,n} = \{f \in \mathscr{S}(\mathbb{Q}_p) : \mathrm{spt} f \subset p^{-m} \mathbb{Z}_p, \ f(x + u) = f(x) \text{ for all } x,$$

$$\text{for all } u \in p^n \mathbb{Z}_p\}$$

In particular, $\mathscr{S}(\mathbb{Q}_p)$ consists of finite linear combinations of characteristic functions of sets $x_o + p^n \mathbb{Z}_p$.

Proof: The lemma asserts that $\mathscr{S}(\mathbb{Q}_p) = \bigcup_{m,n} V_{m,n}$. Since $p^{-m} \mathbb{Z}_p$ is the disjoint union of p^{m+n} distinct cosets $x_o + p^n \mathbb{Z}_p$, the subspace $V_{m,n}$ is the collection of linear combinations of characteristic functions of these sets. ///

Thus, the Schwartz space $\mathscr{S}(\mathbb{Q}_p)$ is not Fréchet but is the simplest type of LF-space, namely, a strict colimit of finite-dimensional spaces (and finite-dimensional spaces have unique topologies [13.4]) like $\mathbb{C}^\infty$ in [13.8].

The following holds for Schwartz functions by direct computation, and then will follow for L^2 functions by denseness of $\mathscr{S}(\mathbb{Q}_p)$ in $L^2(\mathbb{Q}_p)$ and extending by L^2-continuity.

[13.17.4] Theorem: For Schwartz functions, *Fourier inversion* holds:

$$f(x) = \int_{k_v} \psi_\xi(x) \widehat{f}(\xi) \, d\xi \qquad (\text{for } f \in \mathscr{S}(\mathbb{Q}_p))$$

and *Plancherel's theorem* holds:

$$\int_{\mathbb{Q}_p} |f|^2 = \int_{\mathbb{Q}_p} |\widehat{f}|^2 \qquad \text{(for } f \in \mathscr{S}(\mathbb{Q}_p)\text{)}$$

Proof: For Schwartz functions, we prove more by giving sample computations of Fourier transforms which are useful. In particular, we observe simply described functions on $\mathbb{Q}_p$ whose Fourier transforms are of a similar nature. For example, certain natural functions in $\mathscr{S}(\mathbb{Q}_p)$ are their own Fourier transform, analogous to the Gaussian in the archimedean case.

[13.17.5] Claim: The characteristic function f of $\mathbb{Z}_p$ is its own Fourier transform. ///

Proof: Computing directly,

$$\widehat{f}(\xi) = \int_{\mathbb{Q}_p} \overline{\psi}_\xi(x) f(x) \, dx = \int_{\mathbb{Z}_p} \overline{\psi}_1(\xi \cdot x) \, dx = \int_{\mathbb{Z}_p} \psi_1(-\xi \cdot x) \, dx$$

Recall a form of the *cancellation lemma:* (a tiny case of *Schur orthogonality*)

[13.17.6] Lemma: Let $\psi : K \to \mathbb{C}^\times$ be a continuous group homomorphism on a compact group K. Then

$$\int_K \psi(x) \, dx = \begin{cases} \text{meas}\,(K) & \text{(for } \psi = 1\text{)} \\ 0 & \text{(for } \psi \neq 1\text{)} \end{cases}$$

Proof: *(of lemma)* Yes, of course, the measure is a Haar measure on K. Since K is *compact*, it is *unimodular*. For ψ trivial, of course the integral is the total measure of K. For ψ nontrivial, there is $y \in K$ such that $\psi(y) \neq 1$. Using the invariance of the measure, change variables by replacing x by xy:

$$\int_K \psi(x) \, dx = \int_K \psi(xy) \, d(xy) = \int_K \psi(x) \, \psi(y) \, dx = \psi(y) \int_K \psi(x) \, dx$$

Since $\psi(y) \neq 1$, the integral is 0. ///

Apply the lemma to the integrals computing the Fourier transform of the characteristic function f of $\mathbb{Z}_p$. Since $\mathbb{Z}_p$ has measure 1,

$$\widehat{f}(\xi) = \int_{\mathbb{Z}_p} \psi_1(-\xi \cdot x) \, dx = \begin{cases} 1 & (\psi_1(-\xi x) = 1 \text{ for } x \in \mathbb{Z}_p) \\ 0 & \text{(otherwise)} \end{cases}$$

On one hand, for $\xi \in \mathbb{Z}_p$, certainly $\psi_1(\xi x) = 1$ for $x \in \mathbb{Z}_p$. On the other hand, for $\xi \notin \mathbb{Z}_p$, there is $x \in \mathbb{Z}_p$ such that, for example, $\xi \cdot x = 1/p$. Then

$$\psi_1(-\xi \cdot x) = \psi_1(\tfrac{-1}{p}) = e^{+2\pi i \cdot \frac{1}{p}} \neq 1$$

Thus, ψ_ξ is not trivial on $\mathbb{Z}_p$, so the integral is 0. Thus, the characteristic function of $\mathbb{Z}_p$ is its own Fourier transform. ///

[13.17.7] Claim: The Fourier transform of the characteristic function of $p^k\mathbb{Z}_p$ is p^{-k} times the characteristic function of $p^{-k}\mathbb{Z}_p$.

Proof: Let f be the characteristic function of $p^k\mathbb{Z}_p$, so

$$\widehat{f}(\xi) = \int_{\mathbb{Q}_p} \overline{\psi}_\xi(x)\, f(x)\, dx = \int_{p^k\mathbb{Z}_p} \overline{\psi}_1(\xi \cdot x)\, dx$$

$$= |p^k|_p \cdot \int_{\mathbb{Z}_p} \psi_1(-\xi \cdot x/p^k)\, dx = p^{-k} \cdot \int_{\mathbb{Z}_p} \psi_1(-\xi \cdot x/p^k)\, dx$$

This reduces to the previous computation: by *cancellation*, for $\xi/p^k \notin \mathbb{Z}_p$ the character $x \to \psi_1(-\xi x/p^k)$ is nontrivial, so the integral is 0. Otherwise, the integral is 1. ///

[13.17.8] Claim: The Fourier transform of the characteristic function of $\mathbb{Z}_p + y$ is ψ_y times the characteristic function of $\mathbb{Z}_p$.

Proof: Let f be the characteristic function of $\mathbb{Z}_p + y$, so

$$\widehat{f}(\xi) = \int_{\mathbb{Q}_p} \overline{\psi}_\xi(x)\, f(x)\, dx = \int_{\mathbb{Z}_p+y} \overline{\psi}_1(\xi \cdot x)\, dx$$

$$= \int_{\mathbb{Z}_p} \psi_1(-\xi \cdot (x+y))\, dx = \psi_1(-\xi \cdot y) \int_{\mathbb{Z}_p} \psi_1(-\xi \cdot x)\, dx$$

$$= \psi_1(-\xi \cdot y) \cdot f(\xi)$$

by the previous computation. ///

Combining the two computations above,

$$\mathscr{F}\left(\text{char fcn } p^k\mathbb{Z}_p + y\right) = \psi_y \cdot p^{-k} \cdot (\text{char fcn } p^{-k}\mathbb{Z}_p)$$

Conveniently, products $\psi_y \cdot (\text{char fcn } p^{-k}\mathbb{Z}_p)$ are in the same class of functions, since ψ_y has a kernel which is an open (and compact) neighborhood of 0, so we *this class of functions is mapped to itself under Fourier transform.*

We have essentially proven Fourier inversion, in the preceding computations, as follows. Let f^o be the characteristic function of $\mathbb{Z}_p$. We computed $\widehat{f^o} = f$. Let δ_t be the dilation operator $\delta_t f(x) = f(t \cdot x)$ for $t \in \mathbb{Q}_p^\times$. We computed, by changing variables in the integral defining the Fourier transform, that

$$\mathscr{F}(\delta_t f) = \frac{1}{|t|_p} \cdot \delta_{1/t}(\mathscr{F}f)$$

Let τ_y be the translation operator $\tau_y f(x) = f(x+y)$. By changing variables,

$$\mathscr{F}(\tau_y f) = \psi_y \cdot (\mathscr{F}f)$$

It is convenient to also compute that

$$\mathscr{F}(\psi_y \cdot f)(\xi) = \int_{\mathbb{Q}_p} \overline{\psi}_\xi(x) \cdot \psi_y(x) f(x) \, dx = \int_{\mathbb{Q}_p} \overline{\psi}_{\xi-y}(x) f(x) \, dx$$

$$= \widehat{f}(\xi - y) = \tau_{-y}(\mathscr{F}f)$$

Let $\mathscr{F}^*$ be the integral for Fourier inversion, namely,

$$\mathscr{F}^* f(x) = \int_{\mathbb{Q}_p} \psi_\xi(x) f(\xi) \, d\xi$$

Similar computations give

$$\mathscr{F}^*(\delta_t f) = \frac{1}{|t|_p} \delta_{1/t}(\mathscr{F}^* f) \qquad \mathscr{F}^*(\tau_y f) = \psi_{-y}(\mathscr{F}^* f)$$

and

$$\mathscr{F}^*(\psi_y f) = \tau_y(\mathscr{F}^* f)$$

Since every element of $\mathscr{S}(\mathbb{Q}_p)$ is a linear combination of images of f^o under dilation and translation, it suffices to give a sort of inductive proof of Fourier inversion:

$$\mathscr{F}^* \mathscr{F}(\tau_y f) = \mathscr{F}^* \psi_y \mathscr{F} f = \tau_y \mathscr{F}^* \mathscr{F} f$$

$$\mathscr{F}^* \mathscr{F}(\delta_t f) = \mathscr{F}^* \frac{1}{|t|_p} \delta_{1/t} \mathscr{F} f = \frac{1}{|t|_p} \frac{1}{|1/t|_p} \delta_t \mathscr{F}^* \mathscr{F} f = \delta_t \mathscr{F}^* \mathscr{F} f$$

Similarly for multiplication by ψ_y. Since $\mathscr{F}^* \mathscr{F} f^o = \mathscr{F}^* f^o = f^o$, we have Fourier inversion on $\mathscr{S}(\mathbb{Q}_p)$.

The surjectivity of $\mathscr{F} : \mathscr{S}(\mathbb{Q}_p) \to \mathscr{S}(\mathbb{Q}_p)$ is made explicit in the foregoing computations. Then we have the Plancherel theorem on $\mathscr{S}(\mathbb{Q}_p)$:

$$\int_{\mathbb{Q}_p} f \cdot \overline{g} = \int_{\mathbb{Q}_p} f \cdot \overline{\mathscr{F}^{-1} g} = \int_{\mathbb{Q}_p} \int_{\mathbb{Q}_p} f(x) \cdot \psi_1(-\xi x) \cdot \overline{\widehat{g}}(\xi) \, d\xi \, dx$$

$$= \int_{\mathbb{Q}_p} \left(\int_{\mathbb{Q}_p} f(x) \cdot \psi_1(-\xi x) \, dx \right) \cdot \overline{\widehat{g}}(\xi) \, d\xi = \int_{\mathbb{Q}_p} \widehat{f} \cdot \overline{\widehat{g}}$$

This proves the theorem for Schwartz functions.   ///

Similarly, and as for $C_c^o(\mathbb{R})$, the space $C_c^o(\mathbb{Q})$ of compactly supported, continuous, $\mathbb{C}$-valued functions on $\mathbb{Q}_p$ is an LF-space, the strict colimit of the spaces of continuous functions supported on $p^{-n}\mathbb{Z}_p$. Much as in [6.2] and [13.9], we have

[13.17.9] Claim: The translation action $\mathbb{Q}_p \times C_c^o(\mathbb{Q}) \longrightarrow C_c^o(\mathbb{Q})$ by $(x \cdot f)(y) = f(y + x)$ is (jointly) continuous.

Proof: Since $\mathbb{Q}_p$ itself is the colimit of $p^{-m}\mathbb{Z}_p$ (as additive topological group), and $C_c^o(\mathbb{Q}_p)$ is a colimit, it suffices to show that $p^{-m}\mathbb{Z}_p \times C^o(p^{-n}\mathbb{Z}_p) \longrightarrow C_c^o(\mathbb{Q}_p)$ is continuous for all m, n. Indeed, $p^{-m}\mathbb{Z}_p \times C^o(p^{-n}\mathbb{Z}_p)$ maps to $C^o(p^{-\max(m,n)}\mathbb{Z}_p)$, and sup norms are preserved.   ///

[13.17.10] Claim: $\mathscr{S}(\mathbb{Q}_p)$ is *dense* in $C_c^o(\mathbb{Q}_p)$.

Proof: This is a simple p-adic analogue of the smoothing of distributions [14.5], and of Gårding's theorem [14.6], asserting that smooth vectors are dense in a representation, also following from the basic result [14.1.4] about approximate identities and Gelfand-Pettis integrals. Namely, let $\varphi_n = p^n \cdot \chi_n$, where χ_n is the characteristic function of $p^n\mathbb{Z}_p$. These are continuous, compactly supported functions and form an *approximate identity* in $C_c^o(\mathbb{Q}_p)$ in the sense that they are non-negative, their integrals are all 1, and their supports shrink to $\{0\}$. By the previous claim, $\mathbb{Q}_p$ acts continuously on $C_c^o(\mathbb{Q}_p)$, giving integral operators

$$(\varphi_n \cdot f)(x) = \int_{\mathbb{Q}_p} \varphi_n(y)\, f(x + y)\, dy$$

on $f \in C_c^o(\mathbb{Q}_p)$. By [14.1.4], $\varphi_n \cdot f \longrightarrow f$ in $C_c^o(\mathbb{Q}_p)$.

Analogous to the archimedean discussion in the proof of smoothing of distributions theorem [14.5], we check that each $\varphi_n \cdot f$ is locally constant and compactly supported, so is in $\mathscr{S}(\mathbb{Q}_p)$. The compact support is clear, since the support of $\varphi_n \cdot f$ is contained in $\mathrm{spt}(\varphi_n) + \mathrm{spt}(f)$, which is compact, being the image of the compact $\mathrm{spt}(\varphi_n) \times \mathrm{spt}(f) \subset \mathbb{Q}_p \times \mathbb{Q}_p$ under the continuous map $x \times y \to x + y$. For local constancy, for $u \in p^n\mathbb{Z}_p$,

$$(\varphi_n \cdot f)(x + u) = \int_{\mathbb{Q}_p} \varphi_n(y)\, f(x + u + y)\, dy = p^n \int_{p^n\mathbb{Z}_p} f(x + u + y)\, dy$$

$$= p^n \int_{p^n\mathbb{Z}_p} f(x + y)\, dy = (\varphi_n \cdot f)(x)$$

by changing variables $y \to y - u$, since $p^n\mathbb{Z}_p$ is a group.   ///

[13.17.11] Corollary: $\mathscr{S}(\mathbb{Q}_p)$ is dense in $L^1(\mathbb{Q}_p)$ and in $L^2(\mathbb{Q}_p)$.

Proof: $C_c^o(\mathbb{Q}_p)$ is dense in both $L^1(\mathbb{Q}_p)$ and $L^2(\mathbb{Q}_p)$, essentially by Urysohn's lemma [9.E.2], as in [6.1] and [6.2], so $\mathscr{S}(\mathbb{Q}_p)$ is dense in both, by the previous.   ///

[13.17.12] Corollary: $\mathscr{F}$ extends to $L^2(\mathbb{Q}_p)$ *by continuity*, giving the *Fourier-Plancherel* transform $\mathscr{F} : L^2(\mathbb{Q}_p) \longrightarrow L^2(\mathbb{Q}_p)$, no longer defined literally by the integrals but still satisfying Fourier inversion and Plancherel theorem.   ///

14

Vector-Valued Integrals

Quasi-complete, locally convex topological vector spaces V have the useful property that continuous compactly supported V-valued functions have *integrals* with respect to finite, regular Borel measures. Rather than being *constructed* as limits, these vector-valued integrals are *characterized*. Uniqueness follows from the Hahn-Banach theorem, and existence follows from a construction.

An immediate application is justification of differentiation with respect to a parameter inside an integral, under mild, easily understood hypotheses, a special case of the general assertion that Gelfand-Pettis integrals commute with continuous operators, as in the first section. A subtler application is passage of compactly supported distributions inside the integrals expressing Fourier inversion, as in [14.3]. Uniqueness of group-invariant measures, distributions, and other functionals is another corollary. Other applications are to holomorphic vector-valued functions, to holomorphically parametrized families of *generalized functions* (distributions), as in Chapter 14. Many distributions that are not classical functions appear naturally as residues or analytic continuations of meromorphic families of classical functions.

14.1 Characterization and Basic Results

For a topological vectorspace V over $\mathbb{C}$ and for f a continuous V-valued function on a topological space X with a regular Borel measure, a *Gelfand-Pettis integral* of f is a vector $I_f \in V$ so that

$$\lambda(I_f) = \int_X \lambda \circ f \qquad \text{(for all } \lambda \in V^*)$$

If it exists and is unique, this vector I_f is reasonably denoted

$$I_f = \int_X f$$

In contrast to *construction* of integrals as limits, this *characterization* surely should apply to any reasonable notion of integral, without asking how the property comes to be. Since the property of allowing continuous linear functionals to pass inside the integral is an irreducible minimum, the Gelfand-Pettis integral is sometimes called a *weak integral*.

We only consider *locally convex* vectorspaces, so *uniqueness* of the integral is immediate, since V^* *separates points* on V, by Hahn-Banach. Similarly, for such V, *linearity* of $f \to I_f$ follows by Hahn-Banach. The issue is *existence*.[1] We only consider V-valued functions that are *continuous* on *compact* measure spaces with *regular Borel* measures. Under these assumptions, all the $\mathbb{C}$-valued integrals

$$f \longrightarrow f \circ \lambda \longrightarrow \int_X \lambda \circ f \qquad \text{(for } \lambda \in V^*)$$

exist for elementary reasons, being integrals of compactly supported $\mathbb{C}$-valued continuous functions on compact sets with respect to a regular Borel measure.

For *existence* of Gelfand-Pettis integrals of compactly supported, continuous V-valued functions, the literal requirement on V turns out to be that *the closure of the convex hull of a compact set is compact*. We show here that *local convexity* and *quasi-completeness* suffice. For the following, a *probability measure* is a positive, regular Borel measure with total measure 1.

[14.1.1] Theorem: Let X be a compact Hausdorff topological space with a probability measure. Let V be a quasi-complete, locally convex vectorspace. Then *continuous* V-valued functions f on X have Gelfand-Pettis integrals. The

---

[1] We want the integral to be in V itself, rather than in a larger space containing V, such as a double dual V^{**}, for example, to make *existence* trivial, but then leaving technical issues. Some discussions of vector-valued integration do allow integrals to exist in larger spaces, but this only delays certain issues, rather than resolving them directly.

basic estimate holds:

$$\int_X f \in (\text{closure of convex hull of } f(X))$$

substituting for the estimate of a $\mathbb{C}$-valued integral by the integral of its absolute value. *(Proof in [14.8].)*

[14.1.2] Corollary: In the situation of the theorem, but when the total measure of X is *finite* but not necessarily 1, the *basic estimate* becomes

$$\int_X f \in (\text{closure of convex hull of } f(X)) \cdot \int_X 1$$

(Replace the measure by a constant multiple.)   ///

[14.1.3] Corollary: For a continuous linear map of locally convex, quasi-complete topological vectorspaces $T : V \to W$, and f a continuous, compactly supported V-valued function on a finite, regular, positive Borel measure space X. Then

$$T\left(\int_X f \right) = \int_X T \circ f$$

Proof: To verify that the left-hand side of the asserted equality is a Gelfand-Pettis integral of $T \circ f$, show that

$$\mu(\text{left-hand side}) = \int_X \mu \circ (T \circ f) \qquad (\text{for all } \mu \in W^*)$$

Starting with the left-hand side, by associativity

$$\mu\left(T\left(\int_X f \right) \right) = (\mu \circ T)\left(\int_X f \right)$$

Since $\mu \circ T \in V^*$ and $\int_X f$ is a weak integral, this is

$$\int_X (\mu \circ T) \circ f$$

and again by associativity it is

$$\int_X \mu \circ (T \circ f)$$

proving that $T(\int_X f)$ is a weak integral of $T \circ f$.   ///

A *representation* of G on a locally convex, quasi-complete topological vec-torspace V is a continuous map $G \times V \to V$ that is linear in V, has the associa-tivity $(gh) \cdot v = g \cdot (h \cdot v)$ for $g, h \in G$, and $1_G \cdot v = v$ for all $v \in V$.

For any $v \in V$ and $\varphi \in C_c^o(G)$, we have a V-valued Gelfand-Pettis integral

$$\varphi \cdot f = \int_G \varphi(g) \, T_g f \, dg \in \text{closed convex hull of } \{\varphi(g)f : g \in G\} \subset V$$

For present purposes, a continuous *approximate identity* on a topological group G is a sequence $\{\varphi_i\}$ of nonnegative, continuous, real-valued functions such that $\int_G \varphi_i = 1$ for all i, and such that the supports shrink to $\{1\}$, in the sense that for every neighborhood N of 1 in G, there is an index i_o so that the support of φ_i is inside N for all $i \geq i_o$. From Urysohn's lemma [9.E.2], there always exists a *continuous* approximate identity. Let $T_g f(y) = f(yg)$ be right translation. With right translation-invariant measure dg on G, since $\int_G \varphi_i(g) \, dg = 1$ and φ_i is nonnegative, $\varphi_i(g) \, dg$ is a probability measure (total mass 1) on the (compact) support of φ_i.

[14.1.4] Corollary: Given a representation of G on a quasi-complete, locally convex topological vector space V, for every approximate identity $\{\varphi_i\}$ on G, $\varphi_i \cdot v \to v$ for every $v \in V$.

Proof: By continuity, given a neighborhood N of 0 in V, we have $\varphi_i \cdot f \in f + N$ for all sufficiently large i. That is, $\varphi_i \cdot f \to f$.   ///

[14.1.5] Corollary: Given a representation of G on a quasi-complete, locally convex topological vector space V, the action of $C_c^o(G)$ on V is *non-degenerate*, in the sense that, for every $0 \neq v \in V$, there exists $\varphi \in C_c^o(G)$ such that $\varphi \cdot v \neq 0$.

Proof: For every approximate identity $\{\varphi_i\}$, for every $v \in V$, $\varphi_i \cdot v \to v$. Thus, for all sufficiently large i, $\varphi_i \cdot v \neq 0$ for $v \neq 0$.   ///

A *G-subrepresentation* $W \subset V$ of a representation of G on a quasi-complete, locally convex topological vector space V is a (topologically) closed G-stable vector subspace of V. Similarly, a $C_c^o(G)$-*subrepresentation* $W \subset V$ of a G-representation is a (topologically) closed $C_c^o(G)$-stable vector subspace of V.

[14.1.6] Corollary: A $C_c^o(G)$-subrepresentation of a G-representation on a quasi-complete, locally convex topological vector space V is a G-representation.

Proof: Let $\{\varphi_i\}$ be an approximate identity, fix w in the subrepresentation W, and take $g \in G$. We will show that $g \cdot w \in W$. On one hand, from [14.1.4], $\varphi_i \cdot (g \cdot w) \to g \cdot w$. On the other hand,

$$\varphi_i \cdot (g \cdot w) = \int_G \varphi_i(h) \, h \cdot (g \cdot v) \, dh = \int_G \varphi_i(hg^{-1}) \, h \cdot v \, dh$$

by changing variables. The function $h \to \varphi_i(hg^{-1})$ is still in W, by assumption, so $\varphi_i \cdot (g \cdot w)$ is a sequence of vectors in W. Since W is closed, and the sequence converges to $g \cdot w$, necessarily $g \cdot w \in W$.   ///

A representation of G on W is (topologically) *G-irreducible* when there is no proper G-subrepresentation. A representation of G on W is (topologically) $C_c^o(G)$-*irreducible* when there is no proper $C_c^o(G)$-subrepresentation.

[14.1.7] Corollary: For a representation of G on a quasi-complete, locally convex topological vector space V, every irreducible $C_c^o(G)$-subrepresentation is an irreducible G-subrepresentation.

Proof: From the previous corollary, every $C_c^o(G)$-subrepresentation $W \subset V$ is a G-subrepresentation. If W had a proper G-subrepresentation W', then W' would be a proper $C_c^o(G)$-subrepresentation, as well.   ///

14.2 Differentiation of Parametrized Integrals

Differentiation under the integral is an immediate corollary, in many useful situations.

[14.2.1] Claim: A $\mathbb{C}$-valued C^k function F on $[a, b] \times [c, d]$ gives a *continuous* $C^k[c, d]$-valued function $f(x) = F(x, -)$ of $x \in [a, b]$.

Proof: For each $0 \le i \le k$, the function $(x, y) \to \frac{\partial^i}{\partial y^i} F(x, y)$ is continuous as a function of two variables. For each $\varepsilon > 0$ and each $x_o \in [a, b]$, we want $\delta > 0$ such that

$$|x - x_o| < \delta \quad \Longrightarrow \quad \sup_y \left| \frac{\partial^i}{\partial y^i} F(x, y) - \frac{\partial^i}{\partial y^i} F(x_o, y) \right| < \varepsilon$$

The continuous function $(x, y) \to \frac{\partial^k}{\partial y^i} F(x, y)$ is *uniformly* continuous on the compact $[a, b] \times [c, d]$, so there is $\delta > 0$ such that

$$\left| \frac{\partial^k}{\partial y^i} F(x_1, y_1) - \frac{\partial^i}{\partial y^i} F(x_2, y_2) \right| < \varepsilon$$

for all $(x_1, y_1), (x_2, y_2)$ with $|x_1 - x_2| < \delta$ and $|y_1 - y_2| < \delta$. In particular, this holds for all $y_1 = y_2$, and $x_1 = x$, and $x_2 = x_o$.   ///

[14.2.2] Corollary: For a $\mathbb{C}$-valued C^k function F on $[a, b] \times [c, d]$,

$$\frac{\partial}{\partial y} \int_a^b F(x, y) \, dx = \int_a^b F \frac{\partial}{\partial y}(x, y) \, dx$$

Proof: The function-valued function $x \to (y \to F(x, y))$ is a continuous, $C^k[c, d]$-valued function, and $\frac{\partial}{\partial y}$ is a continuous linear map $C^k[c, d] \to C^{k-1}[c, d]$, so the Gelfand-Pettis property allows interchange of the operator and the integral. ///

14.3 Fourier Transforms

Certainly an integral expressing Fourier inversion [13.15]

$$f(x) = \frac{1}{2\pi} \int_{\mathbb{R}} e^{i\xi x} \, \widehat{f}(\xi) \, d\xi$$

for Schwartz function f cannot converge as a Schwartz-function-valued integral because $x \to e^{i\xi x}$ is in $C^\infty(\mathbb{R})$, but not Schwartz. Multiplying by $\widehat{f}$ does not affect decay in x, so does not alter the situation. Examination of the situation is complicated by the fact that the integrand is not compactly supported, but we can follow Schwartz's device of suitably *compactifying* $\mathbb{R}^n$ to a sphere S^n and then invoke the Gelfand-Pettis property for compactly supported functions. Then we will see that the integral *does* converge as a $C^\infty(\mathbb{R})$-valued Gelfand-Pettis integral. First,

[14.3.1] Claim: For any $\Phi \in C^\infty(\mathbb{R}^2)$, the $C^\infty(\mathbb{R})$-valued function $\xi \to \Phi(-, \xi)$ that is, $\xi \to (x \to \Phi(x, \xi))$ is a *continuous*, $C^\infty(\mathbb{R})$-valued function on $\mathbb{R}$. (Similarly, it is a *smooth* $C^\infty(\mathbb{R})$-valued function, but do not need this.)

Proof: The function $(x, \xi) \to \Phi(x, \xi)$ is C^∞ as a function of two variables. In particular, $(x, \xi) \to \frac{\partial^k}{\partial x^k} \Phi(x, \xi)$ is continuous as a function of two variables. For each k, compact $C \subset \mathbb{R}$, $\varepsilon > 0$ and each $\xi_o \in \mathbb{R}$, we want $\delta > 0$ such that

$$|\xi - \xi_o| < \delta \implies \sup_{x \in C} \left| \frac{\partial^k}{\partial x^k} \Phi(x, \xi) - \frac{\partial^k}{\partial x^k} \Phi(x, \xi_o) \right| < \varepsilon$$

Let I be the interval $[\xi_1, \xi_o + 1]$. The continuous function $(x, \xi) \to \frac{\partial^k}{\partial x^k} \Phi(x, \xi)$ is *uniformly* continuous on the compact $C \times I$ that is, there is $\delta > 0$ such that

$$\left| \frac{\partial^k}{\partial x^k} \Phi(x_1, \xi_1) - \frac{\partial^k}{\partial x^k} \Phi(x_2, \xi_2) \right| < \varepsilon$$

for all $(x_1, \xi_1), (x_2, \xi_2) \in I$ with $|x_1 - x_2| < \delta$ and $|\xi_1 - \xi_2| < \delta$. In particular, this holds for all $x_1 = x_2$, and $\xi = \xi_o \in I$, and $\xi_2 = \xi_o$, giving the desired continuity. This previous applies to $\Phi(x, \xi) = e^{i\xi x}$. Since $\xi \to F(\xi)$ is a continuous $\mathbb{C}$-valued function, the product $x \to F(\xi) \cdot e^{i\xi x}$ is a continuous $C^\infty(\mathbb{R})$-valued function of $\xi \in \mathbb{R}$.

Compactify $\mathbb{R}$ to the circle $\mathbb{T} \subset \mathbb{R}^2$ via by stereographic projection

$$\sigma : x \longrightarrow \left(\frac{x}{\sqrt{1 + x^2}}, \frac{1}{\sqrt{1 + x^2}} \right)$$

and adding the point $\infty = (0, 1)$.

[14.3.2] Claim: $\xi \to F(\xi) \cdot \psi_\xi$ extends (by $0 \in C^\infty(\mathbb{R})$) to a continuous, $C^\infty(\mathbb{R})$-valued function on the compactification $\mathbb{T}$ of $\mathbb{R}$.

Proof: We must check continuity in ξ near ∞. That is, for each k, compact $C \subset \mathbb{R}$, and $\varepsilon > 0$, we want (large) B such that

$$|\xi| > B \Longrightarrow \sup_{x \in C} \left| F(\xi) \cdot \frac{\partial^k}{\partial x^k} e^{i\xi x} - 0 \right| < \varepsilon$$

The exponential function is easy to estimate: for example, with M a bound so that $|(1 + \xi^2)^k \cdot F(\xi)| \le M$,

$$\sup_{x \in C} \left| F(\xi) \cdot \frac{\partial^k}{\partial x^k} e^{i\xi x} \right| = \left| F(\xi) \cdot (i\xi)^k \right| \cdot 1 \le \frac{M \cdot |\xi|^k}{(1 + \xi^2)^k}$$

Take B large enough so that $M \cdot B^k / (1 + B^2)^k < \varepsilon$. For any continuous linear functional, $\xi \to \lambda \circ (\psi_\xi \cdot F(\xi))$ is a continuous scalar-valued function on the compact set $\mathbb{T}$, so is *bounded*. The same is true of any $\xi \to \lambda \circ (\psi_\xi \cdot (1 + \xi^2)^N F(\xi))$, so $\xi \to \lambda \circ (\psi_\xi \cdot F(\xi))$ is *rapid decreasing*. Adjust the measure on $\mathbb{R}$ to give total measure 1:

$$\int_{\mathbb{R}} \psi_\xi \cdot F(\xi) \, d\xi = \int_{\mathbb{R}} \psi_\xi \cdot \pi (1 + \xi)^2 F(\xi) \, \frac{d\xi}{\pi(1 + \xi^2)}$$

The function $\pi(1 + \xi)^2 F(\xi)$ is still continuous and of rapid decay. Being continuous and compactly supported on a measure space with total measure 1, with values in a quasi-complete, locally convex topological vector space, $\xi \to \pi(1 + \xi^2) F(\xi) \cdot \psi_\xi$ has a *Gelfand-Pettis integral J* with respect to the measure $d\xi / \pi(1 + \xi^2)$, lying inside the closed convex hull of the image. That is,

$$\lambda(J) = \int_{\mathbb{R}} \lambda(\psi_\xi) \cdot \pi(1 + \xi)^2 F(\xi) \, \frac{d\xi}{\pi(1 + \xi^2)}$$

for every continuous linear functional λ. In the the latter scalar-valued integral the adjustment factors cancel:

$$\int_{\mathbb{R}} \lambda(\psi_\xi \cdot \pi(1 + \xi)^2 F(\xi)) \, \frac{d\xi}{\pi(1 + \xi^2)}$$

$$= \int_{\mathbb{R}} \lambda(\psi_\xi) \cdot \pi(1 + \xi)^2 F(\xi) \, \frac{d\xi}{\pi(1 + \xi^2)}$$

$$= \int_{\mathbb{R}} \lambda(\psi_\xi) \cdot F(\xi) \, d\xi$$

That is, $\lambda(J) = \int_\mathbb{R} \lambda(\psi_\xi) \cdot F(\xi) \, d\xi$, and the Gelfand-Pettis integral J of the mutually adjusted function and measure is the Gelfand-Pettis integral of the original. ///

[14.3.3] Corollary: For rapidly decreasing $F \in C^o(\mathbb{R})$, for any continuous linear $T : C^\infty \to V$ for another topological vector space V,

$$T\left(\int_\mathbb{R} \psi_\xi \cdot F(\xi) \, d\xi \right) = \int_\mathbb{R} T\big(\psi_\xi \cdot F(\xi)\big) \, d\xi = \int_\mathbb{R} T(\psi_\xi) \cdot F(\xi) \, d\xi$$

as V-valued Gelfand-Pettis integral. ///

[14.3.4] Corollary: For rapidly decreasing $F \in C^o(\mathbb{R})$, for any continuous, for any compactly supported distribution u,

$$u\left(\int_\mathbb{R} \psi_\xi \cdot F(\xi) \, d\xi \right) = \int_\mathbb{R} u\big(\psi_\xi \cdot F(\xi)\big) \, d\xi = \int_\mathbb{R} u(\psi_\xi) \cdot F(\xi) \, d\xi$$

with absolutely convergent integral. ///

[14.3.5] Corollary: For rapidly decreasing $F \in C^o(\mathbb{R})$, the Fourier transform is a C^∞ function on $\mathbb{R}$, and its derivative is computed by the expected expression

$$\frac{\partial}{\partial x}\left(\int_\mathbb{R} \psi_\xi \cdot F(\xi) \, d\xi \right) = \int_\mathbb{R} \frac{\partial \psi_\xi}{\partial x} \cdot F(\xi) \, d\xi = i \int_\mathbb{R} \psi_\xi \cdot \xi F(\xi) \, d\xi$$

since $\partial/\partial x$ is a continuous map of $C^\infty(\mathbb{R})$ to itself. ///

14.4 Uniqueness of Invariant Distributions

We prove uniqueness of invariant *functionals* on suitable function spaces V on topological spaces X on which a topological group acts transitively. This includes uniqueness of invariant (Haar) *measures*, and uniqueness of invariant *distributions*, as special cases.

A translation-invariant function f on the real line, that is, a function with $f(x + y) = f(x)$ for all $x, y \in \mathbb{R}$, is *constant*, by a pointwise argument:

$$f(x) = (T_x f)(0) = f(0)$$

where $T_x f(y) = f(x + y)$ is translation. The same conclusion holds for translation-invariant *distributions*, but we cannot argue in terms of pointwise values.

Let G be a *topological group*,[2] with right translation-invariant measure dg, meaning that

$$\int_G f(g \cdot h)\, dg = \int_G f(g)\, dg \qquad \text{(for all } h \in G)$$

We assume only *existence* of a right translation-invariant measure. The theorem proves uniqueness.

For present purposes, a continuous *approximate identity* on a topological group G is a sequence $\{\varphi_i\}$ of nonnegative, continuous, real-valued functions such that $\int_G \varphi_i = 1$ for all i, and such that the supports shrink to $\{1\}$, in the sense that for every neighborhood N of 1 in G, there is an index i_o so that the support of φ_i is inside N for all $i \geq i_o$. From Urysohn's lemma [9.E.2], there always exists a *continuous* approximate identity. Not all classes of functions contain an approximate identity in this strict sense: (real-) analytic functions on a non-compact group, such as $\mathbb{R}$, cannot be compactly supported, so a compromise notion would be needed. The following theorem refers to the strict sense that supports shrink to $\{1\}$:

[14.4.1] Theorem: Let $V \subset C_c^o(G)$ be a quasi-complete, locally convex topological vector space of complex-valued functions on G stable under left and right translations, so that $G \times V \to V$ is continuous and containing an *approximate identity* $\{\varphi_i\}$. Then there is a unique *right G-invariant* element of the dual space V^* (up to constant multiples), and it is

$$f \to \int_G f(g)\, dg \qquad \text{(with right translation-invariant measure)}$$

Proof: Let $T_g f(y) = f(yg)$ be right translation. With right translation-invariant measure dg on G, since $\int_G \varphi_i(g)\, dg = 1$ and φ_i is nonnegative, $\varphi_i(g)\, dg$ is a probability measure (total mass 1) on the (compact) support of φ_i. Thus, for any $f \in V$, we have a V-valued Gelfand-Pettis integral

$$T_{\varphi_i} f = \int_G \varphi_i(g)\, T_g f\, dg \in \text{closure of convex hull of } \{\varphi_i(g)f : g \in G\} \subset V$$

By continuity, given a neighborhood N of 0 in V, we have $T_{\varphi_i} f \in f + N$ for all sufficiently large i. That is, $T_{\varphi_i} f \to f$. For a right-invariant (continuous) functional $u \in V^*$,

$$u(f) = \lim_i u\left(g \to \int_G \varphi_i(h)\, f(gh)\, dh\right)$$

---

[2] A *topological group* is usually understood to be locally compact and Hausdorff, and multiplication and inversion are continuous. To avoid measure-theoretic pathologies, a *countable basis* is often assumed. Perhaps oddly, the local compactness excludes most topological vector spaces.

This is

$$u\left(g \to \int_G f(hg)\,\varphi_i(h^{-1})\,dh\right) = u\left(g \to \int_G f(h)\,\varphi_i(gh^{-1})\,dh\right)$$

by replacing h by hg^{-1}. By properties of Gelfand-Pettis integrals, and since f is guaranteed to be a compactly supported continuous function, we can move the functional u inside the integral: the foregoing becomes

$$\int_G f(h)\,u\left(g \to \varphi_i(gh^{-1})\right)\,dh$$

Using the *right* G-invariance of u the evaluation of u with right translation by h^{-1} gives

$$\int_G f(h)\,u(g \to \varphi_i(g))\,dh = u(\varphi_i) \cdot \int_G f(h)\,dh$$

By assumption the latter expressions approach $u(f)$ as $i \to \infty$. For f so that the latter integral is nonzero, we see that the limit of the $u(\varphi_i)$ exists, and then we conclude that $u(f)$ is a constant multiple of the indicated integral with right Haar measure.   ///

14.5 Smoothing of Distributions

Every locally integrable[3] function f on $\mathbb{R}^n$, for example, gives a *distribution* u_f by *integrating against* it:

$$u_f(\varphi) = \int_{\mathbb{R}^n} \varphi \cdot f \qquad (\text{for } \varphi \in \mathcal{D}(\mathbb{R}^n))$$

Conversely, we prove here that the distributions u_f from $f \in \mathcal{D} = C_c^\infty(\mathbb{R}^n)$ are *dense* in the whole space $\mathcal{D}^*$ of distributions, with the weak dual topology. Further, a *sequence* of such smooth functions approaching a given distribution can be expressed in terms of *smoothing* or *mollifying* u.

Let $g \to T_g$ be the *regular representation* of $\mathbb{R}^n$ on test functions $f \in \mathcal{D} = \mathcal{D}(\mathbb{R}^n)$ by $(T_g f)(x) = f(x+g)$, for $x, g \in \mathbb{R}^n$. The map, $x \times f \to T_x f$ gives a continuous map $\mathbb{R}^n \times \mathcal{D} \longrightarrow \mathcal{D}$. The corresponding *adjoint* action of $\mathbb{R}^n$ on distributions u is

$$(T_g^* u)(f) = u(T_g^{-1} f)$$

For the usual reasons, this gives a continuous map $x \times u \longrightarrow x \cdot u = T_x^* u$ with the *weak dual* topology: for $f \in \mathcal{D}$, let v_f be the seminorm $v_f(u) = |u(f)|$ on

---

[3] Again, *locally integrable* means that $|f|$ is in $L^1(K)$ for every compact K. This makes best sense for positive regular Borel measures, so that the measures of compact sets are finite.

$\mathcal{D}^*$, and then

$$
\begin{aligned}
v\big(T_g^* u - T_h^* v\big) &= \big|u\big(T_g^{-1} f\big) - v\big(T_h^{-1} f\big)\big| \\
&\leq \big|u\big(T_g^{-1} f\big) - u\big(T_h^{-1} f\big)\big| + \big|u\big(T_h^{-1} f\big) - v\big(T_h^{-1} f\big)\big| \\
&= \big|u\big(T_g^{-1} f - T_h^{-1} f\big)\big| + v_{T_h^{-1} f}(u - v)
\end{aligned}
$$

For g close to h, since the translation action of $\mathbb{R}^n$ on $\mathcal{D}$ is continuous and u is a continuous functional, $|u(T_g^{-1} f) - u(T_h^{-1} f)|$ is small. And for u close to v in the weak dual topology, the second term is small. This proves the continuity.

As earlier and throughout, the action of a function $\varphi \in C_c^o(\mathbb{R}^n)$ on distributions u is by *integrating* the group action

$$
T_\varphi^* u = \int_{\mathbb{R}^n} \varphi(x)\, T_x^* u \, dx \in \mathcal{D}^*
$$

Suppressing the T^*, this is

$$
\varphi \cdot u = \int_{\mathbb{R}^n} \varphi(x)\, x \cdot u \, dx \in \mathcal{D}^*
$$

A *smooth* approximate identity on $\mathbb{R}^n$ is a sequence $\{\psi_i\} \subset \mathcal{D}$ which are nonnegative, real-valued, have $\int_{\mathbb{R}^n} \psi_i = 1$ and supports shrink to $\{0\} \subset \mathbb{R}^n$.

[14.5.1] Theorem: For a smooth approximate identity $\{\psi_i\}$ and distribution u, the distributions $T_{\psi_i}^* u$ go to u in the weak dual topology on $\mathcal{D}^*$ and are (integration against) the functions $x \to u(T_x^{-1} \psi_i)$, which are *smooth functions*.

Proof: That $T_{\psi_i}^* u \to u$ as distributions is an instance of a general property of such Gelfand-Pettis integrals from [14.1.4]. To prove that every $T_f u$ for $f \in \mathcal{D}$ is (integration against) a *continuous* or *smooth* function, we first guess what that continuous function is, by determining its pointwise values. Indeed, if $u = u_\varphi$ were known to be integration against a continuous function φ, then with an approximate identity $\{\psi_i\}$

$$
\lim_i u_\varphi(\psi_i) = \lim_i \int_{\mathbb{R}^n} \varphi(x)\, \psi_i(x) \, dx = \varphi(0)
$$

Thus, we anticipate determining values of the alleged continuous function $f \cdot u$ by computing

$$
\text{alleged value } (f \cdot u)(0) = \lim_i (f \cdot u)(\psi_i)
$$

For a continuous function F on $\mathbb{R}^n$, let $F^\vee(x) = F(-x)$. For for f and ψ in $\mathcal{D}$, since Gelfand-Pettis integrals commute with continuous linear maps,

$$(T_f^* u)(\psi) = \left(\int_{\mathbb{R}^n} f(x) \, T_x^* u \, dx \right)(\psi) = \int_{\mathbb{R}^n} f(x) \, (T_x^* u)(\psi) \, dx$$

$$= \int_{\mathbb{R}^n} f(x) \, u(T_x^{-1} \cdot \psi) \, dx = u \left(\int_{\mathbb{R}^n} f(x) \, (T_x^{-1} \cdot \psi) \, dx \right)$$

$$= u \left(\int_{\mathbb{R}^n} f(-x) \, (T_x \psi) \, dx \right) = u(T_{f^\vee} \psi)$$

The function $T_{f^\vee} \psi$ admits a rewriting that reverses the roles of f and ψ, namely

$$(T_{f^\vee} \psi)(y) = \int_{\mathbb{R}^n} f(-x) \, \psi(y + x) \, dx = \int_{\mathbb{R}^n} f(y - x) \, \psi(x) \, dx$$

$$= \int_{\mathbb{R}^n} f(y + x) \, \psi(-x) \, dx = \int_{\mathbb{R}^n} f(y + x) \, \psi^\vee(x) \, dx = (T_{\psi^\vee} f)(y)$$

Thus,

$$(T_f^* \cdot u)(\psi) = u(T_{f^\vee} \psi) = u(T_{\psi^\vee} f) = (T_\psi^* u)(f)$$

We already know that $T_{\psi_i}^* u \to u$ for an approximate identity ψ_i, so the limit exists and has an understandable value:

$$(T_f^* u)(\psi_i) = (T_{\psi_i}^* u)(f) \longrightarrow u(f) = \text{supposed value of } f \cdot u \text{ at } 0$$

Thus, we would guess that $T_f^* u$ should be a function with value $u(f)$ at 0. More generally, for the distribution u_φ given by integration against φ, we have

$$(T_z^* u_\varphi)(\psi_i) = u_\varphi \left(T_z^{-1} \psi_i \right) = \int_{\mathbb{R}^n} \varphi(x) \, \psi_i(x - z) \, dx$$

$$= \int_{\mathbb{R}^n} \varphi(x + z) \, \psi_i(x) \, dx \to \varphi(z)$$

The analogous computation suggests the values of the function $T_f^* u$ at z. First, a more elaborate version of the identity reverses the roles of test functions f and φ, namely

$$\left(T_{f^\vee} T_z^{-1} \psi \right)(y) = \int_{\mathbb{R}^n} f(-x) \, \psi(y + x - z) \, dx = \int_{\mathbb{R}^n} f(y - x - z) \, \psi(x) \, dx$$

$$= \int_{\mathbb{R}^n} f(y + x - z) \, \psi(-x) \, dx = \int_{\mathbb{R}^n} (T_z^{-1} f)(y + x) \, \psi^\vee(x) \, dx$$

$$= \left(T_{\psi^\vee} T_z^{-1} f \right)(y)$$

The same sort of computation gives

$$(T_y^*(T_f^*u))(\psi_i) = (T_f^*u))(T_y^{-1}\psi_i) = u(T_{f^\vee}T_y^{-1}\psi_i) = u(T_{\psi_i^\vee}T_y^{-1}f)$$
$$= (T_y^*(T_{\psi_i}^*u))(f) \to (T_y^*u)(f) = u(T_y^{-1}f)$$
$$= \text{supposed value of } f \cdot u \text{ at } y$$

Since $\mathbb{R}^n \times \mathcal{D} \to \mathcal{D}$ is continuous, and u is continuous, the composition

$$y \times f \longrightarrow T_y^{-1}f \longrightarrow u(T_y^{-1}f)$$

is indeed *continuous* as a function of $y \in \mathbb{R}^n$.

Now we check that the distribution $f \cdot u$ is truly given by integration against the continuous function

$$\varphi(y) = u(T_y^{-1}f)$$

that *apparently* gives the pointwise values of T_f^*u. Letting $h \in \mathcal{D}$,

$$\int_{\mathbb{R}^n} \varphi(x)\,h(x)\,dx = \int_{\mathbb{R}^n} u(T_x^{-1}f)\,h(x)\,dx$$
$$= \left(\int_{\mathbb{R}^n} h(x)\,x \cdot u\,dx\right)(f) = (T_h^*u)(f)$$

We already computed directly that

$$(T_h^*u)(f) = u(T_{h^\vee}f) = u(T_{f^\vee}h) = (T_f^*u)(h)$$

which shows that integration against the continuous function $\varphi(y) = u(T_y^{-1}f)$ gives the distribution T_f^*u.

Smoothness of $y \to u(T_y^{-1}f)$ would follow from the assertion that $y \to T_y^{-1}f$ is a smooth, $\mathcal{D}$-valued function, since u is a continuous linear functional on $\mathcal{D}$. The latter assertion is existence of limits

$$\lim_{t \to 0} \frac{T_{y+tX}^{-1}f - T_y^{-1}f}{t} \qquad (\text{for } X \in \mathbb{R}^n \text{ and } y \in \mathbb{R}^n)$$

in $\mathcal{D}$ for each $X \in \mathbb{R}^n$, and iterates thereof. It suffices to consider $y = 0$. By design, differentiation (such as this directional derivative in the X direction) is a continuous map of $\mathcal{D}$ to itself [13.9]. This gives the *smoothness* of $y \to u(T_y^{-1}f)$. ///

[14.5.2] Remark: That is, given the idea that $f \cdot u$ has been smoothed, *determination* of it as a classical function is straightforward. The proof that $T_{\psi_i}^*u \to u$ did not use the specifics of the situation: the same argument applies to representations of *Lie groups*.

14.6 Density of Smooth Vectors

Let G be a Lie group, so that the notion of C^∞ function on G makes sense. A *representation* of G on a locally convex, quasi-complete topological vectorspace V is a continuous map $G \times V \to V$ that is linear in V and has the associativity $(gh) \cdot v = g \cdot (h \cdot v)$ for $g, h \in G$. The subspace V^∞ of *smooth vectors* is

$$V^\infty = \{v \in V : g \to g \cdot v \text{ is a } C^\infty \text{ } V\text{-valued function on } G\}$$

It suffices to consider derivatives associated to the Lie algebra $\mathfrak{g}$ of G:

$$(x \cdot f)(g) = \frac{\partial}{\partial t}\Big|_{t=0} \left((ge^{tx}) \cdot v\right) \qquad \text{(for } x \in \mathfrak{g})$$

where $x \to e^x$ is the exponential map $\mathfrak{g} \to G$.

Note that in the representation of $\mathbb{R}^n$ on distributions $\mathcal{D}^*$ *every* distribution is a smooth vector, since every distribution is infinitely differentiable as a distribution. Thus, smooth *vectors* are not necessarily smooth *functions*. Nevertheless, as in the previous section, distributions are approximable by smooth functions. For general representations $G \times V \to V$, the following is the appropriate corollary of [14.1.4]:

[14.6.1] Theorem: *(Gårding)* For quasi-complete, locally convex V with a continuous action of a real Lie group G, V^∞ is dense in V.

Proof: Let $\{\psi_i\}$ be an approximate identity in $\mathcal{D}(G)$. On one hand, by [14.1.4], for each $v \in V$, $T_{\psi_i} v \to v$. On the other hand, we claim that $T_{\psi_i} v$ is a smooth vector in V. That is, for any $\psi \in \mathcal{D}$, we claim that $g \to T_g(T_\psi v)$ is a smooth function of $g \in G$. By the weak-to-strong result [15.1.1], it suffices to show that, for all $\lambda \in V^*$, $g \to \lambda(T_g T_\psi v)$ is a smooth scalar-valued function. By properties of Gelfand-Pettis integrals,

$$\lambda(T_g T_\psi v) = \lambda T_g \int_G \psi(h) T_h v \, dh = \int_G \psi(h) \lambda(T_g T_h v) \, dh$$
$$= \int_G \psi(g^{-1}h) \lambda(T_h v) \, dh$$

To show differentiability near a given g_o, without loss of generality we can multiply by a smooth, compactly supported cutoff function $\eta(g)$ which is identically 1 near g_o. Then $h \to (g \to \eta(g)\psi(g^{-1}h))$ is a smooth, compactly supported function on $G \times G$, and the integrand $h \to (g \to \eta(g)\psi(g^{-1}h)\lambda(T_h v))$ is a continuous, compactly supported, $\mathcal{D}(G)$-valued function on G. Thus, it admits a Gelfand-Pettis integral $g \to \int_G \psi(g^{-1}h)\lambda(T_h v) \, dh$ that is a smooth function. This holds for every $\lambda \in V^*$, so by [15.1.1] shows that $g \to T_g T_\psi v$ is a smooth V-valued function on G. ///

14.7 Quasi-Completeness and Convex Hulls of Compacts

A subset E of a *complete metric space* X is *totally bounded* if, for every $\varepsilon > 0$ there is a covering of E by *finitely many* open balls of radius ε. The property of *total boundedness* in a metric space is generally stronger than mere *boundedness*. It is immediate that any subset of a totally bounded set is totally bounded. Recall:

[14.7.1] Proposition: A subset of a complete metric space has compact closure if and only if it is *totally bounded*.

Proof: Certainly if a set has compact closure then it admits a finite covering by open balls of arbitrarily small (positive) radius. On the other hand, suppose that a set E is totally bounded in a complete metric space X. To show that E has compact closure, it suffices to show that any sequence $\{x_i\}$ in E has a Cauchy subsequence.

Choose such a subsequence as follows. Cover E by finitely many open balls of radius 1. In at least one of these balls there are infinitely many elements from the sequence. Pick such a ball B_1, and let i_1 be the smallest index so that x_{i_1} lies in this ball.

The set $E \cap B_1$ is still totally bounded and contains infinitely many elements from the sequence. Cover it by finitely many open balls of radius $1/2$, and choose a ball B_2 with infinitely many elements of the sequence lying in $E \cap B_1 \cap B_2$. Choose the index i_2 to be the smallest one so that both $i_2 > i_1$ and so that x_{i_2} lies inside $E \cap B_1 \cap B_2$.

Inductively, suppose that indices $i_1 < \cdots < i_n$ have been chosen, and balls B_i of radius $1/i$, so that

$$x_i \in E \cap B_1 \cap B_2 \cap \cdots \cap B_i$$

Cover $E \cap B_1 \cap \cdots \cap B_n$ by finitely many balls of radius $1/(n+1)$ and choose one, call it B_{n+1}, containing infinitely many elements of the sequence. Let i_{n+1} be the first index so that $i_{n+1} > i_n$ and so that

$$x_{n+1} \in E \cap B_1 \cap \cdots \cap B_{n+1}$$

For $m < n$ we have $d(x_{i_m}, x_{i_n}) \leq \frac{1}{m}$ so this subsequence is Cauchy. ///

In a not-necessarily-metric *topological vectorspace* V, a subset E is *totally bounded* if, for every neighborhood U of 0, there is a finite subset F of V so that $E \subset F + U$, where

$$F + U = \bigcup_{v \in F} v + U = \{v + u : v \in F, u \in U\}$$

[14.7.2] Proposition: A totally bounded subset E of a *locally convex* topological vectorspace V has totally bounded *convex hull*.

Proof: First, recall that the convex hull of a *finite* set $F = \{x_1, \ldots, x_n\}$ in a topological vectorspace is *compact*, since it is the continuous image of the compact set $\{(c_1, \ldots, c_n) \in \mathbb{R}^n : \sum_i c_i = 1, \ 0 \le c_i \le 1, \ \text{for all } i\} \subset \mathbb{R}^n$ under $(c_1, \ldots, c_n) \to \sum_i c_i x_i$.

Given a neighborhood U of 0 in V, let U_1 be a *convex* neighborhood of 0 so that $U_1 + U_1 \subset U$. For some finite subset F, we have $E \subset F + U_1$, by total boundedness. The convex hull K of F is *compact*. Then $E \subset K + U_1$, and the latter is *convex*. Therefore, the convex hull H of E lies inside $K + U_1$. Since K is compact, it is totally bounded, so can be covered by a finite union $\Phi + U_1$ of translates of U_1. Thus, since $U_1 + U_1 \subset U$, $H \subset (\Phi + U_1) + U_1 \subset \Phi + U$. Thus, H lies inside this finite union of translates of U. This holds for any open U containing 0, so H is totally bounded. ///

[14.7.3] Corollary: In a Fréchet space, the closure of the convex hull of a compact set is compact.

Proof: A compact set in a Fréchet space (or in any complete metric space) is totally bounded, as recalled earlier. By the previous, the convex hull of a totally bounded set in a Fréchet space is totally bounded. Thus, this convex hull has compact closure, since totally bounded sets in complete metric spaces have compact closure. ///

The general case reduces to the case of Fréchet spaces.

[14.7.4] Proposition: In a *quasi-complete*, locally convex topological vectorspace X, the closure of the convex hull of a compact set is *compact*.

Proof: Since X is locally convex, its topology is given by a collection of seminorms v. For each seminorm v, let X_v be the completion of the quotient $X/\{x \in X : v(x) = 0\}$ with respect to the *metric* that v induces on the latter quotient. Thus, X_v is a Banach space. Consider $Z = \prod_v X_v$ with product topology, with the natural injection $j : X \to Z$, and with projection p_v to the vth factor. By construction, and by definition of the topology given by the seminorms, j is a (linear) homeomorphism to its image. That is, X is homeomorphic to the subset jX of Z, given the subspace topology.

Let $K \subset X$ be compact, with convex hull H, and C the closure of H. The continuous image $p_v jK$ of compact K is compact. Since X_v is Fréchet, the convex hull H_v of $p_v jK$ has compact closure C_v. The convex hull jH of jK is contained in the product $\prod_v H_v$ of the convex hulls H_v of the projections $p_v jK$. By Tychonoff's theorem, the product $\prod_v C_v$ is *compact*.

Since jC is contained in the compact set $\prod_v C_v$, to prove that the closure $\overline{jC}$ of jH in jX is compact, it suffices to prove that jC is closed in Z. Since jC is a subset of the compact set $\prod_v C_v$, it is totally bounded and so is certainly *bounded* (in Z, hence in $X \approx jX$). By the quasi-completeness, a Cauchy net in jC is necessarily *bounded* and converges to a point in jC. Since any point in the closure of jC in Z has a Cauchy net in jC converging to it, jC is closed in Z.   ///

14.8 Existence Proof

To simplify, divide by a constant to make X have total measure 1. The closure H of the convex hull of $f(X)$ in V is *compact* by hypothesis. We will show that there is an integral of f inside H.

For *finite* $L \subset V^*$, let

$$V_L = \{v \in V : \lambda v = \int_X \lambda \circ f, \ \forall \lambda \in L\} \qquad \text{and} \qquad I_L = H \cap V_L$$

Since H is compact and V_L is closed, I_L is *compact*. Certainly $I_L \cap I_{L'} = I_{L \cup L'}$ for two finite subsets L, L' of V^*. If all the I_L are *nonempty*, then the intersection of *all* these compact sets I_L is nonempty, by the *finite intersection property*, giving *existence*.

To prove that each I_L is nonempty for finite subsets L of V^*, choose an ordering $\lambda_1, \ldots, \lambda_n$ of the elements of L. Make a continuous linear mapping $\Lambda = \Lambda_L$ from V to $\mathbb{R}^n$ by $\Lambda(v) = (\lambda_1 v, \ldots, \lambda_n v)$. Since this map is continuous, the image $\Lambda(f(X))$ is compact in $\mathbb{R}^n$.

For a finite set L of functionals, the integral $y = y_L = \int_X \Lambda f(x)\,dx$ is readily defined by component-wise integration. Take y in the convex hull of $\Lambda(f(X))$. Since Λ_L is linear, $y = \Lambda_L v$ for some v in the convex hull of $f(X)$. Then

$$\Lambda_L v = y = (\ldots, \int \lambda_i f(x)\,dx, \ldots)$$

Thus, $v \in I_L$ as desired. It remains to show that y lies in the convex hull of $\Lambda_L(f(x))$.

Suppose *not*. From the following lemma, in a *finite-dimensional* space the convex hull of a compact set is still compact, *without* taking closure. By the finite-dimensional case of the Hahn-Banach theorem, there would be a linear functional η on $\mathbb{R}^n$ so that $\eta y > \eta z$ for all z in this convex hull. That is, letting $y = (y_1, \ldots, y_n)$, there would be real $c_1, \ldots, c_n$ so that for all $(z_1, \ldots, z_n)$ in the convex hull $\sum_i c_i z_i < \sum c_i y_i$. In particular, for all $x \in X$

$$\sum_i c_i \lambda_i(f(x)) < \sum_i c_i y_i$$

Integration of both sides of this over X *preserves ordering*, giving the impossible $\sum_i c_i y_i < \sum_i c_i y_i$. Thus, y *does* lie in this convex hull. ///

[14.8.1] Lemma: The convex hull of a compact set K in $\mathbb{R}^n$ is compact.

Proof: First claim that, for $E \subset \mathbb{R}^n$ and for any x a point in the convex hull of E, there are $n+1$ points $x_0, x_1, \ldots, x_n$ in E of which x is a convex combination.

By induction, it suffices to consider a convex combination $v = c_1 v_1 + \cdots + c_N v_N$ of vectors v_i with $N > n+1$ and show that v is actually a convex combination of $N-1$ of the v_i. Further, without loss of generality that all the coefficients c_i are nonzero. Define a linear map

$$L : \mathbb{R}^N \longrightarrow \mathbb{R}^n \times \mathbb{R} \qquad \text{by} \qquad L(x_1, \ldots, x_N) \longrightarrow \left(\sum_i x_i v_i, \sum_i x_i \right)$$

By dimension-counting, since $N > n+1$, the kernel of L is nontrivial. Let $(x_1, \ldots, x_N)$ be a nonzero vector in the kernel. Since $c_i > 0$ for every index, and since there are only finitely many indices altogether, there is a constant c so that $|c x_i| \leq c_i$ for every index i and so that $c x_{i_o} = c_{i_o}$ for at least one index i_o. Then

$$v = v - 0 = \sum_i c_i v_i - c \cdot \sum_i x_i v_i = \sum_i (c_i - c x_i) v_i$$

Since $\sum_i x_i = 0$ this is still a convex combination, and since $c x_{i_o} = c_{i_o}$, at least one coefficient has become zero. This is the induction proving the claim.

By this claim, a point v in the convex hull of K is a convex combination $c_o v_o + \cdots + c_n v_n$ of $n+1$ points $v_o, \ldots, v_n$ of K. Let σ be the compact set $(c_o, \ldots, c_n)$ with $0 \leq c_i \leq 1$ and $\sum_i c_i = 1$. The convex hull of K is the image of the compact set $\sigma \times K^{n+1}$ under the continuous map

$$L : (c_o, \ldots, c_n) \times (v_o, v_1, \ldots, v_n) \longrightarrow \sum_i c_i v_i$$

so is compact. This proves the lemma, finishing the proof of the theorem. ///

14.A Appendix: Hahn-Banach Theorems

For a locally convex vectorspace V, functionals $\lambda \in V^*$ separate points, and convex sets can be separated by linear functionals. Continuous linear functionals on arbitrary subspaces have continuous extensions to the whole space. In contrast, in general, linear maps from subspaces W to not-finite-dimensional topological vectorspaces need not extend to V. Indeed, if the identity map

$T : W \to W$ extended to $T' : V \to W$, then $\ker T'$ would be a *complementary subspace*, which need not exist even for *closed* subspaces W.

Let k be either $\mathbb{R}$ or $\mathbb{C}$, and let V be a k-vectorspace, without any assumptions about topologies for the moment. A k-linear k-valued function on V is a *linear functional*. A linear functional λ on V is *bounded* when there is a neighborhood U of 0 in V and constant c so that $|\lambda x| \leq c$ for $x \in U$, where $|\ |$ is the usual absolute value on k. The following proposition is the general analogue of the corresponding assertion for Banach spaces, in which *boundedness* has a different sense.

[14.A.1] Proposition: The following conditions on a linear functional λ on a topological vectorspace V over k are equivalent: (i) λ is continuous, (ii) λ is continuous at 0, (iii) λ is bounded.

Proof: The first assertion certainly implies the second. Assume the second. Then, given $\varepsilon > 0$, there is a neighborhood U of 0 so that $|\lambda|$ is bounded by ε on U. This proves boundedness. Finally, suppose that $|\lambda(x)| \leq c$ on a neighborhood U of 0. Then given $x \in V$ and given $\varepsilon > 0$, we *claim* that for $y \in x + \frac{\varepsilon}{2c}U$ we have $|\lambda(x) - \lambda(y)| < \varepsilon$. Indeed, letting $x - y = \frac{\varepsilon}{2c}u$ with $u \in U$, we have

$$|\lambda(x) - \lambda(y)| = \frac{\varepsilon}{2c}|\lambda(u)| \leq \frac{\varepsilon}{2c} \cdot c = \frac{\varepsilon}{2} < \varepsilon$$

This proves the proposition. ///

The immediate goal is to *extend* a linear functional while preserving a comparison to another function (denoted p below). For this, we need *not* suppose that the vectorspaces involved are *topological* vectorspaces. Let V be a *real* vectorspace, without any assumption about topologies. Let $p : V \to \mathbb{R}$ be a *nonnegative* real-valued function on V so that

$$p(tv) = t \cdot p(v) \quad \text{(for } t \geq 0\text{)} \quad \textit{(positive-homogeneity)}$$

$$p(v + w) \leq p(v) + p(w) \quad \textit{(triangle inequality)}$$

Lacking a description of $p(tv)$ for $t < 0$, p is not quite a *seminorm*.

[14.A.2] Theorem: Let λ be a real-linear function on a real vector subspace W of V, so that $\lambda(w) \leq p(w)$ for all $w \in W$. There is an extension of λ to a real-linear function Λ on all of V, so that $-p(-v) \leq \Lambda(v) \leq p(v)$ for all $v \in V$.

Proof: The key issue is extending the functional *one step*. That is, for $v_o \in V$, attempt to extend λ' of λ to $W + \mathbb{R}v_o$ by $\lambda'(w + tv_o) = \lambda(w) + ct$ and examine the resulting conditions on c.

For all $w, w' \in W$

$$\lambda(w) - p(w - v_o) = \lambda(w + w') - \lambda(w') - p(w - v_o)$$
$$\leq p(w + w') - \lambda(w') - p(w - v_o)$$
$$= p(w - v_o + w' + v_o) - \lambda(w') - p(w - v_o)$$
$$\leq p(w - v_o) + p(w' + v_o) - \lambda(w') - p(w - v_o)$$
$$= p(w' + v_o) - \lambda(w')$$

That is,

$$\lambda(w) - p(w - v_o) \leq p(w' + v_o) - \lambda(w') \qquad \text{(for all } w, w' \in W)$$

Let σ be the sup of all the left-hand sides as w ranges over W. Since the right-hand side is finite, this sup is finite. With μ, the inf of the right-hand side as w' ranges over W,

$$\lambda(w) - p(w - v_o) \leq \sigma \leq \mu \leq p(w' + v_o) - \lambda(w')$$

Take any real number c so that $\sigma \leq c \leq \mu$ and define $\lambda'(w + tv_o) = \lambda(w) + tc$.

To compare with p is easy: in the inequality $\lambda(w) - p(w - v_o) \leq \sigma$ replace w by w/t with $t > 0$, multiply by t and invoke the positive-homogeneity to obtain $\lambda(w) - p(w - tv_o) \leq t\sigma$ from which

$$\lambda'(w - tv_o) = \lambda(w) - tc \leq \lambda(w) - t\sigma \leq p(w - tv_o)$$

Likewise, from $\mu \leq p(w + v_o) - \lambda(w)$ a similar trick produces

$$\lambda'(w + tv_o) = \lambda(w) + tc \leq \lambda(w) + t\mu \leq p(w + tv_o)$$

for $t > 0$, the other half of the desired inequality. Thus, for all $v \in W + Rv_o$, we have $\lambda'(v) \leq p(v)$. Using the linearity of λ', $\lambda'(v) = -\lambda'(-v) \geq -p(-v)$ giving the bottom half of the comparison of λ' and p.

Extend to a functional on the *whole* space dominated by p by transfinite induction, as follows. Let $\mathcal{X}$ be the collection of all pairs (X, μ), where X is a subspace of V (containing W) and where μ is real-linear real-valued function on X so that μ restricted to W is λ, and so that $-p(-x) \leq \mu(x) \leq p(x)$ for all $x \in X$. *Order* these by writing $(X, \mu) \leq (Y, \nu)$ when $X \subset Y$ and $\nu|_X = \mu$. By the Hausdorff maximality principle, there is a *maximal* totally ordered subset $\mathcal{Y}$ of $\mathcal{X}$. Let

$$V' = \bigcup_{(X, \mu) \in \mathcal{Y}} X$$

be the ascending union of all the subspaces in $\mathcal{Y}$. Define a linear functional λ' on this union as follows: for $v \in V'$, take any X so that $(X, \mu) \in \mathcal{Y}$ and $v \in X$

and define $\lambda'(v) = \mu(v)$. The total ordering on $\mathcal{Y}$ makes the choice of (X, μ) not affect the definition of λ'. If V' were not the whole space V the first part of the proof would create an extension to a properly larger subspace, contradicting the maximality.   ///

[14.A.3] Theorem: For a nonempty convex open subset X of a *locally convex* topological vectorspace V, and a nonempty convex set Y in V with $X \cap Y = \phi$, there is a *continuous* real-linear real-valued functional λ on V and a constant c so that $\lambda(x) < c \leq \lambda(y)$ for all $x \in X$ and $y \in Y$.

Proof: Fix $x_o \in X$ and $y_o \in Y$. Since X is open, $X - x_o$ is open, and thus

$$U = (X - x_o) - (Y - y_o) = \{(x - x_o) - (y - y_o) : x \in X, \ y \in Y\}$$

is open. Further, since $x_o \in X$ and $y_o \in Y$, U contains 0. Since X, Y are convex, U is convex. The *Minkowski functional* $p = p_U$ attached to U is $p(v) = \inf\{t > 0 : v \in tU\}$. The convexity ensures that this function p has the *positive-homogeneity* and *triangle-inequality* properties of the auxiliary functional p above.

Let $z_o = -x_o + y_o$. Since $X \cap Y = \phi$, $z_o \notin U$, so $p(z_o) \geq 1$. Define a linear functional λ on $\mathbb{R}z_o$ by $\lambda(tz_o) = t$. Check that λ is dominated by p in the sense of the previous section:

$$\lambda(tz_o) = t \leq t \cdot p(z_o) = p(tz_o) \qquad \text{(for } t \geq 0)$$

while

$$\lambda(tz_o) = t < 0 \leq p(tz_o) \qquad \text{(for } t < 0)$$

Thus, $\lambda(tz_o) \leq p(tz_o)$ for all real t, and λ extends to a real-linear real-valued functional Λ on V, still so that $-p(-v) \leq \Lambda(v) \leq p(v)$ for all $v \in V$. From the definition of p, $|\Lambda| \leq 1$ on U. Thus, on $\frac{\varepsilon}{2}U$ we have $|\Lambda| < \varepsilon$. That is, the linear functional Λ is *bounded*, so is *continuous* at 0, so is *continuous* on V.

For arbitrary $x \in X$ and $y \in Y$,

$$\Lambda x - \Lambda y + 1 = \Lambda(x - y + z_o) \leq p(x - y + z_o) < 1$$

since $x - y + z_o \in U$. Thus, $\Lambda x - \Lambda y < 0$ for all such x, y. Therefore, $\Lambda(X)$ and $\Lambda(Y)$ are *disjoint* convex subsets of $\mathbb{R}$. Since Λ is not the zero functional, it is *surjective* to $\mathbb{R}$, and so is an *open* map. Thus, $\Lambda(X)$ is open, and $\Lambda(X) < \sup \Lambda(X) \leq \Lambda(Y)$ as desired.   ///

The analogous results for complex scalars are corollaries of the real-scalar cases, as follows. Let V be a complex vectorspace. Given a complex-linear complex-valued functional λ on V, let its real part be

$$u(v) = \operatorname{Re}\lambda(v) = \frac{\lambda(v) + \overline{\lambda(v)}}{2}$$

where the overbar denotes complex conjugation. On the other hand, given a *real*-linear *real*-valued functional u on V, its *complexification* is $Cu(x) = u(x) - iu(ix)$ where $i = \sqrt{-1}$.

[14.A.4] Lemma: For a real-linear functional u on the complex vectorspace V, the complexification Cu is a complex-linear functional so that $\operatorname{Re}(Cu) = u$ and for a complex-linear functional λ $C(\operatorname{Re}\lambda) = \lambda$. *(Straightforward computation).*   ///

[14.A.5] Corollary: Let p be a *seminorm* on the complex vectorspace V. Let λ be a complex-linear function on a complex vector subspace W of V, so that $|\lambda(w)| \le p(w)$ for all $w \in W$. Then there is an extension of λ to a complex-linear function Λ on all of V, so that $|\Lambda(v)| \le p(v)$ for all $v \in V$.

Proof: Certainly if $|\lambda| \le p$ then $|\operatorname{Re}\lambda| \le p$. By the theorem for *real*-linear functionals, there is an extension u of $\operatorname{Re}\lambda$ to a *real*-linear functional so that still $|u| \le p$. Let $\Lambda = Cu$. In light of the lemma, it remains to show that $|\Lambda| \le p$. To this end, given $v \in V$, let μ be a complex number of absolute value 1 so that $|\Lambda(v)| = \mu\Lambda(v)$. Then

$$|\Lambda(v)| = \mu\Lambda(v) = \Lambda(\mu v) = \operatorname{Re}\Lambda(\mu v) \le p(\mu v) = p(v)$$

using the seminorm property of p. Thus, the complex-linear functional made by complexifying the *real*-linear extension of the real part of λ satisfies the desired bound.   ///

[14.A.6] Corollary: Let X be a nonempty convex open subset of a locally convex topological vectorspace V, and let Y be an arbitrary nonempty convex set in V so that $X \cap Y = \phi$. Then there is a *continuous* complex-linear complex-valued functional λ on V and a constant c so that

$$\operatorname{Re}\lambda(x) < c \le \operatorname{Re}\lambda(y) \qquad \text{(for all } x \in X \text{ and } y \in Y)$$

Proof: Invoke the real-linear version of the theorem to make a *real*-linear functional u so that $u(x) < c \le u(y)$ for all $x \in X$ and $y \in Y$. By the lemma, u is the real part of its own complexification.   ///

[14.A.7] Corollary: Let V be a locally convex topological vectorspace. Let K and C be *disjoint* sets, where K is a *compact* convex nonempty subset of V, and

C is a *closed* convex subset of V. Then there is a continuous linear functional λ on V and there are real constants $c_1 < c_2$ so that

$$\mathrm{Re}\,\lambda(x) \le c_1 < c_2 \le \mathrm{Re}\,\lambda(y) \qquad \text{(for all } x \in K \text{ and } y \in C)$$

Proof: Take a small-enough convex neighborhood U of 0 in V so that $(K + U) \cap C = \phi$. Apply the separation theorem to $X = K + U$ and $Y = C$. The constant c_2 can be taken to be $c_2 = \sup \mathrm{Re}\,\lambda(K + U)$. Since $\mathrm{Re}\,\lambda(K)$ is a compact subset of $\mathrm{Re}\,\lambda(K + U)$, its sup c_1 is strictly less than c_2.   ///

[14.A.8] Corollary: Let V be a locally convex topological vectorspace, W a subspace, and $v_o \in V$. Let $\overline{W}$ be the topological closure of W. Then $v_o \notin \overline{W}$ if and only if there is a *continuous* linear functional λ on V so that $\lambda(W) = 0$ while $\lambda(v) = 1$.

Proof: On one hand, if v_o lies in the closure of W, then any continuous function that is 0 on W must be 0 on v_o, as well. On the other hand, suppose that v_o does *not* lie in the closure of W. Then apply the previous corollary with $K = \{v_o\}$ and $C = \overline{W}$. We find that

$$\mathrm{Re}\,\lambda(\{v_o\}) \cap \mathrm{Re}\,\lambda(\overline{W}) = \phi$$

Since $\mathrm{Re}\,\lambda(\overline{W})$ is a vector subspace of the real line and is not the whole real line, it is just $\{0\}$, and $\mathrm{Re}\,\lambda(v_o) \ne 0$. Divide λ by the constant $\mathrm{Re}\,\lambda(v_o)$ to obtain a continuous linear functional zero on W but 1 on v_o.   ///

[14.A.9] Corollary: Let V be a locally convex topological (real) vectorspace. Let λ be a continuous linear functional on a subspace W of V. Then there is a continuous linear functional Λ on V extending λ.

Proof: Without loss of generality, take $\lambda \ne 0$. Let W_o be the kernel of λ (on W), and pick $w_1 \in W$ so that $\lambda w_1 = 1$. Evidently w_1 is not in the closure of W_o, so there is Λ on the whole space V so that $\Lambda|_{W_o} = 0$ and $\Lambda w_1 = 1$. It is easy to check that this Λ is an extension of λ.   ///

[14.A.10] Corollary: Let V be a locally convex topological vector-space. Given two distinct vectors $x \ne y$ in V, there is a continuous linear functional λ on V so that $\lambda(x) \ne \lambda(y)$

Proof: The set $\{x\}$ is compact convex nonempty, and the set $\{y\}$ is closed convex nonempty, so we can apply a corollary just above.   ///

15

Differentiable Vector-Valued Functions

15.1 Weak-to-Strong Differentiability

A V-valued function $f : [a, c] \to V$ on an interval $[a, c] \subset \mathbb{R}$ is *differentiable* if for every $x_o \in [a, c]$

$$f'(x_o) = \lim_{x \to x_o} (x - x_o)^{-1} \left(f(x) - f(x_o) \right)$$

exists. The function f is *continuously differentiable* when it is differentiable and f' is continuous. A k-times continuously differentiable function is C^k, and a continuous function is C^o.

A V-valued function f is *weakly C^k* when for every $\lambda \in V^*$ the scalar-valued function $\lambda \circ f$ is C^k. This sense of *weak differentiability* of a function f does *not* refer to distributional derivatives but to differentiability of every scalar-valued function $\lambda \circ f$ where $\lambda \in V^*$ for V-valued f.

[15.1.1] Theorem: For quasi-complete, locally convex V, a *weakly C^k* V-valued function f on an interval $[a, c]$ is *strongly C^{k-1}*.

271

Proof: This is a corollary of [15.7.1]. To have f be (strongly) differentiable at fixed $b \in [a, c]$ is to have (strong) *continuity* at b of

$$g(x) = \frac{f(x) - f(b)}{x - b} \qquad (\text{for } x \neq b)$$

Weak C^2-ness of f implies that every $\lambda \circ g$ extends to a C^1 scalar-valued function on $[a, c]$. We need to get from this to a (strongly) continuous extension of g to the whole interval.

The (strong) continuity of f' will follow from consideration of the function of two variables (initially for $x \neq y$)

$$g(x, y) = \frac{f(x) - f(y)}{x - y}$$

The weak C^2-ness of f ensures that g extends to a weakly C^1 function on $[a, c] \times [a, c]$. In particular, the function $x \to g(x, x)$ of (the extended) g is weakly C^1, and $x \to g(x, x)$ is $f'(x)$, so f' is weakly C^1. By [15.7.1], f' is (strongly) C^o. Suppose that we already know that f is C^ℓ, for $\ell < k - 1$. As the ℓth derivative $g = f^{(\ell)}$ of f is weakly C^2, it is (strongly) C^1 by the first part of the argument. That is, f is $C^{\ell+1}$. ///

15.2 Holomorphic Vector-Valued Functions

Let V be a quasi-complete, locally convex topological vector space. A V-valued function f on a nonempty open set $\Omega \subset \mathbb{C}$ is (strongly) *complex-differentiable* when $\lim_{z \to z_o} (f(z) - f(z_o))/(z - z_o)$ exists (in V) for all $z_o \in \Omega$, where $z \to z_o$ specifically means for *complex z* approaching z_o. The function f is *weakly holomorphic* when the $\mathbb{C}$-valued functions $\lambda \circ f$ are holomorphic for all λ in V^*. The useful version of vector-valued *meromorphy* of f at z_o is that $(z - z_o)^n \cdot f(z)$ extends to a vector-valued *holomorphic* function at z_o for some n. After some preparation, we will prove

[15.2.1] Theorem: *Weakly* holomorphic V-valued functions f are *continuous*. *(Proof in [15.8.1].)* ///

[15.2.2] Corollary: Weakly holomorphic V-valued functions are (strongly) holomorphic. The Cauchy integral formula applies: as Gelfand-Pettis V-valued integral,

$$f(z) = \frac{1}{2\pi i} \int_\gamma \frac{f(w)}{w - z} \, dw$$

Proof: Since $f(z)$ is continuous, the integral

$$I(z) = \frac{1}{2\pi i} \int_\gamma \frac{f(w)}{w - z} \, dw$$

exists as a Gelfand-Pettis integral [14.1]. Thus, for any $\lambda \in V^*$

$$\lambda(I(z)) = \frac{1}{2\pi i} \int_\gamma \frac{(\lambda \circ f)(w)}{w - z} \, dw = (\lambda \circ f)(z)$$

by the holomorphy of $\lambda \circ f$. By Hahn-Banach, linear functionals separate points, so $I(z) = f(z)$, giving the Cauchy integral formula for f itself.

To prove (strong) complex-differentiability of f at z_o, take $z_o = 0$ and use $f(0) = 0$, for convenience. There is a disk $|z| < 3r$ such that for every $\lambda \in V^*$

$$F_\lambda(z) = \frac{(\lambda \circ f)(z)}{z} \qquad \text{(on } 0 < |z| < r\text{)}$$

extends to a holomorphic function on $|z| < r$. Continuity of f assures existence of

$$\frac{1}{2\pi i} \int_\gamma \frac{f(w)}{w} \frac{dw}{w - z}$$

By Cauchy theory for $\mathbb{C}$-valued functions, and Gelfand-Pettis,

$$\lambda\left(\frac{f(z)}{z}\right) = F_\lambda(z) = \frac{1}{2\pi i} \int_\gamma \frac{(\lambda \circ f)(w)}{w} \frac{dw}{w - z}$$

$$= \lambda\left(\frac{1}{2\pi i} \int_\gamma \frac{f(w)}{w} \frac{dw}{w - z}\right)$$

Since functionals separate points,

$$\frac{f(z)}{z} = \frac{1}{2\pi i} \int_\gamma \frac{f(w)}{w} \frac{dw}{w - z}$$

From

$$\frac{1}{w(w - z)} = \frac{1}{w^2} + \frac{z}{w^2(w - z)}$$

we have

$$\frac{f(z)}{z} = \frac{1}{2\pi i} \int_\gamma \frac{f(w)}{w^2} \, dw + z \cdot \frac{1}{2\pi i} \int_\gamma \frac{f(w)}{w^2(w - z)} \, dw$$

Using the continuity of f, given a convex balanced neighborhood U of 0 in V, the compact set $K = \{f(w) : |w| = 2r\}$ is contained in some multiple $t_o U$ of U. Thus, for $|z| < r$,

$$\frac{f(z)}{z} - \frac{1}{2\pi i} \int_\gamma \frac{f(w)}{w^2} \, dw \quad \in \quad |z| \cdot \frac{1}{(2r)^2 \, r} \cdot t_o U$$

so $\lim_{z\to 0} f(z)/z$ exists. Since $f(0) = 0$,

$$\lim_{z\to z_o} \frac{f(z) - f(z_o)}{z - z_o} = \frac{1}{2\pi i} \int_\gamma \frac{f(w)\, dw}{(w - z_o)^2}$$

giving the complex differentiability of f.   ///

[15.2.3] Corollary: The usual Cauchy-theory integral formulas apply. In particular, weakly holomorphic f is (strongly) infinitely differentiable, in fact expressible as a convergent power series with coefficients given by Cauchy's formulas:

$$f(z) = \sum_{n \geq 0} c_n \, (z - z_o)^n$$

with

$$c_n = \frac{f^{(n)}(z_o)}{n!} = \frac{1}{2\pi i} \int_\gamma \frac{f(w)}{(w - z_o)^{n+1}} \, dw$$

for γ a path with winding number $+1$ around z_o.

Proof: Without loss of generality, treat $z_o = 0$, and $|z| < \rho|w|$ with $\rho < 1$, and $|w| = r$. The expansion

$$\frac{1}{w - z} = \frac{1}{w} \frac{1}{1 - \frac{z}{w}} = \frac{1}{w}\left(1 + \frac{z}{w} + \left(\frac{z}{w}\right)^2 + \cdots + \left(\frac{z}{w}\right)^N + \frac{(z/w)^{N+1}}{1 - \frac{z}{w}}\right)$$

combined with an integration around γ against $f(w)$, and the basic Cauchy integral formula, give

$$f(z) = \sum_{n=0}^{N} \frac{1}{2\pi i} \int_\gamma \frac{f(w)\, dw}{w^{n+1}} \cdot z^n + \frac{1}{2\pi i} \int_\gamma \frac{1}{w^{N+1}} \frac{f(w)\, dw}{w - z} \cdot z^{N+1}$$

Much as in the previous proof, given a convex balanced neighborhood U of 0 in V, the compact set $K = \{f(w) : |w| = r\}$ is contained in some multiple $t_o U$ of U, and

$$\frac{1}{2\pi i} \int_\gamma \frac{1}{w^{N+1}} \frac{f(w)\, dw}{w - z} \cdot z^{N+1} \in \frac{1}{r^{N+1}} \cdot t_o U \cdot \frac{1}{r(1 - \rho)} \cdot (\rho r)^{N+1}$$

$$= U \frac{t_o}{r(1 - \rho)} \rho^{N+1}$$

Since $0 < \rho < 1$, $\rho^{N+1}/r(1 - \rho) < 1$ for sufficiently large N, so the leftover term is inside given U.   ///

Appendix [15.A] discusses the differentiability of power series with coefficients in topological vector spaces.

The next section collects some important corollaries of the main result, before preparation for the proof that weak holomorphy implies *continuity*,

15.3 Holomorphic Hol(Ω, V)-Valued Functions

The vector-valued versions of Cauchy's formulas have useful corollaries. First, recall some aspects of the classical scalar-valued case.

For open $\phi \neq \Omega \subset \mathbb{C}$, give the space Hol(Ω) of holomorphic functions on Ω the topology given by the seminorms $\mu_K(f) = \sup_{z \in K} |f(z)|$ for compacts $K \subset \Omega$.

[15.3.1] Claim: Hol(Ω) is a Fréchet space.

Proof: Let $\{f_n\}$ be a Cauchy sequence in that topology. As in [13.5], the pointwise limit $f(z) = \lim_n f_n(z)$ is at least *continuous*. Then, for a small circle γ inside Ω and enclosing z,

$$f(z) = \lim_n f_n(z) = \lim_n \frac{1}{2\pi i} \int_\gamma \frac{f_n(w)}{w - z} \, dw$$

Since γ is compact and the limit is uniformly approached on compacts, this gives

$$f(z) = \frac{1}{2\pi i} \int_\gamma \lim_n \frac{f_n(w)}{w - z} \, dw = \frac{1}{2\pi i} \int_\gamma \frac{f(w)}{w - z} \, dw$$

Direct estimates (simpler than in the previous section) show that the latter integral is complex-differentiable in w. ///

Let V be quasi-complete, locally convex, with topology given by seminorms $\{v\}$. The space Hol(Ω, V) of holomorphic V-valued functions on Ω has the natural topology given by seminorms

$$\mu_{v,K}(f) = \sup_{z \in K} v(f(z)) \qquad \text{(compacts } K \subset \Omega, \text{ seminorms } v \text{ on } V)$$

This topology is obviously the analogue of the sups-on-compacts seminorms on scalar-valued holomorphic functions, and there is the analogous corollary of the vector-valued Cauchy formulas:

[15.3.2] Corollary: Hol(Ω, V) is locally convex, quasi-complete. ///

Proof: Let $\{f_n\}$ be a bounded Cauchy net. Just as in the scalar case, the pointwise limits $\lim_n f_n(z)$ exist. The same three-epsilon argument as for scalar-valued functions will show that the pointwise limit exists and is continuous, as follows. First, using compact $K = \{z\}$, the value $\mu_{\{z\},v}(f)$ is just $v(f(z))$.

Thus, by quasi-completeness of V, for each fixed z the bounded Cauchy net $f_n(z)$ converges to a value $f(z)$. Given $\varepsilon > 0$ and $z_o \in \Omega$, let K be a compact neighborhood of z_o, and take N sufficiently large so that $\nu(f_m(z) - f_n(z')) < \varepsilon$ for all $z, z' \in K$ and all $m, n \geq N$. Then

$$\mu_{K,\nu}(f(z) - f(z_o))$$
$$\leq \mu_{K,\nu}(f(z) - f_n(z)) + \mu_{K,\nu}(f_n(z) - f_n(z_o)) + \mu_{K,\nu}(f_n(z_o) - f(z_o))$$
$$\leq 3\varepsilon$$

proving the continuity of the pointwise limit. Then, as in the previous scalar-valued argument, the vector-valued Cauchy formula gives, for a small circle γ inside Ω and enclosing z,

$$f(z) = \lim_n f_n(z) = \lim_n \frac{1}{2\pi i} \int_\gamma \frac{f_n(w)}{w - z} \, dw$$

with Gelfand-Pettis integrals. Since γ is compact and the limit is uniformly approached on compacts, this gives

$$f(z) = \frac{1}{2\pi i} \int_\gamma \lim_n \frac{f_n(w)}{w - z} \, dw = \frac{1}{2\pi i} \int_\gamma \frac{f(w)}{w - z} \, dw$$

Again, the differentiability of latter integral is directly verifiable, and f is holomorphic. ///

It is occasionally useful to iterate the previous ideas: A V-valued function $f(z, w)$ on a nonempty open subset $\Omega \subset \mathbb{C}^2$ is *complex analytic* when it is locally expressible as a convergent power series in z and w, with coefficients in V. The two-variable version of the discussion of convergence of power series with coefficients in V in Appendix [15.A] succeeds without incident in the two-variable case.[1]

[15.3.3] Corollary: Let $f(z, w)$ be complex-analytic $\mathbb{C}$-valued in two variables, on a domain $\Omega_1 \times \Omega_2 \subset \mathbb{C}^2$. Then the function $w \longrightarrow (z \to f(z, w))$ is a holomorphic $\mathrm{Hol}(\Omega_1)$-valued function on Ω_2.

Proof: The issue is the uniformity in z in compacts K of the limit

$$\lim_{h \to 0} \frac{f(z, w + h) - f(z, w)}{h}$$

---

[1] We have no immediate need of subtleties concerning functions of several complex variables, such as Hartogs's theorem that separate analyticity implies joint analyticity.

Using the scalar-valued Cauchy integral, for a small circle γ about w, letting f_2 be the partial derivative of f with respect to its second argument,

$$\frac{f(z, w + h) - f(z, w)}{h} - f_2(z, w)$$

$$= \frac{1}{2\pi i} \int_\gamma f(z, \zeta) \left(\frac{\frac{1}{\zeta - (w+h)} - \frac{1}{\zeta - w}}{h} - \frac{1}{(\zeta - w)^2} \right) d\zeta$$

$$= \frac{1}{2\pi i} \int_\gamma f(z, \zeta) \left(\frac{1}{(\zeta - (w + h))(\zeta - h)} - \frac{1}{(\zeta - w)^2} \right) d\zeta$$

The two-variable analytic function $z, \zeta \to f(z, \zeta)$ is certainly *continuous* as a function of two variables, so is *uniformly* continuous on compacts $K \times \gamma$. Thus, the limit as $h \to 0$ is approached uniformly. ///

Application of the vector-valued form of Cauchy's integrals gives the same result for $f(z, w)$ taking values in a quasi-complete, locally convex V:

[15.3.4] Corollary: Let V be quasi-complete, locally convex. Let $f(z, w)$ be complex-analytic V-valued in two variables, on a domain $\Omega_1 \times \Omega_2 \subset \mathbb{C}^2$. Then the function $w \longrightarrow (z \to f(z, w))$ is a holomorphic $\mathrm{Hol}(\Omega_1, V)$-valued function on Ω_2. ///

15.4 Banach-Alaoglu: Compactness of Polars

The *polar* U^o of an open neighborhood U of 0 in a topological vector space V is

$$U^o = \{\lambda \in V^* : |\lambda u| \leq 1, \text{ for all } u \in U\}$$

[15.4.1] Theorem: *(Banach-Alaoglu)* In the weak dual topology on V^* the polar U^o of an open neighborhood U of 0 in V is *compact*.

Proof: For every v in V there is real t_v sufficiently large such that $v \in t_v \cdot U$, and $|\lambda v| \leq t_v$ for $\lambda \in U^o$. Tychonoff gives compactness of the product

$$P = \prod_{v \in V} \{z \in \mathbb{C} : |z| \leq t_v\} \subset \prod_{v \in V} \mathbb{C}$$

Map V^* to $\prod_{v \in V} C$ by $j(\lambda) = \{\lambda(v) : v \in V\}$. By design, $j(U^o) \subset P$. To prove the compactness of U^o it suffices to show that the weak dual topology on U^o is identical to the subspace topology on $j(U^o)$ inherited from P and that $j(U^o)$ is closed in P.

The sub-basis sets

$$\{\lambda \in V^* : |\lambda v - \lambda_o v| < \delta\} \qquad (\text{for } v \in V \text{ and } \delta > 0)$$

for V^* are mapped by j to the sub-basis sets

$$\{p \in P : |p_v - \lambda_o v| < \delta\} \qquad (\text{for } v \in V \text{ and } \delta > 0)$$

for the product topology on P. That is, j maps U^o with the weak star-topology homeomorphically to $j(U^o)$.

To show that $j(U^o)$ is closed in P, consider L in the closure of U^o in P. Given $x, y \in V$, $a, b \in \mathbb{C}$, the sets

$$\{p \in P : |(p - L)_x| < \delta\} \qquad \{p \in P : |(p - L)_y| < \delta\}$$
$$\{p \in P : |(p - L)_{ax+by}| < \delta\}$$

are open in P and contain L, so meet $j(U^o)$. Let $\lambda \in j(U^o)$ lie in the intersection of these three sets and $j(U^o)$. Then

$$|aL_x + bL_y - L_{ax+by}| \leq |a| \cdot |L_x - \lambda x| + |b| \cdot |L_y - \lambda y|$$
$$+ |L_{ax+by} - \lambda(ax + by)| + |a\lambda x + b\lambda y - \lambda(ax + by)|$$
$$\leq |a| \cdot \delta + |b| \cdot \delta + \delta + 0 \qquad (\text{for every } \delta > 0)$$

so L is *linear*. Given $\varepsilon > 0$, for N be a neighborhood of 0 in V such that $x - y \in N$ implies $\lambda x - \lambda y \in N$,

$$|L_x - L_y| = |L_x - \lambda x| + |L_y - \lambda y| + |\lambda x - \lambda y| \delta + \delta + \varepsilon$$

Thus, L is *continuous*. Also, $|L_x - \lambda x| < \delta$ for all $x \in U$ and all $\delta > 0$, so $L \in j(U^o)$, and $j(U^o)$ is *closed*, giving compactness.   ///

15.5 Variant Banach-Steinhaus/Uniform Boundedness

This variant of the Banach-Steinhaus (uniform boundedness) theorem is used with Banach-Alaoglu to show that weak boundedness implies boundedness in a locally convex space, the starting point for *weak-to-strong principles*. It uses the version of Baire category for locally compact Hausdorff spaces, rather than complete metric spaces.

[15.5.1] Theorem: *(Variant Banach-Steinhaus)* Let K be a compact convex set in a topological vectorspace X, and $\mathscr{T}$ a set of continuous linear maps $X \to Y$ from X to another topological vectorspace Y. Suppose that for every *individual* $x \in K$ the collection of images $\mathscr{T}x = \{Tx : T \in \mathscr{T}\}$ is *bounded* in Y. Then $B = \bigcup_{x \in K} \mathscr{T}x$ is bounded in Y.

Proof: Let U, V be balanced neighborhoods of 0 in Y so that $\overline{U} + \overline{U} \subset V$, and let

$$E = \bigcap_{T \in \mathscr{T}} T^{-1}(\overline{U})$$

By the boundedness of $\mathscr{T}x$, there is a positive integer n such that $\mathscr{T}x \subset nU$, and then $x \in nE$. For every $x \in K$ there is such n, so

$$K = \bigcup_{n} (K \cap nE)$$

Since E is closed, the version of the Baire category theorem for locally compact Hausdorff spaces implies that at least one set $K \cap nE$ has nonempty interior in K. For such n, let x_o be an interior point of $K \cap nE$. Pick a balanced neighborhood W of 0 in X such that

$$K \cap (x_o + W) \subset nE$$

Since K is compact, it is bounded, so $K - x_o$ is bounded, and $K \subset x_o + tW$ for large enough positive real t. Since K is convex, $(1 - t^{-1})x + t^{-1}x \in K$ for any $x \in K$ and $t \geq 1$. At the same time,

$$z - x_o = t^{-1}(x - x_o) \in W \qquad \text{(for large enough } t\text{)}$$

by the boundedness of K, so $z \in x_o + W$. Thus, $z \in K \cap (x_o + V) \subset nE$. From the definition of $E, TE \subset \overline{U}$, so $T(nE) = nT(E) \subset n\overline{U}$. And $x = tz - (t - 1)x_o$ yields

$$Tx \in tn\overline{U} - (t - 1)n\overline{U} \subset tn(\overline{U} + \overline{U})$$

by the balanced-ness of U. Since $\overline{U} + \overline{U} \subset V$, we have $B \subset tnV$. Since V was arbitrary, this proves the boundedness of B.   ///

15.6 Weak Boundedness Implies (strong) Boundedness

[15.6.1] Theorem: Let V be a locally convex topological vectorspace. A subset E of V is bounded if and only if it is weakly bounded.

Proof: For the proof, we need the notion of *second polar* N^{oo} of an open neighborhood N of 0 in a topological vector space V:

$$N^{oo} = \{v \in V : |\lambda v| \leq 1 \text{ for all } \lambda \in N^o\}$$

where N^o is the polar of N. Conveniently,

[15.6.2] Claim: *(On second polars)* For V a locally convex topological vectorspace and N a convex, balanced neighborhood of 0, the second polar N^{oo} of N is the closure $\overline{N}$ of N.

Proof: Certainly N is contained in N^{oo}, and in fact $\overline{N}$ is contained in N^{oo} since N^{oo} is closed. By the local convexity of V, Hahn-Banach implies that for $v \in V$ but $v \notin \overline{N}$ there is $\lambda \in V^*$ such that $\lambda v > 1$ and $|\lambda v'| \leq 1$ for all $v' \in \overline{N}$. Thus, λ is in N^o, and every element $v \in N^{oo}$ is in $\overline{N}$, so $N^{oo} = \overline{N}$. ///

Returning to the proof of the theorem: clearly boundedness implies weak boundedness. On the other hand, take E weakly bounded, and U be a neighborhood of 0 in V in the original topology. By local convexity, there is a convex (and balanced) neighborhood N of 0 such that the closure $\overline{N}$ is contained in U.

By the weak boundedness of E, for each $\lambda \in V^*$ there is a bound b_λ such that $|\lambda x| \leq b_\lambda$ for $x \in E$. By Banach-Alaoglu the polar N^o of N is compact in V^*. The functions $\lambda \to \lambda x$ are continuous, so by variant Banach-Steinhaus there is a uniform constant $b < \infty$ such that $|\lambda x| \leq b$ for $x \in E$ and $\lambda \in N^o$. Thus, $b^{-1}x$ is in the second polar N^{oo} of N, shown by the previous proposition to be the closure $\overline{N}$ of N. That is, $b^{-1}x \in \overline{N}$. By the balanced-ness of N, $E \subset t\overline{N} \subset tU$ for any $t > b$, so E is bounded. ///

15.7 Proof That Weak C^1 Implies Strong C^0

The claim below, needed to complete the proof of [15.1.1] that *weak C^k* implies (strong) C^{k-1}, is an application of the fact that weak boundedness implies boundedness.

[15.7.1] Claim: Let V be a quasi-complete locally convex topological vector space. Fix real numbers $a \leq b \leq c$. Let g be a V-valued function defined on $[a, b) \cup (b, c]$. Suppose that for $\lambda \in V^*$ the scalar-valued function $\lambda \circ g$ extends to a C^1 function F_λ on the whole interval $[a, c]$. Then $g(b)$ can be chosen such that the extended $g(x)$ is (strongly) continuous on $[a, c]$.

Proof: For each $\lambda \in V^*$, let F_λ be the extension of $\lambda \circ g$ to a C^1 function on $[a, c]$. The differentiability of F_λ implies that for each λ the function

$$\Phi_\lambda(x, y) = \frac{F_\lambda(x) - F_\lambda(y)}{x - y} \qquad (\text{for } x \neq y)$$

has a continuous extension $\tilde{\Phi}_\lambda$ to the compact set $[a, c] \times [a, c]$. The image C_λ of $[a, c] \times [a, c]$ under this continuous map is compact in $\mathbb{R}$, so bounded. Thus, the subset

$$\left\{ \frac{\lambda f(x) - \lambda f(y)}{x - y} : x \neq y \right\} \subset C_\lambda$$

is *bounded* in $\mathbb{R}$. That is,

$$E = \left\{ \frac{g(x) - g(y)}{x - y} : x \neq y \right\} \subset V$$

is *weakly* bounded. Because weakly bounded implies (strongly) bounded, E is (strongly) bounded. That is, for a balanced, convex neighborhood N of 0 in V, there is t_o such that $(g(x) - g(y))/(x - y) \in tN$ for $x \neq y$ in $[a, c]$ and $t \geq t_o$. That is, $g(x) - g(y) \in (x - y)tN$. Given N and t_o as earlier, $g(x) - g(y) \in N$ for $|x - y| < \frac{1}{t_o}$. That is, as $x \to b$ the collection $g(x)$ is a bounded Cauchy net. By quasi-completeness, define $g(b) \in V$ as the limit of the values $g(x)$. For $x \to y$ the values $g(x)$ approach $g(y)$, so this extension of g is continuous on $[a, c]$. ///

15.8 Proof That Weak Holomorphy Implies Continuity

With the above preparation, we prove that *weak holomorphy* implies (strong) *continuity*, completing the proof of [15.2.2], as another application of the fact that *weak* boundedness implies boundedness, by an argument parallel to that of [15.7] that weak C^1 implies C^o for vector-valued functions on $[a, b]$.

[15.8.1] Claim: *Weak holomorphy* implies (strong) *continuity*.

Proof: To show that weak holomorphy of f implies $f : D \to V$ is (strongly) *continuous*, without loss of generality prove continuity at $z = 0$ and suppose $f(0) = 0 \in V$. Since $\lambda \circ f$ is holomorphic for each $\lambda \in V^*$ and vanishes at 0, each function $(\lambda \circ f)(z)/z$ initially defined on a punctured disk at 0 extends to a holomorphic function on a full disk at 0. By Cauchy theory for the scalar-valued holomorphic function $z \to \frac{\lambda(f(z))}{z}$,

$$\frac{(\lambda \circ f)(z)}{z} = \frac{1}{2\pi i} \int_\gamma \frac{1}{w - z} \cdot \frac{(\lambda \circ f)(w)}{w} \, dw$$

where γ is a circle of radius $2r$ centered at 0, and $|z| < r$. With M_λ the sup of $|\lambda \circ f|$ on γ,

$$\left| \frac{(\lambda \circ f)(z)}{z} \right| \leq \frac{\text{length } \gamma}{2\pi} \cdot \frac{1}{2r - r} \cdot \frac{M_\lambda}{2r} = \frac{1}{2\pi} \cdot (2\pi \cdot 2r) \cdot \frac{1}{r} \cdot \frac{M_\lambda}{2r} = \frac{M_\lambda}{r}$$

Thus, the set of values

$$S = \left\{ \frac{f(z)}{z} : |z| \leq r \right\}$$

is *weakly* bounded. Weak boundedness implies (strong) boundedness, so S is *bounded*. That is, given a balanced convex neighborhood N of 0 in V, there is $t_o > 0$ such that for complex w with $|w| \geq t_o$, the set S lies inside wN. Then $f(z) \in zwN$ and $f(z) \in N$ for $|z| < |w|$. As $f(0) = 0$, we have proven that, given N, for z sufficiently near 0 $f(z) - f(0) \in N$. This is (strong) continuity.   ///

15.A Appendix: Vector-Valued Power Series

We should confirm that power series with values in a quasi-complete, locally compact vectorspace V behave essentially as well as scalar-valued ones. First,

[15.A.1] Lemma: Let c_n be a *bounded* sequence of vectors in the locally convex, quasi-complete topological vector space V. Let z_n be a sequence of complex numbers, let $0 \leq r_n$ be real numbers such that $|z_n| \leq r_n$, and suppose that $\sum_n r_n < +\infty$. Then $\sum_n c_n z_n$ *converges* in V. Further, given a convex balanced neighborhood U of 0 in V let t be a positive real such that for all complex w with $|w| \geq t$ we have $\{c_n\} \subset tU$. Then

$$\sum_n c_n z_n \in \left(\sum_n |z_n| \right) \cdot tU \subset \left(\sum_n r_n \right) \cdot tU$$

Proof: For convex balanced neighborhood N of 0 in the topological vector space, with complex numbers z and w such that $|z| \leq |w|$, then $zN \subset wN$, since $|z/w| \leq 1$ implies $(z/w)N \subset N$, or $zN \subset wN$. Further, for an absolutely convergent series $\sum_n \alpha_n$ of complex numbers, for any n_o

$$\sum_{n \leq n_o} (\alpha_n \cdot V) = \sum_{n \leq n_o} (|\alpha_n| \cdot V) \subset \left(\sum_{n \leq n_o} |\alpha_n| \right) \cdot N$$

$$\subset \left(\sum_{n < \infty} |\alpha_n| \right) \cdot N$$

For a balanced open U containing 0, let t be large enough such that for any complex w with $|w| \geq t$ the sequence c_n is contained in wU. The previous discussion shows that

$$\sum_{m \leq \ell \leq n} c_\ell z_\ell \in (|z_m| + \cdots + |z_n|) \cdot tU$$

Given $\varepsilon > 0$, invoking absolute convergence, take m sufficiently large such that $|z_m| + \cdots + |z_n| < t \cdot \varepsilon$ for all $n \geq m$. Then

$$\sum_{m \leq \ell \leq n} c_\ell z_\ell \in t \cdot (\varepsilon/t) \cdot U = U$$

Thus, the original series is convergent. Since X is quasi-complete the limit exists in V. The last containment assertion follows from this discussion as well.   ///

[15.A.2] Corollary: Let c_n be a *bounded* sequence of vectors in a locally convex quasi-complete topological vector space V. Then on $|z| < 1$ the series $f(z) = \sum_n c_n z^n$ converges and gives a *holomorphic* V-valued function. That is, the function is infinitely many times complex-differentiable.

Proof: The lemma shows that the series expressing $f(z)$ and its apparent k^{th} derivative $\sum_n c_n \binom{n}{k} z^{n-k}$ all converge for $|z| < 1$. The usual direct proof of Abel's theorem on the differentiability of (scalar-valued) power series can be adapted to prove the infinite differentiability of the X-valued function given by this power series, as follows. Let

$$g(z) = \sum_{n \geq 0} n c_n z^{n-1}$$

Then

$$\frac{f(z) - f(w)}{z - w} - g(w) = \sum_{n \geq 1} c_n \left(\frac{z^n - w^n}{z - w} - n w^{n-1} \right)$$

For $n = 1$, the expression in the parentheses is 1. For $n > 1$, it is

$$(z^{n-1} + z^{n-2}w + \cdots + zw^{n-2} + w^{n-1}) - nw^{n-1}$$

$$= (z^{n-1} - w^{n-1}) + (z^{n-2}w - w^{n-1}) + \cdots + (z^2 w^{n-3} - w^{n-1})$$

$$+ (zw^{n-2} - w^{n-1}) + (w^{n-1} - w^{n-1})$$

$$= (z - w)[(z^{n-2} + \cdots + w^{n-2}) + w(z^{n-3} + \cdots + w^{n-3})$$

$$+ \cdots + w^{n-3}(z + w) + w^{n-2} + 0]$$

$$= (z - w) \sum_{k=0}^{n-2} (k+1) z^{n-2-k} w^k$$

For $|z| \leq r$ and $|w| \leq r$ the latter expression is dominated by

$$|z - w| \cdot r^{n-2} \frac{n(n-1)}{2} < |z - w| \cdot n^2 r^{n-2}$$

Let U be a balanced neighborhood of 0 in X, and t a sufficiently large real number such that for all complex w with $|w| \geq t$ all c_n lie in wU. For $|z| \leq r < 1$

and $|w| \leq r < 1$, by the lemma,

$$\frac{f(z) - f(w)}{z - w} - g(w) = (z - w) \sum_{n \geq 2} c_n \cdot \left(\sum_{k=0}^{n-2} (k+1) z^{n-2-k} w^k \right)$$

$$\in (z - w) \cdot \left(\sum_n n^2 r^{n-2} \right) \cdot tU$$

Thus, as $z \to w$, eventually $\frac{f(z) - f(w)}{z - w} - g(w)$ lies in U. ///

[15.A.3] Corollary: Let c_n be a sequence of vectors in a Banach space X such that for some $r > 0$ the series $\sum |c_n| \cdot r^n$ converges in X. Then for $|z| < r$ the series $f(z) = \sum c_n z^n$ converges and gives a holomorphic (infinitely many times complex-differentiable) X-valued function. ///

15.B Appendix: Two Forms of the Baire Category Theorem

This standard result is indispensable and mysterious. We give the more typical version for *complete metric* spaces in parallel with the argument for *locally compact Hausdorff* spaces.

A set E in a topological space X is *nowhere dense* if its closure $\bar{E}$ contains no nonempty open. A *countable union* of nowhere dense sets is said to be *of first category*, while every other subset is *of second category*. The idea, not suggested by this traditional terminology, is that first category sets are *small*, while second category sets are *large*. The theorem asserts that (nonempty) *complete metric* spaces and *locally compact Hausdorff* spaces are *of second category*.

[15.B.1] Theorem: For X be a *complete metric space* or a *locally compact Hausdorff topological space*, the intersection of a *countable* collection $U_1, U_2, \ldots$ of *dense open subsets* U_i of X is still *dense* in X.

Proof: Let B_o be a nonempty open set in X and show that $\bigcap_i U_i$ meets B_o. Suppose that we have inductively chosen an open ball B_{n-1}. By the denseness of U_n, there is an open ball B_n whose closure $\overline{B_n}$ satisfies

$$\overline{B_n} \subset B_{n-1} \cap U_n$$

Further, for complete metric spaces, take B_n to have radius less than $1/n$ (or any other sequence of reals going to 0), and in the locally compact Hausdorff case take B_n to have compact closure.

Let

$$K = \bigcap_{n \geq 1} \overline{B_n} \subset B_o \cap \bigcap_{n \geq 1} U_n$$

For complete metric spaces, the centers of the nested balls B_n form a Cauchy sequence (since they are nested and the radii go to 0). By completeness, this Cauchy sequence *converges*, and the limit point lies inside each *closure* $\overline{B_n}$, so lies in the intersection. In particular, K is nonempty. For locally compact Hausdorff spaces, the intersection of a nested family of nonempty compact sets is nonempty, so K is nonempty, and B_o necessarily meets the intersection of the U_n.   ///

15.C Appendix: Hartogs's Theorem on Joint Analyticity

This proof roughly follows that in [Hörmander 1973] which roughly follows Hartogs' original argument [Hartogs 1906].

[15.C.1] Theorem: Let f be a $\mathbb{C}$-valued function defined in a nonempty open set $U \subset \mathbb{C}^n$. If f is analytic in each variable z_j when the other coordinates z_k for $k \neq j$ are fixed, then f is analytic as a function of all n coordinates.

[15.C.2] Remark: It is striking that no additional hypothesis on f is used beyond its separate analyticity: there is no assumption of continuity, nor even of measurability. Indeed, the beginning of the proof illustrates the fact that an assumption of continuity trivializes things. The strength of the theorem is that no hypothesis whatsoever is necessary.

Proof: The assertion is local, so it suffices to prove it when the open set U is a polydisk. The argument approaches the full assertion in stages.

First, suppose that f is *continuous* on the closure $\bar{U}$ of a polydisk U, and separately analytic. Even without continuity, simply by separate analyticity, an n-fold iterated version of Cauchy's one-variable integral formula is valid, namely

$$f(z) = \frac{1}{(2\pi i)^n} \int_{C_1} \cdots \int_{C_n} \frac{f(\zeta)}{(\zeta_1 - z_1)\ldots(\zeta_n - z_n)} \, d\zeta_1 \ldots d\zeta_n$$

where C_j is the circle bounding the disk in which z_j lies, traversed in the positive direction. The integral is a compactly supported integral of the function

$$(\zeta_1, \ldots, \zeta_n) \to \frac{f(\zeta_1, \ldots, \zeta_n)}{(\zeta_1 - z_1)\ldots(\zeta_n - z_n)}$$

For $|z_j| < |\zeta_j|$, the geometric series expansion

$$\frac{1}{\zeta_j - z_j} = \sum_{n \geq 0} \frac{z_j^n}{\zeta_j^{n+1}}$$

can be substituted into the latter integral. Fubini's theorem justifies interchange of summation and integration, yielding a (convergent) power series for $f(z)$. Thus, continuity of $f(z)$ (with separate analyticity) implies joint continuity.

Note that if we could be sure that *every* conceivable integral of analytic functions were analytic, then this iterated one-variable Cauchy formula would prove (joint) analyticity immediately. However, it is not obvious that *separate* analyticity implies continuity, for example.

Next we see that *boundedness* of a separately analytic function on a closed polydisk implies continuity, using Schwarz's lemma and its usual corollary:

[15.C.3] Lemma: (*Schwarz*) Let $g(z)$ be a holomorphic function on $\{z \in \mathbb{C} : |z| < 1\}$, with $g(0) = 0$ and $|g(z)| \le 1$. Then $|g(z)| \le |z|$ and $|g'(0)| \le 1$. (*Proof:* Apply the maximum modulus principle to $f(z)/z$ on disks of radius less than 1.)

[15.C.4] Corollary: Let $g(z)$ be a holomorphic function on $\{z \in \mathbb{C} : |z| < r\}$, with $|g(z)| \le B$ for a bound B. Then for z, ζ in that disk,

$$|g(z) - g(\zeta)| \le 2 \cdot B \cdot \left| \frac{r(z - \zeta)}{r^2 - \bar{\zeta}z} \right|$$

Proof: (*of corollary*) The linear fractional transformation

$$\mu : z \to r \cdot \begin{pmatrix} 1 & \zeta/r \\ \bar{\zeta}/r & 1 \end{pmatrix} (rz) = r \cdot \frac{z + r\zeta}{\bar{\zeta}z + r}$$

sends the disk of radius 1 to the disk of radius r, and sends 0 to ζ. Then the function

$$z \to \frac{g(\mu(z)) - g(\zeta)}{2B}$$

is normalized to match Schwarz's lemma, namely, that it vanishes at 0 and is bounded by 1 on the open unit disk. Thus, we conclude that for $|z| < 1$

$$\left| \frac{g(\mu(z)) - g(\zeta)}{2B} \right| \le |z|$$

Replace z by

$$\mu^{-1}(z) = \frac{r(z - \zeta)}{r^2 - \bar{\zeta}z}$$

to obtain

$$\left| \frac{g(z) - g(\zeta)}{2B} \right| \le \left| \frac{r(z - \zeta)}{r^2 - \bar{\zeta}z} \right|$$

as asserted in the corollary. ///

Now let f be separately analytic and *bounded* on the closure of the polydisk $\{(z_1, \ldots, z_n) : |z_j| < r_j\}$. We show that f is (jointly) analytic by proving it is continuous, invoking the first part of the proof (above). Let B be a bound for $|f|$ on the closed polydisk. We claim that the inequality

$$|f(z) - f(\zeta)| \leq 2B \sum_{1 \leq j \leq n} \frac{r_j |z_j - \zeta_j|}{|r_j^2 - \bar{\zeta}_j z_j|}$$

holds, which would prove continuity. Because of the telescoping expression

$$
f(z) - f(\zeta)
$$
$$
= \sum_{1 \leq j \leq n} \left(f(\zeta_1, \ldots, \zeta_{j-1}, z_j, \ldots, z_n) - f(\zeta_1, \ldots, \zeta_j, \zeta_j, z_{j+1}, \ldots, z_n) \right)
$$

it suffices to prove the inequality in the single-variable case, which is the immediate corollary to Schwarz's lemma as above. Thus, a *bounded* separately analytic f is continuous and (from earlier) jointly analytic.

Now we do induction on the dimension n: suppose that Hartogs's theorem is proven on $\mathbb{C}^{n-1}$, and prove it for $\mathbb{C}^n$. Here the Baire category theorem intervenes, getting started on the full statement of the theorem by first showing that a separately analytic function must be bounded on *some* polydisk, hence (from earlier) *continuous* on that polydisk, hence (from above) *analytic* on that polydisk.

Let f be separately analytic on a (nonempty) closed polydisk $D = \prod_{1 \leq j \leq n} D_j$, where D_j is a disk in $\mathbb{C}$. We claim that there exist nonempty closed disks $E_j \subset D_j$ with $E_n = D_n$ such that f is bounded on $E = \prod_{1 \leq j \leq n} E_j$ (and, hence, f is analytic in E).

To see this, for each bound $B > 0$ let

$$\Omega_B = \{z' \in \prod_{1 \leq j < n-1} E_j : |f(z', z_n)| \leq B \text{ for all } z_n \in E_n\}$$

By induction, for fixed z_n the function $z' \to f(z', z_n)$ is analytic, so continuous, so Ω_B is closed. For any fixed z', the function $z_n \to f(z', z_n)$ is assumed analytic, so is continuous on the closed disk $E_n = D_n$, hence bounded. Thus

$$\bigcup_{B=1}^{\infty} \Omega_B = \prod_{1 \leq j \leq n-1} D_j$$

Then the Baire category theorem shows that some Ω_B must have nonempty interior, so must contain a (nonempty) closed polydisk, as claimed.

Now let f be separately analytic in a polydisk

$$D = \{(z_1, \ldots, z_n) : |z_j| < r\} \subset \mathbb{C}^n$$

analytic in $z' = (z_1, \ldots, z_n - 1)$ for fixed z_n, and suppose that f is analytic in a smaller (nonempty) polydisk

$$E = \left(\prod_{1 \le j \le n-1} \{z_j \in \mathbb{C} : |z_j| < \varepsilon\} \right) \times \{z_n \in \mathbb{C} : |z_n| < r\}$$

inside D. Then we claim that f is analytic on the original polydisk D.

By the iterated form of Cauchy's formula, the function $z' \to f(z', z_n)$ has a Taylor expansion in z'

$$f(z', z) = \sum_\alpha c_\alpha(z_n) z'^\alpha$$

where the coefficients depend upon z_n, given by the usual formula

$$c_\alpha(z_n) = \frac{\partial^\alpha}{\partial z'^\alpha} f(0, z_n) / \alpha!$$

using multi-index notation. Cauchy's integral formula in z' for derivatives

$$\frac{\partial^\alpha}{\partial z'^\alpha} f(0, z_n) = \frac{\alpha!}{(2\pi i)^{n-1}} \int_{C_1} \cdots \int_{C_{n-1}} \frac{f(\zeta) \, d\zeta_1 \ldots d\zeta_{n-1}}{(\zeta_1 - z_1)^{\alpha_1+1} \ldots (\zeta_{n-1} - z_{n-1})^{\alpha_{n-1}+1}}$$

shows that $c_\alpha(z_n)$ is analytic in z_n, again by expanding convergent geometric series and their derivatives, and interchanging summation and integration.

Fix $0 < r_1 < r_2 < r$ and fix z_n with $|z_n| < r$. Then

$$|c_\alpha(z_n)| \cdot r_2^{|\alpha|} \to 0$$

as $|\alpha| \to \infty$, by the convergence of the power series. Let B be a bound for $|f|$ on the smaller polydisk E. Then on that smaller polydisk the Cauchy integral formula for the derivative gives

$$|c_\alpha(z_n)| \le B/\varepsilon^{|\alpha|}$$

Therefore, the subharmonic functions

$$u_\alpha(z_n) = \frac{1}{|\alpha|} \log |c_\alpha(z_n)|$$

are uniformly bounded from above for $|z_n| < r$. And the property $|c_\alpha(z_n)| \cdot r_2^{|\alpha|} \to 0$ shows that for fixed z_n $\log(1/r_2)$ is an upper bound for these subharmonic functions as $|\alpha| \to \infty$. Thus, Hartogs's lemma (recalled below) on subharmonic functions implies that for large $|\alpha|$, *uniformly* in $|z_n| < r_1$

$$\frac{1}{|\alpha|} \log |c_\alpha(z_n)| \leq \log(1/r_1)$$

Thus, for large $|\alpha|$

$$|c_\alpha(z_n)| \cdot r_1^{|\alpha|} \leq 1$$

uniformly in $|z_n| < r_1$. Therefore, since the summands $c_\alpha(z_n) z'^\alpha$ are analytic, the series

$$f(z', z) = \sum_\alpha c_\alpha(z_n) z'^\alpha$$

converges to a function analytic in the polydisk D.

Thus, in summary, given $z \in U$, choose $r > 0$ so that the polydisk of radius $2r$ centered at z is contained in U. The foregoing Baire category argument shows that there is w such that z is inside a polydisk D of radius r centered at w, and such that f is holomorphic on some smaller polydisk E inside D (still centered at w). Finally one uses Hartogs's lemma on subharmonic functions (below) to see that the power series for f on the small polydisk E at w converges on the larger polydisk D at w. Since D contains the given point z, f is analytic on a neighborhood of z. Thus, f is analytic throughout U. ///

[15.C.5] Lemma: (*Hartogs*) Let u_j be a sequence of real-valued subharmonic functions in an open set U in $\mathbb{C}$. Suppose that the functions are uniformly bounded from above and that

$$\limsup_k u_k(z) \leq C$$

for every $z \in U$. Then, given $\varepsilon > 0$ and compact $K \subset U$ there exists k_o such that for $z \in K$ and $k \geq k_o$

$$u_k(z) \leq C + \varepsilon$$

Proof: Without loss of generality, replacing U by an open subset with compact closure contained inside U, we may suppose that the functions u_k are uniformly bounded in U, for example, $u_k(z) \leq 0$ for all $z \in U$. Let $r > 0$ be small enough so that the distance from K to every point of the complement of U is more than $3r$. Using the proposition below characterizing subharmonic functions, we have, for every $z \in K$,

$$\pi r^2 u_k(z) \leq \int_{|z-\zeta|<r} u_k(\zeta) \, d\zeta$$

By Fatou's lemma, the lim sup of the right hand side is at most $\pi r^2 C$ as $k \to \infty$. Thus, for every $z \in K$ there is k_o such that for $k \geq k_o$

$$\int_{|z-\zeta|<r} u_k(\zeta)\,d\zeta \leq \pi r^2 (C + \varepsilon/2)$$

Since $u_k(z) \leq 0$, for $|z - w| < \delta < r$

$$\pi (r+\delta)^2 u_k(w) \leq \int_{|\zeta-w|<r+\delta|} u_k(\zeta)\,d\zeta \leq \int_{|\zeta-z|\leq r} u_k(\zeta)\,d\zeta$$

Thus, for $\delta > 0$ sufficiently small, for $k \geq k_o$ and $|w - z| < \delta$,

$$u_k(w) < C + \varepsilon$$

Since K is compact the lemma follows. ///

For convenience, recall the following basic property of subharmonic functions.

Proposition: For a real-valued subharmonic function u bounded above on an open set U, for every positive measure μ on $[0, \delta]$, and for $z \in U$ of distance more than δ from the complement of U,

$$u(z) \cdot 2\pi \cdot \int d\mu \leq \int_0^{2\pi} \int u(z + re^{i\theta})\,d\theta\,d\mu(r)$$

Proof: The definition of a function u being subharmonic on an open set Ω is that u is upper semicontinuous (that is, $\{z \in \Omega : u(z) < c\}$ is open for every constant c), and for every compact $K \subset \Omega$, for every continuous function h on K harmonic on K and $h(\beta) \geq u(\beta)$ for β on the boundary of K, $u(z) \leq h(z)$ throughout K. The condition may be vacuous unless u is assumed bounded from above.

Let $z \in U$ be distance more than δ away from the complement of U, and fix r with $0 < r \leq \delta$. Let D be the closed disk of radius r about z. Since $r \leq \delta$, $D \subset U$. For a trigonometric polynomial

$$g(\theta) = \sum_k c_k e^{i\theta}$$

with real coefficients c_k with $u(z + re^{i\theta}) \leq g(\theta)$, the polynomial

$$G(\zeta) = c_0 + \sum_{k>0} (c_k + c_{-k}) \frac{(\zeta - z)^k}{r^k}$$

has real part $\operatorname{Re} G$ which is an upper bound for u on the boundary of the disk D. Thus, $u \leq \operatorname{Re} G$ on D by the subharmonicness of u and in particular at the center of D, at z,

$$u(z) \le c_o + \frac{1}{2\pi} \int_0^{2\pi} g(\theta) \, d\theta$$

Then for an arbitrary continuous real-valued function h on the boundary of D and with $u(z + re^{i\theta}) \le h(\theta)$ (by Weierstrass approximation, for example), given $\varepsilon > 0$, we can find a trigonometric polynomial g so that $\sup |g(\theta) - h(\theta)| < \varepsilon$. Thus, for every $\varepsilon > 0$,

$$u(z) \le c_o + \frac{1}{2\pi} \int_0^{2\pi} h(\theta) \, d\theta + \varepsilon$$

Thus, the latter inequality must hold with $\varepsilon = 0$, for continuous h. Since the integral of an upper-semicontinuous function is the infimum of the integrals of continuous functions dominating it, we have the same inequality with u in place of h. Integration with respect to the radius r gives the result. ///

In fact, suppose that for every $\delta > 0$ and for every z at distance more than δ from the complement of U there exists a positive measure μ on $[0, \delta]$ with support not just $\{0\}$ and

$$u(z) \cdot 2\pi \cdot \int d\mu \le \int_0^{2\pi} \int u(z + re^{i\theta}) \, d\theta \, d\mu(r)$$

Then u is subharmonic. To see this, let K be a compact subset of U, h a continuous function on K which is harmonic in the interior of K and such that $u \le g$ on the boundary of K. If the supremum of $u - h$ over K is strictly positive, the upper semicontinuity of $u - h$ implies that $u - h$ attains its sup S on a nonempty compact subset M of the interior of K. Let z_o be a point of M closest to the boundary of K. If the distance is greater than δ, then *every* circle $|z - z_o| = r$ with $0 < r \le \delta$ contains a nonempty arc of points where $u - h < S$. Then

$$\int (u - h)(z_o + re^{i\theta})) \, d\theta \, d\mu(r) < S \cdot 2\pi \cdot \int d\mu(r)$$

$$= (u - h)(z_o) \cdot 2\pi \cdot \int d\mu(r)$$

when μ is a measure not supported just at $\{0\}$. The mean value property for harmonic functions gives

$$\int h(z_o + re^{i\theta})) \, d\theta \, d\mu(r) = h(z_o) \cdot 2\pi \cdot \int d\mu(r)$$

Thus,

$$\int u(z_o + re^{i\theta})) \, d\theta \, d\mu(r) < u(z_o) \cdot 2\pi \cdot \int d\mu(r)$$

contradicting the hypothesis. Thus, $\sup_K (u - h) \leq 0$, which proves that u is subharmonic.   ///

16

Asymptotic Expansions

The simplest notion of *asymptotic* of $f(s)$ as s goes to $+\infty$ on $\mathbb{R}$ is a simpler function $F(s)$ such that $\lim_s f(s)/F(s) = 1$, written $f \sim F$. One might require an error estimate, for example,

$$f \sim F \iff f(s) = F(s) \cdot (1 + O(\frac{1}{|s|}))$$

A more precise form is to say that

$$f(s) \sim f_0(s) \cdot (\frac{c_0}{s^\alpha} + \frac{c_1}{s^{\alpha+1}} + \frac{c_2}{s^{\alpha+2}} + \cdots)$$

293

(with any auxiliary function f_0) is an *asymptotic expansion* for f when

$$f = f_0(s) \cdot (\frac{c_0}{s^\alpha} + \frac{c_1}{s^{\alpha+1}} + \ldots + \frac{c_n}{s^{\alpha+n}} + O(\frac{1}{|s|^{\alpha+n+1}}))$$

The exposition is revisionist: Laplace's method is proven by reducing it to Watson's lemma.

16.1 Heuristic for Stirling's Asymptotic

First we give a heuristic and mnemonic for the main term of the Laplace-Stirling asymptotic, namely

$$\Gamma(s) \sim e^{-s} \cdot s^{s-\frac{1}{2}} \cdot \sqrt{2\pi}$$

Using Euler's integral,

$$s \cdot \Gamma(s) = \Gamma(s+1) = \int_0^\infty e^{-u} u^{s+1} \frac{du}{u}$$

$$= \int_0^\infty e^{-u} u^s \, du = \int_0^\infty e^{-u+s\log u} \, du$$

The trick is to replace the exponent $-u + s\log u$ by the quadratic polynomial in u best approximating it near its maximum and evaluate the resulting integral. This replacement is justified via Watson's lemma and Laplace's method, shown subsequently, but the heuristic is simpler than the justification. The exponent takes its maximum where its derivative vanishes, at the unique solution $u_o = s$ of

$$-1 + \frac{s}{u} = 0$$

The second derivative in u of the exponent is $-s/u^2$, which takes value $-1/s$ at $u_o = s$. Thus, near $u_o = s$, the quadratic Taylor-Maclaurin polynomial in u approximating the exponent is

$$-s + s\log s - \frac{1}{2!\,s} \cdot (u-s)^2$$

We imagine that

$$s \cdot \Gamma(s) \sim \int_0^\infty e^{-s+s\log s - \frac{1}{2s} \cdot (u-s)^2} \, du = e^{-s} \cdot s^s \cdot \int_{-\infty}^\infty e^{-\frac{1}{2s} \cdot (u-s)^2} \, du$$

The latter integral is taken over the whole real line. Evaluation of the integral over the whole line, and simple estimates on the integral over $(-\infty, 0]$, show that the integral over $(-\infty, 0]$ is of a lower order of magnitude than the whole. Thus, the leading term of the asymptotics of the integral over the whole line is the same than the integral from 0 to $+\infty$. To simplify the remaining integral, replace u by su and cancel a factor of s from both sides,

$$\Gamma(s) \sim e^{-s} \cdot s^s \cdot \int_{-\infty}^{\infty} e^{-s(u-1)^2/2} \, du$$

Replace u by $u + 1$, and u by $u \cdot \sqrt{2\pi/s}$, obtaining

$$\int_{-\infty}^{\infty} e^{-s(u-1)^2/2} \, du = \int_{-\infty}^{\infty} e^{-su^2/2} \, du = \frac{\sqrt{2\pi}}{\sqrt{s}} \int_{-\infty}^{\infty} e^{-\pi u^2} \, du$$

$$= \frac{\sqrt{2\pi}}{\sqrt{s}}$$

and

$$\Gamma(s) \sim e^{-s} \cdot s^{s-\frac{1}{2}} \cdot \sqrt{2\pi}$$

It is striking that this heuristic can be made rigorous, as shown subsequently.

16.2 Watson's Lemma

Watson's lemma gives an asymptotic expansion for certain Laplace transforms, valid in half-planes in $\mathbb{C}$. For example, let h be a smooth function on $(0, +\infty)$ all whose derivatives are of polynomial growth, and expressible for small $x > 0$ as

$$h(x) = x^\alpha \cdot g(x)$$

for some $\alpha \in \mathbb{C}$, where $g(x)$ is differentiable on $\mathbb{R}$ near 0. Thus, $h(x)$ has an expression

$$h(x) = x^\alpha \cdot \sum_{n=0}^{\infty} c_n x^n \qquad \text{(for } 0 < x \text{ sufficiently small)}$$

Then there is an *asymptotic expansion* of the Laplace transform of h,

$$\int_0^{\infty} e^{-xs} h(x) \frac{dx}{x} \sim \frac{\Gamma(\alpha) c_0}{s^\alpha} + \frac{\Gamma(\alpha+1) c_1}{s^{\alpha+1}} + \frac{\Gamma(\alpha+2) c_2}{s^{\alpha+2}} + \cdots$$

for $\text{Re}(s) > 0$. A simple corollary of the error estimates given below is that, letting $\text{Re}(\alpha) + 1 - \varepsilon$ be the greatest integer less than or equal $\text{Re}(\alpha) + 1$,

$$\int_0^{\infty} e^{-xs} h(x) \frac{dx}{x} = \int_0^{\infty} e^{-xs} x^\alpha g(x) \frac{dx}{x}$$

$$= \frac{\Gamma(\alpha) g(0)}{s^\alpha} + O\left(\frac{1}{|s|^{\text{Re}(\alpha)+1-\varepsilon}} \right)$$

Since $\mathrm{Re}\,(\alpha) + 1 - \varepsilon > \mathrm{Re}\,(\alpha)$, the error term is of strictly smaller order of magnitude in s.

The idea of the proof is straightforward: the expansion is obtained from

$$
\int_0^\infty e^{-xs}\, h(x)\, \frac{dx}{x} = \int_0^\infty e^{-xs}\, x^\alpha (c_0 + \cdots + c_n x^n)\, \frac{dx}{x}
$$
$$
+ \int_0^\infty e^{-xs}\, x^\alpha \left(g(x) - (c_0 + \cdots + c_n x^n)\right) \frac{dx}{x}
$$

The first integral gives the asymptotic expansion, and for $\mathrm{Re}\,(s) > 0$ the second integral can be integrated by parts essentially $\mathrm{Re}\,(\alpha) + n$ times and trivially bounded to give a $O(1/s^{\alpha+n-\varepsilon})$ error term for some small $\varepsilon \geq 0$. Note that for the integration by parts the denominator x in the measure must be moved into the integrand proper, accounting for a slight reduction of the order of vanishing of the integrand at 0.

To understand the error, let $\varepsilon \geq 0$ be the smallest such that

$$
N = \mathrm{Re}\,(\alpha) + n - \varepsilon \in \mathbb{Z}
$$

The subtraction of the initial polynomial and re-allocation of the $1/x$ from the measure makes $x^{\alpha-1}(g(x) - (c_0 + \cdots + c_n x^n))$ vanish to order N at 0. This, with the exponential e^{-sx} and the presumed polynomial growth of h and its derivatives, allows integration by parts N times without boundary terms, giving

$$
\int_0^\infty e^{-xs}\, h(x)\, dx = \frac{\Gamma(\alpha)\, c_0}{s^\alpha} + \frac{\Gamma(\alpha+1)\, c_1}{s^{\alpha+1}} + \cdots + \frac{\Gamma(\alpha+n)\, c_n}{s^{\alpha+n}}
$$
$$
+ \frac{1}{s^N} \int_0^\infty e^{-sx} \left(\frac{\partial}{\partial x}\right)^N (x^\alpha \cdot (g(x) - (c_0 + \cdots + c_n x^n)))\, dx
$$

The last error-like term is $O(s^{-[\mathrm{Re}\,(\alpha)+n-\varepsilon]})$. That is, computing in this fashion, the error term swallows up the last term in the asymptotic expansion. Visibly, this argument applies to more general sorts of expansions near 0.

16.3 Watson's Lemma Illustrated on the Beta Function

Here is an important example of an asymptotic result nontrivial to derive from Stirling's formula for $\Gamma(s)$ but easy to obtain from Watson's lemma. Euler's beta integral is

$$
B(s, a) = \int_0^1 x^{s-1} (1 - x)^{a-1}\, dx = \frac{\Gamma(s)\,\Gamma(a)}{\Gamma(s+a)}
$$

We recall how to express Beta in terms of Gamma: with $x = u/(u+1)$ in the beta integral,

$$B(s, a) = \int_0^\infty u^{s-1} (u+1)^{-(s-1)-(a-1)-2} \, du = \int_0^\infty u^{s-1} (u+1)^{-s-a} \, du$$

$$= \frac{1}{\Gamma(s+a)} \int_0^\infty \int_0^\infty u^s \, e^{-v(u+1)} \, v^{s+a} \, \frac{dv}{v} \frac{du}{u}$$

using $\int_0^\infty e^{-vy} v^b \, dv/v = \Gamma(b)/y^b$. Replacing u by u/v gives $B(s, a) = \Gamma(s)\Gamma(a)/\Gamma(s+a)$.

Fix a with $\operatorname{Re}(a) > 0$, and consider this integral as a function of s. Letting $x = e^{-u}$ gives an integrand fitting Watson's lemma,

$$B(s, a) = \int_0^\infty e^{-su} (1 - e^{-u})^{a-1} \, du = \int_0^\infty e^{-su} (u - \frac{u^2}{2!} + \ldots)^{a-1} \, du$$

$$= \int_0^\infty e^{-su} u^a \cdot (1 - \frac{u}{2!} + \ldots)^{a-1} \frac{du}{u} \sim \frac{\Gamma(a)}{s^a}$$

taking just the first term in an asymptotic expansion, using Watson's lemma. Thus,

$$\frac{\Gamma(s) \, \Gamma(a)}{\Gamma(s+a)} \sim \frac{\Gamma(a)}{s^a}$$

giving

$$\frac{\Gamma(s)}{\Gamma(s+a)} \sim \frac{1}{s^a} \qquad \text{(for } a \text{ fixed)}$$

16.4 Simple Form of Laplace's Method

A simple version of Laplace's method obtains asymptotics in s for certain integrals of the form

$$\int_0^\infty e^{-s \cdot f(u)} \, du$$

with f real-valued. The idea is that the *minimum values* of $f(u)$ should dominate, and the leading term of the asymptotics should be

$$\int_0^\infty e^{-s \cdot f(u)} \, du \sim e^{-sf(u_o)} \cdot \frac{\sqrt{2\pi}}{\sqrt{f''(u_o)}} \cdot \frac{1}{\sqrt{s}}$$

for $|s| \to \infty$, with $\operatorname{Re}(s) \geq \delta > 0$. To reduce this to Watson's lemma, break the integral at points where the derivative f' changes sign, and change variables to convert each fragment to a Watson-lemma integral. For Watson's lemma to be legitimately applied, we will find that f must be smooth with all derivatives of at most polynomial growth *and* at most polynomial *decay*, as $u \to +\infty$. For simplicity *assume* that there is exactly *one* point u_o at which $f'(u_o) = 0$, and that $f''(u_o) > 0$. Further, assume that $f(u)$ goes to $+\infty$ at 0^+ and at $+\infty$. Since $f'(u) > 0$ for $u > u_o$ and $f'(u) < 0$ for $0 < u < u_o$, on each of these two intervals there is a smooth square root $\sqrt{f(u) - f(u_o)}$ and there are smooth functions F, G such that

$$\begin{cases} F(\sqrt{f(u) - f(u_o)}) = u & \text{(for } u_o < u < +\infty) \\[2mm] G(\sqrt{f(u) - f(u_o)}) = u & \text{(for } 0 < u < u_o) \end{cases}$$

Then

$$\int_0^\infty e^{-s\,f(u)}\,du = e^{-sf(u_o)} \int_0^{u_o} e^{-s\,(f(u)-f(u_o))}\,du + e^{-sf(u_o)} \int_{u_o}^\infty e^{-s\,(f(u)-f(u_o))}\,du$$

$$= e^{-sf(u_o)} \left(\int_0^\infty e^{-sx^2}\,F'(x)\,dx + \int_0^\infty e^{-sx^2}\,G'(x)\,dx \right)$$

by letting $x = \sqrt{f(u) - f(u_o)}$ in the two intervals. In both integrals, replacing x by $\sqrt{x}$ gives Watson's-lemma integrals

$$\int_0^\infty e^{-s\,f(u)}\,du$$

$$= e^{-sf(u_o)} \left(\int_0^\infty e^{-sx} \tfrac{1}{2} x^{1/2}\, F'(\sqrt{x})\, \frac{dx}{x} + \int_0^\infty e^{-sx} \tfrac{1}{2} x^{1/2}\, G'(\sqrt{x})\, \frac{dx}{x} \right)$$

At this point the needed conditions on F, hence, on f, become clear: since F must be smooth with all derivatives of at most polynomial growth, direct chain-rule computations show that it suffices that no derivative of f increases *or decreases* faster than polynomially as $u \to +\infty$. The assumptions $f'(u_o) = 0$ and $f''(u_o) > 0$ assure that F has a Taylor series expansion near 0, giving a suitable expansion

$$\tfrac{1}{2} x^{1/2} F'(x)$$

$$= \tfrac{1}{2} F'(0) x^{1/2} + \frac{\tfrac{1}{2} F^{(2)}(0)}{1!} x^{3/2} + \frac{\tfrac{1}{2} F^{(3)}(0)}{2!} x^{5/2} + \frac{\tfrac{1}{2} F^{(4)}(0)}{3!} x^{7/2} + \cdots$$

for small $x > 0$. From this, the main term of the Watson's lemma asymptotics for the integral involving F would be

$$\int_0^\infty e^{-sx} \tfrac{1}{2} x^{1/2}\, F'(\sqrt{x})\, \frac{dx}{x} \sim \frac{\Gamma(\tfrac{1}{2})\, F'(0)}{2} \cdot \frac{1}{\sqrt{s}}$$

To determine $F'(0)$, or any higher coefficients, from $F(x) = u$, we have

$F'(x) \cdot \frac{dx}{du} = 1$. Since

$$x = \sqrt{f(u) - f(u_o)} = \sqrt{(u - u_o)^2 \cdot \frac{f''(u_o)}{2!} + \cdots}$$

$$= \sqrt{\frac{f''(u_o)}{2}} \cdot ((u - u_o) + \cdots)$$

the derivative is

$$\frac{dx}{du} = \sqrt{\frac{f''(u_o)}{2}} \cdot (1 + O(u - u_o))$$

Thus,

$$F'(x) = \frac{1}{\frac{dx}{du}} = \sqrt{\frac{2}{f''(u_o)}} \cdot (1 + O(u - u_o))$$

which allows evaluation at $x = 0$, namely

$$F'(0) = \sqrt{\frac{2}{f''(u_o)}}$$

The same argument applied to G gives $G'(0) = F'(0)$. Thus,

$$\int_0^\infty e^{-s\,f(u)}\,du \sim e^{-sf(u_o)} \cdot \frac{\Gamma(\frac{1}{2}) \cdot 2 \cdot \sqrt{\frac{2}{f''(u_o)}}}{2\sqrt{s}}$$

$$= e^{-sf(u_o)} \cdot \frac{\sqrt{2\pi}}{\sqrt{f''(u_o)}} \cdot \frac{1}{\sqrt{s}}$$

Last, we verify that this outcome is what would be obtained by replacing $f(u)$ by its quadratic approximation

$$f(u_o) + \frac{f''(0)}{2!} \cdot (u - u_o)^2$$

in the exponent in the original integral, integrated over the whole line. The latter would be

$$\int_{-\infty}^\infty e^{s \cdot \left(f(u_o) + \frac{1}{2}f''(u_o)(u-u_o)^2\right)}\,du = e^{sf(u_o)} \int_{-\infty}^\infty e^{s \cdot \frac{1}{2}f''(u_o)(u-u_o)^2}\,du$$

$$= e^{sf(u_o)} \int_{-\infty}^\infty e^{s \cdot \frac{1}{2}f''(u_o)u^2}\,du$$

$$= e^{sf(u_o)} \cdot \frac{\sqrt{\pi}}{\sqrt{\frac{1}{2}f''(u_o)}} \cdot \frac{1}{\sqrt{s}}$$

$$= e^{sf(u_o)} \cdot \frac{\sqrt{2\pi}}{\sqrt{f''(u_o)}} \cdot \frac{1}{\sqrt{s}}$$

This does indeed agree. Last, verify that the integral of the exponentiated quadratic approximation over $(-\infty, 0]$ is of a lower order of magnitude. Indeed, for $u \leq 0$ and $u_o > 0$ we have $(u - u_o)^2 \geq u^2 + u_o^2$, and $f''(u_o) < 0$ by assumption, so

$$
e^{sf(u_o)} \int_{-\infty}^{0} e^{s \cdot \left(\frac{1}{2} f''(u_o)(u - u_o)^2 \right)} \, du
$$

$$
\leq e^{sf(u_o)} \cdot e^{s \cdot \frac{1}{2} f''(u_o) \cdot u_o^2} \int_{-\infty}^{0} e^{s \cdot \frac{1}{2} f''(u_o) u^2} \, du
$$

$$
\leq e^{sf(u_o)} \cdot e^{s \cdot \frac{1}{2} f''(u_o) \cdot u_o^2} \int_{-\infty}^{\infty} e^{s \cdot \frac{1}{2} f''(u_o) u^2} \, du
$$

$$
= e^{sf(u_o)} \cdot e^{s \cdot \frac{1}{2} f''(u_o) \cdot u_o^2} \cdot \frac{\sqrt{2\pi}}{\sqrt{f''(u_o)}} \cdot \frac{1}{\sqrt{s}}
$$

Thus, the integral over $(-\infty, 0]$ has an additional exponential decay by comparison to the integral over the whole line, so the leading term of the asymptotics of the integral from 0 to $+\infty$ is the same as those of the integral from $-\infty$ to $+\infty$.

The case of $\Gamma(s)$ can be converted to this situation as follows. For *real s* > 0, in the integral

$$
s \cdot \Gamma(s) = \Gamma(s + 1) = \int_0^{\infty} e^{-u} u^s \, du = \int_0^{\infty} e^{-u + s \log u} \, du
$$

can replace u by su, to put the integral into the desired form

$$
s \cdot \Gamma(s) = \int_0^{\infty} e^{-su + s \log u + s \log s} s \, du = s \cdot e^{s \log s} \int_0^{\infty} e^{-s(u + \log u)} \, du
$$

For complex s with $\mathrm{Re}(s) > 0$, both $s \cdot \Gamma(s)$ and the integral $s \cdot e^{s \log s} \int_0^{\infty} e^{-s(u + \log u)} \, du$ are holomorphic in s, and they *agree* for *real s*. Thus, by the identity principle, they are equal for $\mathrm{Re}(s) > 0$.

16.5 Laplace's Method Illustrated on Bessel Functions

Consider the standard integral

$$
K_v(y) = \frac{1}{2} \int_0^{\infty} e^{(u + \frac{1}{u})y/2} u^v \frac{du}{u}
$$

The function K_v is variously called a *Bessel function of imaginary argument* or *MacDonald's function* or *modified Bessel function of third kind*. Being interested mainly in the case that the parameter v here is purely imaginary, in the text we replace v by iv. The leading factor of $\sqrt{y}$ arises in applications, where such an integral appears as a *Whittaker function*.

$$f(y) = \sqrt{y} \int_0^\infty e^{-(u+\frac{1}{u})y} u^{iv} \frac{du}{u}$$

The exponent $-(u + \frac{1}{u})y$ is of the desired form, with the earlier s replaced by y, but the u^{iv} in the integrand does not fit into the simpler Laplace' method. Thus, consider integrals

$$\int_0^\infty e^{-sf(u)} g(u) \, du$$

where f is real-valued, but g may be complex-valued. The idea still is that the minimum values of $f(u)$ should dominate, and the leading term of the asymptotics should be (assuming a *unique* minimum at u_o)

$$\int_0^\infty e^{-s \cdot f(u)} g(u) \, du \sim e^{-sf(u_o)} \cdot \frac{\sqrt{2\pi} \cdot g(u_o)}{\sqrt{f''(u_o)}} \cdot \frac{1}{\sqrt{s}}$$

As in the simpler case, reduce this to Watson's lemma by breaking the integral where f' changes sign, and change variables to convert each fragment to a Watson-lemma integral. Thus, the first part of the story is much as the simple case of Laplace's method. As in the simple case of Laplace's method, the course of the argument reveals conditions on f and g.

For simplicity, assume that there is exactly *one* point u_o at which $f'(u_o) = 0$, that $f''(u_o) > 0$ and that $f(u)$ goes to $+\infty$ at 0^+ and at $+\infty$. Thus, on these intervals there are smooth square roots $\sqrt{f(u) - f(u_o)}$ and smooth functions F, G such that

$$\begin{cases} F(\sqrt{f(u) - f(u_o)}) = u & \text{(for } u_o < u < +\infty) \\[2mm] G(\sqrt{f(u) - f(u_o)}) = u & \text{(for } 0 < u < u_o) \end{cases}$$

Then, letting $x = \sqrt{f(u) - f(u_o)}$ in each of the two intervals, so that $F(x) = u$ and $G(x) = u$, respectively,

$$\int_0^\infty e^{-sf(u)} g(u) \, du$$

$$= e^{-sf(u_o)} \int_0^{u_o} e^{-s(f(u) - f(u_o))} g(u) \, du + e^{-sf(u_o)} \int_{u_o}^\infty e^{-s(f(u) - f(u_o))} g(u) \, du$$

$$= e^{-sf(u_o)} \left(\int_0^\infty e^{-sx^2} g(F(x)) F'(x) \, dx + \int_0^\infty e^{-sx^2} g(G(x)) G'(x) \, dx \right)$$

In both integrals, replacing x by $\sqrt{x}$ gives Watson's lemma integrals

$$
\int_0^\infty e^{-sf(u)} g(u)\, du = e^{-sf(u_o)} \left(\int_0^\infty e^{-sx} \frac{\sqrt{x}}{2} g(F(\sqrt{x})) F'(\sqrt{x}) \frac{dx}{x} \right.
$$
$$
\left. + \int_0^\infty e^{-sx} \frac{\sqrt{x}}{2} g(G(\sqrt{x})) G'(\sqrt{x}) \frac{dx}{x} \right)
$$

The assumptions $f'(u_o) = 0$ and $f''(u_o) > 0$ ensure that F has a Taylor series expansion near 0, which gives an expansion

$$
\frac{\sqrt{x}}{2} g(F(\sqrt{x})) F'(\sqrt{x}) = x^{1/2} \cdot \left(\frac{g(F(0)) F'(0)}{2} + \cdots \right)
$$

for small $x > 0$. From this, the main term of the Watson's lemma asymptotics for the integral involving F would be

$$
\int_0^\infty e^{-sx} \frac{\sqrt{x}}{2} g(F(\sqrt{x})) F'(\sqrt{x}) \frac{dx}{x} \sim \frac{\Gamma(\frac{1}{2}) g(F(0)) F'(0)}{2} \cdot \frac{1}{\sqrt{s}}
$$

Note that $F(0) = u_o$. Determine $F'(0)$ from $F(x) = u$. First, $F'(x) \cdot \frac{dx}{du} = 1$. Since

$$
x = \sqrt{f(u) - f(u_o)} = \sqrt{(u - u_o)^2 \cdot \frac{f''(u_o)}{2!} + \cdots}
$$
$$
= \sqrt{\frac{f''(u_o)}{2}} \cdot ((u - u_o) + \cdots)
$$

the derivative is

$$
\frac{dx}{du} = \sqrt{\frac{f''(u_o)}{2}} \cdot (1 + O(u - u_o))
$$

Thus,

$$
F'(x) = \frac{1}{\frac{dx}{du}} = \sqrt{\frac{2}{f''(u_o)}} \cdot (1 + O(u - u_o))
$$

which allows evaluation at $x = 0$, namely

$$
F'(0) = \sqrt{\frac{2}{f''(u_o)}}
$$

The same argument applied to G gives $G'(0) = F'(0)$. Thus,

$$\int_0^\infty e^{-s\,f(u)}\,du \sim e^{-sf(u_o)} \cdot \frac{\Gamma(\frac{1}{2}) \cdot 2 \cdot g(u_o) \cdot \sqrt{\frac{2}{f''(u_o)}}}{2\sqrt{s}}$$

$$= e^{-sf(u_o)} \cdot \frac{\sqrt{2\pi} \cdot g(u_o)}{\sqrt{f''(u_o)}} \cdot \frac{1}{\sqrt{s}}$$

Returning to

$$f(y) = \sqrt{y} \int_0^\infty e^{-(u+\frac{1}{u})y}\, u^{iv}\, \frac{du}{u}$$

we have critical point $u_o = 1$, and $f(u_o) = 2$ and $f''(u_o) = 2$. Applying the just derived asymptotic,

$$f(y) \sim \sqrt{y} \cdot \left(e^{-2y} \cdot \frac{\sqrt{2\pi} \cdot 1^{iv}}{\sqrt{2}} \cdot \frac{1}{\sqrt{y}} \right) = \sqrt{\pi} \cdot e^{-2y}$$

as $y \to +\infty$. Even though the exponent is plausible, it would be easy to lose track of the power of y, which might matter. Also, the leading constant does not depend on the index v.

16.6 Regular Singular Points Heuristic: Freezing Coefficients

Differential equations

$$x^2 u'' + bxu' + cu = 0 \qquad \text{(with constants } b, c)$$

have easy-to-understand solutions on $(0, +\infty)$: linear combinations of x^α, x^β for α, β solutions of the indicial equation

$$X(X-1) + bX + c = 0$$

when the roots are distinct. Therefore, it is reasonable to imagine that a differential equation

$$x^2 u'' + xb(x)u' + c(x)u = 0$$

with b, c *analytic* near 0 has solutions *asymptotic*, as $x \to 0^+$, to solutions of the differential equation $x^2 u'' + b(0)xu' + c(0)u = 0$ obtained by *freezing* the coefficients $b(x)$, $c(x)$ of the original at $x = 0^+$. That is, solutions of the variable-coefficient equation should be asymptotic to x^α for solutions α to the *indicial equation* $X(X-1) + b(0)X + c(0) = 0$. An equation of that form, with b, c analytic near 0, is said to have a *regular singular point* at 0. The discussion below explains the behavior of solutions to such equations.

We give a useful example from the non-Euclidean geometry on the upper half-plane. Recall the $SL_2(\mathbb{R})$-invariant Laplacian on the upper half-plane $\mathfrak{H}$ is

$$\Delta^{\mathfrak{H}} = y^2 \left(\frac{\partial^2}{\partial x^2} + \frac{\partial^2}{\partial y^2} \right)$$

[16.6.1] Translation-Equivariant Eigenfunctions: We ask for $\Delta^{\mathfrak{H}}$-eigen-functions $f(z)$ of the special form

$$f(x + iy) = e^{2\pi i x} u(y)$$

That is, such an eigenfunction is *equivariant* under *translations*: with $t \in \mathbb{R}$ and $z \in \mathfrak{H}$,

$$f(z + t) = e^{2\pi i(x+t)} u(y) = e^{2\pi i t} \cdot \left(e^{2\pi i x} u(y) \right) = e^{2\pi i t} \cdot f(z)$$

The eigenfunction condition is the partial differential equation

$$(\Delta^{\mathfrak{H}} - \lambda) \, e^{2\pi i x} u(y) = 0$$

Since the dependence on x is completely specified, this partial differential equation simplifies to the ordinary differential equation[1]

$$y^2 u'' - (4\pi^2 y^2 + \lambda)u = 0$$

The point $y = 0$ is *not* an *ordinary point* for this equation because in the form

$$u'' - \left(4\pi^2 + \frac{\lambda}{y^2} \right) u = 0$$

the coefficient of u has a pole at 0. But $y = 0$ is a *regular singular point* because that pole is of order at most 2. Thus, following the idea to freeze $y^2 u'' + y b(y) u' + c(y)$ to $y^2 u'' + y b(0) u' + c(0)u$, the indicial equation of the frozen equation is

$$X(X - 1) - \lambda = 0$$

Expressing λ as $\lambda = s(s - 1)$, the roots of the indicial equation are $s, 1 - s$. The frozen equation has distinct solutions y^s and y^{1-s} for $s \neq \frac{1}{2}$. Thus, we could hope that solutions would have asymptotics as $y \to 0^+$ beginning

$$u(y) = Ay^s(1 + O(y)) + By^{1-s}(1 + O(y)) \qquad (\text{as } y \to 0^+)$$

Indeed, this is the case, as we see below. It seems more difficult to obtain the asymptotics at 0^+ from *integral representations* of solutions of the differential equation.

---

[1] This equation is a type of *Bessel* equation, with solutions that are K-type and I-type Bessel functions.

[16.6.2] Remark: As we discuss subsequently, $y^2 u'' - (4\pi^2 y^2 + \lambda)u = 0$ has an *irregular* singular point at $+\infty$, so other methods are needed to obtain asymptotics for solutions as $y \to +\infty$.

[16.6.3] Remark: Up to choices of normalizations, the function u above, depending on the spectral parameter λ or s, is called a *Whittaker function* or *Bessel function*, enjoying an enormous literature. Here, we wish to have direct access to their properties, as instances of general phenomena.

[16.6.4] An Irregular Singular Point: For the translation-equivariant eigenfunctions on $\mathfrak{H}$, we check that $y = +\infty$ is *not* an ordinary point nor a regular singular point: given

$$u'' - \left(4\pi^2 + \frac{\lambda}{y^2}\right)u = 0$$

again let $u(x) = v(1/x)$ and put $z = 1/x$, obtaining

$$\left(z^4 v'' + 2z^3 v'\right) - (4\pi^2 + \lambda z^2)v = 0$$

or

$$z^2 v'' + 2z v' - \left(\frac{4\pi^2}{z^2} + \lambda\right)v = 0$$

Since the coefficient of v has a pole at $z = 0$, this equation falls outside the present discussion. Instead, a different *freezing* idea succeeds: letting $y \to +\infty$ freezes the original equation at $+\infty$, giving a *constant-coefficient* equation

$$u'' - 4\pi^2 u = 0$$

with easily understood solutions $e^{\pm 2\pi y}$. Happily the solutions to the original equation *do* have asymptotics with main terms $e^{\pm 2\pi y}$. Details and proofs are given later, in a discussion of *irregular singular points*.

16.7 Regular Singular Points

A homogeneous ordinary differential equation of the form

$$x^2 u'' + x b(x) u' + c(x) u = 0 \qquad \text{(with } b, c \text{ analytic near 0)}$$

is said to have a *regular singular point* at 0. Similarly, with b, c analytic near x_o,

$$(x - x_o)^2 u'' + (x - x_o)b(x)u' + c(x)u = 0$$

has a regular singular point at x_o. Obviously it suffices to treat $x_o = 0$ and is notationally convenient. The coefficients in an expansion of the form

$$u(x) = x^\alpha \cdot \sum_{n=0}^{\infty} a_n x^n \qquad \text{(with } a_0 \neq 0, \alpha \in \mathbb{C})$$

are determined recursively, but we see subsequently that this recursion succeeds only when α satisfies the *indicial equation*

$$\alpha(\alpha - 1) + b(0)\alpha + c(0) = 0$$

Further, when the two roots α, α' of the indicial equation have a relation $n + \alpha - \alpha' = 0$ for $0 < n \in \mathbb{Z}$, the recursion for α may fail, although the recursion for α' will succeed. These conditions are easily discovered, as in the following discussion. The convergence of the recursively defined series is important both because it produces a genuine function, and because it can be differentiated termwise, by Abel's theorem.

[16.7.1] The Recursion: The equation is

$$x^{\alpha+2} \cdot \sum_{n=0}^{\infty} (n + \alpha)(n + \alpha - 1)a_n x^{n-2} + b(x)x^{\alpha+1} \sum_{n=0}^{\infty} (n + \alpha)a_n x^{n-1}$$

$$+ c(x)x^\alpha \sum_{n=0}^{\infty} a_n x^n = 0$$

Dividing through by x^α and grouping,

$$\sum_{n=0}^{\infty} (n + \alpha)(n + \alpha - 1)a_n x^n + b(x) \sum_{n=0}^{\infty} (n + \alpha)a_n x^n + c(x) \sum_{n=0}^{\infty} a_n x^n = 0$$

The vanishing of the sum of coefficients of x^0, and $a_0 \neq 0$, give the *indicial equation*. The coefficients a_n with $n > 0$ are obtained recursively, from the expected

$$\left[(n + \alpha)(n + \alpha - 1) + b(0)(n + \alpha) + c(0)\right] \cdot a_n$$
$$= \text{(in terms of } a_0, a_1, \ldots, a_{n-1})$$

The coefficient of a_n simplifies by invoking the indicial equation and the fact that the sum of the two roots α, α' is $1 - b(0)$:

$$(n + \alpha)(n + \alpha - 1) + b(0)(n + \alpha) + c(0) = n(n + (2\alpha - 1) + b(0))$$
$$= n(n + \alpha - \alpha')$$

That is, for $n > 0$,

$$n(n + \alpha - \alpha') \cdot a_n = \text{(in terms of } a_0, a_1, \ldots, a_{n-1})$$

Since $n > 0$, the recursion can fail only when

$$n + \alpha - \alpha' = 0 \qquad \text{(for some } 0 < n \in \mathbb{Z})$$

[16.7.2] Convergence: To complete the proof of existence, we prove convergence. Let $A, M \geq 1$ be large enough so that

$$b(x) = \sum_{n \geq 0} b_n x^n \qquad \text{(with } |b_n| \leq A \cdot M^n)$$

$$c(x) = \sum_{n \geq 0} c_n x^n \qquad \text{(with } |c_n| \leq A \cdot M^n)$$

Inductively, suppose that $|a_\ell| \leq (CM)^\ell$, with a constant $C \geq 1$ to be determined in the following. Then

$$|n(n + \alpha - \alpha') \cdot a_n| \leq A \sum_{i=1}^{n} |n - i + \alpha| M^i \cdot (CM)^{n-i} + A \sum_{i=1}^{n} M^i \cdot (CM)^{n-i}$$

$$\leq AM^n C^{n-1} \left(\frac{n(n+1)}{2} + n|\alpha| + n \right)$$

Dividing through by $n|n + \alpha - \alpha'|$, this is

$$|a_n| \leq AM^n \cdot C^{n-1} \frac{(n+1) + 2|\alpha| + 2}{2|n + \alpha - \alpha'|}$$

This motivates the choice

$$C \geq \sup_{1 \leq n \in \mathbb{Z}} \frac{(n+1) + 2|\alpha| + 2}{2|n + \alpha - \alpha'|}$$

which gives $|a_n| \leq A(CM)^n$, and a positive radius of convergence.

16.8 Regular Singular Points at Infinity

With $u(x) = v(1/x)$,

$$u'(x) = \frac{-1}{x^2} v'(1/x) \quad \text{and} \quad u''(x) = \frac{1}{x^4} v''(1/x) + \frac{2}{x^3} v'(1/x)$$

Putting $z = 1/x$, this is

$$u' = -z^2 v' \quad \text{and} \quad u'' = z^4 v'' + 2z^3 v'$$

A differential equation $u'' + p(x)u' + q(x)u = 0$ becomes

$$(z^4 v'' + 2z^3 v') + p(x)(-z^2 v') + q(x)v = 0$$

or

$$z^2 v'' + z\left(2 - \frac{p(1/z)}{z}\right)v' + \frac{q(1/z)}{z^2}v = 0$$

The point $z = 0$ is a *regular singular point* when the coefficients

$$2 - \frac{p(1/z)}{z} \qquad \frac{q(1/z)}{z^2}$$

are analytic at 0. That is, $z = 0$ is a regular singular point when p, q have expansions of the forms

$$\begin{cases} p\left(\dfrac{1}{z}\right) = p_1 z + p_2 z^2 + \cdots \\[2mm] q\left(\dfrac{1}{z}\right) = q_2 z^2 + q_3 z^3 + \cdots \end{cases}$$

or, equivalently,

$$\begin{cases} p(x) = \dfrac{p_1}{x} + \dfrac{p_2}{x^2} + \cdots \\[2mm] q(x) = \dfrac{q_2}{x^2} + \dfrac{q_3}{x^3} + \cdots \end{cases}$$

16.9 Example Revisited

Returning to the earlier example from the upper half-plane: we ask for $\Delta = \Delta^{\mathfrak{H}}$ eigenfunctions $f(z)$ of the special form

$$f(x + iy) = e^{2\pi i x} u(y)$$

The equation $(\Delta - \lambda)f = 0$ simplifies to the ordinary differential equation

$$y^2 u'' - (4\pi^2 y^2 + \lambda)u = 0$$

with regular singular point at $y = 0$. The indicial equation is

$$X(X - 1) - \lambda = 0$$

With $\lambda = s(s - 1)$, the roots of the indicial equation are $s, 1 - s$. By now we know that, unless $s - (1 - s)$ is an integer, the equation has solutions of the form

$$u_s(y) = y^s \cdot \sum_{\ell \geq 0} a_\ell y^\ell \qquad\qquad u_{1-s}(y) = y^{1-s} \cdot \sum_{\ell \geq 0} b_\ell y^\ell$$

with coefficients a_ℓ and b_ℓ determined by the natural recursions. We emphasize that these power series *have positive radius of convergence*, so certainly give asymptotics as $y \to 0^+$. Further, convergent series can be *differentiated* termwise, by Abel's theorem.

We execute a few steps of the recursion for the coefficients for y^s. The equation

$$\sum_{\ell \geq 0} (\ell + s)(\ell + s - 1) a_\ell \, y^\ell - (4\pi^2 y^2 + \lambda) \sum_{\ell \geq 0} a_\ell \, y^\ell = 0$$

simplifies to

$$\ell(\ell + 2s - 1)\, a_\ell = 4\pi^2 a_{\ell-2} \qquad (\text{for } \ell \geq 1)$$

with $a_{-1} = 0$ by convention, and $a_0 = 1$. Thus, the odd-degree terms are all 0, and

$$u_s(y) = y^s \cdot \left(1 + \frac{4\pi^2 y^2}{2(1 + 2s)} + \frac{(4\pi^2)^2 \, y^4}{2(1 + 2s) \cdot 4(3 + 2s)} + \cdots \right)$$

Similarly, replacing s by $1 - s$,

$$u_{1-s}(y) = y^{1-s} \cdot \left(1 + \frac{4\pi^2 y^2}{2(3 - 2s)} + \frac{(4\pi^2)^2 \, y^4}{2(3 - 2s) \cdot 4(5 - 2s)} + \cdots \right)$$

For $\mathrm{Re}\,(s) \neq \frac{1}{2}$, one of these solutions is obviously asymptotically larger than the other. For $\mathrm{Re}\,(s) = \frac{1}{2}$, they are the same size, so some cancellation can occur. Write $s = \frac{1}{2} + iv$, so $1 - s = \frac{1}{2} - iv$, and rewrite the expansions in those coordinates:

$$\begin{cases} u_{\frac{1}{2}+iv}(y) = y^{\frac{1}{2}+iv} \cdot \left(1 + \dfrac{\pi^2 y^2}{(1 + iv)} + \dfrac{\pi^4 y^4}{(1 + iv) \cdot 2(2 + iv)} + \cdots \right) \\[2ex] u_{\frac{1}{2}-iv}(y) = y^{\frac{1}{2}-iv} \cdot \left(1 + \dfrac{\pi^2 y^2}{(1 - iv)} + \dfrac{\pi^4 y^4}{(1 - iv) \cdot 2(2 - iv)} + \cdots \right) \end{cases}$$

For example,

$$\begin{cases} u_{\frac{1}{2}+iv} + u_{\frac{1}{2}-iv} = 2y^{\frac{1}{2}} \cos(\log y) + O(y^{\frac{3}{2}}) \\[2ex] u_{\frac{1}{2}+iv} - u_{\frac{1}{2}-iv} = 2y^{\frac{1}{2}} \sin(\log y) + O(y^{\frac{3}{2}}) \end{cases}$$

Further, behavior of the higher terms as functions of v is clear.

16.10 Irregular Singular Points

According to Erdélyi, Thomé found that differential equations with *finite rank* irregular singular points have asymptotic expansions given by the expected recursions. Thus, although the irregularity typically precludes *convergence* of the series expression for solutions, the series is still a legitimate *asymptotic expansion*.

We approximately follow Erdélyi in treating a rank-one irregular singular point of a second-order differential equation: after normalization to get rid of the first-derivative term, these are of the form

$$u'' - q(x)\,u = 0 \quad \text{with} \quad q(x) \sim q_o + \frac{q_1}{x} + \frac{q_2}{x^2} + \cdots$$

as $x \to +\infty$, with $q_o \neq 0$, with q continuous in some range $x \geq a$. The series expression for $q(x)$ need not be convergent: it suffices that it be an *asymptotic* expansion of $q(x)$ at $+\infty$. *Freezing* the coefficient q to its value at $+\infty$, gives the *constant-coefficient* equation

$$u'' - q_o\,u = 0$$

and *suggests* that the solutions $e^{\pm\sqrt{q_o}\,x}$ of the constant-coefficient equation should give the leading term in the asymptotics of solutions of the original equation. This is approximately true: there is an adjustment by a power of x. Solutions have asymptotics of the form

$$u(x) \sim e^{\pm\sqrt{q_o}\,x} \cdot x^\rho \cdot \left(1 + \sum_{n \geq 1} \frac{a_n}{x^n} \right)$$

with $\rho = \dfrac{q_1}{\pm 2\sqrt{q_o}}$, as $x \to +\infty$, with coefficients a_n obtained by a natural recursion. However, the series rarely converges.

The loss of convergence is not a trifling matter. The termwise differentiability of convergent power series is extremely useful. In contrast, termwise differentiation of *asymptotic* series

$$f(x) \sim \sum_{n \geq 0} \frac{a_n}{x^n} \qquad (\text{as } x \to +\infty)$$

for differentiable f produces an asymptotic series for f' only under additional hypotheses, for example, that f' admits such an asymptotic series. (See the Appendix.) While a function admitting an asymptotic expansion of this form determines that expansion uniquely, the expansion does not uniquely determine the function. For example, as $x \to +\infty$, $e^{-x} = o(x^{-N})$ for all N, so e^{-x} has the 0 asymptotic expansion but is not the 0 function.

[16.10.1] Example: Rotationally Symmetric Eigenfunctions on $\mathbb{R}^n$: A natural example arises from the eigenvalue equation for the radial component of the Euclidean Laplacian on $\mathbb{R}^n$:

$$v'' + \frac{n-1}{r}v' - \lambda v = 0$$

For large r, this equation resembles the constant-coefficient equation $v'' - \lambda v = 0$, with solutions $e^{\pm r\sqrt{\lambda}}$. Heuristically, we should have solutions with behavior $v \sim e^{\pm r\sqrt{\lambda}}$ as $v \to +\infty$. This is not quite right: the true asymptotic expansions have main terms

$$v \sim \frac{e^{\pm r\sqrt{\lambda}}}{r^{\frac{n-1}{2}}}$$

That is, the differences between the actual equation and the constant-coefficient approximation do not alter the constant in the exponential, but do have a significant impact, as we see subsequently.

A natural recursion produces an apparent solution to differential equations in this class, of the form

$$e^{\omega x} x^{-\rho} \sum_{n \geq 0} \frac{c_n}{x^n}$$

However, unlike the regular singular point situation, the series is *not convergent*! The relation of this nonconvergent series to any genuine solution is a priori unclear. It is natural to suppose that this non-convergent series is an *asymptotic expansion*, but this is not obvious. A genuine solution must be identified by other means, must be proven to *have* an asymptotic expansion, and the latter must be compared with the series obtained by the recursion. All this will occupy us in following sections.

[16.10.2] Recursion: In more detail, the heuristic recursion is as follows, as applied to the eigenvalue equation for the radial component of the Laplacian on $\mathbb{R}^n$. First, simplify by employing the standard device to eliminate the v' term[2]: take $v = u/r^{(n-1)/2}$, and then

$$u'' - \left(\lambda + \frac{(n-1)(n-3)}{4r^2} \right) u = 0$$

The singular point at infinity is *irregular*, unless $n = 1$ or 3. Nevertheless, intuitively, for $x \to +\infty$, a differential equation of the shape

$$u'' - (\lambda + \frac{C}{x^2})u = 0$$

is approximately a constant-coefficient differential equation, suggesting a

---

[2] Let $v = u \cdot w$ and set the u' term equal to 0 in the left-hand side. This gives $2u'w' + \frac{n-1}{r}u'w = 0$, which gives the differential equation $2w' + \frac{n-1}{r}w = 0$ for w.

solution of the form[3]

$$u(x) = e^{\pm x\sqrt{\lambda}} \cdot \sum_{\ell=0}^{+\infty} \frac{c_\ell}{x^\ell} \qquad (c_0 = 1 \text{ without loss of generality})$$

Substituting the latter into the differential equation and dividing through by $e^{\pm x\sqrt{\lambda}}$, letting $s = \pm\sqrt{\lambda}$, simplifies to

$$\sum_{\ell=0}^{+\infty} ((\ell - 2)(\ell - 1)c_{\ell-2} - 2s(\ell - 1)c_{\ell-1})\frac{1}{x^\ell} - \sum_{\ell=0}^{+\infty} c_{\ell-2}\frac{C}{x^\ell} = 0$$

where by convention $c_{-1} = c_{-2} = 0$. The case $\ell = 0$ is vacuous, as is $\ell = 1$. The case $\ell \geq 2$ determines $c_{\ell-1}$, assuming $s \neq 0$:

$$(\ell - 2)(\ell - 1)c_{\ell-2} - 2s(\ell - 1)c_{\ell-1} - C \cdot c_{\ell-2} = 0$$

or

$$c_{\ell+1} = \frac{\ell(\ell + 1) - C}{2s(\ell + 1)} \cdot c_\ell$$

[16.10.3] Remark: That recursion causes the coefficients to grow approximately as factorials, and the resulting series *does not converge* for any finite nonzero value of $1/x$, unless the constant C happens to be of the form $(\ell - 1)(\ell - 2)$ for some positive integer ℓ, in which case the series *terminates*, and *is* convergent.

Our later discussion will show that the preceding recursion *does* correctly determine *asymptotic expansions* for solutions. In particular, the leading part of the asymptotic is

$$v = \frac{e^{\pm r\sqrt{\lambda}}}{r^{\frac{n-1}{2}}} \cdot \left(1 + O\left(\frac{1}{r}\right)\right) \qquad (\text{as } r \to +\infty, \text{ in } \mathbb{R}^n)$$

The denominator $r^{(n-1)/2}$ might be hard to anticipate. The symmetry $r \to -r$ imposes a further requirement, and for $\sqrt{\lambda}$ not purely imaginary one of the two solutions swamps the other. Indeed, for $\sqrt{\lambda}$ not purely imaginary, the asymptotic components of the large solution are all larger than the main part of the smaller solution. Further, this is an *asymptotic* and not merely a *bound*.

In fact, for n odd, the asymptotic is *finite*: the recursion for coefficients terminates, so gives a convergent series: we obtain not merely *asymptotics*, but

---

[3] Anticipating the adjustment by x^ρ in general, with ρ determined by the asymptotics $q_o + \frac{q_1}{x} + \cdots$ of the coefficient of u by $\rho = q_1/2\sqrt{q_o}$, in the present example we are fortunate that $q_1 = 0$, so the idea of *freezing* is exactly right.

equalities. Thus, in odd-dimensional $\mathbb{R}^n$ the solutions to the differential equation for rotationally invariant λ-eigenfunctions have elementary expressions. For example,

$$\begin{cases} v = e^{\pm r\sqrt{\lambda}} & \text{(on } \mathbb{R}^1) \\[2mm] v = \dfrac{e^{\pm r\sqrt{\lambda}}}{r} & \text{(on } \mathbb{R}^3) \\[2mm] v = e^{\pm r\sqrt{\lambda}}\left(\dfrac{1}{r^2} - \dfrac{1}{\pm r^3\sqrt{\lambda}}\right) & \text{(on } \mathbb{R}^5) \\[2mm] v = e^{\pm r\sqrt{\lambda}}\left(\dfrac{1}{r^3} - \dfrac{3}{\pm r^5\sqrt{\lambda}} + \dfrac{3}{r^7\lambda}\right) & \text{(on } \mathbb{R}^7) \end{cases}$$

[16.10.4] Remark: The same technique applies to differential equations

$$u'' - q(x)\,u = 0$$

with $q(x)$ continuous in some range $x > a$ and admitting an *asymptotic expansion* at infinity of the form

$$q(x) \sim \sum_{\ell \geq 0} \frac{q_\ell}{x^\ell} \qquad \text{(with } q_o \neq 0)$$

The condition $q_o \neq 0$ is essential[4] for the recursion to succeed. Adjustment by x^ρ with $\rho = q_1/2\sqrt{q_o}$ would be found necessary when $q_1 \neq 0$. In any case, the recursion *rarely* produces a convergent power series!

[16.10.5] Comparison to Regular Singular Points: The behavior of the above recursion in [16.10.2] is much different from that resulting from a regular singular point. A power series in $z = 1/x$ behaves differently under d/dx than under d/dz. Indeed, as in the foregoing example, the power series in $1/x$ often diverges, while at a regular singular point the analogous power series has a positive radius of convergence. For $u'' - q(x)u = 0$ to have a regular singular point at infinity, changing variables to $u(x) = v(1/x)$ and $z = 1/x$,

$$u'(x) = \frac{-1}{x^2}v'(1/x) \qquad \text{and} \qquad u''(x) = \frac{1}{x^4}v''(1/x) + \frac{2}{x^3}v'(1/x)$$

Putting $z = 1/x$, this is

$$u' = -z^2 v' \qquad \text{and} \qquad u'' = z^4 v'' + 2z^3 v'$$

---

[4] The condition $q_o \neq 0$ and the assumption that q has the indicated asymptotic at $+\infty$ together imply that there is x_o such that $q(x) \neq 0$ for $x \geq x_o$. That is, in the regime $x \geq x_o$ there are no *transition points.*

Thus, in the coordinate z at infinity, the differential equation becomes

$$(z^4 v'' + 2z^3 v') - q\left(\frac{1}{z}\right) v = 0$$

or

$$v'' + \frac{2}{z} v' - \frac{q(1/z)}{z^4} v = 0$$

The point $z = 0$ is *never* an *ordinary point*, because of the pole in the coefficient of v'. The point $z = 0$ is a regular singular point only when $q(1/z)/z^2$ is analytic at $z = 0$, that is, when $x^2 q(x)$ is analytic at ∞. This requires that $q(x)$ have the form

$$q(x) = \frac{q_2}{x^2} + \frac{q_3}{x^3} + \cdots$$

16.11 Example: Translation-Equivariant Eigenfunctions on $\mathfrak{H}$

Another example of irregular singular point arises from the $SL_2(\mathbb{R})$-invariant Laplacian on the upper half-plane $\mathfrak{H}$:

$$\Delta^{\mathfrak{H}} = y^2 \left(\frac{\partial^2}{\partial x^2} + \frac{\partial^2}{\partial y^2} \right)$$

We ask for $\Delta^{\mathfrak{H}}$-eigenfunctions $f(z)$ of the special form

$$f(x + iy) = e^{2\pi i x} u(y)$$

that is, *equivariant* under *translations*: with $t \in \mathbb{R}$ and $z \in \mathfrak{H}$,

$$f(z + t) = e^{2\pi i (x+t)} u(y) = e^{2\pi i t} \cdot (e^{2\pi i x} u(y)) = e^{2\pi i t} \cdot f(z)$$

The eigenfunction condition

$$(\Delta^{\mathfrak{H}} - \lambda) e^{2\pi i x} u(y) = 0$$

simplifies to the ordinary differential equation

$$y^2 u'' - (4\pi^2 y^2 + \lambda) u = 0$$

This equation has an *irregular* singular point at $+\infty$, seen by changing coordinates, as follows. Let $u(x) = v(1/x)$ and put $z = 1/x$, obtaining

$$(z^4 v'' + 2z^3 v') - (4\pi^2 + \lambda z^2) v = 0$$

or

$$z^2 v'' + 2z v' - \left(\frac{4\pi^2}{z^2} + \lambda \right) v = 0$$

Since the coefficient of v has a pole at $z = 0$, the singular point of this equation in the new coordinate z at 0 is *irregular*.

[16.11.1] Recursion: Happily, following our present prescription, in the form

$$u'' - \left(4\pi^2 + \frac{\lambda}{y^2}\right)u = 0$$

the coefficient

$$q(y) = q_o + \frac{q_1}{y} + \frac{q_2}{y^2} + \ldots = 4\pi^2 + \frac{\lambda}{y^2}$$

is analytic at $y = \infty$, and $q(\infty) = q_o = 4\pi^2 \neq 0$ while $q_1 = 0$, so our later discussion justifies *freezing* y at $+\infty$, obtaining the constant-coefficient equation

$$u'' - 4\pi^2 u = 0$$

with solutions $e^{\pm 2\pi y}$, and assuring existence of solutions of the original equation with asymptotics of the form

$$u(y) = e^{\pm 2\pi y} \cdot \sum_{\ell \geq 0} \frac{a_\ell}{y^\ell}$$

Substituting this into the differential equation and dividing through by $e^{\pm 2\pi y}$ gives

$$\sum_{\ell=0}^{+\infty}((\ell - 2)(\ell - 1)a_{\ell-2} \mp 2\pi(\ell - 1)a_{\ell-1})\frac{1}{y^\ell} - \sum_{\ell=0}^{+\infty} a_{\ell-2}\frac{\lambda}{y^\ell} = 0$$

or

$$\pm 2\pi(\ell - 1)a_{\ell-1} = ((\ell - 2)(\ell - 1) - \lambda)a_{\ell-2}$$

or

$$a_\ell = \frac{(\ell - 1)\ell - \lambda}{\pm 2\pi \ell}a_{\ell-1} = \left(\ell - 1 - \frac{\lambda}{\ell}\right)\frac{a_\ell}{\pm 2\pi}$$

As usual, $a_{-2} = a_{-1} = 0$ by convention, and $a_0 = 1$. The cases $\ell \leq 0$ are vacuous. With $a_0 = 1$, the recursion begins

$$a_1 = \frac{-\lambda}{\pm 2\pi}$$

$$a_2 = \left(1 - \frac{\lambda}{2}\right)\frac{a_1}{\pm 2\pi} = \left(1 - \frac{\lambda}{2}\right)(-\lambda)\frac{1}{(\pm 2\pi)^2}$$

$$a_3 = \left(2 - \frac{\lambda}{3}\right)\frac{a_2}{\pm 2\pi} = \left(2 - \frac{\lambda}{3}\right)\left(1 - \frac{\lambda}{2}\right)(-\lambda)\frac{1}{(\pm 2\pi)^3}$$

[16.11.2] Remark: If λ is of the form $\lambda = \ell(\ell - 1)$ for $0 < \ell \in \mathbb{Z}$, the recursion *terminates*. Then the asymptotic expansion is *convergent*, and produces an elementary solution to the eigenfunction equation.[5]

16.12 Beginning of Construction of Solutions

[16.12.1] Heuristic for Asymptotic Expansion: Consider the equation

$$u'' - q(x)\,u = 0$$

as $x \to +\infty$, where q is continuous in some range $x > a$ and itself admits an asymptotic expansion

$$q(x) \sim \sum_{n \geq 0} \frac{q_n}{x^n} \qquad (\text{as } x \to +\infty, \text{ with } q_o \neq 0)$$

The $q_o \neq 0$ condition is essential. We look for a solution of the form

$$u(x) \sim e^{\omega x} \cdot x^{-\rho} \cdot \sum_{n \geq 0} \frac{c_o}{x^n} \qquad (\text{with } c_o \text{ nonzero})$$

Substituting this expansion in the differential equation and dividing through by $e^{\omega x} x^{-\rho}$, setting the coefficient of $1/x^n$ to 0,

$$((\rho + n - 2)(\rho + n - 1)c_{n-2} - 2\omega(\rho + n - 1)c_{n-1} + \omega^2 c_n)$$
$$- (q_o c_n + q_1 c_{n-1} + \cdots + q_{n-1} c_1 + q_n c_o) = 0$$

By convention, $c_{-2} = c_{-1} = 0$ and $q_{-2} = q_{-1} = 0$. For $n = 0$, the relation is

$$\omega^2 c_o - q_o c_o = 0$$

so $\omega = \pm\sqrt{q_o} \neq 0$, since $c_o \neq 0$. For $n = 1$,

$$(-2\omega\rho c_o + \omega^2 c_1) - (q_o c_1 + q_1 c_o) = 0$$

so, using $\omega^2 = q_o$ and $\omega \neq 0$, this is

$$-2\omega\rho - q_1 = 0$$

so $\rho = -q_1/(2\omega)$. Thus, the choice of $\pm\omega$ is reflected in the choice of $\pm\rho$. For $n \geq 2$, using $\omega^2 = q_o$,

$$(-2\omega(\rho + n - 1) - q_1)c_{n-1}$$
$$= -(\rho + n - 2)(\rho + n - 1)c_{n-2} + (q_2 c_{n-2} + \cdots + q_{n-1} c_1 + q_n c_o)$$

---

[5] These elementary solutions arise from the *finite-dimensional* representations of $SL_2(\mathbb{R})$.

and using $-2\omega\rho - q_1 = 0$,

$$-2\omega(n-1)c_{n-1}$$
$$= -(\rho + n - 2)(\rho + n - 1)c_{n-2} + (q_2 c_{n-2} + \cdots + q_{n-1}c_1 + q_n c_0)$$

Since $\omega \neq 0$, this gives a successful recursion. The following discussion will show that the two asymptotics, with $\pm\omega$ and corresponding $\pm\rho$, are asymptotic expansions of two solutions of the differential equation $u'' - q(x)u = 0$.

[16.12.2] Remark: However, since the preceding expansions usually do not converge, *genuine* solutions must be constructed by other means, and must be shown to *have* asymptotic expansions at $+\infty$.

[16.12.3] Small renormalization: For a solution u to $u'' - q(x)u = 0$, let

$$u(x) = e^{\omega x} \cdot x^{-\rho} \cdot v(x)$$

with ω and ρ determined as earlier. Then

$$u' = e^{\omega x} x^{-\rho} \left(\left(\omega - \frac{\rho}{x} \right)v + v' \right)$$

and

$$u'' = e^{\omega x} x^{-\rho} \left(\omega - \frac{\rho}{x} \right)^2 v + e^{\omega x} x^{-\rho} \frac{\rho}{x^2} v + 2e^{\omega x} x^{-\rho} \left(\omega - \frac{\rho}{x} \right)v' + e^{\omega x} x^{-\rho} v''$$

Dividing through by $e^{\omega x} x^{-\rho}$ gives the differential equation for v, namely,

$$v'' + 2\left(\omega - \frac{\rho}{x} \right)v' + \left(\omega^2 - \frac{2\omega\rho}{x} + \frac{\rho^2 + \rho}{x^2} - q(x) \right)v = 0$$

Unsurprisingly, the ω^2 and $-2\omega\rho/x$ cancel the first two terms of $q(x)$. Thus, the function

$$F(x) = x^2 \cdot \left(\omega^2 - \frac{2\omega\rho}{x} + \frac{\rho^2 + \rho}{x^2} - q(x) \right)$$

is *bounded*. The differential equation is

$$v'' + 2\left(\omega - \frac{\rho}{x} \right)v' + \frac{F(x)}{x^2} v = 0$$

Rewrite the equation as

$$\frac{d}{dx}\left(e^{2\omega x} x^{-2\rho} \frac{dv}{dx} \right) + e^{2\omega x} x^{-2\rho-2} F(x) v(x) = 0$$

Integrate this from $b \geq a$ to $x \geq b$, and multiply through by $e^{-2\omega x} x^{2\rho}$, to obtain

$$\frac{dv}{dx} + e^{-2\omega x} x^{2\rho} \int_b^x e^{2\omega t} t^{-2\rho-2} F(t) v(t) \, dt = \text{const} \cdot e^{-2\omega x} x^{2\rho}$$

Take the constant of integration to be 0 and integrate from a to x, to obtain

$$v(x) + \int_a^x e^{-2\omega s} s^{2\rho} \left(\int_b^s e^{2\omega t} t^{-2\rho-2} F(t) v(t) \, dt \right) ds = \text{const}$$

Rearrange the double integral:

$$\int_a^x e^{-2\omega s} s^{2\rho} \left(\int_b^s e^{2\omega t} t^{-2\rho-2} F(t) v(t) \, dt \right) ds$$

$$= \int_b^x \left(\int_t^x e^{2\omega(t-s)} \left(\frac{s}{t} \right)^{2\rho} ds \right) F(t) v(t) \frac{dt}{t^2}$$

Let $K(x, t)$ denote the inner integral

$$K(x, t) = \int_t^x e^{2\omega(t-s)} \left(\frac{s}{t} \right)^{2\rho} ds$$

Then the equation is

$$v(x) - \int_b^x K(x, t) F(t) v(t) \frac{dt}{t^2} = \text{const}$$

Take the constant to be 1. With $b = +\infty$, this gives an integral equation

$$v(x) = 1 + \int_x^\infty K(x, t) F(t) v(t) \frac{dt}{t^2}$$

We claim that this equation can be solved by *successive approximations*. With the obvious operator

$$T f(x) = \int_x^\infty K(x, t) F(t) f(t) \frac{dt}{t^2}$$

take $w_o(x) = 1$, $w_{n+1} = T w_n$, and then show that the limit

$$v(x) = w_o(x) + w_1(x) + w_2(x) + \cdots = (1 + T + T^2 + \cdots) w_o$$

exists *pointwise*, is *twice differentiable*, and satisfies the differential equation.

16.13 Boundedness of $K(x, t)$

We claim that, with correct choice of $\pm\omega$, the kernel

$$-K(x, t) = \int_x^t e^{2\omega(t-s)} \left(\frac{s}{t} \right)^{2\rho} ds$$

is bounded for $t \geq x \geq a$. Choose $\pm\omega$ so that either $\text{Re}(\omega) < 0$, or $\text{Re}(\omega) = 0$ and $\text{Re}(\rho) \geq 0$.

[16.13.1] Very Easy Case $\rho = 0$: To illustrate the reasonableness of the boundedness assertion, consider the special case $\rho = 0$, where the integral can be computed explicitly:

$$-K(x, t) = \int_x^t e^{2\omega(t-s)}\, ds = \frac{1}{-2\omega}(1 - e^{2\omega(t-x)})$$

Since Re $\omega \leq 0$ and $\omega \neq 0$, this is bounded, for $a \leq x \leq t$.

[16.13.2] Easy Case Re $\omega < 0$**:** When Re $\omega < 0$, absolute value estimates suffice to prove boundedness of $K(x, t)$.

$$|K(x, t)| \leq \int_x^t e^{2\,\mathrm{Re}\,\omega(t-s)} \left(\frac{s}{t}\right)^{2\,\mathrm{Re}\,\rho}\, ds \leq \int_a^t e^{2\,\mathrm{Re}\,\omega(t-s)} \left(\frac{s}{t}\right)^{2\,\mathrm{Re}\,\rho}\, ds$$

Lighten the notation by taking ω, ρ real. For $\rho \geq 0$,

$$\int_a^t e^{2\omega(t-s)} \left(\frac{s}{t}\right)^{2\rho}\, ds \leq \int_0^t e^{2\omega(t-s)}\, ds = e^{2\omega t} \cdot \frac{e^{-2\omega t} - 1}{2|\omega|} \leq \frac{1}{2|\omega|}$$

For $\rho < 0$, still with $\omega < 0$,

$$\int_a^t e^{2\omega(t-s)} \left(\frac{s}{t}\right)^{2\rho}\, ds \leq \int_{t/2}^t e^{2\omega(t-s)} \left(\frac{t/2}{t}\right)^{2\rho}\, ds + \int_0^{t/2} e^{2\omega(t-s)} \left(\frac{1}{t}\right)^{2\rho}\, ds$$

The two integrals are bounded in $t \geq a$, for elementary reasons. Thus, for Re $(\omega) < 0$, the kernel $K(x, t)$ is bounded.

[16.13.3] Re $(\omega) = 0$ and Cancellation: When Re $(\omega) = 0$, absolute value estimates no longer suffice to prove boundedness. Cancellation must be exploited by an integration by parts. Choose $\pm\omega$ so that Re $(\rho) \geq 0$. One integration by parts gives

$$\int_x^t e^{2\omega(t-s)} \left(\frac{s}{t}\right)^{2\rho}\, ds = \left[\frac{e^{2\omega(t-s)}}{-2\omega} \left(\frac{s}{t}\right)^{2\rho}\right]_x^t + \int_x^t \frac{e^{2\omega(t-s)}}{2\omega} \frac{2\rho}{s} \left(\frac{s}{t}\right)^{2\rho}\, ds$$

$$= \frac{1}{-2\omega} - \frac{e^{2\omega(t-x)}}{-2\omega} \left(\frac{x}{t}\right)^{2\rho} + \int_x^t \frac{e^{2\omega(t-s)}}{2\omega} \frac{2\rho}{s} \left(\frac{s}{t}\right)^{2\rho}\, ds$$

The leading terms are bounded for $t \geq x \geq a$. The latter integral can be estimated by absolute values, for Re $\rho \neq 0$:

$$\left| \int_x^t e^{2\omega(t-s)} \frac{1}{s} \left(\frac{s}{t}\right)^{2\rho}\, ds \right| \leq \int_x^t \frac{1}{s} \left(\frac{s}{t}\right)^{2\,\mathrm{Re}\,\rho}\, ds$$

$$= \frac{1}{2\,\mathrm{Re}\,\rho} \left(1 - \left(\frac{x}{t}\right)^{2\,\mathrm{Re}\,\rho}\right)$$

When $\operatorname{Re}\rho = 0$, a second integration by parts gives

$$\int_x^t e^{2\omega(t-s)} \frac{1}{s}\left(\frac{s}{t}\right)^{2\rho} ds$$

$$= \left[\frac{e^{2\omega(t-s)}}{-2\omega}\frac{1}{s}\left(\frac{s}{t}\right)^{2\rho}\right]_x^t + \frac{2\rho-1}{2\omega}\int_x^t e^{2\omega(t-s)}\frac{1}{s^2}\left(\frac{s}{t}\right)^{2\rho} ds$$

The latter integral is estimated by

$$\left|\int_x^t e^{2\omega(t-s)}\frac{1}{s^2}\left(\frac{s}{t}\right)^{2\rho} ds\right| \le \int_x^t \frac{ds}{s^2} = \frac{1}{x} - \frac{1}{t} \le \frac{1}{a}$$

Thus, in all cases, $K(x,t)$ is bounded on $t \ge x \ge a > 0$.

16.14 End of Construction of Solutions

[16.14.1] Bound for T: As we have observed, there is a bound A so that $|F(x)| \le A$ for $x \ge a$. Let $|K(x,t)| \le B$. For f satisfying a bound $|f(x)| \le x^{-\lambda}$ for $x \ge a$, with $\lambda > -1$,

$$|(Tf)(x)| \le \frac{AB}{\lambda+1}x^{-(\lambda+1)} \qquad \text{(for } x \ge a)$$

Indeed,

$$|Tf(x)| = \left|\int_x^\infty K(x,t)\,F(t)\,f(t)\,\frac{dt}{t^2}\right| \le AB\int_x^\infty t^{-(\lambda+2)}\,dt$$

[16.14.2] Bound on f_n: With $f_0 = 1$ and $f_{n+1} = Tf_n$, we claim that

$$|f_n(x)| \le \frac{(AB)^n}{n!}x^{-n} \qquad \text{(for } n = 0, 1, 2, \dots \text{ and } x \ge a)$$

This holds for $n = 0$, and induction using the bound on T gives the result.

[16.14.3] Convergence of the Series: Now we show that the series

$$f(x) = \sum_{n\ge 0} f_n(x) = \sum_{n\ge 0} T^n f_0(x)$$

converges uniformly absolutely and satisfies the integral equation

$$f(x) = 1 + \int_x^\infty K(x,t)\,F(t)\,f(t)\,\frac{dt}{t^2}$$

Uniform absolute convergence in $C^o[a, +\infty)$ follows from the previous estimate. This justifies interchange of summation and integration:

$$Tf(x) = \int_x^\infty K(x, t) F(t) f(t) \frac{dt}{t^2} = \sum_{n \geq 0} \int_x^\infty K(x, t) F(t) T^n f_0(t) \frac{dt}{t^2}$$

$$= \sum_{n \geq 0} T^{n+1} f_0(x) = -1 + \sum_{n \geq 0} T^n f_0(x) = -1 + f(x)$$

Thus, f satisfies the integral equation. Since $K(x, t)$ is differentiable in x, and since the integral for T converges well, the expression

$$f(x) = 1 + \int_x^\infty K(x, t) F(t) f(t) \frac{dt}{t^2}$$

demonstrates the differentiability of f. Further, since $K(x, x) = 0$, the derivative is

$$f'(x) = \int_x^\infty \frac{\partial K(x, t)}{\partial x} F(t) f(t) \frac{dt}{t^2}$$

$$= \int_x^\infty e^{2\omega(t-x)} \left(\frac{x}{t}\right)^{2\rho} F(t) f(t) \frac{dt}{t^2}$$

The integral is again continuously differentiable in x, so f is in C^2.

[16.14.4] Back to the Differential Equation: From the integral expression,

$$f''(x) = -\frac{F(x)}{x^2} f(x) + \int_x^\infty \left(-2\omega + \frac{2\rho}{x}\right) e^{2\omega(t-x)} \left(\frac{x}{t}\right)^{2\rho} F(t) f(t) \frac{dt}{t^2}$$

Substituting into the differential equation,

$$f'' + 2\left(\omega - \frac{\rho}{x}\right) f' + \frac{F}{x^2} f$$

$$= -\frac{F}{x^2} f + \int_x^\infty \left(-2\omega + \frac{2\rho}{x}\right) e^{2\omega(t-x)} \left(\frac{x}{t}\right)^{2\rho} F(t) f(t) \frac{dt}{t^2}$$

$$+ 2\left(\omega - \frac{\rho}{x}\right) \int_x^\infty e^{2\omega(t-x)} \left(\frac{x}{t}\right)^{2\rho} F(t) f(t) \frac{dt}{t^2} + \frac{F}{x^2} f = 0$$

Then

$$u(x) = e^{\omega x} x^{-\rho} f(x)$$

satisfies the original equation

$$u'' - q(x) u = 0$$

[16.14.5] Two Independent Solutions: In the special case that $q_o < 0$ and $q_1 \in \mathbb{R}$, $\omega = \sqrt{\omega}$ has Re $\omega = 0$ and Re $\rho = 0$. In that case, the successive approximation solution to the integral equation can proceed with either values $\pm \omega$, $\pm \rho$, and two linearly independent solutions are obtained.

In all other cases, the successive approximation argument succeeds for only one choice of sign, producing a solution u as we saw earlier. Nevertheless, a second solution can be constructed as follows, by a standard device. Since $f(x) = 1 + O(1/x)$, there is $b \geq a$ large enough so that $u(x) \neq 0$ for $x \geq b$. Then let $v = u \cdot w$, require that v satisfy $v'' - q v = 0$, and see what condition this imposes on w. From

$$v'' - q v = u'' w + 2u' w' + u w'' - q u w = 0$$

using $u'' - q u = 0$, we obtain

$$\frac{w''}{w'} = \frac{-2u'}{u}$$

Then

$$\log w' = -2 \log u + C$$

and

$$w(x) = \int_b^x u(t)^{-2} \, dt$$

Thus, a second solution is

$$u(x) \cdot \int_b^x u(t)^{-2} \, dt$$

That integral is not constant, so the two solutions are linearly independent.

16.15 Asymptotics of Solutions

We show that the solutions on $x \geq a$ have the same asymptotics as the heuristic indicated earlier.

[16.15.1] Some Elementary Asymptotics: Use the standard device $(\rho)_\ell = \rho(\rho + 1) \ldots (\rho + \ell - 1)$ and $(\rho)_0 = 1$. Let $0 \neq \omega \in \mathbb{C}$ with Re $\omega \leq 0$. If Re $\omega = 0$, require that Re $\rho > 1$. Repeated integration by parts and easy estimates yield asymptotic expansions,

$$\begin{cases} \int_x^\infty e^{\omega t} \, t^{-\rho} \, dt & \sim & e^{\omega x} \cdot \sum_{\ell \geq 0} \frac{(\rho)_\ell}{(-\omega)^{\ell+1}} \frac{1}{x^{\rho+\ell}} \\[2ex] \int_b^x e^{-\omega t} \, t^{-\rho} \, dt & \sim & e^{-\omega x} \cdot \sum_{\ell \geq 0} \frac{(\rho)_\ell}{\omega^{\ell+1}} \frac{1}{x^{\rho+\ell}} \end{cases}$$

Since the sup of $|e^{\omega t} t^{-\rho}|$ occurs farther to the right for larger $\mathrm{Re}\,(\rho) < 0$, these asymptotics are *not uniform* in ρ. Note that the boundedness of the kernel $K(x, t)$ proven earlier has a weaker hypothesis than the second asymptotic assertion, requires a slightly more complicated argument and has a weaker conclusion.

[16.15.2] Asymptotics of $T^n f_0$: With $f_0 = 1$, we claim that $f_n = T^n f_0$ has an asymptotic expansion at $+\infty$, of the form

$$ f_n \sim \sum_{\ell \geq n} c_{n\ell}\, x^{-\ell} $$

This holds for $f_0 = 1$. To do the induction step, assume f_n has such an asymptotic expansion. Then $F(x) \cdot f_n(x)$ has a similar expansion

$$ F f_n \sim \sum_{\ell \geq n} b_\ell\, x^{-\ell} $$

because (as the product of two asymptotic expansions in $1/x^n$ is readily shown to be an asymptotic expansion for the product function)

$$ F(x) = x^2 \cdot \left(\omega^2 - \frac{2\omega\rho}{x} + \frac{\rho^2 + \rho}{x^2} - q(x) \right) $$

and q is assumed to have an asymptotic expansion in the functions $1/x^n$ at $+\infty$. We want to insert the asymptotic expansion for $F f_n$ into the integral in the differentiated form of $f_{n+1} = T f_n$, namely, into the equation

$$ f'_{n+1}(x) = \int_x^\infty e^{2\omega(t-x)} \left(\frac{x}{t} \right)^{2\rho} F(t) f_n(t)\, \frac{dt}{t^2} $$

Indeed, from

$$ F(x) f_n(x) - \sum_{n \leq \ell \leq N} b_\ell\, x^{-\ell} = O(x^{-(N+1)}) $$

and from the boundedness of $K(x, t)$ we have

$$ \left| \int_x^\infty e^{2\omega(t-x)} \left(\frac{x}{t} \right)^{2\rho} \left(F(t) f_n(t) - \sum_{n \leq \ell \leq N} b_\ell\, t^{-\ell} \right) \frac{dt}{t^2} \right| $$

$$ = \left| \int_x^\infty e^{2\omega(t-x)} \left(\frac{x}{t} \right)^{2\rho} O(t^{-(N+1)})\, \frac{dt}{t^2} \right| $$

$$ \ll_{\omega, \rho, N} x^{-(N+1)} \int_x^\infty \frac{dt}{t^2} = O(x^{-(N+2)}) = o(x^{-(N+1)}) $$

Thus, the desired asymptotics for f'_{n+1} would follow from asymptotics for the collection

$$\int_x^\infty e^{2\omega(t-x)} \left(\frac{x}{t}\right)^{2\rho} \left(\sum_{n\le\ell\le N} b_\ell\, t^{-\ell}\right) \frac{dt}{t^2} \qquad \text{(for } N \ge n\text{)}$$

As noted earlier,

$$\int_x^\infty e^{\omega t}\, t^{-\rho}\, dt \sim e^{\omega x} \cdot \sum_{\ell\ge 0} \frac{(\rho)_\ell}{(-\omega)^{\ell+1}} \frac{1}{x^{\rho+\ell}}$$

Note that for each N only finitely many asymptotic expansions are used, so *uniformity* is not an issue. After some preliminary rearrangements, this gives

$$\int_x^\infty e^{2\omega(t-x)} \left(\frac{x}{t}\right)^{2\rho} \left(\sum_{n\le\ell\le N} b_\ell\, t^{-\ell}\right) \frac{dt}{t^2}$$

$$= \sum_{n\le\ell\le N} b_\ell \int_x^\infty e^{2\omega(t-x)} \left(\frac{x}{t}\right)^{2\rho} t^{-\ell} \frac{dt}{t^2}$$

$$= \sum_{n\le\ell\le N} b_\ell\, e^{-2\omega x} x^{2\rho} \int_x^\infty e^{2\omega t}\, t^{-(2\rho+\ell+2)}\, dt$$

$$= \sum_{n\le\ell\le N} b_\ell\, e^{-2\omega x} x^{2\rho} \cdot e^{2\omega x} \sum_{0\le m\le N-(2+\ell)} \left(\frac{(\rho+\ell+2)_m}{(-2\omega)^{m+1}} \frac{1}{x^{2\rho+\ell+2+m}} + O\left(\frac{1}{x^{2\rho+N+1}}\right)\right)$$

$$= \sum_{n\le\ell\le N} b_\ell \sum_{0\le m\le N-(2+\ell)} \frac{(\rho+\ell+2)_m}{(-2\omega)^{m+1}} \frac{1}{x^{\ell+2+m}} + O\left(\frac{1}{x^{N+1}}\right)$$

This holds for all N, so we have an asymptotic expansion for f'_{n+1}:

$$f'_{n+1}(x) \sim \sum_{k\ge n+1} \left(\sum_{\ell\,:\,n\le\ell\le k} b_\ell \frac{(\rho+\ell+2)_{k-\ell}}{(-2\omega)^{m+1}}\right) \frac{1}{x^{k+2}}$$

Integrating this in x gives the asymptotic expansion of f_{n+1}. (See the Appendix.)

[16.15.3] Asymptotics of the Solution f: Obviously we expect the asymptotic expansion of $f = \sum_n f_n$ to be the sum of those of f_n, all the more so since the $1/x^m$ terms in the expansion of f_n vanish for $m < n$. The uniform pointwise bound

$$|f_n(x)| \le \frac{(AB)^n}{n!} x^{-n} \qquad \text{(for } n = 0, 1, 2, \ldots \text{ and } x \ge a\text{)}$$

proven earlier legitimizes this. Thus, the solution f has an asymptotic expansion of the desired type.

To prove that this asymptotic expansion is the same as the expansion obtained by a recursion earlier, we show that the coefficients satisfy the same recursion.

The integral expression for f' in terms of f (above) proves that f' has an asymptotic expansion, and similarly for f''. As proven in the appendix, this justifies two termwise differentiations of the asymptotic for f.

The asymptotics for f, f', and f'' can be inserted in the differential equation

$$f'' + 2\left(\omega - \frac{\rho}{x}\right)f' + \left(\omega^2 - \frac{2\omega\rho}{x} + \frac{\rho^2 + \rho}{x^2} - q(x)\right)f = 0$$

for f. We have assumed that the coefficient of f has an asymptotic expansion, and this equation gives the expected recursive relation on the coefficients of the asymptotic for f. Therefore, the solution

$$u(x) = e^{\omega x} x^{-\rho} f(x)$$

to the original differential equation has the asymptotics inherited from f, which match the heuristic asymptotics from the earlier formal/heuristic solution.

[16.15.4] The Second Solution: Now we show that the second solution

$$v(x) = u(x) \cdot \int_b^x u(t)^{-2}\, dt$$

to the original differential equation has the asymptotics given by the heuristic recursion, but with the opposite choice of $\pm\omega$ and $\pm\rho$. In terms of f,

$$v(x) = u(x) \cdot \int_b^x u(t)^{-2}\, dt = e^{\omega x} x^{-\rho} f(x) \int_b^x e^{-2\omega t} x^{-2\rho} f(t)\, dt$$

$$= e^{-\omega x} x^\rho f(x) \int_b^x e^{2\omega(x-s)} \left(\frac{s}{x}\right)^{2\rho} f(s)^{-2}\, ds$$

This motivates taking

$$g(x) = f(x) \int_b^x e^{2\omega(x-s)} \left(\frac{s}{x}\right)^{2\rho} f(s)^{-2}\, ds$$

The lower bound b has been chosen large enough so that $f(x)$ is bounded away from 0 for $x \geq b$. Since f has an asymptotic expansion with leading coefficient 1, it is elementary that there are coefficients a_n so that $1/f^2$ has asymptotics

$$\frac{1}{f(x)^2} = 1 + \sum_{1 \leq n \leq N} \frac{a_n}{x^n} + O\left(\frac{1}{x^{N+1}}\right) \qquad \text{(with } a_0 = 1\text{)}$$

Then

$$\frac{g(x)}{f(x)} = \sum_{0 \leq n \leq N} a_n \int_b^x e^{2\omega(x-s)} \left(\frac{s}{x}\right)^{2\rho} \frac{1}{s^n}\, ds + \int_b^x e^{2\omega(x-s)} \left(\frac{s}{x}\right)^{-2\rho} O\left(\frac{1}{s^{N+1}}\right) ds$$

The last integral is $O(1/x^{N+1})$, from the elementary asymptotics. For each fixed N, the finitely many integrals inside the summation have elementary asymptotics. Since for fixed N there are only finitely many such asymptotics, they are trivially *uniform*, so the asymptotics can be added. The asymptotic expansion for

$$\int_b^x e^{2\omega(x-s)} \left(\frac{s}{x}\right)^{2\rho} \frac{1}{s^n}\, ds$$

begins with $1/x^n$, so the coefficient of each $1/x^n$ is a finite sum, and there is no issue of convergence. Multiplying this asymptotic by that of $f(g)$ give the asymptotic expansion of $g(x)$.

As with f, the derivatives g' and g'' of g have integral representations that yield asymptotic expansions. Thus, as in the appendix, the asymptotic expansion for g can be twice differentiated termwise to give those of g' and g''. Thus, their asymptotic expansions can be inserted in the differential equation. Their coefficients must satisfy the same recursion with some choice of $\pm\omega$ and corresponding $\pm\rho$. Arguing that the asymptotic for g cannot be identical to that of f, we infer that the recursion for the coefficients of g uses the opposite choice $-\omega, -\rho$ from the choice ω, ρ used to construct f.

[16.15.5] Remark: When ω and ρ are both purely imaginary, u and v are bounded, neither approaches 0, and they are uniquely determined up to constant factors. In all other cases, one solution approaches 0, and is uniquely determined up to a constant, while the other is unbounded and ambiguous by multiples of the first, insofar as it depends on the choice of lower bound b in the integral above.

[16.15.6] Remark: *(Stokes's phenomenon)* When the coefficient $q(x)$ of the differential equation $u'' - q(x)u = 0$ is *analytic* in a sector in $\mathbb{C}$, and when q admits the same sort of asymptotic expansion

$$q(e^{i\theta}x) \sim \sum_{n\geq 0} \frac{q_n\, e^{-in\theta}}{x^n} \qquad \text{(uniformly in } \theta\text{)}$$

in that sector, *uniformly in the argument* θ, with $q_o \neq 0$, the preceding discussion still applies. In the real-variable discussion, with $\omega = \pm\sqrt{q_o}$, the case $\text{Re } \omega = 0$ was at the interface between the regimes $\text{Re } \omega \leq 0$ and $\text{Re } \omega \geq 0$ in which behaviors of solutions differed. Similarly, in the complex-variable situation the line $\text{Re}(z \cdot \sqrt{q_o}) = 0$ is the boundary between regimes of different behavior. *On* that line, the behavior is as in the $\text{Re } \omega = 0$ case. On either side of that line, one solution is exponentially larger than the other, etc. This is *Stokes's phenomenon*.

16.A Appendix: Manipulation of Asymptotic Expansions

To say that φ_ℓ is an *asymptotic sequence* at x_o means that $\varphi_{\ell+1}(x) = o(\varphi_\ell(x))$ as $x \to x_o$, for all ℓ. A function f has an *asymptotic expansion* in terms of the φ_n, expressed with coefficients c_n as

$$f(x) \sim \sum_{n \geq 0} c_n \varphi_n$$

when, for all $N \geq 0$,

$$f(x) - \sum_{0 \leq n \leq N} c_n \varphi_n = o(\varphi_N)$$

It is not surprising that a sum or integral of asymptotic expansions *uniform* in a parameter has the expected asymptotics. Circumstances under which an asymptotic expansion can be *differentiated* are more special.

[16.A.1] Summing Asymptotic Expansions: Let functions f_n have asymptotic expansions $f_n \sim \sum_{\ell \geq 0} c_{n\ell} \, \varphi_\ell$, *uniform* in n, meaning that

$$f_n(x) - \sum_{\ell \leq N} c_{n\ell} \, \varphi_\ell = o(\varphi_N)$$

with implied constant and neighborhood of x_o *uniform* in n. Let a_n be coefficients such that $\sum_n a_n \cdot c_{n\ell}$ is convergent and $\sum_n a_n$ is *absolutely* convergent. We claim that $\sum_n a_n f_n$ converges in a neighborhood of x_o and has the expected asymptotic expansion

$$\sum_n a_n \, f_n \sim \sum_\ell \left(\sum_n a_n \, c_{n\ell} \right) \varphi_n$$

The *uniformity* of the asymptotic expansions, and $\sum_n |a_n| < \infty$, give

$$\sum_{n \geq 1} a_n \, (f_n(x) - c_{n1} \, \varphi_1(x)) = o(\varphi_1(x)) \qquad \text{(uniformly in } x)$$

In particular, the sum on the left-hand side converges for fixed x. Since $\sum_n a_n c_{n1}$ converges, $\sum_{n \geq 1} a_n \, f_n(x)$ converges. Similarly,

$$\sum_n a_n f_n(x) - \sum_{\ell \leq N} \left(\sum_n a_n c_{n\ell} \right) \varphi_\ell = o(\varphi_N)$$

[16.A.2] Integrals: The general case is readily extrapolated from the example of an infinite sum. Namely, let $f(x, y) \sim \sum_\ell c_\ell(y) \, \varphi_\ell$ be asymptotic expansions *uniform* in a parameter $y \in Y$, where Y is a measure space. Suppose that $y \to f(x, y)$ is measurable for each x, and that every $c_\ell(y)$ is measurable. Let $a(y)$ be absolutely integrable on Y, and assume that the integrals

$$\int_Y a(y) \, c_\ell(y) \, dy$$

converge for all n. Then

$$\int_Y a(y) \, f(x, y) \, dy$$

exists for x close to x_o and has asymptotic expansion

$$\int_Y a(y) \, f(x, y) \, dy \sim \sum_\ell \left(\int_Y a(y) \, c_\ell(y) \, dy \right) \varphi_\ell$$

[16.A.3] Differentiation of Asymptotics in $1/x^n$: *Asymptotic power series* are asymptotic expansions

$$f(x) \sim c_o + \frac{c_1}{x} + \frac{c_2}{x^2} + \cdots \qquad (\text{as } x \to +\infty)$$

Unlike general situations, two such asymptotic expansions can be *multiplied*. A special property of asymptotic power series is the *absolute integrability* of $f(x) - c_o - c_1/x = O(x^{-2})$ on intervals $[a, +\infty)$. Let

$$F(x) = \int_x^\infty \left(f(t) - c_o - \frac{c_1}{t} \right) dt$$

We claim that F has an asymptotic expansion obtained from that of $f(x) - c_o - c_1/x$ by integrating termwise, namely,

$$F(x) \sim \frac{c_2}{t} + \frac{c_3}{2t^2} + \frac{c_4}{3t^3} + \cdots$$

To prove this, use

$$f(x) - \left(c_o + \frac{c_1}{x} + \cdots + \frac{c_N}{x^N} \right) = O(x^{-(N+1)})$$

Then

$$F(x) - \left(\frac{c_2}{t} + \frac{c_3}{2t^2} + \cdots + \frac{c_N}{Nx^{N-1}} \right)$$
$$= \int_x^\infty \left(f(t) - c_o - \frac{c_1}{t} \right) dt - \int_x^\infty \left(\frac{c_2}{t^2} + \frac{c_3}{t^3} + \cdots + \frac{c_N}{x^N} \right) dt$$
$$= \int_x^\infty O(t^{-(N+1)}) \, dt = O(x^{-N}) = o(x^{-(N-1)})$$

This has a surprising corollary about differentiation: for f with an asymptotic power series at $+\infty$ as above, if f is differentiable, and if f' has an asymptotic power series at $+\infty$, then the asymptotics of f' are obtained by differentiating that of f termwise:

$$f'(x) \sim -\frac{c_1}{x^2} - \frac{2c_2}{x^3} - \frac{3c_3}{x^4} - \cdots$$

When f is *holomorphic* in a region in which the asymptotic holds *uniformly* in the argument of x, Cauchy's integral formula for f' produces an asymptotic for f' from that for f, thus avoiding the need to make a hypothesis that f' admits an asymptotic expansion.

16.B Appendix: Ordinary Points

The following discussion is well known, although the convergence discussion is often omitted. This is the simpler case extended by the discussion of the regular singular points. A homogeneous ordinary differential equation of the form

$$u'' + b(x)u' + c(x)u = 0 \qquad \text{(with } b, c \text{ analytic near } 0\text{)}$$

is said to have an *ordinary point* at 0. The coefficients in a proposed expansion of the form

$$u(x) = \sum_{n=0}^{\infty} a_n x^n \qquad \text{(with } a_0 \neq 0\text{)}$$

are determined recursively from a_0 and a_1, as follows. The equation is

$$\sum_{n=0}^{\infty} n(n-1)a_n x^{n-2} + b(x) \sum_{n=0}^{\infty} n a_n x^{n-1} + c(x) \sum_{n=0}^{\infty} a_n x^n = 0$$

or

$$\sum_{n=0}^{\infty} n(n-1)a_n x^{n-2} + b(x) \sum_{n=0}^{\infty} (n-1)a_{n-1} x^{n-2} + c(x) \sum_{n=0}^{\infty} a_{n-2} x^{n-2} = 0$$

The coefficients a_n with $n \geq 2$ are obtained recursively, from the expected

$$n(n-1) \cdot a_n = (\text{in terms of } a_0, a_1, \ldots, a_{n-1})$$

To complete the proof of existence, we prove *convergence*. Take $A, M \geq 1$ large enough so that

$$\begin{cases} b(x) = \sum_{n \geq 0} b_n x^n & \text{(with } |b_n| \leq A \cdot M^n\text{)} \\ c(x) = \sum_{n \geq 0} c_n x^n & \text{(with } |c_n| \leq A \cdot M^n\text{)} \end{cases}$$

Inductively, suppose that $|a_\ell| \leq (CM)^\ell$, with a constant $C \geq 1$ to be determined in the following. Then

$$n(n-1) \cdot |a_n| \leq A \sum_{i=1}^{n} (n-i)M^{i-1} \cdot (CM)^{n-i} + A \sum_{i=2}^{n} M^{i-2} \cdot (CM)^{n-i}$$

$$\leq AM^{n-1} \cdot C^{n-1}\left(\frac{n(n+1)}{2} + n - 1\right)$$

Dividing through by $n(n-1)$, this is

$$|a_n| \leq AM^{n-1}C^{n-1}\frac{n^2 + 3n - 2}{n(n-1)}$$

This motivates taking

$$C \geq A \sup_{2 \leq n \in \mathbb{Z}} \frac{n^2 + 3n - 2}{n(n-1)}$$

which gives $|a_n| \leq (CM)^n$. In particular, for arbitrary a_0 and a_1 the resulting power series has a positive radius of convergence. In particular, these series can be differentiated termwise, by Abel's theorem.

[16.B.1] Ordinary Points at Infinity: Let $u(x) = v(1/x)$ and $z = 1/x$. Then

$$u'(x) = \frac{-1}{x^2}v'(1/x) \qquad \text{and} \qquad u''(x) = \frac{1}{x^4}v''(1/x) + \frac{2}{x^3}v'(1/x)$$

or

$$u' = -z^2 v' \qquad \text{and} \qquad u'' = z^4 v'' + 2z^3 v'$$

A differential equation $u'' + b(x)u' + c(x)u = 0$ becomes

$$\left(z^4 v'' + 2z^3 v'\right) + b(x)\left(-z^2 v'\right) + c(x)v = 0$$

or

$$v'' + \frac{2z - b\left(\frac{1}{z}\right)}{z^2}v' + \frac{c\left(\frac{1}{z}\right)}{z^4}v = 0$$

The point $z = 0$ is an *ordinary point* when the coefficients of v' and v are analytic at 0. That is, $z = 0$ is an ordinary point when b, c have expansions at infinity of the form

$$\begin{cases} b\left(\dfrac{1}{z}\right) &= 2z + b_2 z^2 + b_3 z^3 \ldots \\[2mm] c\left(\dfrac{1}{z}\right) &= c_4 z^4 + c_5 z^5 + \ldots \end{cases}$$

[16.B.2] Not-Quite-Ordinary Points: Consider a differential equation with coefficients having poles of at most first order at 0:

$$u'' + \frac{b(x)}{x}u' + \frac{c(x)}{x}u = 0$$

with b, c analytic at 0. The coefficients in a proposed expansion of the form

$$u(x) = \sum_{n=0}^{\infty} a_n x^n \qquad \text{(with } a_0 \neq 0\text{)}$$

are determined recursively as follows. The equation is

$$\sum_{n=0}^{\infty} n(n-1)a_n x^{n-2} + b(x)\sum_{n=0}^{\infty} na_n x^{n-2} + c(x)\sum_{n=0}^{\infty} a_n x^{n-1} = 0$$

or

$$\sum_{n=0}^{\infty} n(n-1)a_n x^{n-2} + b(x)\sum_{n=0}^{\infty} na_n x^{n-2} + c(x)\sum_{n=0}^{\infty} a_{n-1} x^{n-2} = 0$$

We expect to determine the coefficients a_n with $n \geq 2$ recursively, from

$$\big(n(n-1) + b(0)n\big) \cdot a_n = (\text{in terms of } a_0, a_1, \ldots, a_{n-1})$$

for $n \geq 1$. For $b(0)$ not a nonpositive integer, the recursion succeeds, and a_0 determines all the other coefficients a_n.

For $b(0) = 0$, so that the coefficient of v' has no pole, the relation from the coefficient of x^{-1},

$$b(0)a_1 + c(0)a_0 = 0$$

implies that *either* $c(0) = 0$ and the coefficient of v has no pole, returning us to the ordinary-point case, *or* $a_0 = 0$, and there is no nonzero solution of this form.

For $b(0)$ a negative integer $-\ell$, the recursion for a_ℓ gives a_ℓ the coefficient 0, and imposes a nontrivial relation on the prior coefficients a_n.

To complete the proof of existence, we prove *convergence*, assuming $b(0)$ is not a nonpositive integer. Dividing through by a constant if necessary, we can take $M \geq 1$ large enough so that

$$\begin{cases} b(x) & = \quad \sum_{n\geq 0} b_n x^n \quad \text{(with } |b_n| \leq M^n\text{)} \\[2mm] c(x) & = \quad \sum_{n\geq 0} c_n x^n \quad \text{(with } |c_n| \leq M^n\text{)} \end{cases}$$

Inductively, suppose that $|a_\ell| \leq (CM)^\ell$, with a constant $C \geq 1$ to be determined in the following. Then

$$\left(n(n-1) + b(0)n\right) \cdot |a_n|$$

$$= \left| \sum_{i=1}^{n} (n-i)M^{i-1}(CM)^{n-i} + \sum_{i=1}^{n} M^{i-1}(CM)^{n-i} \right|$$

$$\leq M^{n-1}C^{n-1}\left(\frac{n(n+1)}{2} + n\right)$$

Dividing through by $n(n-1) + b(0)n$, this is

$$|a_n| \leq M^{n-1}C^{n-1}\frac{n^2 + 3n}{n(n-1) + b(0)n}$$

This motivates taking

$$C \geq \sup_{2 \leq n \in \mathbb{Z}} \frac{n^2 + 3n}{n(n-1) + b(0)n}$$

which gives $|a_n| \leq (CM)^n$. In particular, for arbitrary a_0 the resulting power series has a positive radius of convergence. For example, the series can be differentiated termwise, by Abel's theorem.

Bibliography

[Arthur 1978] J. Arthur, *A trace formula for reductive groups, I. Terms associated to classes in* $G(\mathbb{Q})$, Duke. Math. J. **45** (1978), 911–952.

[Arthur 1980] J. Arthur, *A trace formula for reductive groups, II. Application of a truncation operator*, Comp. Math. J. **40** (1980), 87–121.

[Avakumović 1956] V.G. Avakumović, *Über die Eigenfunktionen auf geschlossenen Riemannschen Mannigfaltigkeiten*, Math. Z. **65** (1956), 327–344.

[Bargmann 1947] V. Bargmann, *Irreducible unitary representations of the Lorentz group*, Ann. Math. **48** (1947), 568–640.

[Berezin 1956] F.A. Berezin, *Laplace operators on semisimple Lie groups*, Dokl. Akad. Nauk SSSR **107** (1956), 9–12.

[Berezin-Faddeev 1961] F.A. Berezin, L.D. Faddeev, *Remarks on Schrödinger's equation with a singular potential*, Soviet Math. Dokl. **2** (1961), 372–375.

[Bethe-Peierls 1935] H. Bethe, R. Peierls, *Quantum theory of the diplon*, Proc. Royal Soc. London **148a** (1935), 146–156.

[Bianchi 1892] L. Bianchi, *Sui gruppi di sostituzioni lineari con coefficienti appartenenti a corpi quadratici immaginari*, Math. Ann. **40** (1892), 332–412.

[Birkhoff 1908] G.D. Birkhoff, *On the asymptotic character of the solutions of certain linear differential equations containing a parameter*, Trans. Amer. Math. Soc. **9** (1908), 219–231.

[Birkhoff 1909] G.D. Birkhoff, *Singular points of ordinary linear differential equations*, Trans. Amer. Math. Soc. **10** (1909), 436–470.

[Birkhoff 1913] G.D. Birkhoff, *On a simple type of irregular singular point*, Trans. Amer. Math. Soc. **14** (1913), 462–476.

[Birkhoff 1935] G. Birkhoff, *Integration of functions with values in a Banach space*, Trans. AMS **38** (1935), 357–378.

[Blaustein-Handelsman 1975] N. Blaustein, R.A. Handelsman, *Asymptotic Expansions of Integrals*, Holt, Rinehart, Winston, 1975, reprinted 1986, Dover.

[Blumenthal 1903/1904] O. Blumenthal, *Über Modulfunktionen von mehreren Veränderlichen*, Math. Ann. Bd. **56** (1903), 509–548; **58** (1904), 497–527.

[Bôcher 1898/1899] M. Bôcher, *The theorems of oscillation of Sturm and Klein*, Bull. AMS **4** (1898), 295–313; **5** (1899), 22–43.

[Bochner 1932] S. Bochner, *Vorlesungen über Fouriersche Integrale*, Akademie-Verlag, 1932.

[Bochner 1935] S. Bochner, *Integration von Funktionen deren Werte die Elemente eines Vektorraumes sind*, Fund. Math. **20** (1935), 262–276.

[Bochner-Martin 1948] S. Bochner, W.T. Martin, *Several Complex Variables*, Princeton University Press, Princeton, 1948.

333

[Borel 1962] A. Borel, *Ensembles fondamentaux pour les groupes arithmétiques*, Colloque sur la théorie des groupes algébriques, Bruxelles, 1962, 23–40.

[Borel 1963] A. Borel, *Some finiteness properties of adele groups over number fields*, IHES Sci. Publ. Math. **16** (1963), 5–30.

[Borel 1965/1966a] A. Borel, *Introduction to automorphic forms*, in *Algebraic Groups and Discontinuous Subgroups, Boulder, 1965*, Proc. Symp. Pure Math. **9**, AMS, New York, 1966, 199–210.

[Borel 1965/1966b] A. Borel, *Reduction theory for arithmetic groups*, in *Algebraic Groups and Discontinuous Subgroups, Boulder, 1965*, Proc. Symp. Pure Math. **9**, AMS, New York 1966, 20–25.

[Borel 1969] A. Borel, *Introductions aux groupes arithmeétiques*, Publ. l'Inst. Math. Univ. Strasbourg, XV, Actualités Sci. et Industrielles, no. 1341, Hermann, Paris, 1969.

[Borel 1976] A. Borel, *Admissible representations of a semi-simple group over a local field with vectors fixed under an Iwahori subgroup*, Inv. Math. **35** (1976), 233–259.

[Borel 1997] A. Borel, *Automorphic Forms on $SL_2(\mathbb{R})$*, Cambridge Tracts in Math. **130**, Cambridge University Press, Cambridge, 1997.

[Borel 2007] A. Borel, *Automorphic forms on reductive groups*, in *Automorphic Forms and Applications*, eds. P. Sarnak and F. Shahidi, IAS/ParkCity Math Series **12**, AMS, 2007.

[Borel-HarishChandra 1962] A. Borel, Harish-Chandra, *Arithmetic subgroups of algebraic groups*, Ann. Math. **75** (1962), 485–535.

[Bourbaki 1987] N. Bourbaki, *Topological Vector Spaces*, ch. 1–5, Springer-Verlag, Berlin-Heidelberg 1987.

[Braun 1939] H. Braun, *Konvergenz verallgemeinerter Eisensteinscher Reihen*, Math. Z. **44** (1939), 387–397.

[Brooks 1969] J.K. Brooks, *Representations of weak and strong integrals in Banach spaces*, Proc. Nat. Acad. Sci. U.S.A., 1969, 266–270.

[Casselman 1978/1980] W. Casselman, *Jacquet modules for real reductive groups*, Proc. Int. Cong. Math. (1978), 557–563, Acad. Sci. Fennica, Helskinki, 1980.

[Casselman 1980] W. Casselman, *The unramified principal series of p-adic groups. I. The spherical function*, Comp. Math. **40** (1980), no. 3, 387–406.

[Casselman 2005] W. Casselman, *A conjecture about the analytical behavior of Eisenstein series*, Pure and Applied Math. Q. **1** (2005) no. 4, part 3, 867–888.

[Casselman-Miličić 1982] W. Casselman, D. Miličić, *Asymptotic behavior of matrix coefficients of admissible representations*, Duke J. Math. **49** (1982), 869–930.

[Casselman-Osborne 1975] W. Casselman, M.S. Osborne, *The n-cohomology of representations with an infinitesimal character*, Comp. Math **31** (1975), 219–227.

[Casselman-Osborne 1978] W. Casselman, M.S. Osborne, *The restriction of admissible representations to n*, Math. Ann. **233** (1978), 193–198.

[Cogdell–Li-PiatetskiShapiro-Sarnak 1991] J. Cogdell, J.-S. Li, I.I. Piatetski-Shapiro, P. Sarnak, *Poincaré series for SO(n, 1)*, Acta Math. **167** (1991), 229–285.

[Cogdell-PiatetskiShapiro 1990] J. Cogdell, I. Piatetski-Shapiro, *The Arithmetic and Spectral Analysis of Poincaré Series*, Perspectives in Mathematics, Academic Press, San Diego, 1990.

[Cohen-Sarnak 1980] P. Cohen, P. Sarnak, *Selberg Trace Formula*, ch. 6 and 7, *Eisenstein series for hyperbolic manifolds*, www.math.princeton.edu/sarnak/

[Colin de Verdière 1981] Y. Colin de Verdière, *Une nouvelle démonstration du prolongement méromorphe des séries d'Eisenstein*, C. R. Acad. Sci. Paris Sér. I Math. **293** (1981), no. 7, 361–363.

[Colin de Verdière 1982/1983] Y. Colin de Verdière, *Pseudo-laplaciens, I, II*, Ann. Inst. Fourier (Grenoble) **32** (1982) no. 3, xiii, 275–286; ibid, **33** (1983) no. 2, 87–113.

[Conway-Smith 2003] J. Conway, D. Smith, *On Quaternions and Octonians*, A. K. Peters, Natick, MA, 2003.

[DeCelles 2012] A. DeCelles, *An exact formula relating lattice points in symmetric spaces to the automorphic spectrum*, Illinois J. Math. **56** (2012), 805–823.

[DeCelles 2016] A. DeCelles, *Constructing Poincaré series for number-theoretic applications*, New York J. Math. **22** (2016) 1221–1247.

[Dirac 1928a/1928b] P.A.M. Dirac, *The quantum theory of the electron*, Proc. R. Soc. Lond. A **117** (1928), 610–624; *II*, ibid, **118** (1928), 351–361.

[Dirac 1930] P.A.M. Dirac, *Principles of Quantum Mechanics*, Clarendon Press, Oxford, 1930.

[Dirichlet 1829] P.G.L. Dirichlet, *Sur la convergence des séries trigonométriques qui servent à représenter une fonction arbitraire entre des limites données*, J. Reine Angew. Math **4** (1829), 157–169 (Werke I, 117–132).

[Douady 1963] A. Douady, *Parties compactes d'un espace de fonctions continues a support compact*, C. R. Acad. Sci. Paris **257** (1963), 2788–2791.

[Elstrodt 1973] J. Elstrodt, *Die Resolvente zum Eigenwertproblem der automorphen Formen in der hyperbolischen Ebene, I*, Math. Ann. **203** (1973), 295–330, *II*, Math. Z. **132** (1973), 99–134.

[Elstrodt-Grunewald-Mennicke 1985] J. Elstrodt, E. Grunewald, J. Mennicke, *Eisenstein series on three-dimensional hyperbolic spaces and imaginary quadratic fields*, J. reine und angew. Math. **360** (1985), 160–213.

[Elstrodt-Grunewald-Mennicke 1987] J. Elstrodt, E. Grunewald, J. Mennicke, *Zeta functions of binary Hermitian forms and special values of Eisenstein series on three-dimensional hyperbolic space*, Math. Ann. **277** (1987), 655–708.

[Epstein 1903/1907] P. Epstein, *Zur Theorie allgemeiner Zetafunktionen*, Math. Ann. **56** (1903), 615–644; ibid, **65** (1907), 205–216.

[Erdélyi 1956] A. Erdélyi, *Asymptotic Expansions*, Technical Report 3, Office of Naval Research Reference No. NR-043-121, reprinted by Dover, 1956.

[Estermann 1928] T. Estermann, *On certain functions represented by Dirichlet series*, Proc. London Math. Soc. **27** (1928), 435–448.

[Faddeev 1967] L. Faddeev, *Expansion in eigenfunctions of the Laplace operator on the fundamental domain of a discrete group on the Lobacevskii plane*, AMS Transl. Trudy (1967), 357–386.

[Faddeev-Pavlov 1972] L. Faddeev, B.S. Pavlov, *Scattering theory and automorphic functions*, Seminar Steklov Math. Inst. **27** (1972), 161–193.

[Fay 1977] J.D. Fay, *Fourier coefficients of the resolvent for a Fuchsian group*, J. für reine und angewandte Math. (Crelle) **293–294** (1977), 143–203.

[Fourier 1822] J. Fourier, *Théorie analytique de la chaleur*, Firmin Didot Père et Fils, Paris, 1822.

[Friedrichs 1934/1935] K.O. Friedrichs, *Spektraltheorie halbbeschränkter Operatoren*, Math. Ann. **109** (1934), 465–487, 685–713; *II*, Math. Ann. **110** (1935), 777–779.

[Gårding 1947] L. Gårding, *Note on continuous representations of Lie groups*, Proc. Nat. Acad. Sci. USA **33** (1947), 331–332.

[Garrett vignettes] P. Garrett, *Vignettes*, www.math.umn.edu/~garrett/m/v/

[Garrett mfms-notes] P. Garrett, *Modular forms notes*, www.math.umn.edu/~garrett/m/mfms/

[Garrett fun-notes] P. Garrett, *Functional analysis notes*, www.math.umn.edu/~garrett/m/fun/

[Garrett alg-noth-notes] P. Garrett, *Algebraic number theory notes*, www.math.umn.edu/~garrett/m/number_theory/

[Gelfand 1936] I.M. Gelfand, *Sur un lemme de la theorie des espaces lineaires*, Comm. Inst. Sci. Math. de Kharkoff, no. 4, **13** (1936), 35–40.

[Gelfand 1950] I.M. Gelfand, *Spherical functions in Riemannian symmetric spaces*, Dokl. Akad. Nauk. SSSR **70** (1950), 5–8.

[Gelfand-Fomin 1952] I.M. Gelfand, S.V. Fomin, *Geodesic flows on manifolds of constant negative curvature*, Uspekh. Mat. Nauk. **7** (1952), no. 1, 118–137. English translation, AMS ser. 2, **1** (1965), 49–65.

[Gelfand-Graev 1959] I.M. Gelfand, M.I. Graev, *Geometry of homogeneous spaces, representations of groups in homogeneous spaces, and related problems of integral geometry*, Trudy Moskov. Obshch. **8** (1962), 321–390.

[Gelfand-Graev-PiatetskiShapiro 1969] I. Gelfand, M. Graev, I. Piatetski-Shapiro, *Representation Theory and Automorphic Functions*, W.B. Saunders Co., Philadelphia, 1969.

[Gelfand-Kazhdan 1975] I.M. Gelfand, D. Kazhdan, *Representations of the group GL(n, k) where k is a local field*, in *Lie Groups and Their Representations*, Halsted, New York, 1975, pp. 95–118.

[Gelfand-PiatetskiShapiro 1963] I.M. Gelfand, I.I. Piatetski-Shapiro, *Automorphic functions and representation theory*, Trudy Moskov. Obshch. **8** (1963), 389–412. [Trans.: Trans. Moscow Math. Soc. **12** (1963), 438–464.]

[Gelfand-Shilov 1964] I.M. Gelfand, G.E. Shilov, *Generalized Functions, I: Properties and Operators*, Academic Press, New York, 1964.

[Gelfand-Vilenkin 1964] I.M. Gelfand, N.Ya. Vilenkin, *Generalized Functions, IV: Applications of Harmonic Analysis*, Academic Press, NY, 1964.

[Godement 1962–1964] R. Godement, *Domaines fondamentaux des groupes arithmetiques*, Sem. Bourb. **257** (1962–3).

[Godement 1966a] R. Godement, *Decomposition of $L^2(\Gamma\backslash G)$ for $\Gamma = SL(2, Z)$*, in Proc. Symp. Pure Math. **9** (1966), 211–224.

[Godement 1966b] R. Godement, *The spectral decomposition of cuspforms*, in Proc. Symp. Pure Math. **9** (1966), AMS 225–234.

[Green 1828] G. Green, *An essay on the application of mathematical analysis to the theories of electricity and magnetism*, arXiv:0807.0088 [physics.hist-ph]. Re-typeset 2008 from Crelle's J. reprint 1850–1854.

[Green 1837] G. Green, *On the Laws of Reexion and Refraction of Light at the Common Surface of Two Non-crystallized Media* Trans. Camb. Phil. Soc. **68** (1837), 457–462.

[Grothendieck 1950] A. Grothendieck, *Sur la complétion du dual d'un espace vectoriel localement convexe*, C. R. Acad. Sci. Paris **230** (1950), 605–606.

[Grothendieck 1953a,1953b] A. Grothendieck, *Sur certaines espaces de fonctions holomorphes, I*, J. Reine Angew. Math. **192** (1953), 35–64; *II*, **192** (1953), 77–95.

[Haas 1977] H. Haas, *Numerische Berechnung der Eigenwerte der Differentialgleichung $y^2 \Delta u + \lambda u = 0$ für ein unendliches Gebiet im $\mathbb{R}^2$*, Diplomarbeit, Universität Heidelberg (1977), 155 pp.

[Harish-Chandra 1954] Harish-Chandra, *Representations of semisimple Lie groups, III*, Trans. AMS **76** (1954), 234–253.

[Harish-Chandra 1959] Harish-Chandra, *Automorphic forms on a semi-simple Lie group*, Proc. Nat. Acad. Sci. **45** (1959), 570–573.

[Harish-Chandra 1968] Harish-Chandra, *Automorphic forms on a semi-simple Lie groups*, notes G.J.M. Mars., SLN **62**, Springer-Verlag, 1968.

[Hartogs 1906] F. Hartogs, *Zur Theorie der analytischen Funktionen mehrerer unabhängiger Veränderlichen, insbesondere über die Darstellung derselben durch Reihen, welche nach Potenzen einer Veränderlichen fortschreiten*, Math. Ann. **62** (1906), 1–88.

[Hejhal 1976/1983] D. Hejhal, *The Selberg trace formula for $PSL_2(\mathbb{R})$, I*, SLN **548**, Springer-Verlag, 1976; *II*, ibid, **1001**, Springer-Verlag, 1983.

[Hejhal 1981] D. Hejhal, *Some observations concerning eigenvalues of the Laplacian and Dirichlet L-series*, in *Recent Progress in Analytic Number Theory*, ed. H. Halberstam and C. Hooley, vol. 2, Academic Press, New York, 1981, pp. 95–110.

[Hilbert 1909] D. Hilbert, *Wesen und Ziele einer Analysis der unendlich vielen unabhängigen Variablen*, Rendiconti Circolo Mat. Palermo **27** (1909), 59–74.

[Hilbert 1912] D. Hilbert, *Grundzüge einer allgemeinen Theorie der linearen Integralgleichungen*, Teubner, Leipzig-Berlin, 1912.

[Hildebrandt 1953] T.H. Hildebrandt, *Integration in abstract spaces*, Bull. AMS, **59** (1953), 111–139.

[Hille-Phillips 1957] E. Hille with R. Phillips, *Functional Analysis and Semigroups*, AMS Coll. Pub., 2nd edition, Providence, RI, 1957.

[Hörmander 1973] L. Hörmander, *An Introduction to Complex Analysis in Several Variables*, 2nd edition, North-Holland, 1973.

[Horvath 1966] J. Horvath, *Topological Vector Spaces and Distributions*, Addison-Wesley, Boston, 1966.

[Huber 1955] H. Huber, *Über eine neue Klasse automorpher Funktionen und ein Gitterpunkt Problem in der hypbolischen Ebene I*, Comm. Math. Helv. **30** (1955), 20–62.

[Hurwitz 1898] A. Hurwitz, *Über die Komposition der quadratische Formen von beliebig vielen Variabeln*, Nachr. König. Gesellschaft der Wiss. zu Göttingen (1898), 309–316.

[Hurwitz 1919] A. Hurwitz, *Vorlesungen über die Zahlentheorie der Quaternionen*, Springer, Berlin, 1919.

[Iwaniec 2002] H. Iwaniec, *Spectral Methods of Automorphic Forms*, 2nd edition, AMS, Providence, 2002. [First edition, Revisto Mathematica Iberoamericana, 1995.]

[Jacquet 1982/1983] H. Jacquet, *On the residual spectrum of GL(n)*, in *Lie group representations, II, College Park, MD*, 185–208, SLN 1041, Springer, Berlin, 1984.

[Kodaira 1949] K. Kodaira, *The eigenvalue problem for ordinary differential equations of the second order and Heisenberg's theory of S-matrices*, Amer. J. Math. **71** (1949), 921–945.

[Krein 1945] M.G. Krein, *On self-adjoint extension of bounded and semi-bounded Hermitian transformations*, Dokl. Akad. Nauk. SSSR **48** (1945), 303–306 [Russian].

[Krein 1947] M.G. Krein, *The theory of self-adjoint extension of semi-bounded Hermitian transformations and its applications*, I. Mat. Sbornik **20** (1947), 431–495 [Russian].

[Kurokawa 1985a,b] N. Kurokawa, *On the meromorphy of Euler products, I*, Proc. London Math. Soc. **53** (1985) 1–49; *II*, ibid **53** (1985) 209–236.

[Lang 1975] S. Lang, $SL_2(\mathbb{R})$, Addison-Wesley, Boston, 1975.

[Langlands 1971] R. Langlands, *Euler Products*, Yale University Press, New Haven, 1971.

[Langlands 1967/1976] R.P. Langlands, *On the functional equations satisfied by Eisenstein series*. Lecture Notes in Mathematics, vol. 544, Springer-Verlag, Berlin and New York, 1976.

[Laplace 1774] P.S. Laplace, *Mémoir on the probability of causes of events*, Mémoires de Mathématique et de Physique, Tome Sixième. (English trans. S.M. Stigler, 1986. Statist. Sci., 1 **19**, 364–378).

[Lax-Phillips 1976] P. Lax, R. Phillips, *Scattering theory for automorphic functions*, Ann. Math. Studies, Princeton, 1976.

[Levi 1906] B. Levi, *Sul Principio di Dirichlet*, Rend. del Circolo Mat. di Palermo **22** (1906), 293–300.

[Lindelöf 1908] E. Lindelöf, *Quelques remarques sur la croissance de la fonction $\zeta(s)$*, Bull. Sci. Math. **32** (1908), 341–356.

[Liouville 1837] J. Liouville, J. Math. Pures et Appl. **2** (1837), 16–35.

[Lützen 1984] J. Lützen, *Sturm and Liouville's work on ordinary differential equations. The emergence of Sturm-Liouville theory*, Arch. Hist. Exact Sci **29** (1984) no. 4, 309–376. Retrieved July 16, 2013, from www.math.ku.dk/~lutzen/

[Maaß 1949] H. Maaß, *Über eine neue Art von nicht analytischen automorphen Funktionen und die Bestimmung Dirichletscher Reihen durch Funktionalgleichungen*, Math. Ann. **121** (1949), 141–183.

[MSE 2017] Math Stack Exchange, *Residual spectrum of a Hermitian operator*, retrieved July 19, 2017, from https://math.stackexchange.com/questions/2363904/

[Matsumoto 1977] H. Matsumoto, *Analyse harmonique dans les systèmes de Tits bornologiques de type affine*, SLN 590, Springer, Berlin, 1977.

[Minakshisundaram-Pleijel 1949] S. Minakshisundaram, Å. Pleijel, *Some properties of the eigenfunctions of the Laplace-operator on Riemannian manifolds*, Canadian J. Math. **1** (1949), 242–256.

[Moeglin-Waldspurger 1989] C. Moeglin, J.-L. Waldspurger, *Le spectre résiduel de GL_n*, with appendix *Poles des fonctions L de pairs pour GL_n*, Ann. Sci. École Norm. Sup. **22** (1989), 605–674.

[Moeglin-Waldspurger 1995] C. Moeglin, J.-L. Waldspurger, *Spectral Decompositions and Eisenstein series*, Cambridge University Press, Cambridge, 1995.

[Müller 1996] W. Müller, *On the analytic continuation of rank one Eisenstein series*, Geom. Fun. Ann. **6** (1996), 572–586.

[Myller-Lebedev 1907] Wera Myller-Lebedev, *Die Theorie der Integralgleichungen in Anwendung auf einige Reihenentwicklungen*, Math. Ann. **64** (1907), 388–416.

[Neunhöffer 1973] H. Neunhöffer, *Über die analytische Fortsetzung von Poincaréreihen*, Sitzungsberichte Heidelberg Akad. Wiss. (1973), no. 2.

[Niebur 1973] D. Niebur, *A class of nonanalytic automorphic functions*, Nagoya Math. J. **52** (1973), 133–145.

[Olver 1954] F.W.J. Olver, *The asymptotic solution of linear differential equations of the second order for large values of a parameter*, Phil. Trans. **247** (1954), 307–327.

[Paley-Wiener 1934] R. Paley, N. Wiener, *Fourier transforms in the complex domain*, AMS Coll. Publ. XIX, New York, 1934.

[Pettis 1938] B.J. Pettis, *On integration in vector spaces*, Trans. AMS **44** (1938), 277–304.

[Phragmén-Lindelöf 1908] L.E. Phragmén, E. Lindelöf, *Sur une extension d'un principe classique de l'analyse et sur quelques propriétés des fonctions monogènes dans le voisinage d'un point singuliere*, Acta Math. **31** (1908), 381–406.

[Piatetski-Shapiro 1979] I.I. Piatetski-Shapiro, Multiplicity-one theorems, in *Automorphic Forms, Representations, and L-functions*, Proc. Symp. Pure Math. XXXIII, part 1, 315–322, AMS, 1979.

[Picard 1882] E. Picard, *Sur une classe de groupes discontinus de substitutions linéaires et sur les fonctions de deux variables indépendantes restant invariables par ces substitutions*, Acta Math. **1** no. 1 (1882), 297–320.

[Picard 1883] E. Picard, *Sur des fonctions de deux variables indépendantes analogues aux fonctions modulaires* Acta Math. no. 1 **2** (1883), 114–135.

[Picard 1884] E. Picard, *Sur un groupe de transformations des points de l'espace situés du même coîé d'un plan*, Bull. Soc. Math. France **12** (1884), 43–47.

[Povzner 1953] A. Povzner, *On the expansion of arbitrary functions in characteristic functions of the operator* $-\Delta u + cu$, Math. Sb. **32** no. 74 (1953), 109–156.

[Rankin 1939] R. Rankin, *Contributions to the theory of Ramanujan's function* $\tau(n)$ *and similar arithmetic functions, I*, Proc. Cam. Phil. Soc. **35** (1939), 351–372.

[Riesz 1907] F. Riesz, *Sur les systèmes orthogonaux de fonctions*, C.R. de l'Acad. des Sci. **143** (1907), 615–619.

[Riesz 1910] F. Riesz, *Untersuchungen über Systeme integrierbarer Funktionen*, Math. Ann. **69** (1910), 449–497.

[Roelcke 1956a] W. Roelcke, *Über die Wellengleichung bei Grenzkreisgruppen erster Art*, S.-B. Heidelberger Akad. Wiss. Math. Nat. Kl. 1953/1955 (1956), 159–267.

[Roelcke 1956b] W. Roelcke, *Analytische Fortsetzung der Eisensteinreihen zu den parabolischen Spitzen von Grenzkreisgruppen erster Art*, Math. Ann. **132** (1956), 121–129.

[Rudin 1991] W. Rudin, *Functional Analysis*, second edition, McGraw-Hill, New York, 1991.

[Schaefer-Wolff 1999] H.H. Schaefer, with M.P. Wolff, *Topological Vector Spaces*, 2nd edition, Springer, Tübingen, 1999.

[Schmidt 1907] E. Schmidt, *Zur Theorie der linearen und nichtlinearen Integralgleichungen. Teil I: Entwicklung wilkürlicher Funktionen nach Systemen vorgeschriebener*, Math. Ann. **63** (1907), 433–476.

[Schwartz 1950] L. Schwartz, *Théorie des noyaux*, Proc. Int. Cong. Math. Cambridge 1950, I, 220–230.

[Schwartz 1950/1951] L. Schwartz, *Théorie des Distributions*, I, II Hermann, Paris, 1950/1951, 3rd edition, 1965.

[Schwartz 1952] L. Schwartz, *Transformation de Laplace des distributions*, Comm. Sém. Math. Univ. Lund (1952), Tome Supplémentaire, 196–206.

[Schwartz 1953/1954] L. Schwartz, *Espaces de fonctions différentiables à valeurs vectorielles* J. d'Analyse Math. **4** (1953/1954), 88–148.

[Selberg 1940] A. Selberg, *Bemerkungen über eine Dirichletsche Reihe, die mit der Theorie der Modulformen nahe verbunden ist*, Arch. Math. Naturvid **43** (1940), 47–50.

[Selberg 1954] A. Selberg, *Harmonic Analysis, 2. Teil*, Vorlesung Niederschrift, Göttingen, 1954; *Collected Papers I*, Springer, Heidelberg, 1988, 626–674.

[Selberg 1956] A. Selberg, *Harmonic analysis and discontinuous groups in weakly symmetric spaces, with applications to Dirichlet series*, J. Indian Math. Soc. **20** (1956), 47–87.

[Shahidi 1978] F. Shahidi, *Functional equation satisfied by certain L-functions*, Comp. Math. **37** (1978), 171–208.

[Shahidi 1985] F. Shahidi, *Local coefficients as Artin factors for real groups*, Duke Math. J. **52** (1985), 973–1007.

[Shahidi 2010] F. Shahidi, *Eisenstein series and automorphic L-functions*, AMS Coll. Publ. **58**, AMS, 2010.

[Shalika 1974] J.A. Shalika, *The multiplicity one theorem for GLn*, Ann. Math. 100 (1974), 171–193.

[Sobolev 1937] S.L. Sobolev, *On a boundary value problem for polyharmonic equations (Russian)*, Mat. Sb. (NS) **2** (44) (1937), 465–499.

[Sobolev 1938] S.L. Sobolev, *On a theorem of functional analysis (Russian)*, Mat. Sb. (NS) **4** (1938), 471–497.

[Sobolev 1950] S.L. Sobolev, *Some Applications of Functional Analysis to Mathematical Physics* [Russian], Paul: Leningrad, 1950.

[Speh 1981/1982] B. Speh, *The unitary dual of $GL_3(\mathbb{R})$ and $GL_4(\mathbb{R})$*, Math. Ann. **258** (1981/1982), 113–133.

[Stone 1929] M.H. Stone, *I, II: Linear transformations in Hilbert space*, Proc. Nat. Acad. Sci. **16** (1929), 198–200, 423–425; *III: operational methods and group theory*, ibid, **16** (1930), 172–175.

[Stone 1932] M.H. Stone, *Linear transformations in Hilbert space*, AMS, New York, 1932.

[Steklov 1898] W. Steklov, *Sur le problème de refroidissement d'une barre hétérogène*, C. R. Acad. Sci. Paris **126** (1898), 215–218.

[Sturm 1836] C. Sturm, *Mémoire sur les équations différentielles du second ordre*, J. de Maths. Pure et Appl. **1** (1836), 106–186.

[Sturm 1833a/1836a] C. Sturm, *Analyse d'un mémoire sur les propriétés générales des fonctions qui dépendent d'équations différentielles linéares du second ordre*, L'Inst. J. Acad. et Soc. Sci. Nov. 9 (1833) 219–223, summary of *Mémoire sur les Équations différentielles linéares du second ordre*, J. Math. Pures Appl. **1** (1836), 106–186 [Sept. 28, 1833].

[Sturm 1833b/1836b] C. Sturm [unnamed note], L'Inst. J. Acad. et Soc. Sci. Nov. 9 (1833) 219–223, summary of *Mémoire sur une classe d'Équations à différences partielles*, J. Math. Pures Appl. **1** (1836), 373–444.

[Thomas 1935] L.H. Thomas, *The interaction between a neutron and a proton and the structure of H^3*, Phys. Rev. **47** (1935), 903–909.

[Varadarajan 1989] V.S. Varadarajan, *An Introduction to Harmonic Analysis on Semisimple Lie Groups*, Cambridge University Press, Cambridge, 1989.

[Venkov 1971] A. Venkov, *Expansion in automorphic eigenfunctions of the Laplace operator and the Selberg Trace Formula in the space $SO(n, 1)/SO(n)$*, Dokl. Akad. Nauk. SSSR **200** (1971); Soviet Math. Dokl. **12** (1971), 1363–1366.

[Venkov 1979] A. Venkov, *Spectral theory of automorphic functions, the Selberg zeta-function, and some problems of analytic number theory and mathematical physics*, Russian Math. Surveys **34** no. 3 (1979), 79–153.

[vonNeumann 1929] J. von Neumann, *Allgemeine eigenwerttheorie Hermitescher Funktionaloperatoren*, Math. Ann. **102** (1929), 49–131.

[vonNeumann 1931] J. von Neumann, *Die Eindeutigkeit der Schrödingersche Operatoren*, Math. Ann. **104** (1931), 570–578.

[Watson 1918] G.N. Watson, *Asymptotic expansions of hypergeometric functions*, Trans. Cambridge Phil. Soc. **22** (1918), 277–308.

[Weyl 1910] H. Weyl, *Über gewöhnliche Differentialgleichungen mit Singularitäten and die zugehörigen Entwicklungen wilkürlicher Funktionen*, Math. Ann. **68** (1910), 220–269.

[Weyl 1925/1926] H. Weyl, *Theorie der Darstellung kontinuierlicher half-einfacher Gruppen durch lineare Transformationen, I*, Math. Z. **23** (1925), 271–309; *II*, ibid, **24** (1926), 328–376; *III (und Nachtrag)*, ibid, 377–395, 789–791.

[Wiener 1933] N. Wiener, *The Fourier Integral and Certain of Its Applications*, Cambridge University Press, Cambridge, 1933.

[Wigner 1939] E. Wigner, *On unitary representations of the inhomogeneous Lorentz group*, Ann. Math. **40** (1939), 149–204.

[Wong 1990] S.-T. Wong, *The meromorphic continuation and functional equations of cuspidal Eisenstein series for maximal cuspidal groups*, Memoirs of AMS, **83** (1990), no. 423.

Index

Printed in the United States
By Bookmasters